Local Bifurcations, Center Manifolds, and Normal Forms in Infinite-Dimensional Dynamical Systems

Universitext

For other titles published in this series, go to
www.springer.com/series/223

Mariana Haragus · Gérard Iooss

Local Bifurcations, Center Manifolds, and Normal Forms in Infinite-Dimensional Dynamical Systems

Mariana Haragus
Université de Franche-Comté
Laboratoire de Mathématiques
25030 Besançon Cedex
France
mariana.haragus@univ-fcomte.fr

Gérard Iooss
IUF, Université de Nice
Laboratoire J.A.Dieudonné
06108 Nice Cedex 02
France
gerard.iooss@unice.fr

A co-publication with EDP Sciences, France, licensed for sale in all countries outside of France. Sold and distributed within France by EDP Sciences, 17, av. du Hoggar F-91944 Les Ulis, France

EDP Sciences ISBN 978-2-7598-0009-4

ISBN 978-0-85729-111-0 e-ISBN 978-0-85729-112-7
DOI 10.1007/978-0-85729-112-7
Springer London Dordrecht Heidelberg New York

British Library Cataloguing in Publication Data
A catalogue record for this book is available from the British Library

Library of Congress Control Number: 2010938140

Mathematics Subject Classification (2010): 34C20, 34C37, 34C45, 35B32, 35C07, 35Q35, 37G40, 37L10, 76B15, 76D05

Cover design: deblik

Printed on acid-free paper

Springer is part of Springer Science+Business Media (www.springer.com)

Preface

This book is an extension of different lectures given by the authors during many years at the University of Nice, at the University of Stuttgart in 1990, and the University of Bordeaux in 2000 and 2001. Large parts of the first four chapters are of master level and contain various examples and exercises, partly posed at exams. However, the infinite-dimensional set-up in Chapter 2 requires several tools and results from the theory of linear operators. A brief description of these tools and results is given in Appendix A.

Bifurcation theory forms the object of many different books over the past 30 years. We refer, for instance, to [4, 58, 17, 38, 29, 30, 39, 51, 110, 84, 16, 10, 79] for some references covering various topics, going from elementary local bifurcations to global bifurcations and applications to partial differential equations. In this book we restrict our attention to the study of local bifurcations. Starting with the simplest bifurcation problems arising for ordinary differential equations in one and two dimensions, the purpose of this book is to describe several tools from the theory of infinite-dimensional dynamical systems, allowing to treat more complicated bifurcation problems, as for instance bifurcations arising in partial differential equations. Such tools are extensively used to solve concrete problems arising in physics and natural sciences.

In a parameter-dependent physical system, for example, modelized by a differential equation, the presence of a bifurcation corresponds to a topological change in the structure of the solution set (which may break its symmetry in the case of a system invariant under some symmetry group). Such a change may imply the occurrence of new solutions, or the disappearance of certain solutions, or may indicate a change of stability of certain solutions. Local bifurcation theory allows one to detect solutions and to describe their geometric (including symmetries) and dynamic properties. During the last decades the use of bifurcation theory, and in particular of the methods presented in this book, led to significant progress in the understanding of nonlinear phenomena in partial differential equations, including hydrodynamic problems, structural mechanics, but also pattern formation, population dynamics, or questions in biophysics. For instance, in the classical Couette–Taylor problem describing flows between two coaxial rotating cylinders (briefly presented in Section 5.1.2), the theory was not only a qualitative one, but also sufficiently quantitative to allow prediction of numerical values of the parameters, where new flows, such as "ribbons," were expected to be observed. These were indeed later observed experimentally [117]. This predictive power of the local theory appeared again in water wave theory, where new forms of "solitary waves," with damping oscillations at infinity, were found (see Section 5.2.1), or in the propagation of interfaces between metastable states, where new types of fronts were constructed (see Section 5.2.2).

In this book we focus on two specific methods that arise in the analysis of local bifurcations in infinite-dimensional systems, namely the center manifold reduction and the normal form theory. Center manifolds provide a powerful method of analysis of such systems, as they allow one to reduce, under certain conditions, the infinite-dimensional dynamics near a bifurcation point to a finite-dimensional dynamics, described by a system of ordinary differential equations. An efficient way of studying the resulting reduced systems is with the help of normal form theory, which consists in suitably transforming a nonlinear system, in order to keep only the relevant nonlinear terms and to allow easier recognition of its dynamics. The combination of these two methods led over the recent years to significant progress in the understanding of various problems arising in applied sciences, and in particular in the study of nonlinear waves. A common feature of many of these problems is the presence of symmetries, as for instance reversibility symmetries. It turns out that both the center manifold reduction and the normal form transformations preserve symmetries, allowing then an efficient treatment of such problems. In addition, they provide a detailed comprehensive study near a singularity in the solution set of the system, which might also orient a numerical treatment of such problems.

The book is organized as follows. We start in Chapter 1 with a presentation of the simplest bifurcations for one- and two-dimensional ordinary differential equations: saddle-node, pitchfork, Hopf, and steady bifurcations in the presence of a simple symmetry group. The purpose of this particular choice is to also introduce the reader to some of the techniques and notations used in the next chapters. Chapter 2 is devoted to the center manifold theory. This is the core tool used all throughout this book. We present the center manifold reduction for infinite-dimensional systems, together with simple examples and exercises illustrating the variety of possible applications. The aim is to allow readers who are not familiar with the subject to use this reduction method simply by checking some clear assumptions. Chapter 3 is concerned with the normal form theory. In particular, we show how to systematically compute the normal forms in concrete situations. We illustrate the general theory on different bifurcation problems, for which we provide explicit formulas for the normal form, allowing one to obtain quantitative results for the resulting systems. In Chapter 4 the normal form theory is applied to the study of reversible bifurcations, which appear to be of particular importance in applications, as this is shown in Chapter 5. We focus on bifurcations of codimension 1, i.e., bifurcations involving a single parameter, which arise generically for systems in dimensions 2, 3, and 4. In all cases, we give the normal forms and collect some known facts on their dynamics. Finally, in Chapter 5 we present some applications of the methods described in the previous chapters. Without going into detail, for which we refer to the literature, we discuss hydrodynamic instabilities arising in the Couette–Taylor and the Bénard–Rayleigh convection problems and the questions of existence of traveling water waves, of almost planar waves in reaction-diffusion systems, and of traveling waves in lattices. The proofs (few being original) of some of the results in Chapters 2 and 3, and some of the normal form calculations in Chapters 3 and 4, are given in the Appendix. The Appendix is completed by a brief collection of results from the theory of linear operators used in Chapters 2, 3, and 5, and a short introduction to basic Sobolev spaces.

Historical Remark

Many authors refer to the work of C. G. J. Jacobi from 1834, on equilibria of self-gravitating rotating ellipsoids [71], as a first reference in the field of bifurcation theory. However, it seems that the first serious works on bifurcation problems were by Archimedes and Apollonios over 200 years BCE. Archimedes studied the equilibria of a floating paraboloid of revolution [107]. In today's terminology his results would correspond to a pitchfork bifurcation which breaks a flip symmetry, or to a steady bifurcation with $O(2)$ symmetry, when taking into account the invariance under rotations about the paraboloid axis. Apollonios studied the extrema of the length of segments joining a point of the plane to a given conic [74]. The number of solutions changes from one to three in crossing the envelope of the normals to the conic. Here again, due to the symmetry of the conic, we have an example of a pitchfork bifurcation. Finally, it seems that the French word "bifurcation" was introduced by Poincaré in 1885 [103].

Notational Remark

We adopt Arnold's notation [4] to distinguish classes of real matrices $\mathbf{L}$ *with the same Jordan form by indicating the eigenvalues of* $\mathbf{L}$ *and the length of their Jordan chain (e.g.,* $i\omega$ *when* $\mathbf{L}$ *has a pair of simple complex eigenvalues* $\pm i\omega$, 0^2 *when* $\mathbf{L}$ *has a double zero eigenvalue with a Jordan block of length 2,* $(i\omega_1)(i\omega_2)$ *when* $\mathbf{L}$ *has two pairs of complex eigenvalues* $\pm i\omega_1$ *and* $\pm i\omega_2$*, and so on).*

Remark on Numbering

Each of the five chapters of this book is numbered with Arabic numerals. Sections and subsections are numbered within chapters. *The sections are identified by two numbers, the number of the chapter and the number of the section in the chapter (e.g., Section 1.2 is the second section in Chapter 1). The subsections are identified by three numbers, the number of the chapter, the number of the section, and the number of the subsection (e.g., Section 1.2.1 is the first subsection in Section 1.2 of Chapter 1).*

Equations are numbered within sections *and identified by only two numbers: the number of the section inside the chapter (omitting the number of the chapter), and the number of the equation inside the section (e.g., equation (2.1) is the first equation in the second section of the current chapter). When referring to an equation, we only give the number, e.g., equation (2.1), if the equation is in the current chapter, but also mention the number of the chapter if the equation is in a different chapter, e.g., equation (2.1) in Chapter 2.*

Definitions, hypotheses, theorems, lemmas, corollaries, remarks, and exercises are numbered together within sections, *and identified by two numbers, just as the equations. Figures are numbered independently* within sections *and identified also by two numbers, just as equations.*

Acknowledgements

The authors gratefully acknowledge the Centre International de Rencontres Mathématiques in Luminy, France, for hosting them in the context of the Research in Pairs program during two weeks in February 2008, at the beginning of this work. We express very warm thanks to Klaus Kirchgässner for his seminal works which were the germ of a large part of this book (Chapters 2 and 4). Both authors kindly received from Klaus continuous invaluable encouragements during the past decades. We thank Pascal Chossat for his encouragements and support during the preparation of this book and Pierre Coullet and Alain Joets for attracting our attention to Archimedes and Apollonios' works.

Besançon and Nice,
March 2009

Mariana Haragus
Gérard Iooss

Contents

Chapter 1
Elementary Bifurcations

In this chapter we discuss typical local bifurcations in one and two dimensions. We restrict our attention to bifurcations of codimension 1, which require only one real parameter in order to generically occur. We include several cases of systems that possess an invariance under some simple symmetry.

1.1 Bifurcations in Dimension 1

We consider in this section two generic bifurcations that are found for scalar differential equations of the form

$$\frac{du}{dt} = f(u,\mu). \tag{1.1}$$

Here the unknown u is a real-valued function of the "time" t, and the vector field f is real-valued depending, besides u, upon a real parameter μ. The parameter μ is the *bifurcation parameter*.

We assume that the vector field f is of class C^k, $k \geq 2$, in a neighborhood of $(0,0)$ satisfying

$$f(0,0) = 0, \quad \frac{\partial f}{\partial u}(0,0) = 0. \tag{1.2}$$

The first condition shows that $u = 0$ is an equilibrium of (1.1) at $\mu = 0$. We are interested in (local) bifurcations that occur in the neighborhood of this equilibrium when we vary the parameter μ. Then the second equality in (1.2) is a necessary, but not sufficient, condition for the appearance of local bifurcations at $\mu = 0$. If $\partial f/\partial u(0,0) \neq 0$, the condition (1.2) is not satisfied and a direct application of the implicit function theorem shows that the equation $f(u,\mu) = 0$ possesses a unique solution $u = u(\mu)$ in a neighborhood of 0, for any μ sufficiently small. In particular, $u = 0$ is the only equilibrium of (1.1) in a neighborhood of the origin when $\mu = 0$, and the same property holds for μ sufficiently small. Furthermore, it is not difficult to show that the dynamics of (1.1) in a neighborhood of the origin is qualitatively

M. Haragus, G. Iooss, *Local Bifurcations, Center Manifolds, and Normal Forms in Infinite-Dimensional Dynamical Systems*, Universitext,
DOI 10.1007/978-0-85729-112-7_1,

the same for all sufficiently small values of the parameter μ. Consequently, in this situation no bifurcation occurs for small values of μ.

1.1.1 Saddle-Node Bifurcation

We discuss in this section the simplest bifurcation that occurs in one dimension, the *saddle-node bifurcation*. Throughout this section we make the following hypothesis.

Hypothesis 1.1 *Assume that the vector field f is of class $\mathscr{C}^k$, $k \geq 2$, in a neighborhood of $(0,0)$, and that it satisfies (1.2) and*

$$\frac{\partial f}{\partial \mu}(0,0) =: a \neq 0, \quad \frac{\partial^2 f}{\partial u^2}(0,0) =: 2b \neq 0. \tag{1.3}$$

An immediate consequence of this hypothesis is that f has the expansion

$$f(u,\mu) = a\mu + bu^2 + o(|\mu| + u^2),$$

as $(u,\mu) \to (0,0)$. It is then natural to start by studying the truncated equation

$$\frac{du}{dt} = a\mu + bu^2, \tag{1.4}$$

for which we expect that the dynamics near 0 are the same as those of (1.1).

Truncated Equation

The equilibria of (1.4) are solutions of the equation $a\mu + bu^2 = 0$, so that the truncated equation has no equilibria if $ab\mu > 0$, one equilibrium $u = 0$ if $\mu = 0$, and a pair of equilibria $u = \pm\sqrt{-a\mu/b}$ if $ab\mu < 0$. As for the dynamics, in the case $ab\mu > 0$ the function $a\mu + bu^2$ has a constant sign for all $u \in \mathbb{R}$, so that the solutions are monotone: increasing when $b > 0$ and decreasing when $b < 0$ (see Figure 1.1(a)). The same property holds for the nonequilibrium solutions in the case $\mu = 0$: They are increasing when $b > 0$ and decreasing when $b < 0$ (see Figure 1.1(b)). Finally, in the case $ab\mu < 0$, the function $a\mu + bu^2$ changes sign at the equilibrium points $u = \pm\sqrt{-a\mu/b}$, and we find that solutions with $|u(t)| < \sqrt{-a\mu/b}$ are decreasing when $b > 0$ and increasing when $b < 0$, whereas solutions with $|u(t)| > \sqrt{-a\mu/b}$ are increasing when $b > 0$ and decreasing when $b < 0$ (see Figure 1.1(c)). In particular, the equilibrium $-\sqrt{-a\mu/b}$ is attractive, asymptotically stable, when $b > 0$, and repelling, unstable, when $b < 0$; whereas, the equilibrium $\sqrt{-a\mu/b}$ has opposite stability properties.

We summarize in Figure 1.2 the dynamics of the truncated equation. In all cases, the qualitative behavior of the solutions changes when μ crosses 0. The value $\mu = 0$ is the bifurcation point. At this value, a pair of equilibria with opposite stability

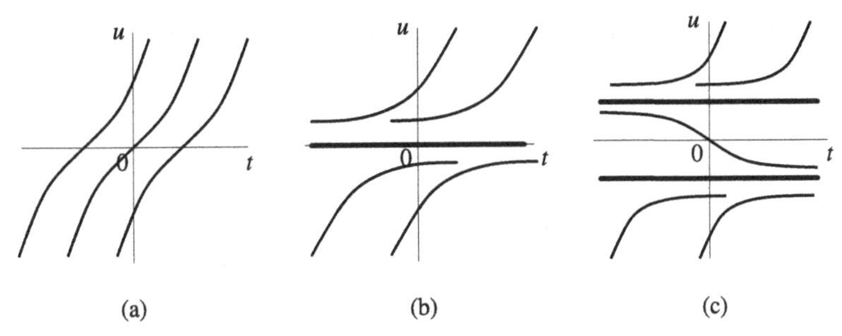

Fig. 1.1 Extended phase portrait, in the (t,u)-plane, of the truncated equation (1.4) for $b>0$ and (a) $a\mu>0$, (b) $\mu=0$, (c) $a\mu<0$.

properties emerges for $\mu>0$ when $ab<0$, and $\mu<0$ when $ab>0$. We are here in the presence of a *saddle-node bifurcation* (also called *fold* or *turning point* bifurcation).

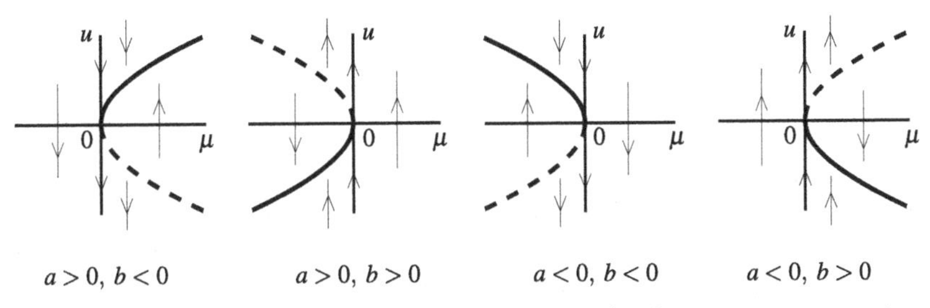

Fig. 1.2 Saddle-node bifurcation: bifurcation diagrams, in the (μ,u)-plane, of the truncated equation (1.4) for different values of a and b. The solid lines represent branches of stable equilibria, the dashed lines branches of unstable equilibria, and the arrows indicate the sense of increasing time t. For the full equation (1.1), under Hypothesis 1.1, the bifurcation diagrams are qualitatively the same in a neighborhood of the origin.

Remark 1.2 (Saddle-node bifurcation) *The names* fold *and* turning point bifurcations *are inspired by the form of the branch of the bifurcating equilibria in the (μ,u)-plane. The name* saddle-node bifurcation *comes from the fact that in the n-dimensional case, when $u(t)\in\mathbb{R}^n$, the two emerging equilibria are typically a saddle point and a node.*

Remark 1.3 (Explicit solutions) *The truncated equation (1.4) can be easily solved explicitly. For $ab\mu>0$ we set*

$$u=\sqrt{\frac{a\mu}{b}}v$$

and obtain the equation

$$\frac{dv}{dt}=\operatorname{sign}(b)\sqrt{ab\mu}\,(1+v^2).$$

The unique solution v of this first order ordinary differential equation (ODE) with initial data $v(0) = v_0$ is then given by

$$v(t) = \tan\left(\operatorname{sign}(b)\sqrt{ab\mu}\,t + \arctan(v_0)\right), \tag{1.5}$$

with $\arctan v_0 \in (-\pi/2, \pi/2)$. *Similarly, for $ab\mu < 0$ we set*

$$u = \sqrt{\frac{-a\mu}{b}}\,v$$

and obtain

$$\frac{dv}{dt} = \operatorname{sign}(b)\sqrt{-ab\mu}\,(v^2 - 1).$$

Hence,

$$\frac{v(t)+1}{v(t)-1} = \frac{v_0+1}{v_0-1}e^{-2\operatorname{sign}(b)\sqrt{-ab\mu}\,t},$$

for any $v_0 \neq \pm 1$. For $v_0 = \pm 1$, we find the constant solutions $v(t) = \pm 1$. Finally, for $\mu = 0$ we have the unique solution

$$u(t) = \frac{u_0}{1 - bu_0 t}$$

for initial data $u(0) = u_0$. These calculations then give the results described above and summarized in Figures 1.1 and 1.2. In addition, they show that the solutions blow up in finite time (either positive or negative), except for initial data $u_0 \in \left[-\sqrt{-a\mu/b}, \sqrt{-a\mu/b}\right]$, when $ab\mu \leq 0$.

Full Dynamics

Let us now consider the full equation (1.1). The equilibria are solutions of the equation $f(u,\mu) = 0$. Since $a \neq 0$ we can apply the implicit function theorem, which shows that this equation possesses a unique solution $\mu = g(u)$ for u close to 0. The map g is of class $\mathscr{C}^k$ in a neighborhood of the origin, and $g(0) = 0$. Moreover, its Taylor expansion is given by

$$\mu = -\frac{b}{a}u^2 + o(u^2).$$

This gives a curve in the (μ,u)-plane, which has a second order tangency at $(0,0)$ to the parabola $\mu = -bu^2/a$ found for the truncated equation (see Figure 1.2). In particular, this shows that the truncated equation and the full equation have the same number of equilibria in a neighborhood of the origin, which are $o(|\mu|^{1/2})$-close to each other. Consequently, the full equation has no equilibria if $ab\mu > 0$, one equilibrium $u = 0$ if $\mu = 0$, and a pair of equilibria $u_\pm(\mu) = \pm\sqrt{-a\mu/b} + o(|\mu|^{1/2})$ if $ab\mu < 0$.

As for the dynamics, the situation is also similar to that for the truncated equation, provided u and μ are sufficiently small. In the case $ab\mu > 0$ the function $f(u,\mu)$ has constant sign for sufficiently small u and μ, so that in a neighborhood of the origin the solutions are monotone: increasing when $b > 0$ and decreasing when $b < 0$ (see Figure 1.1(a)). When $\mu = 0$, the nonequilibrium solutions are monotone: increasing when $b > 0$ and decreasing when $b < 0$ (see Figure 1.1(b)). Finally, in the case $ab\mu < 0$, the function $f(u,\mu)$ changes sign at the equilibrium points $u_\pm(\mu)$, where

$$\frac{\partial f}{\partial u}(u_\pm(\mu),\mu) = 2bu_\pm(\mu) + o(|\mu|^{1/2})$$

has a definite sign. Then the equilibrium $u_-(\mu)$ is attractive, asymptotically stable when $b > 0$, and repelling, unstable when $b < 0$; whereas, the equilibrium $u_+(\mu)$ has opposite stability properties. Further, we find that solutions with $u(t) \in (u_-(\mu), u_+(\mu))$ are decreasing when $b > 0$ and increasing when $b < 0$, whereas solutions outside this interval, with $u(t) > u_+(\mu)$ or $u(t) < u_-(\mu)$ are increasing when $b > 0$ and decreasing when $b < 0$ (see Figure 1.1(c)). Just as for the truncated equation, we have here a *saddle-node bifurcation* (see Figure 1.2). We summarize this result in the following theorem.

Theorem 1.4 (Saddle-node bifurcation) *Assume that the vector field f satisfies Hypothesis 1.1. Then, for the differential equation (1.1) a saddle-node bifurcation occurs at $\mu = 0$. More precisely, the following properties hold in a neighborhood of 0 in $\mathbb{R}$ for sufficiently small μ:*

(i) If $ab < 0$ (resp., $ab > 0$) the differential equation has no equilibria for $\mu < 0$ (resp., for $\mu > 0$).
(ii) If $ab < 0$ (resp., $ab > 0$), the differential equation possesses precisely two equilibria $u_\pm(\varepsilon)$, $\varepsilon = |\mu|^{1/2}$ for $\mu > 0$ (resp., for $\mu < 0$), with opposite stabilities. Furthermore, the map $\varepsilon \mapsto u_\pm(\varepsilon)$ is of class $\mathscr{C}^{k-2}$ in a neighborhood of 0, and $u_\pm(\varepsilon) = O(\varepsilon)$.

Remark 1.5 (Higher orders) *In the case when $b = 0$, but still $a \neq 0$, one has to look for the lowest positive integer n for which the derivative $\partial^n f/\partial u^n(0,0) = bn! \neq 0$. The equilibria are then of order $O(|\mu|^{1/n})$, and for n even the qualitative phase portraits are as in Figure 1.2. When n is odd, the branch of equilibria crosses the u-axis, and on each side the equilibria have the same stability (stable if $b < 0$, or unstable if $b > 0$). If $a = b = 0$, then the situation requires a study of the Newton polygon and enters more into the framework of singularity theory (e.g., see [29]).*

1.1.2 Pitchfork Bifurcation

In many physical situations the problem possesses some symmetry. The simplest one that occurs in one dimension is the *reflection*, or *mirror, symmetry*: $u \mapsto -u$.

In this section we discuss this situation and the corresponding generic bifurcation, which is the *pitchfork bifurcation*.

We consider again the scalar differential equation (1.1) and now make the following assumptions.

Hypothesis 1.6 *Assume that the vector field in (1.1) is of class* $\mathscr{C}^k$, $k \geq 3$, *in a neighborhood of* $(0,0)$, *that it satisfies (1.2), and that it is odd with respect to u, i.e.,*

$$f(-u,\mu) = -f(u,\mu). \tag{1.6}$$

Further assume that

$$\frac{\partial^2 f}{\partial \mu \partial u}(0,0) =: a \neq 0, \quad \frac{\partial^3 f}{\partial u^3}(0,0) =: 6b \neq 0. \tag{1.7}$$

An immediate consequence of the oddness property of f is that

$$f(0,\mu) = 0 \text{ for all } \mu,$$

so that $u = 0$ is an equilibrium of (1.1) for all μ.

Truncated Equation

We start again by studying the truncated equation, which in this case is

$$\frac{du}{dt} = a\mu u + bu^3. \tag{1.8}$$

As for the full equation, $u = 0$ is an equilibrium of this equation for all values of μ. Upon solving the equation $a\mu u + bu^3 = 0$, we find that $u = 0$ is the only equilibrium of (1.8) if $ab\mu \geq 0$, and that for $ab\mu < 0$ there is an additional pair of nontrivial equilibria $u = \pm\sqrt{-a\mu/b}$. As for the dynamics, the nonequilibrium solutions are monotone, with monotonicity determined by the sign of the function $a\mu u + bu^3$. This function changes sign precisely at the equilibrium points, and a direct calculation leads to the diagram in Figure 1.3.

Again, the qualitative behavior of the solutions changes when μ crosses 0, so that $\mu = 0$ is a bifurcation point. At this value, the trivial equilibrium $u = 0$ changes its stability, and a pair of equilibria having the same stability, but opposite to that of the trivial equilibrium, emerges for $\mu > 0$ when $ab < 0$, and $\mu < 0$ when $ab > 0$. Here we are in the presence of a *pitchfork bifurcation*. The cases in which the emerging nontrivial equilibria are stable are called *supercritical*, whereas the cases in which these equilibria are unstable are called *subcritical*.

Remark 1.7 (Pitchfork bifurcation) *The name* pitchfork bifurcation *comes from the form of the branches of equilibria in the bifurcation diagram (even though actual pitchforks in the countryside may look different in various countries).*

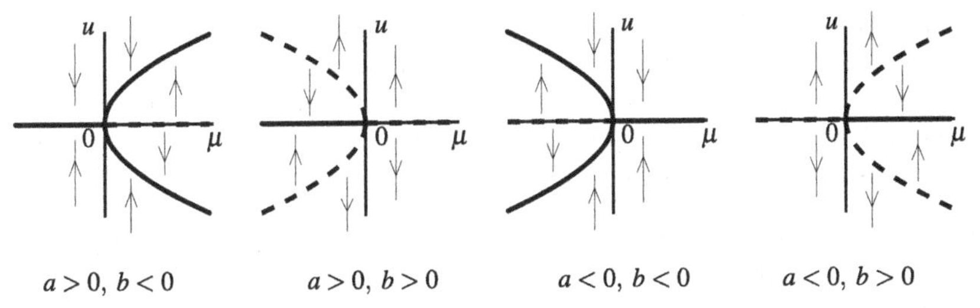

Fig. 1.3 Pitchfork bifurcation: bifurcation diagrams, in the (μ,u)-plane, of the truncated equation (1.8) for different values of a and b. The solid lines represent branches of stable equilibria, the dashed lines branches of unstable equilibria, and the arrows indicate the sense of increasing time t. For the full equation (1.1), under Hypothesis 1.6, the bifurcation diagrams are qualitatively the same.

Remark 1.8 (Explicit solution) *The truncated equation (1.8) can be easily solved explicitly. A direct calculation shows that its unique solution for initial data $u(0) = u_0$ is given by*

$$u^2(t) = \frac{a\mu u_0^2}{a\mu e^{-2a\mu t} + bu_0^2(e^{-2a\mu t} - 1)}.$$

This formula allows us to recover the bifurcation diagrams in Figure 1.3 and shows, in addition, that the unbounded nonequilibrium solutions blow up in either positive or negative finite time.

Full Dynamics

We consider now the full equation (1.1), under Hypothesis 1.6. The equilibria are solutions of $f(u,\mu) = 0$, and as already noticed, $u = 0$ is always an equilibrium because of the oddness of f in u. In addition, a standard analysis argument shows that we can rewrite the vector field f as follows:

$$f(u,\mu) = uh(u^2,\mu), \quad h(u^2,\mu) = a\mu + bu^2 + o(|\mu| + u^2),$$

where h is of class $\mathscr{C}^{(k-1)/2}$ in a neighborhood of $(0,0)$. Since $a \neq 0$ we can apply the implicit function theorem to the equation $h(u^2,\mu) = 0$, which shows that it has a unique solution $\mu = g(u^2)$ with $g(0) = 0$ and g of class $\mathscr{C}^{(k-1)/2}$ in a neighborhood of 0. The Taylor expansion of g is given by

$$\mu = -\frac{b}{a}u^2 + o(u^4).$$

We then conclude that there is a curve of nontrivial equilibria in the (μ,u)-plane that has a second order tangency at $(0,0)$ to the parabola $\mu = -bu^2/a$ found for the truncated equation (see Figure 1.3), and which is symmetric with respect to the μ-axis. Again, this shows that the truncated equation and the full equation have the same number of equilibria in a neighborhood of the origin, which are $o(|\mu|^{1/2})$-close to each other. As for the dynamics, it is here again easy to study by looking at the sign of $f(u,\mu)$. The arguments are analogous to those in the case discussed in Section 1.1.1 and lead to the bifurcation diagrams in Figure 1.3. We summarize these results in the following theorem.

Theorem 1.9 (Pitchfork bifurcation) *Assume that the vector field f satisfies Hypothesis 1.6. Then, for the differential equation (1.1), a supercritical (resp., subcritical) pitchfork bifurcation occurs at $\mu = 0$ when $b < 0$ (resp., $b > 0$). More precisely, the following properties hold in a neighborhood of 0 in $\mathbb{R}$ for sufficiently small μ:*

(i) If $ab < 0$ (resp., $ab > 0$) the differential equation has precisely one trivial equilibrium $u = 0$ for $\mu < 0$ (resp., for $\mu > 0$). This equilibrium is stable when $b < 0$ and unstable when $b > 0$.

(ii) If $ab < 0$ (resp., $ab > 0$), the differential equation possesses, for $\mu > 0$ (resp., for $\mu < 0$), the trivial equilibrium $u = 0$ and two nontrivial equilibria $u_{\pm}(\varepsilon)$, $\varepsilon = |\mu|^{1/2}$, which are symmetric, $u_{-}(\varepsilon) = -u_{+}(\varepsilon)$. The map $\varepsilon \mapsto u_{\pm}(\varepsilon)$ is of class $\mathscr{C}^{k-3}$ in a neighborhood of 0, and $u_{\pm}(\varepsilon) = O(\varepsilon)$. Furthermore, the nontrivial equilibria are stable when $b < 0$ and unstable when $b > 0$, whereas the trivial equilibrium has opposite stability.

Remark 1.10 (Higher orders) *In the case $b = 0$, but still $a \neq 0$, one has to look for the lowest n for which the derivative $\partial^{2n+1} f/\partial u^{2n+1}(0,0) \neq 0$. The equilibria are then of order $O(|\mu|^{1/2n})$ and the qualitative phase portraits are as in Figure 1.3. If $a = b = 0$ then the situation requires a study of the Newton polygon and belongs more in the field of singularity theory (e.g., see [29]).*

1.2 Bifurcations in Dimension 2

In the remainder of this chapter we consider differential equations in $\mathbb{R}^2$,

$$\frac{du}{dt} = \mathbf{F}(u,\mu). \tag{2.1}$$

Now the unknown u takes values in $\mathbb{R}^2$, just as the vector field $\mathbf{F}$, which depends again besides depending on u, upon a real parameter μ.

We assume that the vector field $\mathbf{F}$ is of class $\mathscr{C}^k$, $k \geq 3$, in a neighborhood of $(0,0)$, satisfying

$$\mathbf{F}(0,0) = 0. \tag{2.2}$$

Again, this condition shows that $u = 0$ is an equilibrium of (2.1) at $\mu = 0$. We are interested in (local) bifurcations which occur in the neighborhood of this equilibrium when varying the parameter μ. The appearance, or the absence, of bifurcations is in this case determined by the linearization of the vector field at $(0,0)$,

$$\mathbf{L} := D_u F(0,0),$$

which is a linear map (operator) acting in $\mathbb{R}^2$. In the case when the linear map $\mathbf{L}$ has no eigenvalue on the imaginary axis, the Hartman–Grobman theorem shows that the phase portraits of the equation (2.1) are qualitatively the same upon varying μ in a neighborhood of 0 (e.g., see [32], [46]). In particular, no local bifurcations occur in this case. When $\mathbf{L}$ has eigenvalues on the imaginary axis, bifurcations may occur at $\mu = 0$. The type of these bifurcations depend upon the location of the eigenvalues on the imaginary axis. While we do not attempt to give a complete description of the possible bifurcations for two-dimensional systems, we focus in this section on two cases: $\mathbf{L}$ has a pair of complex conjugated purely imaginary eigenvalues (Hopf bifurcation), and $\mathbf{L}$ has a double zero eigenvalue (steady bifurcation) for a system possessing an $O(2)$-symmetry. The cases in which 0 is a simple eigenvalue of $\mathbf{L}$ and another eigenvalue is real and different from 0, fall in the discussion of Chapter 2, using a center manifold reduction (e.g., see the examples in Sections 2.2.4 and 2.4.2). The case of 0 a double, non-semisimple (with only one eigenvector) eigenvalue of $\mathbf{L}$ is treated in Chapter 4 in the presence of a reversibility symmetry, which makes this bifurcation generically occur with only one real parameter.

1.2.1 Hopf Bifurcation

One generic bifurcation in two dimensions is the *Hopf bifurcation*, which occurs when the linear operator $\mathbf{L}$ possesses a pair of purely imaginary complex conjugated eigenvalues. This bifurcation was first proved in two dimensions by Andronov [3] in 1937; it is therefore also referred to as Andronov–Hopf bifurcation, after it was guessed by H. Poincaré in the early 1900s [102]. The n-dimensional case was proved by Hopf in 1942, using the Lyapunov–Schmidt method [50]. Our analysis relies upon the *normal form* theory that we develop in detail in Chapter 3.

Hypothesis 2.1 *Assume that the vector field* $\mathbf{F}$ *in (2.1) is of class* $\mathscr{C}^k$, $k \geq 5$, *that it satisfies (2.2), and that the two eigenvalues of the linear operator* $\mathbf{L}$ *are* $\pm i\omega$ *for some* $\omega > 0$.

Remark 2.2 *(i) Since the operator* $\mathbf{L}$ *is real, its spectrum is symmetric with respect to the real axis, so that purely imaginary eigenvalues occur in pairs* $\pm i\omega$.

(ii) Hypothesis 2.1 implies that $\mathbf{L}$ *is invertible, since* 0 *is not an eigenvalue of* $\mathbf{L}$. *By arguing using the implicit function theorem, we can then solve the equation* $\mathbf{F}(u,\mu) = 0$ *near* $(0,0)$. *This gives a unique family of steady solutions* $u = u(\mu)$ *for sufficiently small* μ, *with* $u(0) = 0$. *Furthermore, the map* $\mu \mapsto u(\mu)$ *is of class* $\mathscr{C}^k$, *and by making the change of variables* $u \mapsto u(\mu) + u$, *we may replace assumption (2.2) by*

$$\mathbf{F}(0,\mu) = 0. \tag{2.3}$$

In this way, $u = 0$ *becomes an equilibrium of (2.1) for all values of* μ *sufficiently small. We point out that it is then a generic possibility that a pair of complex eigenvalues of the linearized operators* $\mathbf{L}_\mu = D_u\mathbf{F}(0,\mu)$ *crosses the imaginary axis for a critical value of the parameter* μ *(here* $\mu = 0$*).*

(iii) *In contrast to the two bifurcations discussed before, now the number of equilibria of the differential equation stays constant upon varying μ in a neighborhood of* 0. *As we shall see, we have here a different type of bifurcation in which it is the dynamics of the differential equation that change at the bifurcation point $\mu = 0$, and not the number of equilibria. Such bifurcations are also called* dynamic bifurcations, *whereas those in which the number of equilibria changes are also called* steady bifurcations.

Consider the eigenvectors ζ and $\overline{\zeta}$ associated with the eigenvalues $i\omega$ and $-i\omega$ of $\mathbf{L}$, respectively,

$$\mathbf{L}\zeta = i\omega\zeta, \quad \mathbf{L}\overline{\zeta} = -i\omega\overline{\zeta}.$$

A convenient way of looking at equation (2.1) in this case is by representing any $u \in \mathbb{R}^2$ by a complex coordinate $z \in \mathbb{C}$ through

$$u = z\zeta + \overline{z}\overline{\zeta}. \tag{2.4}$$

Adopting the same decomposition for $\mathbf{F}$, we write

$$\mathbf{F}(u,\mu) = f(z,\overline{z},\mu)\zeta + \overline{f}(z,\overline{z},\mu)\overline{\zeta}$$

and then obtain two complex differential equations

$$\frac{dz}{dt} = f(z,\overline{z},\mu), \tag{2.5}$$

together with its complex conjugate. The complex-valued vector field f is of class $\mathscr{C}^k$ in a neighborhood of the origin in $\mathbb{R}^2 \times \mathbb{R}$, where the argument in $\mathbb{R}^2$ is represented by the "diagonal" $(z,\overline{z}) \in \mathbb{C}^2$. (Notice that f is not holomorphic in z.) In these coordinates, the differential of the new vector field $(f,\overline{f})$ at the origin is given by

$$\mathbf{L} = \begin{pmatrix} \frac{\partial f}{\partial z}(0,0,0) & \frac{\partial f}{\partial \overline{z}}(0,0,0) \\ \frac{\partial \overline{f}}{\partial z}(0,0,0) & \frac{\partial \overline{f}}{\partial \overline{z}}(0,0,0) \end{pmatrix} = \begin{pmatrix} i\omega & 0 \\ 0 & -i\omega \end{pmatrix}.$$

Though the linear part $\mathbf{L}$ of (2.5) is now in canonical form, it is still difficult to detect its dynamics in general. Our approach relies upon the normal form theory developed in Chapter 3. Roughly speaking, the idea of normal forms consists in adding a polynomial term to the change of coordinates (2.4), such that the vector field of the resulting system has a simpler, particular form, also at the nonlinear level.

Normal Form

According to the general normal form theorem, Theorem 2.2 in Chapter 3, for any integer $p \leq k$, and any μ sufficiently small, there exists a polynomial Φ_μ of degree p in $(A,\overline{A})$, with complex coefficients, depending upon μ, and taking values in $\mathbb{R}^2$, such that

$$\Phi_0(0,0)=0, \quad \partial_A\Phi_0(0,0)=0, \quad \partial_{\overline{A}}\Phi_0(0,0)=0,$$

and that the (near to identity) change of variables in $\mathbb{R}^2$,

$$u = A\zeta + \overline{A\zeta} + \Phi_\mu(A,\overline{A}), \quad A \in \mathbb{C}, \tag{2.6}$$

transforms the equation (2.1) into a differential equation, or "amplitude equation,"

$$\frac{dA}{dt} = i\omega A + N_\mu(A,\overline{A},) + \rho(A,\overline{A},\mu). \tag{2.7}$$

Here N_μ is a complex polynomial of degree p in $(A,\overline{A})$, with

$$N_0(0,0)=0, \quad \partial_A N_0(0,0)=0, \quad \partial_{\overline{A}} N_0(0,0)=0, \tag{2.8}$$

and the remainder ρ satisfies

$$\rho(A,\overline{A},\mu) = o(|A|^p).$$

Furthermore, the polynomial

$$\mathbf{N}_\mu(A,\overline{A}) = (N_\mu(A,\overline{A}), \overline{N_\mu}(A,\overline{A}))$$

commutes with the mapping

$$(A,\overline{A}) \mapsto (e^{i\omega t}A, e^{-i\omega t}\overline{A}),$$

which implies that

$$N_\mu(e^{i\omega t}A, e^{-i\omega t}\overline{A}) = e^{i\omega t}N_\mu(A,\overline{A}) \text{ for all } A,t. \tag{2.9}$$

Remark 2.3 (Symmetry) *We observe that the transformation (2.6) has the effect of* adding a symmetry for the terms up to degree p *in the expansion of the transformed vector field. The property (2.9) means that the truncation at order p of the vector field is equivariant under rotations in the complex plane, which is a rather strong restriction. We point out that in general this transformation cannot be achieved for $p=\infty$, even when* $\mathbf{F}$ *in (2.1) is analytic.*

The following elementary lemma allows us to describe more precisely the polynomials N_μ satisfying (2.9).

Lemma 2.4 *Let f be a complex-valued function of class $\mathscr{C}^k$, $k \geq 1$, defined in a neighborhood $\mathscr{U}$ of the origin in $\{(z,\overline{z})\ ;\ z \in \mathbb{C}\}$, and which verifies*

$$f(e^{i\omega t}z, e^{-i\omega t}\overline{z}) = e^{i\omega t}f(z,\overline{z}) \text{ for any } t \in \mathbb{R} \text{ and } (z,\overline{z}) \in \mathscr{U}. \tag{2.10}$$

Then there exists an even, complex-valued function g of class $\mathscr{C}^{k-1}$ defined in a neighborhood of 0 *in $\mathbb{R}$, such that*

$$f(z,\overline{z}) = zg(|z|). \tag{2.11}$$

Furthermore, if f is a polynomial, then g is an even polynomial, $g(|z|) = \phi(|z|^2)$, for a polynomial ϕ.

Proof First, choose $t = -\arg z/\omega$ in (2.10), which then reads

$$f(z,\overline{z}) = e^{i\arg z} f(|z|,|z|) \text{ for all } (z,\overline{z}) \in \mathscr{U}.$$

Next, we take $t = \pi/\omega$ and find

$$f(-z,-\overline{z}) = -f(z,\overline{z}).$$

It follows that $f(|z|,|z|)$ is odd in $|z|$, and a standard analysis result implies that there exists an even function g of class $\mathscr{C}^{k-1}$ in a neighborhood of 0 in $\mathbb{R}$, such that

$$f(|z|,|z|) = |z|g(|z|).$$

This proves the first part of the lemma. For a polynomial f, the above identity implies that g is an even polynomial in $|z|$. Hence, there exists a polynomial ϕ such that $g(|z|) = \phi(|z|^2)$, which completes the proof of the lemma. □

Going back to the differential equation (2.7), the above lemma together with the equalities (2.8) show that it is of the form

$$\frac{dA}{dt} = i\omega A + AQ(|A|^2,\mu) + \rho(A,\overline{A},\mu). \tag{2.12}$$

Here Q is a complex-valued polynomial with expansion

$$Q(|A|^2,\mu) = a\mu + b|A|^2 + O((|\mu| + |A|^2)^2), \tag{2.13}$$

in which a and b are complex numbers. We make the following generic assumption on the coefficients a and b.

Hypothesis 2.5 *The complex coefficients a and b in the expansion (2.13) of the polynomial Q have nonzero real parts, $a_r \neq 0$ and $b_r \neq 0$.*

Truncated System

We start again by the study of the truncated system obtained by suppressing the higher order terms ρ in (2.12). We introduce polar coordinates by setting

$$A = re^{i\phi},$$

where $r \in \mathbb{R}^+$ and $\phi \in \mathbb{R}/2\pi\mathbb{Z}$. We obtain the equation

$$\frac{dr}{dt} + ir\frac{d\phi}{dt} = i\omega r + rQ(r^2,\mu),$$

and by taking the real and imaginary parts, we find the system

$$\frac{dr}{dt} = rQ_r(r^2,\mu), \tag{2.14}$$

$$\frac{d\phi}{dt} = \omega + Q_i(r^2,\mu), \tag{2.15}$$

where $Q_r = (Q+\overline{Q})/2$ and $Q_i = (Q-\overline{Q})/2i$ are the real and imaginary parts of the polynomial Q, respectively. Then Q_r and Q_i are polynomials of degree $\leq (p-1)/2$ in r^2, with $Q_r(0,0) = Q_i(0,0) = 0$, and expansions

$$Q_r(r^2,\mu) = a_r\mu + b_r r^2 + O((|\mu|+r^2)^2),$$
$$Q_i(r^2,\mu) = a_i\mu + b_i r^2 + O((|\mu|+r^2)^2).$$

The real coefficients a_r and b_r represent the real parts of a and b, respectively, which are both nonzero, by Hypothesis 2.5, whereas a_i and b_i represent the imaginary parts of a and b, respectively.

The key property of the system (2.14)–(2.15) for r and ϕ is that the *radial equation (2.14) for r* decouples, so that we can solve it separately. Upon comparing (2.14) with the scalar differential equation discussed in Section 1.1.2, we conclude that for this equation a *pitchfork bifurcation* occurs at $\mu = 0$, which is supercritical when $b_r < 0$ and subcritical when $b_r > 0$. The bifurcation diagrams for this equation are the same as those in Figure 1.3 with a and b replaced by a_r and b_r, respectively. Since for the radial equation we restrict ourselves to positive solutions, then for $a_r b_r < 0$ (resp., $a_r b_r > 0$), the radial equation possesses the positive steady solution

$$r^*(\mu) = \sqrt{-\frac{a_r\mu}{b_r}} + O(|\mu|^{3/2}),$$

for $\mu > 0$ (resp., $\mu < 0$). Upon substituting this solution in the equation (2.15) we obtain the derivative of the phase (pulsation),

$$\frac{d\phi^*(\mu)}{dt} = \omega^*(\mu) = \omega + Q_i((r^*(\mu))^2,\mu) = \omega + \left(a_i - b_i\frac{a_r}{b_r}\right)\mu + O(|\mu|^2),$$

and going back to the amplitude A this gives the periodic solutions

$$A^*(t,\mu) = r^*(\mu)e^{i\omega^*(\mu)t}, \quad t \in \mathbb{R}. \tag{2.16}$$

The stability of these periodic solutions is the same as that of the steady solution $r^*(\mu)$ of (2.14): They are stable when $b_r < 0$ and unstable when $b_r > 0$. Figure 2.1 illustrates the bifurcation diagram in the supercritical case $a_r > 0$ and $b_r < 0$. Similar bifurcation diagrams can be easily obtained in the other three cases, just as for the pitchfork bifurcation in Figure 1.3.

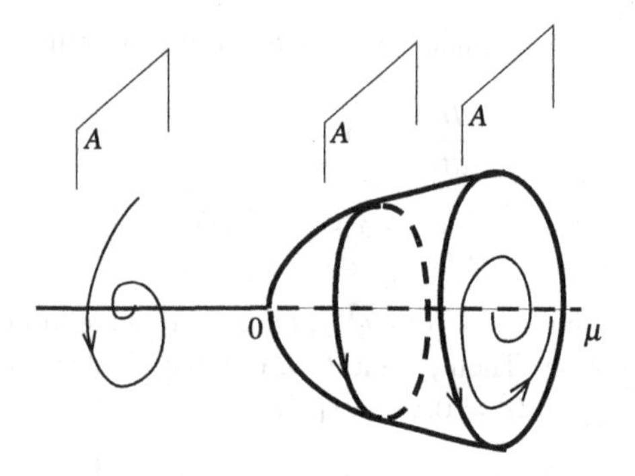

Fig. 2.1 Hopf bifurcation in the case $a_r > 0$, $b_r < 0$.

Persistence of Periodic Solutions

We now turn back to the hardest part of the analysis, that is, the proof of the persistence of such periodic orbits for the full equation (2.12). In what follows, we assume $a_r > 0$, $b_r < 0$, and then also $\mu > 0$, to fix ideas, the proof being analogous in the other cases.

As for the truncated system we introduce polar coordinates by setting

$$A = re^{i\phi}, \quad r \in \mathbb{R}^+, \quad \phi \in \mathbb{R}/2\pi\mathbb{Z},$$

and obtain the system

$$\frac{dr}{dt} = f_r(r,\phi,\mu) = rQ_r(r^2,\mu) + R_r(r,\phi,\mu),$$
$$\frac{d\phi}{dt} = f_\phi(r,\phi,\mu) = \omega + Q_i(r^2,\mu) + R_\phi(r,\phi,\mu),$$

where $R_r = O((r+|\mu|)^{p+1})$ and $R_\phi = O((r+|\mu|)^{p+1}/r)$. We now set

$$r = \mu^{1/2}\left(\sqrt{-\frac{a_r}{b_r}} + v\right),$$

where the new unknown v is supposed to lie in a small interval near 0. In this annular region of the plane, for μ small enough,

$$f_\phi(r,\phi,\mu) = \omega + O(\mu)$$

has a constant sign, and

$$f_r(r,\phi,\mu) = \mu^{3/2}\left(\sqrt{-\frac{a_r}{b_r}} + v\right)\left(2vb_r\sqrt{-\frac{a_r}{b_r}} + b_r v^2\right) + O(\mu^{(p+1)/2}).$$

Using the fact that we can choose $p \geq 4$, this leads to the equation

$$\frac{dv}{d\phi} = -\frac{2a_r\mu}{\omega}v + \rho_1(v,\phi,\mu), \quad \rho_1(v,\phi,\mu) = O(\mu v^2 + \mu^2), \tag{2.17}$$

where ρ_1 is Lipschitz-continuous and bounded for $-\varepsilon < v < \varepsilon$, for ε small enough. We use a fixed point argument to show that this equation possesses a 2π-periodic solution for sufficiently small μ, which then gives the desired result.

By Duhamel's formula, the solution $v(\phi)$, for initial data $v(0) = v_0$, of the differential equation (2.17) satisfies the integral equation

$$v(\phi) = e^{-\frac{2a_r\mu}{\omega}\phi}v_0 + \int_0^\phi e^{-\frac{2a_r\mu}{\omega}(\phi-\theta)}\rho_1(v,\theta,\mu)d\theta.$$

The uniqueness of the solution of the initial value problem, and its differentiability with respect to the initial data v_0, allow us to conclude that, for $|v_0| < \varepsilon$, we have

$$v(\phi) = e^{-\frac{2a_r\mu}{\omega}\phi}v_0 + h(v_0,\phi,\mu), \quad h(v_0,\mu) = O(\mu v_0^2 + \mu^2)$$

for $\phi \in [0,2\pi]$, where the function h is continuously differentiable. Now, if we can find a solution v_0 for the equation

$$v_0 = e^{-\frac{2a_r\mu}{\omega}2\pi}v_0 + h(v_0,2\pi,\mu), \tag{2.18}$$

then the corresponding solution of the integral equation satisfies $v(2\pi) = v_0$, so that we have a *periodic orbit* of (2.12) in a small neighborhood of the circle $|A| = \mu^{1/2}\sqrt{-a_r/b_r}$. Indeed, observe that the *Poincaré map*

$$v_0 \mapsto e^{-\frac{2a_r\mu}{\omega}2\pi}v_0 + h(v_0,2\pi,\mu),$$

is a contraction in a sufficiently small interval $[-\varepsilon,\varepsilon]$, because the derivative of the right hand side with respect to v_0 is

$$1 - \frac{2a_r\mu}{\omega}2\pi + O(\mu^2 + \mu\varepsilon) < 1$$

for ε and $\mu > 0$ small enough. Consequently, this mapping possesses a unique fixed point v_0 solution of (2.18) for sufficiently small $\mu > 0$.

This shows that the full equation (2.12) has a periodic orbit close to the circle of radius $|A| = r^*(\mu)$, and with period approximated by that of the solution (2.16) of the truncated equation. In addition, this proof allows us to conclude that this periodic orbit is attractive for $b_r < 0$. We summarize this result in the following *Hopf bifurcation theorem* (see also Figure 2.1).

Theorem 2.6 (Hopf bifurcation) *Assume that Hypotheses 2.1 and 2.5 hold. Then, for the differential equation (2.1) a supercritical (resp., subcritical) Hopf bifurcation occurs at* $\mu = 0$ *when* $b_r < 0$ *(resp.,* $b_r > 0$*). More precisely, the following properties hold in a neighborhood of* 0 *in* $\mathbb{R}^2$ *for sufficiently small* μ*:*

(i) *If $a_r b_r < 0$ (resp., $a_r b_r > 0$) the differential equation has precisely one equilibrium $u(\mu)$ for $\mu < 0$ (resp., for $\mu > 0$) with $u(0) = 0$. This equilibrium is stable when $b_r < 0$ and unstable when $b_r > 0$.*

(ii) *If $a_r b_r < 0$ (resp., $a_r b_r > 0$), the differential equation possesses for $\mu > 0$ (resp., for $\mu < 0$) an equilibrium $u(\mu)$ and a unique periodic orbit $u^*(\mu) = O(|\mu|^{1/2})$, which surrounds this equilibrium. The periodic orbit is stable when $b_r < 0$ and unstable when $b_r > 0$, whereas the equilibrium has opposite stability.*

Remark 2.7 *The proof in dimension 2 is originally due to Andronov [3]. The n dimensional case is due to Hopf [50]. The present proof using normal form arguments is contained in Ruelle and Takens [108]. We also refer to Marsden and McCracken [94] and Vanderbauwhede [120].*

Remark 2.8 (Higher orders) *In the above proof, we extensively use the assumption that the coefficient b_r is not zero. In the case when this coefficient is zero, one needs to consider the higher order terms, like the term of order $O(A|A|^4)$ in the expansion of the amplitude equation, and so on. If the problem is not completely degenerated, it is then possible to adapt the above proof without difficulty. Of course, this then gives other orders of magnitude for the bifurcating periodic solutions. We shall see in Chapter 4 that in the case of* reversible systems *all terms in Q_r in the radial equation disappear, leading to a degenerated situation.*

How to Compute the Hopf Bifurcation

We show now how to compute the important coefficients a and b in the normal form (2.12), (2.13), starting from the expansion of the vector field $\mathbf{F}$ in (2.1).

Consider the Taylor expansion of the vector field $\mathbf{F}$ in (2.1),

$$\mathbf{F}(u,\mu) = \sum_{1 \le r+q \le k} \mu^q \mathbf{F}_{rq}(u^{(r)}) + o(|\mu| + ||u||)^k, \quad \mathbf{L} = \mathbf{F}_{10}, \tag{2.19}$$

where $\mathbf{F}_{rq}$ is the r-linear symmetric operator from $(\mathbb{R}^2)^r$ to $\mathbb{R}^2$,

$$\mathbf{F}_{rq} = \frac{1}{r!q!} \frac{\partial^q}{\partial \mu^q} D_u^r \mathbf{F}(0,0),$$

and $u^{(r)} := (u, u, ...u)$ for $u \in \mathbb{R}^2$. In particular, the map $u \mapsto \mathbf{F}_{rq}(u^{(r)})$ is homogeneous of degree r in the coordinates of u. Similarly, for $\boldsymbol{\Phi}_\mu$ in (2.6) we write

$$\Phi_\mu(A, \overline{A}) = \sum_{1 \le r+s+q \le p} \Phi_{rsq} A^r \overline{A}^s \mu^q, \tag{2.20}$$

with

$$\Phi_{100} = 0, \quad \Phi_{010} = 0, \quad \Phi_{rsq} = \overline{\Phi}_{srq}.$$

Next, we substitute the change of variables (2.6) into (2.1) and obtain the identity

$$(\zeta + \partial_A \Phi_\mu)\frac{dA}{dt} + (\overline{\zeta} + \partial_{\overline{A}} \Phi_\mu)\frac{d\overline{A}}{dt} = \mathbf{F}(A\zeta + \overline{A}\overline{\zeta} + \Phi_\mu, \mu),$$

in which, according to the normal form (2.12)–(2.13), we have

$$\frac{dA}{dt} = i\omega A + a\mu A + bA|A|^2 + O(\mu^2|A| + |\mu||A|^3 + |A|^5). \tag{2.21}$$

Replacing $\mathbf{F}$ and Φ_μ by the expressions (2.19) and (2.20), we now identify the different powers of $(A, \overline{A}, \mu)$ in the identity above in order to determine the coefficients a, b, and Φ_{rsq} from the known coefficients $\mathbf{F}_{rq}$.

First, at order $O(A)$ we recover the eigenvalue problem

$$i\omega\zeta = \mathbf{L}\zeta,$$

and then successively, respectively at orders $O(\mu)$, $O(\mu A)$, $O(A^2)$, $O(A\overline{A})$, $O(A^3)$, $O(A^2\overline{A})$, we find

$$0 = \mathbf{L}\Phi_{001} + \mathbf{F}_{01} \tag{2.22}$$
$$a\zeta + (i\omega - \mathbf{L})\Phi_{101} = \mathbf{F}_{11}\zeta + 2\mathbf{F}_{20}(\zeta, \Phi_{001}) \tag{2.23}$$
$$(2i\omega - \mathbf{L})\Phi_{200} = \mathbf{F}_{20}(\zeta, \zeta) \tag{2.24}$$
$$-\mathbf{L}\Phi_{110} = 2\mathbf{F}_{20}(\zeta, \overline{\zeta}) \tag{2.25}$$
$$(3i\omega - \mathbf{L})\Phi_{300} = 2\mathbf{F}_{20}(\zeta, \Phi_{200}) + \mathbf{F}_{30}(\zeta, \zeta, \zeta) \tag{2.26}$$
$$b\zeta + (i\omega - \mathbf{L})\Phi_{210} = 2\mathbf{F}_{20}(\overline{\zeta}, \Phi_{200}) + 2\mathbf{F}_{20}(\zeta, \Phi_{110}) + 3\mathbf{F}_{30}(\zeta, \zeta, \overline{\zeta}). \tag{2.27}$$

All these equations are linear, and (2.22), (2.24), (2.25), (2.26) can be easily solved, because the operators $\mathbf{L}, (2i\omega - \mathbf{L}), (3i\omega - \mathbf{L})$ are invertible. This allows us to compute $\Phi_{001}, \Phi_{200}, \Phi_{110}, \Phi_{300}$, and the complex conjugates Φ_{020}, Φ_{030}. The equations (2.23) and (2.27) have the same structure, however, with the noninvertible matrix $(i\omega - \mathbf{L})$. The kernel of this matrix is one-dimensional, since $\pm i\omega$ are simple eigenvalues of $\mathbf{L}$, and one compatibility condition is needed in order to solve each of these equations. A convenient way of computing this compatibility condition is with the help of the eigenvector ζ^* of the adjoint operator satisfying

$$(-i\omega - \mathbf{L}^*)\zeta^* = 0, \quad \langle \zeta, \zeta^* \rangle = 1,$$

where $\langle \cdot, \cdot \rangle$ denotes the Hermitian scalar product in $\mathbb{C}^2$. (For $\zeta = (z_1, z_2) \in \mathbb{C}^2$ and $\zeta^* = (z_1^*, z_2^*) \in \mathbb{C}^2$, we take the Hermitian scalar product defined by

$$\langle \zeta, \zeta^* \rangle = z_1\overline{z_1^*} + z_2\overline{z_2^*}.)$$

Upon computing the Hermitian scalar product of these equations with ζ^* we find

$$a = \langle \mathbf{F}_{11}\zeta + 2\mathbf{F}_{20}(\zeta, \Phi_{001}), \zeta^* \rangle, \tag{2.28}$$

and

$$b = \langle 2\mathbf{F}_{20}(\overline{\zeta}, \Phi_{200}) + 2\mathbf{F}_{20}(\zeta, \Phi_{110}) + 3\mathbf{F}_{30}(\zeta, \zeta, \overline{\zeta}), \zeta^* \rangle, \tag{2.29}$$

in which

$$\begin{aligned}
\Phi_{001} &= -\mathbf{L}^{-1}\mathbf{F}_{01}, \\
\Phi_{200} &= (2i\omega - \mathbf{L})^{-1}\mathbf{F}_{20}(\zeta, \zeta), \\
\Phi_{110} &= -2\mathbf{L}^{-1}\mathbf{F}_{20}(\zeta, \overline{\zeta}),
\end{aligned}$$

are obtained as explained above. We point out that $\Phi_{001} = 0$ in the case when $u = 0$ is a solution for all μ, since then $\mathbf{F}(0, \mu) = 0$, so that $\mathbf{F}_{01} = 0$. In the same way, it is possible to derive formulas for higher order coefficients in (2.21), if needed.

1.2.2 Example: Homogeneous Brusselator

Consider the following system of ODEs:

$$\begin{aligned}
\frac{du_1}{dt} &= -(\beta + 1)u_1 + u_1^2 u_2 + \alpha \\
\frac{du_2}{dt} &= \beta u_1 - u_1^2 u_2,
\end{aligned} \tag{2.30}$$

in which $u(t) = (u_1(t), u_2(t)) \in \mathbb{R}^2$ and α, β are positive constants.

Remark 2.9 *This system, called the* homogeneous Brusselator *[106], arises in the modeling of an autocatalytic chemical reaction ruled by the following reaction mechanism:*

$$\begin{aligned}
A &\overset{k_1}{\rightarrow} X \\
B + X &\overset{k_2}{\rightarrow} Y + D \\
2X + Y &\overset{k_3}{\rightarrow} 3X \\
X &\overset{k_4}{\rightarrow} E.
\end{aligned}$$

Here A, B, D, and E denote different chemical species, X and Y are intermediate products, and k_j represent the speeds of reactions. Denoting by X, Y, A, B the chemical concentrations of the corresponding species, assuming that the concentrations are homogeneous, and that the concentrations of components A and B are maintained constant, one finds that the evolution of X and Y is governed by the system of ODEs

$$\begin{aligned}
\frac{dX}{dt} &= k_1 A - k_2 BX + k_3 X^2 Y - k_4 X \\
\frac{dY}{dt} &= k_2 BX - k_3 X^2 Y.
\end{aligned}$$

Upon setting

$$u_1 = \sqrt{\frac{k_3}{k_4}}X, \quad u_2 = \sqrt{\frac{k_3}{k_4}}Y, \quad \alpha = \sqrt{\frac{k_3}{k_4}}\frac{k_1 A}{k_4}, \quad \beta = \frac{k_2 B}{k_4}, \quad \bar{t} = k_4 t,$$

this leads to the system (2.30), in which we have dropped the bar on t.

The system (2.30) possesses one equilibrium at $(u_1, u_2) = (\alpha, \beta/\alpha)$ for any positive constants α and β. The linearization at this equilibrium has the two eigenvalues

$$\lambda_{\pm} = \frac{1}{2}(\beta - 1 - \alpha^2) \pm \left(-\alpha^2 - \frac{1}{4}(\beta - 1 - \alpha^2)^2\right)^{1/2}.$$

When $\beta < 1 + \alpha^2$, the equilibrium is stable, and it loses its stability at $\beta = 1 + \alpha^2$. At this point, the two eigenvalues are purely imaginary, $\lambda_{\pm} = \pm i\alpha$, and we are in the presence of a Hopf bifurcation.

Computation of the Hopf Bifurcation

In the system (2.30) we set

$$u_1 = \alpha + v_2, \quad u_2 = \frac{\beta}{\alpha} - (v_1 + v_2),$$

and

$$\omega = \alpha, \quad 2\mu = \beta - 1 - \alpha^2.$$

This leads to the system

$$\begin{aligned} \frac{dv_1}{dt} &= v_2 \\ \frac{dv_2}{dt} &= -\omega^2 v_1 + 2\mu v_2 - 2\omega v_1 v_2 + \frac{2\mu + 1 - \omega^2}{\omega} v_2^2 - (v_1 + v_2)v_2^2, \end{aligned} \tag{2.31}$$

in which ω is fixed, μ is a small bifurcation parameter, and $(0,0)$ is a solution for all values of ω and μ. The system (2.31) is of the form

$$\frac{dv}{dt} = \mathbf{L}v + \mathbf{R}(v, \mu), \tag{2.32}$$

where $v(t) = (v_1(t), v_2(t)) \in \mathbb{R}^2$ and

$$\mathbf{L} = \begin{pmatrix} 0 & 1 \\ -\omega^2 & 0 \end{pmatrix}, \quad \mathbf{R}(v, \mu) = \mu \mathbf{R}_{11} v + \mathbf{R}_{20}(v, v) + \mu \mathbf{R}_{21}(v, v) + \mathbf{R}_{30}(v, v, v),$$

with

$$\mathbf{R}_{11}v = \begin{pmatrix} 0 \\ 2v_2 \end{pmatrix}, \quad \mathbf{R}_{21}(u,v) = \begin{pmatrix} 0 \\ \frac{2}{\omega}u_2v_2 \end{pmatrix},$$

$$\mathbf{R}_{20}(u,v) = \begin{pmatrix} 0 \\ -\omega(u_1v_2 + v_1u_2) + \frac{1-\omega^2}{\omega}u_2v_2 \end{pmatrix},$$

$$\mathbf{R}_{30}(u,v,w) = \begin{pmatrix} 0 \\ -\frac{1}{3}(u_1v_2w_2 + u_2v_1w_2 + u_2v_2w_1) - u_2v_2w_2 \end{pmatrix}.$$

Now, the linear operator $\mathbf{L}$ has the pair of simple purely imaginary eigenvalues $\pm i\omega$ with the associated eigenvectors

$$\zeta = \begin{pmatrix} 1 \\ i\omega \end{pmatrix}, \quad \overline{\zeta} = \begin{pmatrix} 1 \\ -i\omega \end{pmatrix}.$$

According to the results in the previous section the system (2.32) has the normal form (2.12). We are interested in computing the coefficients a and b in the expansion (2.13) of the polynomial Q. Of course we can use directly the formulas (2.28) and (2.29) for the coefficients a and b, but for the sake of clarity we prefer to go through the steps of the calculation, again.

Since we restrict ourselves to the terms of order 3 in the expansion of the normal form, it is enough to take $p = 3$ in the expansion (2.20). Then Φ_μ is a polynomial of degree 3,

$$\Phi_\mu(A,\overline{A}) = \sum_{1 \le p+q+r \le 3} \Phi_{pqr} A^p \overline{A}^q \mu^r, \quad \Phi_{100} = \Phi_{010} = 0,$$

such that the change of variables

$$v = A\zeta + \overline{A\zeta} + \Phi_\mu(A,\overline{A}) \tag{2.33}$$

transforms (2.31) into the normal form

$$\frac{dA}{dt} = i\omega A + a\mu A + bA|A|^2 + O(|A|(|\mu|^2 + |\mu||A|^2 + |A|^3)). \tag{2.34}$$

By arguing as explained in the previous section, i.e., substituting (2.33) in (2.32), then replacing dA/dt from (2.34), and finally identifying the different powers of $(A,\overline{A},\mu)$, we find the system (2.22)–(2.27) with $\mathbf{F}_{ij} = \mathbf{R}_{ij}$. Since $\mathbf{R}_{01} = 0$, we have $\Phi_{001} = 0$, and the identity (2.23) becomes

$$a\zeta + (i\omega - \mathbf{L})\Phi_{101} = \mathbf{R}_{11}\zeta = \begin{pmatrix} 0 \\ 2i\omega \end{pmatrix}.$$

The coefficient a is now found from the solvability condition for this equation, obtained by taking the Hermitian scalar product with the vector ζ_* in the kernel of the adjoint operator satisfying

$$(-i\omega - \mathbf{L}^*)\zeta^* = 0, \quad \langle \zeta, \zeta^* \rangle = 1.$$

A direct calculation shows that

$$\mathbf{L}^* = \begin{pmatrix} 0 & -\omega^2 \\ 1 & 0 \end{pmatrix}, \quad \zeta^* = \frac{1}{2i\omega}\begin{pmatrix} i\omega \\ -1 \end{pmatrix},$$

and then

$$a = \langle \mathbf{R}_{11}\zeta, \zeta^* \rangle = 1.$$

Remark 2.10 *Since* $\mathbf{R}(0,\mu) = 0$*, it is not difficult to check in this case that the eigenvalues of the* 2×2*-matrix*

$$\mathbf{L} + \mu\mathbf{R}_{11} = \begin{pmatrix} 0 & 1 \\ -\omega^2 & 2\mu \end{pmatrix},$$

obtained by linearizing the system (2.32) *at* $U = 0$*, are the same as the eigenvalues of the* 2×2*-matrix obtained by linearizing the normal form equation* (2.34)*, together with the complex conjugated equation, at* $(A,\overline{A}) = (0,0)$*. We can use this property to compute the coefficient* a *in a different way. Indeed, this latter matrix is of the form*

$$\begin{pmatrix} i\omega + a\mu & 0 \\ 0 & -i\omega + \overline{a}\mu \end{pmatrix} + O(\mu^2),$$

and since the eigenvalues of $\mathbf{L} + \mu\mathbf{R}_{11}$ *are*

$$\lambda_{\pm} = \mu \pm i\sqrt{\omega^2 - \mu^2} = \pm i\omega + \mu \mp \frac{i\mu^2}{2\omega} + O(\mu^4),$$

we can conclude that $a = 1$.

Next, in order to compute the coefficient b we use the equations (2.24), (2.25), and (2.27), i.e.,

$$\begin{aligned}
(2i\omega - \mathbf{L})\Phi_{200} &= \mathbf{R}_{20}(\zeta,\zeta), \\
-\mathbf{L}\Phi_{110} &= 2\mathbf{R}_{20}(\zeta,\overline{\zeta}), \\
b\zeta + (i\omega - \mathbf{L})\Phi_{210} &= 2\mathbf{R}_{20}(\overline{\zeta},\Phi_{200}) + 2\mathbf{R}_{20}(\zeta,\Phi_{110}) + 3\mathbf{R}_{30}(\zeta,\zeta,\overline{\zeta}).
\end{aligned}$$

Solving the first two equations we find

$$\Phi_{200} = \begin{pmatrix} \frac{1-\omega^2}{3\omega} + \frac{2i}{3} \\ -\frac{4\omega}{3} + \frac{2i}{3}(1-\omega^2) \end{pmatrix}, \quad \Phi_{110} = \begin{pmatrix} \frac{2(1-\omega^2)}{\omega} \\ 0 \end{pmatrix},$$

and then

$$2\mathbf{R}_{20}(\overline{\zeta},\Phi_{200}) = \begin{pmatrix} 0 \\ \frac{4}{3}(1-\omega^2+\omega^4) + 2i\omega(1-\omega^2) \end{pmatrix},$$

$$2\mathbf{R}_{20}(\zeta,\Phi_{110}) = \begin{pmatrix} 0 \\ -4i\omega(1-\omega^2) \end{pmatrix}, \quad 3\mathbf{R}_{30}(\zeta,\zeta,\overline{\zeta}) = \begin{pmatrix} 0 \\ -\omega^2 - 3i\omega^3 \end{pmatrix}.$$

Finally, we compute b from the solvability condition for the third equation,

$$b = \langle 2\mathbf{R}_{20}(\overline{\zeta}, \Phi_{200}) + 2\mathbf{R}_{20}(\zeta, \Phi_{110}) + 3\mathbf{R}_{30}(\zeta, \zeta, \overline{\zeta}), \zeta^* \rangle.$$

We find

$$2\mathbf{R}_{20}(\overline{\zeta}, \Phi_{200}) + 2\mathbf{R}_{20}(\zeta, \Phi_{110}) + 3\mathbf{R}_{30}(\zeta, \zeta, \overline{\zeta}) \\ = \begin{pmatrix} 0 \\ \frac{1}{3}(4 - 7\omega^2 + 4\omega^4) - i\omega(2 + \omega^2) \end{pmatrix},$$

which gives

$$b = -\frac{1}{2}(2 + \omega^2) - \frac{i}{6\omega}(4 - 7\omega^2 + 4\omega^4). \tag{2.35}$$

In particular, this shows that the real part b_r of b is negative, so that we have here a *supercritical Hopf bifurcation*.

Remark 2.11 *In Section 2.4.4, Chapter 2, we discuss the inhomogeneous Brusselator, in which u_1 and u_2 also depend upon a spatial variable x. This is a system of partial differential equations (PDEs), for which we show that a Hopf bifurcation occurs. It is then a Hopf bifurcation in infinite dimensions, and we show in Section 2.4.4 how to compute the coefficients a and b explicitly.*

1.2.3 Hopf Bifurcation with $SO(2)$ Symmetry

We discuss in this section a particular case of a Hopf bifurcation, where the vector field possesses a continuous symmetry. As before, we assume that the vector field $\mathbf{F}$ in (2.1) satisfies Hypotheses 2.1 and 2.5, and now further assume that the following holds.

Hypothesis 2.12 *We assume that the vector field $\mathbf{F}$ is $SO(2)$-equivariant, that is, there exists a one-parameter continuous family of linear maps $\mathbf{R}_\varphi$ on $\mathbb{R}^2$, for $\varphi \in \mathbb{R}/2\pi\mathbb{Z}$, with the following properties:*

(i) $\mathbf{R}_\varphi \circ \mathbf{R}_\psi = \mathbf{R}_{\varphi+\psi}$ *for all* $\varphi,\ \psi \in \mathbb{R}/2\pi\mathbb{Z}$;
(ii) $\mathbf{R}_0 = \mathbb{I}$;
(iii) $\mathbf{F}(\mathbf{R}_\varphi u, \mu) = \mathbf{R}_\varphi \mathbf{F}(u, \mu)$ *for all* $\varphi \in \mathbb{R}/2\pi\mathbb{Z}$.

An immediate consequence of the third property in this hypothesis is that if $u(\mu)$ is a steady solution of (2.1), then $\mathbf{R}_\varphi u(\mu)$ is also a steady solution of (2.1). On the other hand, as already noticed in the Remark 2.2, the system (2.1) has a unique steady solution in a neighborhood of the origin for all sufficiently small μ. Then we necessarily have $\mathbf{R}_\varphi u(\mu) = u(\mu)$, that is, *the steady solution $u(\mu)$ is invariant under the action of* $\mathbf{R}_\varphi$. In addition, notice that

$$\mathbf{L}(\mathbf{R}_\varphi \zeta) = \mathbf{R}_\varphi(\mathbf{L}\zeta) = i\omega(\mathbf{R}_\varphi \zeta),$$

and since the eigenvalue $i\omega$ is simple we have

$$\mathbf{R}_\varphi \zeta = k(\varphi)\zeta \text{ for some } k(\varphi) \in \mathbb{C}.$$

Using the group properties of $\mathbf{R}_\varphi$, Hypothesis 2.12(i)–(ii), we obtain that $k(\varphi + \psi) = k(\varphi)k(\psi)$for all φ, ψ, and that $k(0) = 1$. The fact that k is a continuous function of $\varphi \in \mathbb{R}/2\pi\mathbb{Z}$, now implies that

$$k(\varphi) = e^{im\varphi}, \quad m \in \mathbb{Z}. \tag{2.36}$$

We now distinguish two cases depending upon the value of m in (2.36).

First, assume that $m = 0$, which means that the action of the group $\mathbf{R}_\varphi$ on the eigenvector ζ is trivial, $\mathbf{R}_\varphi \zeta = \zeta$. Then the same also holds for the complex conjugated eigenvector $\overline{\zeta}$, and since $\{\zeta, \overline{\zeta}\}$ forms a basis of $\mathbb{R}^2$, we have in this case $\mathbf{R}_\varphi = \mathbb{I}$for all φ. Consequently, the action of the continuous group $\mathbf{R}_\varphi$ is trivial, so that there is no new fact with respect to Theorem 2.6 in this case, except that all points of the periodic bifurcating orbit are invariant under $\mathbf{R}_\varphi$.

Next, assume that $m \neq 0$. Then in the basis $\{\zeta, \overline{\zeta}\}$ of $\mathbb{R}^2$, the action of $\mathbf{R}_\varphi$ on the coordinates $(z, \overline{z})$, $z \in \mathbb{C}$ is given by

$$\mathbf{R}_\varphi = \begin{pmatrix} e^{im\varphi} & 0 \\ 0 & e^{-im\varphi} \end{pmatrix}.$$

This matrix commutes now with the vector field in equation (2.5), so that we have

$$f(e^{im\varphi}z, e^{-im\varphi}\overline{z}, \mu) = e^{im\varphi} f(z, \overline{z}, \mu)$$

for all $\varphi \in \mathbb{R}/2\pi\mathbb{Z}$ and all z in a neighborhood of 0. Then, by Lemma 2.4, it follows that the differential equation (2.5) is of the form

$$\frac{dz}{dt} = i\omega z + z g(|z|, \mu), \tag{2.37}$$

with g of class $\mathscr{C}^{k-1}$ and even in $|z|$. This means that in this case the *equation is already in the normal form* (2.12), with polynomial Q given by the regular part in the Taylor expansion of g, and the rest, ρ, being of the form z times a function depending only upon $|z|$. The particular form of this part allows to use the same arguments as for the truncated normal form and to show that in this case for the bifurcating periodic solutions $u^*(\cdot; \mu)$ the coordinate $z^*(\cdot, \mu)$ is of the form (2.16). In particular, they describe a "circle" in the plane $\mathbb{C}$. Furthermore, from (2.16) we obtain

$$\mathbf{R}_\varphi u(t; \mu) = u(t + \frac{m\varphi}{\omega^*(\mu)}; \mu).$$

Choosing $\varphi = -\omega^*(\mu)t/m$, we obtain

$$\mathbf{R}_{-\omega^*(\mu)t/m} u(t; \mu) = u(0; \mu),$$

and this gives a new formula for the periodic solutions,

$$u(t;\mu) = \mathbf{R}_{\frac{\omega^*(\mu)t}{m}} u(0;\mu). \tag{2.38}$$

These periodic solutions are *rotating waves*, with wavenumber m thanks to the property

$$u(t;\mu) = \mathbf{R}_{\frac{2\pi}{m}} u(t;\mu).$$

This proves the following result:

Corollary 2.13 (Hopf bifurcation with $SO(2)$ symmetry) *Assume that Hypotheses 2.1, 2.5, and 2.12 hold. Further assume that the action of the group $\mathbf{R}_\varphi$ is not trivial. Then the family of periodic solutions bifurcating in the Hopf bifurcation at $\mu = 0$ are the* rotating waves *(2.38), with wavenumber m given by the action of the group on the eigenvector ζ of $\mathbf{L}$ associated with the purely imaginary eigenvalue $i\omega$.*

1.2.4 Steady Bifurcation with $O(2)$ Symmetry

We end this chapter with a case where the differential equation (2.1) possesses a one-parameter group of symmetries together with one discrete symmetry. More precisely, we make the following assumption.

Hypothesis 2.14 *Assume that the vector field $\mathbf{F}$ in (2.1) is of class $\mathscr{C}^k$, $k \geq 3$, that it satisfies (2.2), and that 0 is an eigenvalue of $\mathbf{L}$. Further assume that $\mathbf{F}$ is $O(2)$-equivariant, that is, there exists a one-parameter continuous family of linear maps $\mathbf{R}_\varphi$ on $\mathbb{R}^2$, for $\varphi \in \mathbb{R}/2\pi\mathbb{Z}$, and a symmetry $\mathbf{S}$ on $\mathbb{R}^2$ with the following properties:*

(i) $\mathbf{R}_\varphi \circ \mathbf{R}_\psi = \mathbf{R}_{\varphi+\psi}$ *and* $\mathbf{S}\mathbf{R}_\varphi = \mathbf{R}_{-\varphi}\mathbf{S}$ *for all* $\varphi,\ \psi \in \mathbb{R}/2\pi\mathbb{Z}$;
(ii) $\mathbf{R}_0 = \mathbb{I}$ *and* $\mathbf{S}^2 = \mathbb{I}$;
(iii) $\mathbf{F}(\mathbf{R}_\varphi u,\mu) = \mathbf{R}_\varphi \mathbf{F}(u,\mu)$ *and* $\mathbf{F}(\mathbf{S}u,\mu) = \mathbf{S}\mathbf{F}(u,\mu)$ *for all* $\varphi \in \mathbb{R}/2\pi\mathbb{Z}$.

Remark 2.15 *This type of symmetry is very frequent in physical examples, particularly in systems of PDEs (infinite-dimensional case) when the system is invariant under translations in one unbounded spatial direction and possesses a reflection symmetry in this direction. When looking for solutions that are periodic in this unbounded spatial direction, the invariance under spatial translations provides the one-parameter group of symmetries, whereas the reflection is the discrete symmetry. We present an example of such a PDE in Section 2.4.3, Chapter 2.*

An important consequence of the $O(2)$-equivariance in this hypothesis is that any eigenvalue of the linear map $\mathbf{L}$ is double, provided the action of the group $\mathbf{R}_\varphi$ is not trivial. Indeed, any eigenvalue of $\mathbf{L}$ is either simple or double. Assume that $\lambda \in \mathbb{C}$ is a simple eigenvalue of $\mathbf{L}$, with associated eigenvector ζ. Then we have

$$\mathbf{L}(\mathbf{R}_\varphi\zeta) = \mathbf{R}_\varphi(\mathbf{L}\zeta) = \lambda(\mathbf{R}_\varphi\zeta),$$

so that $\mathbf{R}_\varphi\zeta = r(\varphi)\zeta$ for some $r(\varphi) \in \mathbb{C}$, and similarly $\mathbf{S}\zeta = s\zeta$ for some $s \in \mathbb{C}$. As for $k(\varphi)$ given by (2.36), in the case of the Hopf bifurcation with $SO(2)$ symmetry discussed in the previous section, we conclude that $r(\varphi) = e^{im\varphi}$, with $m \in \mathbb{Z}$. Moreover, since $\mathbf{S}^2 = \mathbb{I}$, we have that $s = \pm 1$, and from the equality $\mathbf{R}_\varphi \mathbf{S}\zeta = \mathbf{S}\mathbf{R}_{-\varphi}\zeta$, we obtain that $se^{im\varphi}\zeta = se^{-im\varphi}\zeta$ for all φ. Thus $m = 0$, so that $\mathbf{R}_\varphi = \mathbb{I}$ for all φ, which means that the group represented by $\mathbf{R}_\varphi$ reduces to the identity. Consequently, if the action of the group $\mathbf{R}_\varphi$ is not trivial, then λ is a double eigenvalue of $\mathbf{L}$. We shall therefore make the following hypothesis.

Hypothesis 2.16 *Assume that zero is a double eigenvalue of* $\mathbf{L}$ *and that the action of* $\mathbf{R}_\varphi$ *on* $\mathbb{R}^2$ *is not trivial.*

Now we construct a suitable basis for $\mathbb{R}^2$ in which the action of $\mathbf{R}_\varphi$ and $\mathbf{S}$ is given by the 2×2-matrices

$$\mathbf{R}_\varphi = \begin{pmatrix} e^{im\varphi} & 0 \\ 0 & e^{-im\varphi} \end{pmatrix}, \quad \mathbf{S} = \begin{pmatrix} 0 & 1 \\ 1 & 0 \end{pmatrix}. \tag{2.39}$$

First, we claim that the eigenvectors of $\mathbf{R}_\varphi$ are independent of φ, and more precisely, that an eigenvector ζ_0 of $\mathbf{R}_{\varphi_0}$ for some φ_0 is also an eigenvector of $\mathbf{R}_\varphi$ for any φ, namely,

$$\mathbf{R}_\varphi \zeta_0 = r(\varphi)\zeta_0, \tag{2.40}$$

with corresponding eigenvalue $r(\varphi)$ depending continuously upon φ such that

$$r(0) = 1, \quad r(\varphi + \psi) = r(\varphi)r(\psi) \text{ for all } \varphi, \psi \in \mathbb{R}/2\pi\mathbb{Z}. \tag{2.41}$$

Indeed, consider $\varphi_0 \in \mathbb{R}/2\pi\mathbb{Z}$ such that $\varphi_0/2\pi \notin \mathbb{Q}$. Then the integer multiples of φ_0 form a dense set on the circle $\mathbb{R}/2\pi\mathbb{Z}$, so that for any $\varphi \in \mathbb{R}/2\pi\mathbb{Z}$ there exists a sequence of integers $(n_p)_{p\in\mathbb{N}}$ such that

$$\lim_{p\to\infty} (n_p\varphi_0) = \varphi \text{ in } \mathbb{R}/2\pi\mathbb{Z}. \tag{2.42}$$

Take an eigenvector ζ_0 of $\mathbf{R}_{\varphi_0}$,

$$\mathbf{R}_{\varphi_0}\zeta_0 = \lambda_0\zeta_0,$$

for some eigenvalue λ_0 of $\mathbf{R}_{\varphi_0}$. Using successively equality (2.42), the continuity of the map $\varphi \mapsto \mathbf{R}_\varphi$, and Hypothesis 2.14(i), we find

$$\mathbf{R}_\varphi\zeta_0 = \lim_{p\to\infty} \mathbf{R}_{n_p\varphi_0}\zeta_0 = \lim_{p\to\infty} \lambda_0^{n_p}\zeta_0,$$

so that

$$\mathbf{R}_\varphi\zeta_0 = r(\varphi)\zeta_0, \text{ with } r(\varphi) = \lim_{p\to\infty} \lambda_0^{n_p}.$$

This proves that the eigenvectors of $\mathbf{R}_\varphi$ are independent of φ. Finally, the continuity of $\mathbf{R}_\varphi$ in φ and the properties (i) and (ii) in Hypothesis 2.14 imply the claim.

An immediate consequence of the continuity of $r(\varphi)$ in φ and of the equalities (2.41) is that $r(\varphi) = e^{im\varphi}$ for some $m \in \mathbb{R}$ for all φ. Here $m \neq 0$, because $\mathbf{R}_\varphi$ acts nontrivially on $\mathbb{R}^2$. Next, $\mathbf{R}_\varphi \mathbf{S}\zeta_0 = e^{-im\varphi}\mathbf{S}\zeta_0$, so that $\mathbf{S}\zeta_0$ is an eigenvector of $\mathbf{R}_\varphi$ for the eigenvalue $e^{-im\varphi}$. Together with ζ_0, which is an eigenvector of the same operator $\mathbf{R}_\varphi$ for the eigenvalue $e^{im\varphi}$, this provides us with a basis for $\mathbb{R}^2$. In particular, there exists $k \in \mathbb{C}$ such that

$$\mathbf{S}\zeta_0 = k\overline{\zeta_0},$$

and the property $\mathbf{S}^2 = \mathbb{I}$ leads to $|k| = 1$, i.e.,

$$k = e^{i\beta}, \quad \beta \in \mathbb{R}.$$

We set

$$\zeta = e^{-i\beta/2}\zeta_0,$$

for which we find

$$\mathbf{S}\zeta = e^{-i\beta/2}\mathbf{S}\zeta_0 = e^{i\beta/2}\overline{\zeta_0} = \overline{\zeta}.$$

It is then straightforward to conclude that the action of the operators $\mathbf{R}_\varphi$ and $\mathbf{S}$ in the basis $\{\zeta, \overline{\zeta}\}$ is given by (2.39).

We now proceed as for the Hopf bifurcation, and represent any $u \in \mathbb{R}^2$ by a complex coordinate $z \in \mathbb{C}$ through

$$u = z\zeta + \overline{z}\overline{\zeta}.$$

Similarly, for the vector field $\mathbf{F}$ we write

$$\mathbf{F}(u, \mu) = f(z, \overline{z}, \mu)\zeta + \overline{f}(z, \overline{z}, \mu)\overline{\zeta},$$

and then obtain two complex differential equations

$$\frac{dz}{dt} = f(z, \overline{z}, \mu)$$

and its complex conjugate. The equivariance properties in Hypothesis 2.14(iii) and the equalities in (2.39) imply that f satisfies the relations

$$f(e^{im\varphi}z, e^{-im\varphi}\overline{z}, \mu) = e^{im\varphi}f(z, \overline{z}, \mu),$$

and

$$f(\overline{z}, z, \mu) = \overline{f(z, \overline{z}, \mu)}$$

for all z and μ. Using Lemma 2.4, again, the first relation implies that

$$f(z, \overline{z}, \mu) = zg(|z|, \mu),$$

where g is a complex function of class $\mathscr{C}^{k-1}$ in a neighborhood of 0, and even in $|z|$. The second relation implies that, in addition, g is real-valued.

We introduce polar coordinates $A = re^{i\phi}$ and obtain the system

$$\frac{dr}{dt} = rg(r,\mu) = a\mu r + br^3 + o(r|\mu| + r^3) \tag{2.43}$$

$$\frac{d\phi}{dt} = 0, \tag{2.44}$$

in which the coefficients a and b are found from the Taylor expansion of g. Since the function g is even in r, the scalar vector field in (2.43) satisfies Hypothesis 1.6 from the case of a pitchfork bifurcation, provided the coefficients a and b are nonzero. We therefore assume now:

Hypothesis 2.17 *Assume that the coefficients a and b in (2.43) are nonzero,*

$$\frac{\partial g}{\partial \mu}(0,0) =: a \neq 0, \quad \frac{\partial^2 g}{\partial r^2}(0,0) =: 2b \neq 0.$$

Applying the result in Theorem 1.9, we conclude that for the equation (2.43) a *pitchfork bifurcation* occurs at $\mu = 0$, which is supercritical when $b < 0$ and subcritical when $b > 0$. The bifurcation diagrams for this equation are the same as those in Figure 1.3. Since for the radial equation we are restricted to positive solutions, this shows that for $ab < 0$ (resp., $ab > 0$), the radial equation possesses the positive steady solution

$$r^*(\mu) = \sqrt{-\frac{a\mu}{b}} + o(|\mu|^{3/2})$$

for $\mu > 0$ (resp., $\mu < 0$). The dynamics of the second equation (2.44) is trivial, showing that the phase ϕ of the solutions stays constant in time t.

Going back to the two-dimensional equation (2.1), this shows that at the bifurcation point $\mu = 0$, a "circle" of equilibria, parameterized by the phase ϕ,

$$u^*(\mu,\phi) = r^*(\mu)e^{i\phi}\zeta + r^*(\mu)e^{-i\phi}\overline{\zeta},$$

bifurcates for $\mu > 0$ (resp., $\mu < 0$) when $ab < 0$ (resp., $ab > 0$). We have here a steady bifurcation. The stability of the bifurcating equilibria is given by that of $r^*(\mu)$, so that they are stable when $b < 0$ and unstable when $b > 0$. Figure 2.2 illustrates the phase portraits for $\mu < 0$ and $\mu > 0$ in the case $a > 0$, $b < 0$. Similar phase portraits can be obtained in the other cases.

In addition, we have that the bifurcating equilibria are invariant under the rotation $\mathbf{R}_{\frac{2\pi}{m}}$, since

$$\mathbf{R}_{\frac{2\pi}{m}} u^*(\mu,\phi) = u^*(\mu,\phi),$$

and there are two equilibria that are symmetric, i.e., invariant under the symmetry $\mathbf{S}$,

$$\mathbf{S}u^*(\mu,0) = u^*(\mu,0), \quad \mathbf{S}u^*(\mu,\pi) = u^*(\mu,\pi).$$

Moreover, $u^*(\mu,\phi)$ may be obtained from $u^*(\mu,0)$ through

$$u^*(\mu,\phi) = R_{\frac{\phi}{m}} u^*(\mu,0).$$

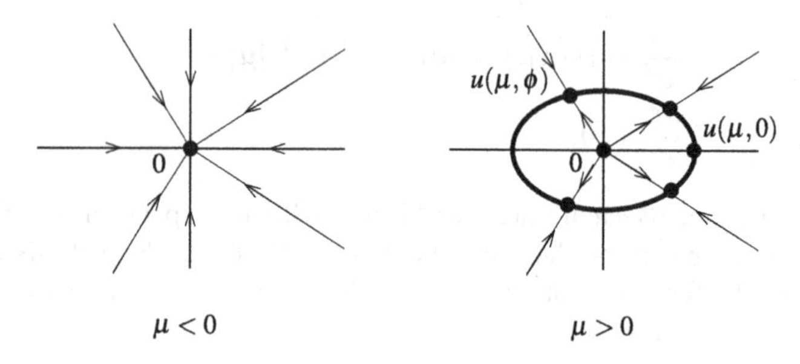

Fig. 2.2 Steady bifurcation with $O(2)$ symmetry: phase portraits in the u-plane for equation (2.1) in the case $a>0$ and $b<0$.

This shows that we have a *group orbit* of equilibria. We summarize these results in the following theorem.

Theorem 2.18 (Steady bifurcation with $O(2)$ symmetry) *Assume that Hypotheses 2.14, 2.16, and 2.17 hold. Then, for the differential equation (2.1) a steady bifurcation occurs at $\mu=0$. More precisely, the following properties hold in a neighborhood of 0 in $\mathbb{R}^2$ for sufficiently small μ:*

(i) If $ab<0$ (resp., $ab>0$) the differential equation has precisely one trivial equilibrium $u=0$ for $\mu<0$ (resp., for $\mu>0$). This equilibrium is stable when $b<0$ and unstable when $b>0$.
(ii) If $ab<0$ (resp., $ab>0$), the differential equation possesses for $\mu>0$ (resp., for $\mu<0$), the equilibrium $u=0$ and a unique closed orbit of equilibria $u^*(\mu,\phi)=O(|\mu|^{1/2})$ for $\phi\in\mathbb{R}/2\pi\mathbb{Z}$, which surrounds this equilibrium. These equilibria are stable when $b<0$ and unstable when $b>0$, whereas the equilibrium $u=0$ has opposite stability.
(iii) The equilibria $u^*(\mu,\phi)$ satisfy

$$u^*(\mu,\phi)=R_{\frac{\phi}{m}}u^*(\mu,0),$$

they are all invariant under the action of $\mathbf{R}_{\frac{2\pi}{m}}$, and there are two equilibria, $u^*(\mu,0)$ and $u^*(\mu,\pi)$, invariant under the symmetry $\mathbf{S}$.

Remark 2.19 (Higher orders) *In the case where the coefficients a or (and) b in Hypothesis 2.17 vanish, one has to consider the next nonzero higher order terms in the expansion of g, just as in the case of the pitchfork bifurcation.*

Chapter 2
Center Manifolds

This chapter is devoted to center manifold theory. We present a general result on the existence of local center manifolds for infinite-dimensional systems in Section 2.2 and then discuss several particular cases and extensions, as, for instance, to parameter-dependent systems and systems possessing different symmetries in Section 2.3. We give a series of examples showing how these results apply to various situations in Section 2.2.4 and in Section 2.4. A brief description of the tools and results from the theory of linear operators needed in this chapter is given in Appendix A.

2.1 Notations

Consider two (complex or real) Banach spaces $\mathscr{X}$ and $\mathscr{Z}$. Throughout this chapter we shall use the following notations:

- $B_\varepsilon(\mathscr{X})$ is the closed ball $\{u \in \mathscr{X}; \|u\|_\mathscr{X} \leq \varepsilon\}$.
- $\mathscr{C}^k(\mathscr{Z},\mathscr{X})$ is the Banach space of k-times continuously differentiable functions $F : \mathscr{Z} \to \mathscr{X}$ equipped with the sup norm on all derivatives up to order k,

$$\|F\|_{\mathscr{C}^k} = \max_{j=0,\dots,k}\left(\sup_{y\in\mathscr{Z}}\left(\|D^jF(y)\|_{\mathscr{L}(\mathscr{Z}^j,\mathscr{X})}\right)\right);$$

  here, and in the following, D denotes the differentiation operator.
- For a positive constant $\eta > 0$, we define the space of exponentially growing functions

$$\mathscr{C}_\eta(\mathbb{R},\mathscr{X}) = \{u \in \mathscr{C}^0(\mathbb{R},\mathscr{X})\ ;\ \|u\|_{\mathscr{C}_\eta} = \sup_{t\in\mathbb{R}}\left(e^{-\eta|t|}\|u(t)\|_\mathscr{X}\right) < \infty\},$$

  which is a Banach space when equipped with the norm $\|\cdot\|_{\mathscr{C}_\eta}$; we also consider the Banach space

M. Haragus, G. Iooss, *Local Bifurcations, Center Manifolds, and Normal Forms in Infinite-Dimensional Dynamical Systems*, Universitext,
DOI 10.1007/978-0-85729-112-7_2, © EDP Sciences 2011

$$\mathscr{F}_\eta(\mathbb{R},\mathscr{X}) = \{u \in \mathscr{C}^0(\mathbb{R},\mathscr{X})\ ;\ \|u\|_{\mathscr{F}_\eta} = \sup_{t\in\mathbb{R}} \left(e^{\eta t}\|u(t)\|_{\mathscr{X}}\right) < \infty\},$$

equipped with the norm $\|\cdot\|_{\mathscr{F}_\eta}$, of functions which may grow exponentially at $-\infty$ and which tend towards 0 exponentially at $+\infty$. Notice that $\mathscr{F}_\eta(\mathbb{R},\mathscr{X}) \subset \mathscr{C}_\eta(\mathbb{R},\mathscr{X})$ with continuous embedding.

- $\mathscr{L}(\mathscr{Z},\mathscr{X})$ is the Banach space of linear bounded operators $\mathbf{L} : \mathscr{Z} \to \mathscr{X}$, equipped with the operator norm

$$\|\mathbf{L}\|_{\mathscr{L}(\mathscr{Z},\mathscr{X})} = \sup_{\|u\|_{\mathscr{Z}}=1} \left(\|\mathbf{L}u\|_{\mathscr{X}}\right).$$

If $\mathscr{Z} = \mathscr{X}$, we write $\mathscr{L}(\mathscr{X}) = \mathscr{L}(\mathscr{X},\mathscr{X})$.

- For a linear operator $\mathbf{L} : \mathscr{Z} \to \mathscr{X}$, we denote by $\mathrm{im}\,\mathbf{L}$ its *range*,

$$\mathrm{im}\,\mathbf{L} = \{\mathbf{L}u \in \mathscr{X}\ ;\ u \in \mathscr{Z}\} \subset \mathscr{X},$$

and by $\ker\mathbf{L}$ its *kernel*,

$$\ker\mathbf{L} = \{u \in \mathscr{Z}\ ;\ \mathbf{L}u = 0\} \subset \mathscr{Z}.$$

- Assume that $\mathscr{Z} \hookrightarrow \mathscr{X}$ with continuous embedding. For a linear operator $\mathbf{L} \in \mathscr{L}(\mathscr{Z},\mathscr{X})$ we denote by $\rho(\mathbf{L})$, or simply ρ, if there is no risk of confusion, the *resolvent set* of $\mathbf{L}$,

$$\rho = \{\lambda \in \mathbb{C}\ ;\ \lambda\mathbb{I} - \mathbf{L} : \mathscr{Z} \to \mathscr{X} \text{ is bijective}\}.$$

The complement of the resolvent set is the *spectrum* $\sigma(\mathbf{L})$, or simply σ,

$$\sigma = \mathbb{C}\setminus\{\rho\}.$$

Notice that when the operator $\mathbf{L}$ is real, the resolvent set and the spectrum of $\mathbf{L}$ are both symmetric with respect to the real axis in the complex plane.

2.2 Local Center Manifolds

In this section we present the main result on the existence of local center manifolds. We discuss the hypotheses in Section 2.2.1, and then in Section 2.2.3, and state the main theorem in Section 2.2.2. The proof of the theorem is given in Appendix B.1.

2.2.1 Hypotheses

Let $\mathscr{X}$, $\mathscr{Z}$, $\mathscr{Y}$ be (real or complex) Banach spaces such that

$$\mathscr{Z} \hookrightarrow \mathscr{Y} \hookrightarrow \mathscr{X},$$

with continuous embeddings. We consider a differential equation in $\mathscr{X}$ of the form

$$\frac{du}{dt} = \mathbf{L}u + \mathbf{R}(u), \tag{2.1}$$

in which we assume that the linear part $\mathbf{L}$ and the nonlinear part $\mathbf{R}$ are such that the following holds.

Hypothesis 2.1 *We assume that* $\mathbf{L}$ *and* $\mathbf{R}$ *in (2.1) have the following properties:*

(i) $\mathbf{L} \in \mathscr{L}(\mathscr{Z}, \mathscr{X})$;
(ii) for some $k \geq 2$, *there exists a neighborhood* $\mathscr{V} \subset \mathscr{Z}$ *of* 0 *such that* $\mathbf{R} \in \mathscr{C}^k(\mathscr{V}, \mathscr{Y})$ *and*

$$\mathbf{R}(0) = 0, \quad D\mathbf{R}(0) = 0.$$

Remark 2.2 *The condition* $\mathbf{R}(0) = 0$ *means that* 0 *is an equilibrium of the differential equation (2.1), and the condition* $D\mathbf{R}(0) = 0$ *then shows that* $\mathbf{L}$ *is the linearization of the vector field about* 0, *so that* $\mathbf{R}$ *represents the nonlinear terms which are* $O(\|u\|^2_{\mathscr{Z}})$. *More generally, for an equation which has a nonzero equilibrium,* u_*, *say, we recover these conditions after replacing* u *by* $u - u_*$ *and then taking for* $\mathbf{L}$ *the differential of the resulting vector field at* 0.

Definition 2.3 *A* solution *of the differential equation (2.1) is a function* $u : \mathscr{I} \to \mathscr{Z} \hookrightarrow X$ *defined on an interval* $\mathscr{I} \subset \mathbb{R}$, *with the following properties:*

(i) the map $u : \mathscr{I} \to \mathscr{Z}$ *is continuous;*
(ii) the map $u : \mathscr{I} \to \mathscr{X}$ *is continuously differentiable;*
(iii) the equality (2.1) holds in $\mathscr{X}$ *for all* $t \in \mathscr{I}$.

Besides Hypothesis 2.1, we make two further assumptions on the linear operator $\mathbf{L}$, which are essential for the center manifold theorem.

Hypothesis 2.4 (Spectral decomposition) *Consider the spectrum* σ *of the linear operator* $\mathbf{L}$, *and write*

$$\sigma = \sigma_+ \cup \sigma_0 \cup \sigma_-,$$

in which

$$\sigma_+ = \{\lambda \in \sigma\,;\, \mathrm{Re}\,\lambda > 0\}, \quad \sigma_0 = \{\lambda \in \sigma\,;\, \mathrm{Re}\,\lambda = 0\}, \quad \sigma_- = \{\lambda \in \sigma\,;\, \mathrm{Re}\,\lambda < 0\}.$$

We assume that

(i) there exists a positive constant $\gamma > 0$ such that

$$\inf_{\lambda \in \sigma_+} (\mathrm{Re}\,\lambda) > \gamma, \quad \sup_{\lambda \in \sigma_-} (\mathrm{Re}\,\lambda) < -\gamma;$$

(ii) the set σ_0 *consists of a finite number of eigenvalues with finite algebraic multiplicities.*

Remark 2.5 *(i) The sets σ_+, σ_0, and σ_- are called* unstable, central, and stable spectrum, *respectively.*

(ii) The hypothesis above implies that the resolvent set ρ of $\mathbf{L}$ is not empty. This further implies that $\mathbf{L}$ is a closed operator in $\mathscr{X}$. Indeed, for some $\lambda \in \rho$, the operator $\lambda\mathbb{I} - \mathbf{L}$ is bijective, and since $\mathbb{I}$ and $\mathbf{L}$ belong to $\mathscr{L}(\mathscr{Z},\mathscr{X})$, by the closed graph theorem the resolvent $(\lambda\mathbb{I} - \mathbf{L})^{-1}$ belongs to $\mathscr{L}(\mathscr{X},\mathscr{Z})$. Now $\mathscr{L}(\mathscr{X},\mathscr{Z}) \subset \mathscr{L}(\mathscr{X})$, so that $(\lambda\mathbb{I} - \mathbf{L})^{-1} \in \mathscr{L}(\mathscr{X})$ and then by the closed graph theorem $\lambda\mathbb{I} - \mathbf{L}$ is closed in $\mathscr{X}$. Consequently, $\mathbf{L}$ is closed in $\mathscr{X}$.

As a consequence of Hypothesis 2.4(ii), we can define the (spectral) projection $\mathbf{P}_0 \in \mathscr{L}(\mathscr{X})$, corresponding to σ_0, by the Dunford integral formula

$$\mathbf{P}_0 = \frac{1}{2\pi i}\int_{\Gamma} (\lambda\mathbb{I} - \mathbf{L})^{-1} d\lambda, \tag{2.2}$$

where Γ is a simple, oriented counterclockwise, Jordan curve surrounding σ_0 and lying entirely in $\{\lambda \in \mathbb{C}\,;\, |\mathrm{Re}\,\lambda| < \gamma\}$. Then

$$\mathbf{P}_0^2 = \mathbf{P}_0, \quad \mathbf{P}_0\mathbf{L}u = \mathbf{L}\mathbf{P}_0 u \text{ for all } u \in \mathscr{Z},$$

and the range $\mathrm{im}\,\mathbf{P}_0$ is finite-dimensional, since σ_0 consists of a finite number of eigenvalues with finite algebraic multiplicities. In particular, it satisfies $\mathrm{im}\,\mathbf{P}_0 \subset \mathscr{Z}$, and

$$\mathbf{P}_0 \in \mathscr{L}(\mathscr{X},\mathscr{Z}),$$

since the map $\lambda \mapsto (\lambda\mathbb{I} - \mathbf{L})^{-1} \in \mathscr{L}(\mathscr{X},\mathscr{Z})$ is analytic in a neighborhood of Γ.

We define a second projection $\mathbf{P}_h : \mathscr{X} \to \mathscr{X}$ by

$$\mathbf{P}_h = \mathbb{I} - \mathbf{P}_0,$$

which then also satisfies

$$\mathbf{P}_h^2 = \mathbf{P}_h, \quad \mathbf{P}_h\mathbf{L}u = \mathbf{L}\mathbf{P}_h u \text{ for all } u \in \mathscr{Z},$$

and

$$\mathbf{P}_h \in \mathscr{L}(\mathscr{X}) \cap \mathscr{L}(\mathscr{Z}) \cap \mathscr{L}(\mathscr{Y}),$$

since $\mathbf{P}_0 \in \mathscr{L}(\mathscr{X},\mathscr{Z})$ and the embeddings $\mathscr{Z} \hookrightarrow \mathscr{Y} \hookrightarrow \mathscr{X}$ are continuous[1].

Next, we consider the spectral subspaces associated with these two projections,

$$\mathscr{E}_0 = \mathrm{im}\,\mathbf{P}_0 = \ker\mathbf{P}_h \subset \mathscr{Z}, \quad \mathscr{X}_h = \mathrm{im}\,\mathbf{P}_h = \ker\mathbf{P}_0 \subset \mathscr{X},$$

which provide a decomposition of $\mathscr{X}$ into invariant subspaces,

$$\mathscr{X} = \mathscr{E}_0 \oplus \mathscr{X}_h.$$

[1] If there is no risk of confusion we shall sometimes use the same notation for an operator $\mathbf{L} \in \mathscr{L}(\mathscr{X})$, say, and its restrictions to $\mathscr{Z}$ and $\mathscr{Y}$, $\mathbf{L}|_{\mathscr{Z}} \in \mathscr{L}(\mathscr{Z})$ and $\mathbf{L}|_{\mathscr{Y}} \in \mathscr{L}(\mathscr{Y})$, respectively.

We also set

$$\mathscr{Z}_h = \mathbf{P}_h \mathscr{Z} \subset \mathscr{Z}, \quad \mathscr{Y}_h = \mathbf{P}_h \mathscr{Y} \subset \mathscr{Y},$$

and denote by $\mathbf{L}_0$ and $\mathbf{L}_h$ the restrictions of $\mathbf{L}$ to $\mathscr{E}_0$ and $\mathscr{Z}_h$, respectively,

$$\mathbf{L}_0 \in \mathscr{L}(\mathscr{E}_0), \quad \mathbf{L}_h \in \mathscr{L}(\mathscr{Z}_h, \mathscr{X}_h).$$

An immediate consequence of these definitions is that the spectrum of $\mathbf{L}_0$ is σ_0 and the spectrum of $\mathbf{L}_h$ is $\sigma_h = \sigma_+ \cup \sigma_-$.

Remark 2.6 *As already noticed, the space $\mathscr{E}_0$ is finite-dimensional by Hypothesis 2.4(ii). Then $\mathbf{L}_0$ acts in a finite-dimensional space, and the exponential $e^{\mathbf{L}_0 t}$ allows us to explicitly solve the linear ordinary differential equation*

$$\frac{du_0}{dt} = \mathbf{L}_0 u_0 + f(t) \tag{2.3}$$

via the variation of constant formula,

$$u_0(t) = e^{\mathbf{L}_0 t} u_0(0) + \int_0^t e^{\mathbf{L}_0 (t-s)} f(s)\, \mathrm{d}s.$$

Our second hypothesis concerns the analogue of this linear problem for the operator $\mathbf{L}_h$.

Hypothesis 2.7 (Linear equation) *For any $\eta \in [0, \gamma]$ and any $f \in \mathscr{C}_\eta(\mathbb{R}, \mathscr{Y}_h)$ the linear problem*

$$\frac{du_h}{dt} = \mathbf{L}_h u_h + f(t), \tag{2.4}$$

has a unique solution $u_h = \mathbf{K}_h f \in \mathscr{C}_\eta(\mathbb{R}, \mathscr{Z}_h)$. Furthermore, the linear map $\mathbf{K}_h$ belongs to $\mathscr{L}(\mathscr{C}_\eta(\mathbb{R}, \mathscr{Y}_h), \mathscr{C}_\eta(\mathbb{R}, \mathscr{Z}_h))$, and there exists a continuous map $C : [0, \gamma] \to \mathbb{R}$ such that

$$\|\mathbf{K}_h\|_{\mathscr{L}(\mathscr{C}_\eta(\mathbb{R}, \mathscr{Y}_h), \mathscr{C}_\eta(\mathbb{R}, \mathscr{Z}_h))} \leq C(\eta).$$

While Hypotheses 2.1 and 2.4 are rather easy to check, in applications it is much more difficult to check Hypothesis 2.7. In Section 2.2.3, we discuss this hypothesis in more detail and give standard results showing how to verify it for a large class of infinite- dimensional systems.

Exercise 2.8 *Prove that Hypothesis 2.7 is satisfied in finite dimensions when $\mathscr{X} = \mathbb{R}^n$.*
Hint: *For the differential equation (2.4) the initial condition $u_h(0)$ is uniquely determined by the exponential growth required for the solution, $u_h \in \mathscr{C}_\eta(\mathbb{R}, \mathscr{Z}_h)$, which is given by*

$$u_h(t) = -\int_t^\infty e^{\mathbf{L}(t-s)} \mathbf{P}_+ f(s) \mathrm{d}s + \int_{-\infty}^t e^{\mathbf{L}(t-s)} \mathbf{P}_- f(s) \mathrm{d}s.$$

Here, $\mathbf{P}_\pm$ are the spectral projections associated to $\sigma_\pm$, which are in this case finite sets, just as σ_0, and the projections can therefore be defined by formulae similar to (2.2).

2.2.2 Main Result

In this section we state the center manifold theorem. This result has been proved for the first time in finite dimensions by Pliss [101] in 1964, in the case where the unstable spectrum σ_+ is empty, and by Kelley [77] in 1967, in the case where σ_+ is not empty. There are several versions of these results in infinite dimensions (e.g., see [47], σ_+ is empty, and [97, 122, 82], and the references therein, σ_+ is not empty), and there are analogous results for mappings (e.g., see [87, 94, 72]).

Theorem 2.9 (Center manifold theorem) *Assume that Hypotheses 2.1, 2.4, and 2.7 hold. Then there exists a map* $\Psi \in \mathscr{C}^k(\mathscr{E}_0, \mathscr{Z}_h)$, *with*

$$\Psi(0) = 0, \quad D\Psi(0) = 0, \tag{2.5}$$

and a neighborhood $\mathscr{O}$ *of* 0 *in* $\mathscr{Z}$ *such that the manifold*

$$\mathscr{M}_0 = \{u_0 + \Psi(u_0)\ ;\ u_0 \in \mathscr{E}_0\} \subset \mathscr{Z} \tag{2.6}$$

has the following properties:

(i) $\mathscr{M}_0$ *is locally invariant, i.e., if* u *is a solution of (2.1) satisfying* $u(0) \in \mathscr{M}_0 \cap \mathscr{O}$ *and* $u(t) \in \mathscr{O}$ *for all* $t \in [0,T]$, *then* $u(t) \in \mathscr{M}_0$ *for all* $t \in [0,T]$.
(ii) $\mathscr{M}_0$ *contains the set of bounded solutions of (2.1) staying in* $\mathscr{O}$ *for all* $t \in \mathbb{R}$, *i.e., if* u *is a solution of (2.1) satisfying* $u(t) \in \mathscr{O}$ *for all* $t \in \mathbb{R}$, *then* $u(0) \in \mathscr{M}_0$.

We give the proof of this theorem in Appendix B.1.

Remark 2.10 *The manifold* $\mathscr{M}_0$ *is called a* local center manifold *of (2.1), and the map* Ψ *is often referred to as the* reduction function. *Notice that* $\mathscr{M}_0$ *has the same* dimension *as* $\mathscr{E}_0$, *so it is finite-dimensional, and that it is tangent to* $\mathscr{E}_0$ *in* 0, *due to (2.5).*

Remark 2.11 *We give in Section 2.3.4 a specific center manifold theorem corresponding to the cases in which the unstable part* σ_+ *of the spectrum of* $\mathbf{L}$ *is empty.*

Center manifolds are fundamental for the study of dynamical systems near "critical situations," and in particular in bifurcation theory. Starting with an infinite-dimensional problem of the form (2.1), the center manifold theorem reduces the study of small solutions, staying sufficiently close to 0, to that of small solutions of a reduced system with finite dimension, equal to the dimension of $\mathscr{E}_0$. Indeed, such solutions belong to the center manifold $\mathscr{M}_0$, and are therefore of the form $u = u_0 + \Psi(u_0)$. The corollary below shows that solutions on the center manifold are described by a finite-dimensional system of ordinary differential equations, also called *reduced system,* which has the same dimension as $\mathscr{E}_0$.

Corollary 2.12 *Under the assumptions in Theorem 2.9, consider a solution* u *of (2.1) which belongs to* $\mathscr{M}_0$ *for* $t \in \mathscr{I}$, *for some open interval* $\mathscr{I} \subset \mathbb{R}$. *Then* $u = u_0 + \Psi(u_0)$, *and* u_0 *satisfies*

$$\frac{du_0}{dt} = \mathbf{L}_0 u_0 + \mathbf{P}_0 \mathbf{R}(u_0 + \Psi(u_0)). \tag{2.7}$$

Furthermore, the reduction function Ψ satisfies the equality

$$D\Psi(u_0)\,(\mathbf{L}_0 u_0 + \mathbf{P}_0 \mathbf{R}(u_0 + \Psi(u_0))) = \mathbf{L}_h \Psi(u_0) + \mathbf{P}_h \mathbf{R}(u_0 + \Psi(u_0)) \text{ for all } u_0 \in \mathscr{E}_0. \tag{2.8}$$

Proof By substituting $u = u_0 + \Psi(u_0)$ into (2.1) we obtain

$$\frac{du_0}{dt} + D\Psi(u_0)\frac{du_0}{dt} = \mathbf{L}_0 u_0 + \mathbf{L}_h \Psi(u_0) + \mathbf{R}(u_0 + \Psi(u_0)).$$

Projecting this equality with $\mathbf{P}_0$ we find that u_0 satisfies (2.7), and then projecting with $\mathbf{P}_h$ we obtain

$$D\Psi(u_0)\frac{du_0}{dt} = \mathbf{L}_h \Psi(u_0) + \mathbf{P}_h \mathbf{R}(u_0 + \Psi(u_0)).$$

Inserting du_0/dt from (2.7) in the equality above gives (2.8). □

Remark 2.13 *In applications it is important to compute the reduced vector field in (2.7), and more precisely its Taylor expansion. Very often it is enough to know the lowest order terms in its Taylor expansion, which can be computed directly from the formula $P_0\mathbf{R}(u_0 + \Psi(u_0))$. However, there are situations in which we need to know the terms at the next orders. This requires the computation of the Taylor expansion of the reduction function Ψ, as well, which can be done with the help of formula (2.8). We point out that one can compute the Taylor expansions of the reduced vector field and of the reduction function up to the order k, but these computations become more involved as k increases. Several examples of such computations are made in Section 2.4.*

Remark 2.14 (i) *Local center manifolds are in general* not unique *even though the Taylor expansion at the origin is unique. This is due to the occurrence in the proof of the theorem of a smooth cut-off function χ_0 on the space $\mathscr{E}_0$, which is not unique (see Appendix B.1). Uniqueness can be achieved under appropriate boundedness conditions on the nonlinearity $\mathbf{R}$: it should be Lipschitzian with sufficiently small Lipschitz constant. We refer to [122, Theorems 1 and 2] for a precise statement of this result. In addition, in this case the resulting center manifold is* global *in the sense that the properties in Theorem 2.9 hold with $\mathscr{O} = \mathscr{X}$.*

(ii) *Center manifolds are in general* not analytic *even when the right hand side of the differential equation (2.1) is analytic in u. We refer to [114, 12, 112], and [94, pp. 44–45], [38, p. 126], [120, p. 123] for examples of analytic vector fields leading to nonanalytic center manifolds.*

(iii) *A crucial hypothesis in the existing proofs on local center manifolds is Hypothesis 2.4(ii) on the set σ_0, which has to be finite. Without this hypothesis one would expect to construct an* infinite- dimensional manifold. *However, this raises a number of difficulties, which, so far, have been overcome in only very* *particular situations [98, 100]. Such a construction would require we first build a "good" projection* $\mathbf{P}_0$ *associated with the infinite spectral set* σ_0, *allowing us to obtain a group property for* $e^{\mathbf{L}_0 t}$ *together with a subexponential growth as* $t \to \pm\infty$, *and then also to construct a smooth cut-off function* χ_0 *on the central space* $\mathscr{E}_0 = \mathbf{P}_0 \mathscr{X}$.

2.2.3 Checking Hypothesis 2.7

We discuss in this section Hypothesis 2.7, and more precisely how to check it in applications. While this hypothesis always holds in finite dimensions (see Exercise 2.8), in infinite dimensions this is not always the case. Here, we distinguish between

- the *semilinear case*, $\mathscr{Y} \subset \mathscr{X}$ with $\mathscr{Y} \neq \mathscr{X}$, and
- the *quasilinear case*, $\mathscr{Y} = \mathscr{X}$.

First, we give some conditions on the resolvent of $\mathbf{L}$ which are sufficient for Hypothesis 2.7 to hold in the semilinear case. In contrast, in the quasilinear case Hypothesis 2.7 is in general not true. We discuss this situation in the second part of this section.

Semilinear Equations in Banach Spaces

We assume that Hypotheses 2.1, 2.4 hold, and show here that we may replace Hypothesis 2.7 by the following one. Though we do not make explicitly the assumption that $\mathscr{Y} \neq \mathscr{X}$, the hypothesis below can only be verified in this case.

Hypothesis 2.15 (Resolvent estimates) *Assume that there exist positive constants* $\omega_0 > 0$, $c > 0$, *and* $\alpha \in [0,1)$ *such that for all* $\omega \in \mathbb{R}$, *with* $|\omega| \geq \omega_0$, *we have that* $i\omega$ *belongs to the resolvent set of* $\mathbf{L}$, *and*

$$\|(i\omega\mathbb{I} - \mathbf{L})^{-1}\|_{\mathscr{L}(\mathscr{X})} \leq \frac{c}{|\omega|}, \tag{2.9}$$

$$\|(i\omega\mathbb{I} - \mathbf{L})^{-1}\|_{\mathscr{L}(\mathscr{Y},\mathscr{Z})} \leq \frac{c}{|\omega|^{1-\alpha}}. \tag{2.10}$$

Remark 2.16 (Hilbert spaces) *Though necessary to show that Hypothesis 2.7 holds, as we shall see in Theorem 2.20, the second inequality (2.10) is not needed for the center manifold Theorem 2.9 to hold when* $\mathscr{X}$, $\mathscr{Z}$, *and* $\mathscr{Y}$ *are Hilbert spaces. We make use of this fact in the examples presented in Section 2.4.*

We prove in Appendix B.2 that Hypothesis 2.15 above implies Hypothesis 2.7, so that the following holds.

Theorem 2.17 (Center manifold theorem in the semilinear case) *Assume that Hypotheses 2.1, 2.4, and 2.15 hold. Then*

(i) Hypothesis 2.7 is satisfied;
(ii) the result in Theorem 2.9 holds.

Remark 2.18 (Parabolic problems) *An important class of problems for which Hypothesis 2.15 usually holds is that of parabolic equations in Hilbert spaces. In such a situation the operator* $\mathbf{L}$ *is typically sectorial and generates an analytic semigroup. In particular, its resolvent satisfies Hypothesis 2.15, so that center manifold Theorem 2.9 applies provided Hypotheses 2.1 and 2.4 hold.*

Remark 2.19 *In Section 5.2.3 of Chapter 5 we give an example (waves in lattices) where (2.9) does not hold, while Hypothesis 2.7 is verified.*

Quasilinear Equations in Hilbert Spaces

We consider now the quasilinear case, $\mathscr{Y} = \mathscr{X}$. In this case Hypothesis 2.7 requires a maximal regularity property for the linear equation (2.4), and it turns out that such a property does not hold in general for spaces of continuous functions such as $\mathscr{C}_\eta(\mathbb{R}, \mathscr{X}_h)$.

Nevertheless, maximal regularity has been shown in Sobolev and Hölder spaces. We mention here the maximal regularity result by da Prato and Grisvard [21] in Sobolev spaces $W^{\theta,p}(\mathbb{R}, \mathscr{X})$, with $\theta \in (0,1)$ and $p \in (1,\infty]$, $\mathscr{X}$ is a Banach space, and the result by Mielke [96] in Sobolev spaces $L^p(\mathbb{R}, \mathscr{X})$, with $p \in (1,\infty)$, $\mathscr{X}$ is a Hilbert space. For both results, the resolvent estimate (2.9) turns out to be a sufficient condition for maximal regularity in these spaces. As for the Hölder spaces, Kirrmann [82] proved a maximal regularity result in $\mathscr{C}^{0,\alpha}(\mathbb{R}, \mathscr{X})$ with $\mathscr{X}$ a Banach space, but under a slightly different resolvent estimate.

Since these maximal regularity results hold in different spaces (Sobolev or Hölder spaces instead of spaces of continuous functions), the proof of the center manifold theorem given in Appendix B.1 does not work anymore, and needs to be adapted. Starting with the result in [96] for Hilbert spaces, Mielke [97] proved a center manifold theorem for quasilinear equations in Hilbert spaces. In Banach spaces, the maximal regularity result by Kirrmann allowed proof of a center manifold theorem [82], with a reduction function Ψ of class $\mathscr{C}^{k-1}$ instead of $\mathscr{C}^k$. We state below the result in Hilbert spaces, which uses our resolvent estimate (2.9), and refer to [97] for its proof and to [82] for the slightly different result in Banach spaces.

Theorem 2.20 (Center manifold theorem in the quasilinear case) *Assume that* $\mathscr{X}$, $\mathscr{Z}$, *and* $\mathscr{Y}$ *are Hilbert spaces, and that Hypotheses 2.1 and 2.4 hold. If the linear operator* $\mathbf{L}_h$ *satisfies (2.9), then the result in Theorem 2.9 holds.*

2.2.4 Examples

We show in this section how to apply the center manifold theorem in two examples. The first one is a fourth order ODE, for which $\mathscr{X} = \mathbb{R}^4$, while the second one is a parabolic PDE, for which $\mathscr{X}$ is a Banach space of continuous functions.

A Fourth Order ODE

Consider the fourth order ODE

$$u^{(4)} - u'' - au^2 = 0, \tag{2.11}$$

where a is a given real number.

Formulation as a First Order System

We start by writing the equation (2.11) in the form (2.1). We set $U = (u, u_1, u_2, u_3)$ with $u_1 = u'$, $u_2 = u'' - u$, $u_3 = u_2'$, and then the equation is equivalent with the system

$$\frac{dU}{dt} = \mathbf{L}U + \mathbf{R}(U), \tag{2.12}$$

in which

$$\mathbf{L} = \begin{pmatrix} 0&1&0&0 \\ 1&0&1&0 \\ 0&0&0&1 \\ 0&0&0&0 \end{pmatrix}, \qquad \mathbf{R}(U) = \begin{pmatrix} 0 \\ 0 \\ 0 \\ au^2 \end{pmatrix}.$$

Here $\mathbf{L}$ is a 4×4-matrix and $\mathbf{R}$ is a smooth vector field in $\mathbb{R}^4$, so that we can choose

$$\mathscr{X} = \mathscr{Y} = \mathscr{Z} = \mathbb{R}^4.$$

Checking the Hypotheses

Clearly, Hypothesis 2.1 is satisfied for $\mathbf{L}$ and $\mathbf{R}$ as above, for any $k \geq 2$ and the neighborhood $\mathscr{V} = \mathbb{R}^4$.

Next, in order to check Hypothesis 2.4 we have to compute the spectrum of $\mathbf{L}$, i.e., the eigenvalues of $\mathbf{L}$. A direct calculation gives

$$\sigma(\mathbf{L}) = \{-1, 0, 1\},$$

with ± 1 simple eigenvalues, and 0 a geometrically simple and algebraically double eigenvalue. Consequently, Hypothesis 2.4 is also satisfied with

$$\sigma_+ = \{1\}, \quad \sigma_0 = \{0\}, \quad \sigma_- = \{-1\}.$$

Finally, according to the result in Exercise 2.8, Hypothesis 2.7 holds in this case since $\mathscr{X}$ is finite-dimensional.

Consequently, we can apply center manifold Theorem 2.9, and conclude the existence of a local center manifold of class $\mathscr{C}^k$ for any arbitrary, but fixed, $k \geq 2$. Since 0 is an algebraically double eigenvalue, the space $\mathscr{E}_0$ is two-dimensional, so that the center manifold is two-dimensional.

Reduced Equation

Our purpose is to compute the Taylor expansion, up to order 2, of the vector field in the reduced equation.

We start by computing a basis for $\mathscr{E}_0$, which is the two-dimensional generalized kernel of $\mathbf{L}$. Solving successively the eigenvalue problem $\mathbf{L}\zeta_0 = 0$ and the generalized eigenvalue problem $\mathbf{L}\zeta_1 = \zeta_0$, we find a basis $\{\zeta_0, \zeta_1\}$ for $\mathscr{E}_0$ given by

$$\zeta_0 = \begin{pmatrix} -1 \\ 0 \\ 1 \\ 0 \end{pmatrix}, \quad \zeta_1 = \begin{pmatrix} 0 \\ -1 \\ 0 \\ 1 \end{pmatrix}.$$

According to the center manifold Theorem 2.9, solutions on the center manifold are of the form

$$U(t) = U_0(t) + \Psi(U_0(t)), \tag{2.13}$$

in which $\Psi(0) = 0$, $D\Psi(0) = 0$, and $U_0(t) \in \mathscr{E}_0$, so that

$$U_0(t) = A(t)\zeta_0 + B(t)\zeta_1, \tag{2.14}$$

where A and B are real-valued functions. The reduced system is an ODE for $U_0 = (A, B)$, and according to Corollary 2.12 it is given by

$$\frac{dU_0}{dt} = \mathbf{L}_0 U_0 + \mathbf{P}_0 \mathbf{R}(U_0 + \Psi(U_0)), \tag{2.15}$$

where $\mathbf{L}_0$ is the restriction of $\mathbf{L}$ to $\mathscr{E}_0$, and $\mathbf{P}_0$ is the spectral projection onto $\mathscr{E}_0$. We compute the expansion, up to order 2, of the vector field in (2.15), by calculating successively the 2×2-matrix $\mathbf{L}_0$, the spectral projector $\mathbf{P}_0$, and the expansion of $\mathbf{P}_0 \mathbf{R}(U_0 + \Psi(U_0))$.

First, since $\mathbf{L}_0$ is the restriction of $\mathbf{L}$ to the space $\mathscr{E}_0$, in the basis $\{\zeta_0, \zeta_1\}$ of $\mathscr{E}_0$ calculated above we find that the 2×2-matrix representing $\mathbf{L}_0$ is given by

$$\mathbf{L}_0 = \begin{pmatrix} 0 & 1 \\ 0 & 0 \end{pmatrix},$$

since $\mathbf{L}\zeta_0 = 0$ and $\mathbf{L}\zeta_1 = \zeta_0$. Next, there are several ways of computing the spectral projection $\mathbf{P}_0$ in finite dimensions. Here, we compute $\mathbf{P}_0$ with the help of the adjoint matrix

$$\mathbf{L}^* = \begin{pmatrix} 0 & 1 & 0 & 0 \\ 1 & 0 & 0 & 0 \\ 0 & 1 & 0 & 0 \\ 0 & 0 & 1 & 0 \end{pmatrix},$$

since this calculation also works in infinite dimensions, provided the operator $\mathbf{L}$ possesses an adjoint $\mathbf{L}^*$. Recall that the adjoint matrix $\mathbf{L}^*$ satisfies

$$\langle \mathbf{L}U, V \rangle = \langle U, \mathbf{L}^* V \rangle \text{ for all } U, V \in \mathbb{R}^4,$$

where $\langle \cdot, \cdot \rangle$ is the usual Euclidean scalar product in $\mathbb{R}^4$.

We claim that the spectral projection $\mathbf{P}_0$ is given by

$$\mathbf{P}_0 U = \langle U, \zeta_0^* \rangle \zeta_0 + \langle U, \zeta_1^* \rangle \zeta_1, \tag{2.16}$$

where $\{\zeta_0^*, \zeta_1^*\}$ is a dual basis satisfying

$$\mathbf{L}^* \zeta_0^* = \zeta_1^*, \quad \mathbf{L}^* \zeta_1^* = 0, \quad \langle \zeta_i, \zeta_j^* \rangle = \delta_{ij} \text{ for all } i, j \in \{0, 1\}. \tag{2.17}$$

Indeed, since $\mathbf{P}_0$ is a linear map from $\mathbb{R}^4$ onto $\mathscr{E}_0$, there exist two vectors $\zeta_0^*, \zeta_1^* \in \mathbb{R}^4$ such that $\mathbf{P}_0 U$ is given by (2.16). Next, since $\mathbf{P}_0$ is a projection, $\mathbf{P}_0^2 = \mathbf{P}_0$, it follows that $\mathbf{P}_0 \zeta_0 = \zeta_0$ and $\mathbf{P}_0 \zeta_1 = \zeta_1$, which implies that the last equality in (2.17) holds for all $i, j \in \{0, 1\}$. Finally, the spectral projection $\mathbf{P}_0$ commutes with $\mathbf{L}$, $\mathbf{P}_0 \mathbf{L} = \mathbf{L} \mathbf{P}_0$, which implies that

$$\langle \mathbf{L}U, \zeta_0^* \rangle = \langle U, \zeta_1^* \rangle, \quad \langle \mathbf{L}U, \zeta_1^* \rangle = 0 \text{ for all } U \in \mathbb{R}^4,$$

and these equalities are equivalent with the first two equalities in (2.17). This proves the claim.

It is now straightforward to compute the vectors ζ_0^* and ζ_1^* in (2.16). We obtain that

$$\zeta_0^* = \begin{pmatrix} 0 \\ 0 \\ 1 \\ 0 \end{pmatrix}, \quad \zeta_1^* = \begin{pmatrix} 0 \\ 0 \\ 0 \\ 1 \end{pmatrix}.$$

Finally, it remains to compute the Taylor expansion up to order 2 of $\mathbf{P}_0 \mathbf{R}(U_0 + \Psi(U_0))$. Notice that since the last component of the vector ζ_0^* vanishes, the scalar product $\langle \mathbf{R}(U_0 + \Psi(U_0)), \zeta_0^* \rangle = 0$, so that

$$\mathbf{P}_0 \mathbf{R}(U_0 + \Psi(U_0)) = \langle \mathbf{R}(U_0 + \Psi(U_0)), \zeta_1^* \rangle \zeta_1.$$

Furthermore, since $\Psi(0) = 0$ and $D\Psi(0) = 0$, we have $\Psi(U_0) = O(\|U_0\|^2)$, which together with the fact that $\mathbf{R}$ is a quadratic map implies that

$$\mathbf{P}_0\mathbf{R}(U_0+\Psi(U_0)) = \langle \mathbf{R}(U_0),\zeta_1^*\rangle\zeta_1 + O(\|U_0\|^3).$$

The explicit formulas for $\mathbf{P}_0$, $\mathbf{R}$, and U_0 give

$$\mathbf{R}(U_0) = \mathbf{R}(A\zeta_0 + B\zeta_1) = \begin{pmatrix} 0 \\ 0 \\ 0 \\ aA^2 \end{pmatrix},$$

so that

$$\mathbf{P}_0\mathbf{R}(U_0+\Psi(U_0)) = \left(aA^2 + O((|A|+|B|)^3)\right)\zeta_1.$$

Together with the explicit formula for $\mathbf{L}_0$ above, this implies that the reduced system (2.15), in the basis $\{\zeta_0,\zeta_1\}$, is

$$\begin{aligned} \frac{dA}{dt} &= B \\ \frac{dB}{dt} &= aA^2 + O((|A|+|B|)^3). \end{aligned}$$

Remark 2.21 *(i) In the calculation of the expansion up to order 2 of the reduced system, it was not necessary to compute the expansion of Ψ. This property is always true because $\Psi(U_0) = O(\|U_0\|^2)$ and $\mathbf{R}(U) = O(\|U\|^2)$. However, the expansion of Ψ is necessary when computing the expansion up to order 3, or higher, of the reduced system. For instance, for a computation up to order 3 one needs to compute the terms of order 2 in the expansion of Ψ. This can be done by substituting the Ansatz*

$$\Psi(A,B) = \Psi_{20}A^2 + \Psi_{11}AB + \Psi_{02}B^2 + O((|A|+|B|)^3) \tag{2.18}$$

in the identity (2.8). Then the vectors Ψ_{20}, Ψ_{11}, and Ψ_{02} are determined by identifying powers of A and B in this identity and taking into account that these vectors belong to the space $(\mathbb{I}-\mathbf{P}_0)\mathbb{R}^4$, i.e., they are orthogonal to both ζ_0^* and ζ_1^*.

(ii) An alternative way of computing the reduced system, is by directly substituting the formulas (2.13), (2.14), and (2.18) into the first order system (2.12) and calculating the Taylor expansions of both sides of the resulting system. We use this alternative approach in most of the examples in Section 2.4. It turns out that such an approach is particularly convenient when the center manifold reduction is followed by a normal form transformation (see Chapter 3, Section 3.4).

A Parabolic PDE

Consider the parabolic boundary value problem

$$\frac{\partial u}{\partial t} = \frac{\partial^2 u}{\partial x^2} + u + g\left(u, \frac{\partial u}{\partial x}\right) \tag{2.19}$$

$$u(0,t) = u(\pi,t) = 0, \tag{2.20}$$

where $u(x,t) \in \mathbb{R}$ for $(x,t) \in (0,\pi) \times \mathbb{R}$, and $g \in \mathscr{C}^k(\mathbb{R}^2, \mathbb{R})$, $k \geq 2$, satisfying

$$g(0,v) = 0 \text{ for all } v \in \mathbb{R}, \text{ and } g(u,v) = O(|u|^2 + |v|^2) \text{ as } (u,v) \to 0.$$

Formulation and Hypothesis 2.1

First we write the problem (2.19)–(2.20) in form (2.1) by setting

$$\mathbf{L}u = \frac{d^2 u}{dx^2} + u, \quad \mathbf{R}(u) = g\left(u, \frac{du}{dx}\right),$$

and choosing the Banach space

$$\mathscr{X} = C^0([0,\pi])$$

of real-valued continuous functions on $[0,\pi]$. Then $\mathbf{L}$ is a closed linear operator in $\mathscr{X}$ with domain

$$\mathscr{Z} = \{u \in C^2([0,\pi])\ ;\ u(0) = u(\pi) = 0\},$$

taken such that $\mathbf{L}u \in \mathscr{X}$ for $u \in \mathscr{Y}$, and such that the functions in $\mathscr{Y}$ satisfy the boundary conditions (2.20). The nonlinear terms $\mathbf{R}$ satisfy $\mathbf{R}(u) \in C^1([0,\pi])$ and $(\mathbf{R}(u))(0) = (\mathbf{R}(u))(\pi) = 0$ for $u \in \mathscr{Y}$. We therefore set

$$\mathscr{Y} = \{u \in C^1([0,\pi])\ ;\ u(0) = u(\pi) = 0\},$$

and then we have $\mathbf{R} \in C^k(\mathscr{Z}, \mathscr{Y})$. In particular, these show that $\mathbf{L}$ and $\mathbf{R}$ satisfy Hypothesis 2.1.

Spectrum and Hypothesis 2.4

Next, we investigate the spectrum of $\mathbf{L}$ and check Hypothesis 2.4. For this we have to solve the linear equation

$$\lambda u - \mathbf{L}u = f$$

for $\lambda \in \mathbb{C}$, $f \in \mathscr{X}$, and $u \in \mathscr{Z}$; that is, we have to find solutions $u \in C^2([0,\pi])$ of the linear problem

$$\lambda u - u - u'' = f \tag{2.21}$$

$$u(0) = u(\pi) = 0 \tag{2.22}$$

for $f \in C^0([0,\pi])$. The second order ODE (2.21) has a unique solution $u \in C^2([0,\pi])$ satisfying the boundary conditions (2.22), for $f \in C^0([0,\pi])$, precisely when the associated homogeneous equation

$$u'' + u - \lambda u = 0 \tag{2.23}$$

possesses no nontrivial solutions. When this is the case, then λ belongs to the resolvent set $\rho(\mathbf{L})$ of $\mathbf{L}$. A direct calculation shows that (2.23) has nontrivial solutions for $\lambda = 1 - n^2$, with n any positive integer. We conclude that the resolvent set and the spectrum of $\mathbf{L}$ are, respectively,

$$\rho(\mathbf{L}) = \mathbb{C} \setminus \sigma(\mathbf{L}), \quad \sigma(\mathbf{L}) = \{\lambda \in \mathbb{C}\,;\, \lambda = 1 - n^2,\ n \in \mathbb{N}^*\};$$

(here, and later in the text, $\mathbb{N}^* = \{n \in \mathbb{N}\,;\, n \geq 1\}$).

With the notations from Hypothesis 2.4 we now have

$$\sigma_+ = \varnothing, \quad \sigma_0 = \{0\}, \quad \sigma_- \subset (-\infty, -3],$$

so that part (i) of this hypothesis holds. Next, the kernel of $\mathbf{L}$ is one-dimensional, spanned by $\xi_0 = \sin x$, so that the eigenvalue $\lambda = 0$ has geometric multiplicity one. A generalized eigenvector v associated to the eigenvalue 0 satisfies the ODE

$$v'' + v = \sin x,$$

and the boundary conditions (2.22). Multiplying this equation by $\sin x$, integrating over $[0,\pi]$, and then integrating twice by parts on the left hand side gives

$$\int_0^\pi v''(x) \sin x\, dx + \int_0^\pi v(x) \sin x\, dx = -\int_0^\pi v(x) \sin x\, dx + \int_0^\pi v(x) \sin x\, dx = 0,$$

while the right hand side is equal to

$$\int_0^\pi \sin^2 x\, dx = \frac{\pi}{2},$$

so that there are no solutions to the ODE above. This proves that 0 is a simple eigenvalue of $\mathbf{L}$, with algebraic multiplicity one, as well, and then shows that part (ii) of Hypothesis 2.4 holds. Notice that the spectral subspace $\mathscr{E}_0$ associated to σ_0 is one-dimensional, spanned by ξ_0, so that we expect in this case to find a one-dimensional center manifold.

Checking Hypothesis 2.7

Finally, we have to check Hypothesis 2.7. For this we use the result in Theorem 2.17, so that we have to verify the estimates on the resolvent (2.9) and (2.10). Since our problem is formulated in Banach spaces we need to check both inequalities (see Remark 2.16).

Consider $\omega \neq 0$. Since $\sigma_0 = \{0\}$, we have that $i\omega$ belongs to the resolvent set of $\mathbf{L}$, so that the equation

$$(i\omega\mathbb{I} - \mathbf{L})u = f$$

has a unique solution $u \in \mathscr{Z}$ for $f \in \mathscr{X}$. This solution satisfies

$$\begin{aligned}(i\omega - 1)u - u'' &= f\\ u(0) = u(\pi) &= 0,\end{aligned}$$

and a direct computation gives

$$u(x) = \frac{1}{\gamma\sinh(\gamma\pi)}\left(\int_0^x \sinh(\gamma\xi)\sinh(\gamma(\pi - x))f(\xi)d\xi \right.$$
$$\left. + \int_x^\pi \sinh(\gamma x)\sinh(\gamma(\pi - \xi))f(\xi)d\xi\right)$$

in which

$$\gamma = \sqrt{i\omega - 1}.$$

We need to show that

$$\|u\|_{C^0} \leq \frac{c}{|\omega|}\|f\|_{C^0}, \qquad \|u\|_{C^2} \leq \frac{c}{|\omega|^{1-\alpha}}\|f\|_{C^1} \tag{2.24}$$

for $|\omega| \geq \omega_0$ and constants $c > 0$ and $\alpha \in [0, 1)$, which then proves that (2.9) and (2.10) hold.

We write

$$u(x) = \frac{1}{\gamma\sinh(\gamma\pi)}\left(\frac{1}{2}\int_0^x \cosh(\gamma(\pi + \xi - x))f(\xi)d\xi \right.$$
$$\left. + \frac{1}{2}\int_x^\pi \cosh(\gamma(\pi + x - \xi))f(\xi)d\xi - \frac{1}{2}\int_0^\pi \cosh(\gamma(x + \xi - \pi))f(\xi)d\xi\right),$$

and $\gamma = \gamma_r + i\gamma_i$, $\gamma_r > 0$. Using the inequalities

$$|\sinh(a + ib)| \geq \sinh(a), \qquad |\cosh(a + ib)| \leq 1 + \sinh(a),$$

which hold for real numbers $a > 0$ and $b \in \mathbb{R}$, we estimate

$$\begin{aligned}|u(x)| \leq{} & \frac{\|f\|_{C^0}}{2|\gamma|\sinh(\gamma_r\pi)}\left(\int_0^x (1 + \sinh(\gamma_r(\pi + \xi - x)))d\xi \right.\\ & + \int_x^\pi (1 + \sinh(\gamma_r(\pi + x - \xi)))d\xi + \int_0^{\pi - x} (1 + \sinh(\gamma_r(\pi - x - \xi)))d\xi\\ & \left. + \int_{\pi - x}^\pi (1 + \sinh(\gamma_r(x + \xi - \pi)))d\xi\right)\\ ={} & \frac{\|f\|_{C^0}}{|\gamma|\gamma_r\sinh(\gamma_r\pi)}(\gamma_r\pi + \cosh(\gamma_r\pi) - 1) \leq \frac{2\|f\|_{C^0}}{|\gamma|\gamma_r}.\end{aligned}$$

This proves the first inequality in (2.24).

Similar calculations show that

$$\|u'\|_{C^0} \leq \frac{c}{|\omega|^{1/2}}\|f\|_{C^0},$$

and it remains to estimate $\|u''\|_{C^0}$. Now we use the fact that $f \in \mathscr{Y}$, in order to obtain the second inequality in (2.24), with $\alpha \neq 0$. (We point out that $\|u''\|_{C^0} \leq c\|f\|_{C^0}$, since $u'' = \gamma^2 u - f$, which gives the second inequality in (2.24) for $\alpha = 1$, only.) Integrating by parts in the formula for u we find, for $f \in C^1([0,\pi])$,

$$\begin{aligned} u''(x) &= \gamma^2 u(x) - f(x) \\ &= \frac{1}{\sinh(\gamma\pi)}\Big(-\sinh(\gamma(\pi - x))f(0) - \sinh(\gamma x)f(\pi) \\ &\qquad - \int_0^x \cosh(\gamma\xi)\sinh(\gamma(\pi - x))f'(\xi)d\xi \\ &\qquad + \int_x^\pi \sinh(\gamma x)\cosh(\gamma(\pi - \xi))f'(\xi)d\xi \Big). \end{aligned}$$

Using the fact that $f(0) = f(\pi) = 0$ for $f \in \mathscr{Y}$, and arguing as above, we find

$$\|u''\|_{C^0} \leq \frac{c}{|\omega|^{1/2}}\|f'\|_{C^0},$$

which completes the proof of (2.24). Notice that the equalities $f(0) = f(\pi) = 0$ were essential in this last part of the proof, taking $f \in C^1([0,\pi])$, only, does not allow us to obtain the second inequality in (2.24) with $\alpha \neq 0$. However, such boundary conditions on f are not necessary when the Banach spaces $C^k([0,\pi])$ are replaced by the Sobolev spaces $H^k(0,\pi)$, for which one can prove the second inequality in (2.24), with $\alpha = 3/4$, without imposing $f(0) = f(\pi) = 0$ (see [122]).

Reduced Equation

Hypotheses 2.1, 2.4, and 2.7 being satisfied, we can now apply center manifold Theorem 2.9. This gives us a one-dimensional center manifold $\mathscr{M}_0$ as in (2.6), parameterized by $u_0 \in \mathscr{E}_0$. Notice that $\mathbf{L}_0 u_0 = 0$ in this case, so that the linear term in the reduced system (2.7) vanishes. Furthermore, since $\mathscr{E}_0$ is spanned by ξ_0, we may write

$$u_0(t) = A(t)\xi_0 \in \mathscr{E}_0, \quad A(t) \in \mathbb{R}.$$

Replacing this formula in the reduced system (2.7) we obtain a first order ODE for A,

$$\frac{dA}{dt} = f_0(A),$$

with $f_0(A) = O(A^2)$ as $A \to 0$. For concrete nonlinear terms g in (2.19), one can compute explicitly the Taylor expansion of f_0 (see Remark 2.13), and then easily determine the dynamics near 0 of the reduced equation, since it is a first order ODE. We present examples of such computations in Section 2.4.

2.3 Particular Cases and Extensions

2.3.1 Parameter-Dependent Center Manifolds

In the same frame as above, we consider a *parameter-dependent* differential equation in $\mathscr{X}$ of the form

$$\frac{du}{dt} = \mathbf{L}u + \mathbf{R}(u, \mu), \tag{3.1}$$

where $\mathbf{L}$ is a linear operator as in Section 2.2 and $\mathbf{R}$ is defined for (u, μ) in a neighborhood of $(0,0)$ in $\mathscr{Z} \times \mathbb{R}^m$. Here $\mu \in \mathbb{R}^m$ is a parameter that we assume to be small. More precisely, we keep Hypotheses 2.4, 2.7, and replace Hypothesis 2.1 by the following:

Hypothesis 3.1 *We assume that* $\mathbf{L}$ *and* $\mathbf{R}$ *in (3.1) have the following properties:*

(i) $\mathbf{L} \in \mathscr{L}(\mathscr{Z}, \mathscr{X})$;
(ii) for some $k \geq 2$, *there exist neighborhoods* $\mathscr{V}_u \subset \mathscr{Z}$ *and* $\mathscr{V}_\mu \subset \mathbb{R}^m$ *of* 0 *such that* $\mathbf{R} \in \mathscr{C}^k(\mathscr{V}_u \times \mathscr{V}_\mu, \mathscr{Y})$ *and*

$$\mathbf{R}(0,0) = 0, \quad D_u\mathbf{R}(0,0) = 0.$$

Remark 3.2 *The equalities above on* $\mathbf{R}$ *imply that* 0 *is an equilibrium of (3.1) for* $\mu = 0$, *and that* $\mathbf{L}$ *represents the linearization of the vector field about this equilibrium at* $\mu = 0$. *Now, if* $\mathbf{L}$ *has a bounded inverse, then this equilibrium persists for small* μ. *More precisely, by arguing with the implicit function theorem, we find that there is a family of stationary solutions* $u = u(\mu)$ *of (3.1) for* μ *close to* 0, *i.e., such that*

$$\mathbf{L}u(\mu) + \mathbf{R}(u(\mu), \mu) = 0.$$

On the contrary, if $\mathbf{L}$ *does not have a bounded inverse, then this equilibrium may not persist for some values of* μ *near* 0.

The analogue of center manifold Theorem 2.9 for the parameter-dependent equation (3.1) is the following result.

Theorem 3.3 (Parameter-dependent center manifolds) *Assume that Hypotheses 3.1, 2.4, and 2.7 hold. Then there exists a map* $\Psi \in \mathscr{C}^k(\mathscr{E}_0 \times \mathbb{R}^m, \mathscr{Z}_h)$, *with*

$$\Psi(0,0) = 0, \quad D_u\Psi(0,0) = 0, \tag{3.2}$$

and a neighborhood $\mathscr{O}_u \times \mathscr{O}_\mu$ *of* $(0,0)$ *in* $\mathscr{Z} \times \mathbb{R}^m$ *such that for* $\mu \in \mathscr{O}_\mu$, *the manifold*

$$\mathscr{M}_0(\mu) = \{u_0 + \Psi(u_0,\mu)\ ;\ u_0 \in \mathscr{E}_0\} \tag{3.3}$$

has the following properties:

(i) $\mathscr{M}_0(\mu)$ *is locally invariant, i.e., if* u *is a solution of (3.1) satisfying* $u(0) \in \mathscr{M}_0(\mu) \cap \mathscr{O}_u$ *and* $u(t) \in \mathscr{O}_u$ *for all* $t \in [0,T]$, *then* $u(t) \in \mathscr{M}_0(\mu)$ *for all* $t \in [0,T]$.
(ii) $\mathscr{M}_0(\mu)$ *contains the set of bounded solutions of (3.1) staying in* $\mathscr{O}_u$ *for all* $t \in \mathbb{R}$, *i.e., if* u *is a solution of (3.1) satisfying* $u(t) \in \mathscr{O}_u$ *for all* $t \in \mathbb{R}$, *then* $u(0) \in \mathscr{M}_0(\mu)$.

Proof We consider (3.1) as a particular case of a system of the form (2.1), namely,

$$\frac{d\widetilde{u}}{dt} = \widetilde{\mathbf{L}}\widetilde{u} + \widetilde{\mathbf{R}}(\widetilde{u}), \tag{3.4}$$

by setting

$$\widetilde{u} = (u,\mu),$$

and

$$\begin{aligned}\widetilde{\mathbf{L}}\widetilde{u} &= (\mathbf{L}u + D_\mu\mathbf{R}(0,0)\mu, 0),\\ \widetilde{\mathbf{R}}(\widetilde{u}) &= (\mathbf{R}(u,\mu) - D_\mu\mathbf{R}(0,0)\mu, 0).\end{aligned}$$

We show that $\widetilde{\mathbf{L}}$ and $\widetilde{\mathbf{R}}$ verify Hypotheses 2.1, 2.4, and 2.7, with Banach spaces

$$\widetilde{\mathscr{X}} = \mathscr{X} \times \mathbb{R}^m, \quad \widetilde{\mathscr{Z}} = \mathscr{Z} \times \mathbb{R}^m, \quad \widetilde{\mathscr{Y}} = \mathscr{Y} \times \mathbb{R}^m,$$

and then the result in the theorem follows from Theorem 2.9.

First, Hypothesis 2.1 is an immediate consequence of Hypothesis 3.1. Next, we show that the spectral sets $\widetilde{\sigma}_\pm$, $\widetilde{\sigma}_0$ of $\widetilde{\mathbf{L}}$ satisfy

$$\widetilde{\sigma}_\pm = \sigma_\pm, \quad \widetilde{\sigma}_0 \setminus \{0\} = \sigma_0 \setminus \{0\}, \tag{3.5}$$

where $\sigma_\pm$, σ_0 are the spectral sets of $\mathbf{L}$, and that $\widetilde{\sigma}_0$ consists of purely imaginary eigenvalues with finite algebraic multiplicities. These properties imply then that Hypothesis 2.4 holds.

Indeed, let us consider the linear equation

$$(\widetilde{\mathbf{L}} - \lambda)\widetilde{u} = \widetilde{v},$$

where $\widetilde{v} = (v, \nu) \in \mathscr{X} \times \mathbb{R}^m$. This means that

$$\begin{aligned}(\mathbf{L} - \lambda)u + D_\mu\mathbf{R}(0,0)\mu &= v,\\ -\lambda\mu &= \nu.\end{aligned}$$

Hence, if $\lambda \neq 0$ we have $\mu = -\nu/\lambda$ and

$$(\mathbf{L} - \lambda)u = v + \lambda^{-1}D_\mu\mathbf{R}(0,0)\nu.$$

Consequently, in $\mathbb{C}\setminus\{0\}$, the resolvent set of $\mathbf{L}$ is identical to the resolvent set of $\widetilde{\mathbf{L}}$. In particular, we have that (3.5) holds. Furthermore, for $\widetilde{\mathbf{L}}$ we can define the spectral projections $\widetilde{\mathbf{P}}_0$, $\widetilde{\mathbf{P}}_h$, and the corresponding spectral spaces $\widetilde{\mathscr{E}_0}$, $\widetilde{\mathscr{X}_h}$ as in Section 2.2.1.

Next, notice that $\mathscr{X}_h\times\{0\}$ is an invariant subspace for $\widetilde{\mathbf{L}}$, since

$$\widetilde{\mathbf{L}}(u_h,0)=(\mathbf{L}_h u_h,0)\in\mathscr{X}_h\times\{0\}\text{ for all }u_h\in\mathscr{Z}_h.$$

From this equality we further deduce that

$$\sigma(\widetilde{\mathbf{L}}|_{\mathscr{X}_h\times\{0\}})=\sigma(\mathbf{L}_h)=\sigma_+\cup\sigma_-=\widetilde{\sigma}_+\cup\widetilde{\sigma}_-.$$

Consequently, $\mathscr{X}_h\times\{0\}\subset\widetilde{\mathscr{X}_h}$, and since

$$\operatorname{codim}\widetilde{\mathscr{X}_h}\leq\operatorname{codim}(\mathscr{X}_h\times\{0\})=\dim\mathscr{E}_0+m<\infty,$$

we conclude that

$$\dim\widetilde{\mathscr{E}_0}=\operatorname{codim}\widetilde{\mathscr{X}_h}<\infty.$$

In particular, this shows that $\widetilde{\sigma}_0$ consists of purely imaginary eigenvalues with finite algebraic multiplicities and proves Hypothesis 2.4.

In order to prove Hypothesis 2.7 it is enough to show that $\widetilde{\mathscr{X}_h}=\mathscr{X}_h\times\{0\}$, and then the conditions on $\widetilde{\mathbf{L}}$ in Hypothesis 2.7 follow from the analogue ones on $\mathbf{L}$. We claim that

$$\widetilde{\mathscr{E}_0}=\{(u_0-\mathbf{L}_h^{-1}D_\mu\mathbf{R}_h(0,0)\mu,\mu)\ ;\ u_0\in\mathscr{E}_0,\ \mu\in\mathbb{R}^m\}=:\mathscr{F}_0.$$

Then this implies that

$$\operatorname{codim}\widetilde{\mathscr{X}_h}=\dim\widetilde{\mathscr{E}_0}=\dim\mathscr{E}_0+m=\operatorname{codim}(\mathscr{X}_h\times\{0\}),$$

and since $\mathscr{X}_h\times\{0\}\subset\widetilde{\mathscr{X}_h}$ we conclude that $\widetilde{\mathscr{X}_h}=\mathscr{X}_h\times\{0\}$.

It remains to prove the claim $\widetilde{\mathscr{E}_0}=\mathscr{F}_0$. First, take $\widetilde{u}=(u,\mu)\in\widetilde{\mathscr{E}_0}\subset\widetilde{\mathscr{Z}}$. We write $u=u_0+u_h$ with $u_0\in\mathscr{E}_0$, $u_h\in\mathscr{Z}_h$, and compute

$$\widetilde{\mathbf{L}}\widetilde{u}=(\mathbf{L}_h u_h+D_\mu\mathbf{R}_h(0,0)\mu,0)+(\mathbf{L}_0 u_0+D_\mu\mathbf{R}_0(0,0)\mu,0),$$

where $\mathbf{R}_h=\mathbf{P}_h\mathbf{R}$ and $\mathbf{R}_0=\mathbf{P}_0\mathbf{R}$. The first term on the right hand side of the above equality belongs to $\mathscr{X}_h\times\{0\}\subset\widetilde{\mathscr{X}_h}$, whereas the second term belongs to $\mathscr{E}_0\times\{0\}\subset\widetilde{\mathscr{E}_0}$. Then, since $\widetilde{\mathbf{L}}\widetilde{u}\in\widetilde{\mathscr{E}_0}$, the first term vanishes, so that

$$\mathbf{L}_h u_h+D_\mu\mathbf{R}_h(0,0)\mu=0.$$

Now $\mathbf{L}_h$ has a bounded inverse because 0 does not belong to its spectrum, so that we find

$$u_h=-\mathbf{L}_h^{-1}D_\mu\mathbf{R}_h(0,0)\mu.$$

Summarizing, for $\widetilde{u} \in \widetilde{\mathscr{E}}_0$, we have

$$\widetilde{u} = (u, \mu) = (u_0 + u_h, \mu) = (u_0 - \mathbf{L}_h^{-1} D_\mu \mathbf{R}_h(0,0)\mu, \mu),$$

which proves that $\widetilde{\mathscr{E}}_0 \subset \mathscr{F}_0$.

Next, notice that

$$\widetilde{\mathbf{L}}(u_0 - \mathbf{L}_h^{-1} D_\mu \mathbf{R}_h(0,0)\mu, \mu) = (\mathbf{L}_0 u_0 + D_\mu \mathbf{R}_0(0,0)\mu, 0) \in \mathscr{E}_0 \times \{0\} \subset \mathscr{F}_0,$$

so that $\mathscr{F}_0$ is an invariant subspace for $\widetilde{\mathbf{L}}$. Consider the bases $\{e_j; j = 1, \ldots, \dim \mathscr{E}_0\}$ and $\{f_k; k = 1, \ldots, m\}$ of $\mathscr{E}_0$ and $\mathbb{R}^m$, respectively. Then the set

$$\{(e_j, 0), (-\mathbf{L}_h^{-1} D_\mu \mathbf{R}_h(0,0) f_k, f_k)\ ;\ j = 1, \ldots, \dim \mathscr{E}_0,\ k = 1, \ldots, m\}$$

is a basis for $\mathscr{F}_0$, in which we find that the matrix of $\widetilde{\mathbf{L}}\big|_{\mathscr{F}_0}$ is of the form

$$\begin{pmatrix} M_0 & M_1 \\ 0 & 0 \end{pmatrix},$$

with M_0 the matrix of $\mathbf{L}_0$ in the basis $\{e_j; j = 1, \ldots, \dim \mathscr{E}_0\}$ and M_1 a matrix of size $m \times \dim \mathscr{E}_0$. The set of eigenvalues of M_0 is precisely the set σ_0, and we then conclude that

$$\sigma(\widetilde{\mathbf{L}}\big|_{\mathscr{F}_0}) = \sigma_0 \cup \{0\} \subset \widetilde{\sigma}_0.$$

In particular, this implies that $\mathscr{F}_0 \subset \widetilde{\mathscr{E}}_0$, which completes the proof of $\widetilde{\mathscr{E}}_0 = \mathscr{F}_0$. □

Remark 3.4 *The analogue of the* reduced equation *(2.7) in this situation is*

$$\frac{du_0}{dt} = \mathbf{L}_0 u_0 + \mathbf{P}_0 \mathbf{R}(u_0 + \Psi(u_0, \mu), \mu) \stackrel{def}{=} f(u_0, \mu), \tag{3.6}$$

where we observe that $f(0,0) = 0$ *and* $D_{u_0} f(0,0) = \mathbf{L}_0$ *has the spectrum* σ_0. *Similarly, we have the analogue of the equality (2.8),*

$$\begin{aligned} D_{u_0}\Psi(u_0, \mu) f(u_0, \mu) &= \mathbf{L}_h \Psi(u_0, \mu) \\ &\quad + \mathbf{P}_h \mathbf{R}(u_0 + \Psi(u_0, \mu), \mu) \text{ for all } u_0 \in \mathscr{E}_0. \end{aligned} \tag{3.7}$$

Exercise 3.5 *Consider a system of the form (3.1) for which* 0 *is a solution for all values of* μ, *i.e., such that* $\mathbf{R}(0, \mu) = 0$ *for all* μ *in a neighborhood of* 0 *in* $\mathbb{R}^m$. *Show that*

$$\Psi(0, \mu) = 0, \quad f(0, \mu) = 0,$$

for μ *sufficiently small. Furthermore, set*

$$\mathbf{L}_\mu = \mathbf{L} + D_u \mathbf{R}(0, \mu) \in \mathscr{L}(\mathscr{Z}, \mathscr{X}) \quad \text{and} \quad \mathbf{A}_\mu = \frac{\partial f}{\partial u_0}(0, \mu).$$

Show that eigenvalues of $\mathbf{A}_\mu$ *are precisely the eigenvalues of* $\mathbf{L}_\mu$, *which are the continuation for small* μ *of the purely imaginary eigenvalues of* $\mathbf{L}$ *(i.e., those of* $\mathbf{L}_0$*)).*

Hint: *Identify the terms linear in* u_0 *in the identity*

$$\left(\mathbb{I}+D_{u_0}\Psi(u_0,\mu)\right)f(u_0,\mu) = \mathbf{L}(u_0+\Psi(u_0,\mu)) + \mathbf{R}(u_0+\Psi(u_0,\mu),\mu) \textit{ for all } u_0 \in \mathscr{E}_0.$$

Remark 3.6 (Case when σ_0 does not lie on the imaginary axis) *A situation arising in some applications is one in which the eigenvalues in σ_0 of the operator* **L** *in (3.1) do not lie on the imaginary axis but stay close to the imaginary axis. More precisely, we still have the spectral decomposition in Hypothesis 2.4, satisfying the properties (i) and (ii), but with σ_0 such that*

$$\sigma_0 = \{\lambda \in \sigma\,;\,|\mathrm{Re}\,\lambda| \leq \delta\} \tag{3.8}$$

for some $\delta \ll \gamma$ sufficiently small. This means that σ_0 consists of a finite number of eigenvalues λ_j, $j=1,\ldots,r$ of **L**, *with real parts that are small but not necessarily* 0*:*

$$\mathrm{Re}\,\lambda_j = \varepsilon_j, \quad |\varepsilon_j| \leq \delta, \quad j=1,\ldots,r.$$

In such a situation we can apply the result in Theorem 3.3 by arguing in the following way:

Consider the bounded linear operator

$$\mathbf{A}_\nu = \sum_{j=1}^{r} \nu_j \mathbf{P}_j \textit{ for } \nu = (\nu_1,\ldots,\nu_r) \in \mathbb{R}^r,$$

where $\mathbf{P}_j$ denotes the spectral projection associated with the eigenvalue $\lambda_j \in \sigma_0$ of **L**. *When $\nu = \varepsilon$, $\varepsilon = (\varepsilon_1,\ldots,\varepsilon_r)$, the operator*

$$\mathbf{L}' = \mathbf{L} - \mathbf{A}_\varepsilon, \quad \varepsilon = (\varepsilon_1,\ldots,\varepsilon_r),$$

satisfies Hypothesis 2.4, the effect of adding $-\mathbf{A}_\varepsilon$ to **L** *being that all eigenvalues in σ_0 are shifted on the imaginary axis. Consequently, we can apply the result in Theorem 3.3 to the modified system*

$$\frac{du}{dt} = \mathbf{L}'u + \mathbf{R}'(u,\mu'),$$

where $\mu' = (\mu,\nu)$ and

$$\mathbf{R}'(u,\mu') = \mathbf{A}_\nu u + \mathbf{R}(u,\mu),$$

which satisfies the hypotheses in Theorem 3.3 with the parameter $\mu' = (\mu,\nu) \in \mathbb{R}^{m+r}$. We recover the original equation by taking $\nu = \varepsilon$, and find the invariant manifolds $\mathscr{M}_0(\mu,\varepsilon)$ for this equation, provided ε is sufficiently small, such that $(0,\varepsilon)$ belongs to the neighborhood $\mathscr{O}_{\mu'}$ of $(0,0)$ in $\mathbb{R}^{m+r}$ given by Theorem 3.3. This latter property is achieved when δ in (3.8) is sufficiently small, i.e., when the eigenvalues in σ_0 are close enough to the imaginary axis.

Remark 3.7 *(i) In (3.1) the parameter μ occurs only in the term* **R**, *which takes values in $\mathscr{Y}$. A more general study would be for cases where μ also occurs in the linear terms which take values in $\mathscr{X}$. Then one would have a family of* *operators $\mathbf{L}_\mu$ with domains which may also depend upon μ. Such a situation requires a more delicate analysis, which does not enter in our setting.*

(ii) *It is possible to develop the theory for a parameter μ lying in a (infinite-dimensional) Banach space instead of $\mathbb{R}^m$. Nevertheless, for such a situation one needs to go back and adapt the proof of the general result in Theorem 2.9. The proof of Theorem 3.3 given above does not extend to this situation, since it relies upon the fact that $\mathbb{R}^m$ is finite-dimensional (one has that $\dim \widetilde{\mathscr{E}_0} = \dim \mathscr{E}_0 + m$, and this quantity is infinite when $\mathbb{R}^m$ is replaced by an infinite- dimensional Banach space, so that the extended system (3.4) does not satisfy Hypothesis 2.4(ii)). We refer the reader to [73] for an example of a problem with a parameter varying in a function space, and for which the continuity of the reduction function Ψ with respect to the parameter, is only valid in $\mathscr{X}$, not in $\mathscr{Z}$.*

2.3.2 Nonautonomous Center Manifolds

We present in this section an extension of the result of center manifold Theorem 2.9 to the case of nonautonomous equations of the form

$$\frac{du}{dt} = \mathbf{L}u + \mathbf{R}(u,t). \tag{3.9}$$

We replace here Hypothesis 2.1 by the following assumptions on $\mathbf{L}$ and $\mathbf{R}$.

Hypothesis 3.8 *We assume that $\mathbf{L}$ and $\mathbf{R}$ in (3.9) have the following properties:*

(i) $\mathbf{L} \in \mathscr{L}(\mathscr{Z}, \mathscr{X})$;

(ii) *for some $k \geq 2$, there exists a neighborhood $\mathscr{V} \subset \mathscr{Z}$ of 0 such that $\mathbf{R} \in \mathscr{C}^k(\mathscr{V} \times \mathbb{R}, \mathscr{Y})$ and*

$$\mathbf{R}(0,t) = 0, \quad D_u\mathbf{R}(0,t) = 0.$$

In addition, we assume that for any sufficiently small ε, there exist positive constants $\delta_0(\varepsilon) = O(\varepsilon^2)$ and $\delta_1(\varepsilon) = O(\varepsilon)$ such that

$$\sup_{u \in B_\varepsilon(\mathscr{Z})} \|\mathbf{R}(u,t)\|_{\mathscr{Y}} = \delta_0(\varepsilon), \qquad \sup_{u \in B_\varepsilon(\mathscr{Z})} \|D_u\mathbf{R}(u,t)\|_{\mathscr{L}(\mathscr{Z},\mathscr{Y})} = \delta_1(\varepsilon). \tag{3.10}$$

The equalities in the formula (3.10) above, show that the nonlinear term $\mathbf{R}$ is bounded with respect to all $t \in \mathbb{R}$, uniformly for u in any sufficiently small closed ball $B_\varepsilon(\mathscr{Z})$. Furthermore, the dependency in t of the system (3.9) is in the nonlinear term $\mathbf{R}$, only. In this sense, the following theorem is a "perturbation" result of center manifold Theorem 2.9.

Theorem 3.9 (Nonautonomous center manifolds) *Assume that Hypotheses 3.8, 2.4, and 2.7 hold. Then, there exist a map $\Psi \in \mathscr{C}^k(\mathscr{E}_0 \times \mathbb{R}, \mathscr{Z}_h)$ and $c > 0$, with*

$$\Psi(0,t) = 0, \quad D_{u_0}\Psi(0,t) = 0,$$

and

$$\sup_{u_0 \in B_\varepsilon(\mathscr{E}_0)} \|\Psi(u_0,t)\|_{\mathscr{Z}} = c\delta_0(\varepsilon), \qquad \sup_{u_0 \in B_\varepsilon(\mathscr{E}_0)} \|D_u\Psi(u_0,t)\|_{\mathscr{L}(\mathscr{Z})} = c\delta_1(\varepsilon),$$

for sufficiently small ε*, and a neighborhood* $\mathscr{O}$ *of* 0 *in* $\mathscr{X}$ *such that the manifold*

$$\mathscr{M}_0(t) = \{u_0 + \Psi(u_0,t)\ ;\ (u_0,t) \in B_\varepsilon(\mathscr{E}_0) \times \mathbb{R}\} \subset \mathscr{X}$$

has the following properties:

(i) the set $\{(t,u(t)) \in \mathbb{R} \times \mathscr{M}_0(t)\}$ *is a local integral manifold of (3.9);*
(ii) any solution u *of (3.9) staying in* $\mathscr{O}$ *for all* $t \in \mathbb{R}$ *satisfies* $u(t) \in \mathscr{M}_0(t)$*.*

We give a brief proof of this result in Appendix B.3 (see also [95] for a complete proof).

Remark 3.10 *The analogue of the reduced equation (2.7) in this situation is*

$$\frac{du_0}{dt} = \mathbf{L}_0 u_0 + \mathbf{P}_0\mathbf{R}(u_0 + \Psi(u_0,t),t) \stackrel{def}{=} f(u_0,t), \tag{3.11}$$

whereas the analogue of the equality (2.8) is

$$\begin{aligned}\partial_t\Psi(u_0,t) + D_{u_0}\Psi(u_0,t)f(u_0,t) &= \mathbf{L}_h\Psi(u_0,t)\\ &\quad + \mathbf{P}_h\mathbf{R}(u_0 + \Psi(u_0,t),t) \text{ for all } u_0 \in \mathscr{E}_0.\end{aligned}$$

There are at least two particular cases of equation (3.9) that are important in applications:

(i) the case in which the map $\mathbf{R}$ is periodic with respect to t, and
(ii) the case in which $\lim_{t\to\infty}\mathbf{R}(u,t) \to \mathbf{R}_\infty(u)$ or $\lim_{t\to-\infty}\mathbf{R}(u,t) \to \mathbf{R}_{-\infty}(u)$.

In these cases the reduction function Ψ, and then also the reduced system, has similar properties. We show in Appendix B.3 that the following result holds.

Corollary 3.11 (Special cases) *Assume that the hypothesis in Theorem 3.9 holds.*

(i) If the map $\mathbf{R}$ *is periodic with respect to* t*,* $\mathbf{R}(u,t) = \mathbf{R}(u,t+\tau)$ *for some* $\tau > 0$*, then one can find a reduction function* Ψ *that is periodic, with the same period, namely* $\Psi(u_0,t) = \Psi(u_0,t+\tau)$ *for any* $(u_0,t) \in B_\varepsilon(\mathscr{E}_0) \times \mathbb{R}$*.*
(ii) Assume that there exist a map $\mathbf{R}_\infty \in \mathscr{C}^k(\mathscr{V},\mathscr{Y})$ *and* $d_0 > 0$ *such that*

$$\|\mathbf{R}(u,t) - \mathbf{R}_\infty(u)\|_{\mathscr{Y}} \leq ce^{-d_0t} \text{ for all } (u,t) \in \mathscr{V} \times \mathbb{R}^+.$$

Then the result in center manifold Theorem 2.9 holds for the autonomous equation

$$\frac{du}{dt} = \mathbf{L}u + \mathbf{R}_\infty(u), \tag{3.12}$$

and there exists $c' > 0$ *such that*

$$\|\Psi(u_0,t) - \Psi_\infty(u_0)\|_{\mathscr{Z}_h} \leq c' e^{-d_0 t} \text{ for all } (u_0,t) \in B_\varepsilon(\mathscr{E}_0) \times \mathbb{R}^+,$$

where Ψ_∞ is the reduction function for the autonomous equation (3.12). A similar result holds when $\|\mathbf{R}(u,t) - \mathbf{R}_{-\infty}(u)\|_{\mathscr{Y}} \leq ce^{d_0 t}$ for all $(u,t) \in \mathscr{V} \times \mathbb{R}^-$.

2.3.3 Symmetries and Reversibility

We discuss in this section three cases of equations possessing a certain symmetry. In each case we show that this symmetry is inherited by both the reduction function Ψ and the reduced system.

Equivariant Systems

We start with the case of an equation that is equivariant under the action of a linear operator. More precisely, we make the following assumptions.

Hypothesis 3.12 (Equivariant equation) *We assume that there exists a linear operator $\mathbf{T} \in \mathscr{L}(\mathscr{X}) \cap \mathscr{L}(\mathscr{Z})$, which commutes with the vector field in equation (2.1),*

$$\mathbf{TL}u = \mathbf{LT}u, \quad \mathbf{TR}(u) = \mathbf{R}(\mathbf{T}u).$$

We further assume that the restriction $\mathbf{T}_0$ of $\mathbf{T}$ to the subspace $\mathscr{E}_0$ is an isometry.

Notice that the fact that the operator $\mathbf{T}$ commutes with the vector field in the equation (2.1) implies that the subspace $\mathscr{E}_0$ is invariant under the action of $\mathbf{T}$, so that the restriction $\mathbf{T}_0$ in the hypothesis above is well defined. Indeed, since $\mathbf{T}$ commutes with $\mathbf{L}$, it also commutes with its resolvent $(\lambda\mathbb{I} - \mathbf{L})^{-1}$, and from the Dunford integral formula (2.2) it follows that $\mathbf{T}$ commutes with the spectral projector $\mathbf{P}_0$. Consequently, the spectral subspace $\mathscr{E}_0$ associated with $\mathbf{P}_0$ is invariant under the action of $\mathbf{T}$.

We show in Appendix B.4 that the following result holds in this situation.

Theorem 3.13 (Center manifold theorem for equivariant equations) *Under the assumptions in Theorem 2.9, we further assume that Hypothesis 3.12 holds. Then one can find a reduction function Ψ in Theorem 2.9 which commutes with $\mathbf{T}$, i.e.,*

$$\mathbf{T}\Psi(u_0) = \Psi(\mathbf{T}_0 u_0) \text{ for all } u_0 \in \mathscr{E}_0,$$

and such that the vector field in the reduced equation (2.7) commutes with $\mathbf{T}_0$.

We point out that analogous results hold for the parameter-dependent equation (3.1) and in the nonautonomous case for the equation (3.9).

Reversible Systems

Next, we consider the case of reversible equations, when the vector field in (2.1) anticommutes with a symmetry $\mathbf{S}$. More precisely, we make the following assumptions.

Hypothesis 3.14 (Reversible equation) *Assume that there exists a linear symmetry* $\mathbf{S} \in \mathscr{L}(\mathscr{X}) \cap \mathscr{L}(\mathscr{Z})$, *with*

$$\mathbf{S}^2 = \mathbb{I}, \quad \mathbf{S} \neq \mathbb{I},$$

and which anticommutes with the vector field in (2.1),

$$\mathbf{SL}u = -\mathbf{LS}u, \quad \mathbf{SR}(u) = -\mathbf{R}(\mathbf{S}u). \tag{3.13}$$

Notice that in this case, if $t \mapsto u(t)$ is a solution of (2.1), then $t \mapsto \mathbf{S}u(-t)$ is also a solution of (2.1). Moreover, the spectrum of the linear operator $\mathbf{L}$ is symmetric with respect to the origin in the complex plane. Indeed, from the first equality in (3.13) we deduce that

$$\mathbf{S}(\lambda\mathbb{I} - \mathbf{L})^{-1} = (\lambda\mathbb{I} + \mathbf{L})^{-1}\mathbf{S},$$

which shows that the resolvent set $\rho(\mathbf{L})$ as well as its complement $\sigma(\mathbf{L})$ are symmetric with respect to the origin. In particular, for real systems, besides the usual symmetry with respect to the real axis, in this case the spectrum of $\mathbf{L}$ is also symmetric with respect to the imaginary axis. We also point out that if λ is an eigenvalue of $\mathbf{L}$ with the associated eigenvector ζ, then $-\lambda$ is an eigenvalue with the associated eigenvector $\mathbf{S}\zeta$.

As in the case of equivariant equations with Hypothesis 3.12, we have that the spectral subspace $\mathscr{E}_0$ is invariant under the action of $\mathbf{S}$. Indeed, since the spectrum of the operator $\mathbf{L}$ is symmetric with respect to the origin in the complex plane, we may choose the curve Γ in the Dunford integral formula (2.2) such that it is also symmetric with respect to the origin in the complex plane. Then a direct calculation shows that the spectral projection $\mathbf{P}_0$ given by (2.2) commutes with $\mathbf{S}$, so that $\mathscr{E}_0$ is invariant under the action of $\mathbf{S}$.

By arguing as in the case of equivariant equations, we obtain here the following result.

Theorem 3.15 (Center manifold theorem for reversible equations) *Under the assumptions of Theorem 2.9, we further assume that Hypothesis 3.14 holds. Then one can find a reduction function Ψ in Theorem 2.9 that commutes with* $\mathbf{S}$,

$$\mathbf{S}\Psi(u_0) = \Psi(\mathbf{S}_0 u_0) \textit{ for all } u_0 \in \mathscr{E}_0,$$

where $\mathbf{S}_0$ is the restriction of $\mathbf{S}$ to the subspace $\mathscr{E}_0$ and such that the reduced equation is reversible, i.e., the vector field in (2.7) anticommutes with $\mathbf{S}_0$.

A similar result holds for the parameter-dependent equation (3.1), whereas in the nonautonomous case for equation (3.9) the following holds.

Corollary 3.16 (Reversible nonautonomous equations) *Under the assumptions of Theorem 3.9, we further assume that the equation (3.9) is reversible, i.e., there exists a symmetry* $\mathbf{S} \in \mathscr{L}(\mathscr{X}) \cap \mathscr{L}(\mathscr{Z})$, *with* $\mathbf{S}^2 = \mathbb{I}$ *and* $\mathbf{S} \neq \mathbb{I}$, *such that*

$$\mathbf{SL}u = -\mathbf{LS}u, \quad \mathbf{SR}(u,t) = -\mathbf{R}(\mathbf{S}u, -t).$$

Then, one can find a reduction function Ψ *in the Theorem 3.9 that satisfies*

$$\mathbf{S}\Psi(u_0,t) = \Psi(\mathbf{S}_0 u_0, -t) \text{ for all } u_0 \in \mathscr{E}_0,$$

and the reduced equation is reversible, i.e., the vector field in (3.11) satisfies

$$\mathbf{S}_0 f(u_0,t) = -f(\mathbf{S}_0 u_0, -t) \text{ for all } u_0 \in \mathscr{E}_0.$$

Continuous Symmetry

We end this section with the case where equation (2.1) is equivariant under a one-parameter group of isometries. We focus on the case of the underlying group $\mathbb{R}$, and, instead of a single equilibrium at the origin, the equation has a "line" of equilibria. This situation is encountered in the applications in Sections 5.1.2, 5.1.3, and 5.2.2 of Chapter 5. Other groups of symmetries can be treated in the same spirit, however, this may require more specific tools and further evolved algebra. We refer the reader to the book [16] for such cases. More precisely, we make here the following hypotheses.

Hypothesis 3.17 (Continuous symmetry) *Assume that there exists a continuous one-parameter group of isometries* $(\mathbf{T}_\alpha)_{\alpha \in \mathbb{R}} \subset \mathscr{L}(\mathscr{Z}) \cap \mathscr{L}(\mathscr{X})$, *which commutes with the vector field in (2.1), that is, such that the following properties hold:*

(i) the map $\alpha \in \mathbb{R} \mapsto \mathbf{T}_\alpha \in \mathscr{L}(\mathscr{Z}) \cap \mathscr{L}(\mathscr{X})$ *is continuous;*
(ii) $\mathbf{T}_0 = \mathbb{I}$ *and* $\mathbf{T}_{\alpha+\beta} = \mathbf{T}_\alpha \mathbf{T}_\beta$ *for all* α, $\beta \in \mathbb{R}$;
(iii) $\mathbf{T}_\alpha \mathbf{L}u = \mathbf{L}\mathbf{T}_\alpha u$ *and* $\mathbf{T}_\alpha \mathbf{R}(u) = \mathbf{R}(\mathbf{T}_\alpha u)$ *for all* $\alpha \in \mathbb{R}$.

Further assume that the infinitesimal generator τ *of the group* $(\mathbf{T}_\alpha)_{\alpha \in \mathbb{R}} \subset \mathscr{L}(\mathscr{X})$ *belongs to* $\mathscr{L}(\mathscr{Z}, \mathscr{Y})$,

$$\tau := \frac{d\mathbf{T}_\alpha}{d\alpha}\Big|_{\alpha=0} \in \mathscr{L}(\mathscr{Z}, \mathscr{Y}).$$

Hypothesis 3.18 (Equilibria) *Assume that equation (2.1) has a nontrivial equilibrium* $u^* \in \mathscr{Z}$,

$$\mathbf{L}u^* + \mathbf{R}(u^*) = 0, \quad u^* \neq 0,$$

satisfying $\tau u^* \in \mathscr{Z} \setminus \{0\}$.

An immediate consequence of the hypotheses above is that equation (2.1) possesses a *line of equilibria* given by $\{\mathbf{T}_\alpha u^* \in \mathscr{Z}; \alpha \in \mathbb{R}\}$. Furthermore, since $\tau u^* \in \mathscr{Z}$, we may differentiate the identity

$$\mathbf{L}\mathbf{T}_\alpha u^* + \mathbf{R}(\mathbf{T}_\alpha u^*) = 0$$

at $\alpha = 0$ and obtain

$$\mathbf{L}\tau u^* + D\mathbf{R}(u^*)\tau u^* = 0. \tag{3.14}$$

This shows that τu^* belongs to the kernel of the linearization $\mathbf{L} + D\mathbf{R}(u^*)$ of the vector field at the equilibrium u^* (this eigenvector is often called the "Goldstone mode" by physicists).

Our purpose is to construct *a local center manifold along this line of equilibria* in $\mathscr{Z}$, taking into account the continuous symmetry of the equation. We make the Ansatz

$$u(t) = \mathbf{T}_{\alpha(t)}(u^* + v(t)), \tag{3.15}$$

replacing the unknown u by the pair (α, v), with $\alpha(t) \in \mathbb{R}$ and $v(t) \in \mathscr{Z}$ satisfying a transversality condition that we define now. For this we decompose the space $\mathscr{X}$ in the subspace spanned by τu^*, parallel to the line of equilibria, and a complementary subspace. Consider the linear form φ^* in the dual space $\mathscr{X}^*$ such that $\langle \tau u^*, \varphi^* \rangle = 1$ (e.g., see [76, p. 135]). We define the subspace $\mathscr{H} \subset \mathscr{X}$ transverse to τu^*,

$$\mathscr{H} = \{v \in \mathscr{X} \; ; \; \langle v, \varphi^* \rangle = 0\},$$

which provides us with a decomposition of $\mathscr{X}$ into two complementary closed subspaces,

$$\mathscr{X} = \{\tau u^*\} \oplus \mathscr{H}.$$

The linear operators

$$\Pi_0 u = \langle u, \varphi^* \rangle \tau u^*, \quad \Pi_{\mathscr{H}} = \mathbb{I} - \Pi_0$$

are projections onto the subspaces $\{\tau u^*\}$ and $\mathscr{H}$, respectively. Since $\tau u^* \in \mathscr{Z}$, we have that $\Pi_{\mathscr{H}} u \in \mathscr{Z}$ (resp., $\Pi_{\mathscr{H}} u \in \mathscr{Y}$) if $u \in \mathscr{Z}$ (resp., $u \in \mathscr{Y}$), so that we have similar decompositions for $\mathscr{Z}$ and $\mathscr{Y}$. We now choose v in (3.15) such that $v(t)$ belongs to $\mathscr{H}$, i.e.,

$$\Pi_0 v(t) = 0 \quad \Longleftrightarrow \quad \langle v(t), \varphi^* \rangle = 0.$$

Next, we substitute the Ansatz (3.15) into the equation (2.1) and obtain the equation

$$\tau \mathbf{T}_\alpha (u^* + v) \frac{d\alpha}{dt} + \mathbf{T}_\alpha \frac{dv}{dt} = \mathbf{L}\mathbf{T}_\alpha v + \mathbf{R}(\mathbf{T}_\alpha (u^* + v)) - \mathbf{R}(\mathbf{T}_\alpha u^*),$$

where we have used the fact that $\mathbf{T}_\alpha u^*$ is an equilibrium of (2.1). Using the equivariance property in Hypothesis 3.17(iii) we find

$$(\tau u^* + \tau v) \frac{d\alpha}{dt} + \frac{dv}{dt} = \mathbf{A} v + \tilde{\mathbf{R}}(v),$$

in which

$$\mathbf{A}v = \mathbf{L}v + D\mathbf{R}(u^*)v, \quad \tilde{\mathbf{R}}(v) = \mathbf{R}(u^* + v) - \mathbf{R}(u^*) - D\mathbf{R}(u^*)v.$$

Projecting successively with Π_0 and $\Pi_{\mathscr{H}}$, this gives the first order system for (α, v),

$$\frac{d\alpha}{dt} = (1 + \langle \tau v, \varphi^* \rangle)^{-1} \langle \mathbf{A}v + \tilde{\mathbf{R}}(v), \varphi^* \rangle \stackrel{\text{def}}{=} g(v) \tag{3.16}$$

$$\frac{dv}{dt} = \Pi_{\mathscr{H}} \mathbf{A}v + \Pi_{\mathscr{H}} \tilde{\mathbf{R}}(v) - g(v) \Pi_{\mathscr{H}} \tau v, \tag{3.17}$$

which holds for $v \in \mathscr{Z}$ sufficiently small.

The key property of the system (3.16)–(3.17) is that the vector field is independent of α, which in particular does not appear in the equation (3.17). This equation decouples, so that we can solve it separately, and once v is known we obtain α from the first equation. The differential equation (3.17) is of the form of (2.1), with the spaces $\mathscr{X}$, $\mathscr{Z}$, $\mathscr{Y}$ replaced by

$$\mathscr{X}' = \mathscr{H}, \quad \mathscr{Z}' = \Pi_{\mathscr{H}} \mathscr{Z}, \quad \mathscr{Y}' = \Pi_{\mathscr{H}} \mathscr{Y},$$

respectively, and operators $\mathbf{L}$ and $\mathbf{R}$ replaced by

$$\mathbf{L}' = \Pi_{\mathscr{H}} \mathbf{A}, \quad \mathbf{R}'(v) = \Pi_{\mathscr{H}} (\tilde{\mathbf{R}}(v) - g(v) \tau v), \tag{3.18}$$

respectively. In particular, this means that thanks to the choice of the Ansatz (3.15), the dimension of the problem is decreased by one, the space $\mathscr{X}$ being replaced by $\mathscr{H}$. In fact we suppressed the direction τu^*, which belongs to the kernel of $\mathbf{A}$ as shown by (3.14). Furthermore, once we obtain a local center manifold for equation (3.17), we have a center manifold for equation (2.1), with one additional dimension, in a neighborhood of the line of stationary solutions $\{\mathbf{T}_\alpha u^* \in \mathscr{Z}; \alpha \in \mathbb{R}\}$. More precisely, we have the following result.

Theorem 3.19 (Center manifolds in presence of continuous symmetry) *Assume that Hypothesis 2.1 holds and that the linear operator $\mathbf{L}' = \Pi_{\mathscr{H}} \mathbf{A}$ in (3.18) acting in $\mathscr{X}'$ satisfies Hypotheses 2.4 and 2.7. Then for the differential equation (3.17) the result in Theorem 2.9 holds.*

Let $\mathscr{O}'$, Ψ', and $\mathscr{E}_0'$ be respectively the neighborhood of the origin in $\mathscr{Z}'$, the reduction function, and the spectral subspace, given by Theorem 2.9 for (3.17). Consider the "tubular" neighborhood

$$\mathscr{O} = \{\mathbf{T}_\alpha (u^* + v) \; ; \; v \in \mathscr{O}', \; \alpha \in \mathbb{R}\} \subset \mathscr{Z}$$

of the line of equilibria $\{\mathbf{T}_\alpha u^* \in \mathscr{Z}; \alpha \in \mathbb{R}\}$, and the manifold

$$\mathscr{M}_0 = \{\mathbf{T}_\alpha (u^* + v_0 + \Psi(v_0)) \; ; \; v_0 \in \mathscr{E}_0', \; \alpha \in \mathbb{R}\} \subset \mathscr{Z}. \tag{3.19}$$

Then for differential equation (2.1) the following properties hold:

(i) The manifold $\mathscr{M}_0$ is locally invariant, i.e., if u is a solution of (2.1) satisfying $u(0) \in \mathscr{M}_0 \cap \mathscr{O}$ and $u(t) \in \mathscr{O}$ for all $t \in [0, T]$, then $u(t) \in \mathscr{M}_0$ for all $t \in [0, T]$.

(ii) $\mathscr{M}_0$ *contains the set of solutions of (2.1) staying in* $\mathscr{O}$ *for all* $t \in \mathbb{R}$*, i.e., if* u *is a solution of (2.1) satisfying* $u(t) \in \mathscr{O}$ *for all* $t \in \mathbb{R}$*, then* $u(0) \in \mathscr{M}_0$*.*

We point out that in this situation the center manifold $\mathscr{M}_0$ contains the solutions which stay close to the line of equilibria for all $t \in \mathbb{R}$. These solutions are of the form

$$u = \mathbf{T}_\alpha(u^* + v_0 + \Psi(v_0)),$$

with α and v_0 satisfying the reduced system

$$\frac{d\alpha}{dt} = g(v_0 + \Psi(v_0)) \tag{3.20}$$

$$\begin{aligned} \frac{dv_0}{dt} &= \Pi_{\mathscr{H}}\mathbf{A}v_0 + \mathbf{P}'_0\left(\Pi_{\mathscr{H}}\tilde{\mathbf{R}}(v_0 + \Psi(v_0))\right) \\ &\quad -\mathbf{P}'_0\left(g(v_0 + \Psi(v_0))\Pi_{\mathscr{H}}\tau(v_0 + \Psi(v_0))\right), \end{aligned} \tag{3.21}$$

in which g is defined in (3.16) and $\mathbf{P}'_0$ is the spectral projector for the linear operator $\mathbf{L}' = \Pi_{\mathscr{H}}\mathbf{A}$ defined as in Section 2.2.1. Furthermore, for such a solution we have that v_0 is a small bounded solution of the equation (3.21), whereas α given by (3.20) has bounded derivative and may grow linearly in t.

Similar results hold for the parameter-dependent equation (3.1) and for the nonautonomous equation (3.9).

2.3.4 Empty Unstable Spectrum

A particular case, which appears in some applications, e.g. in parabolic problems, occurs when the unstable spectrum σ_+ of $\mathbf{L}$ is empty. Then we complete general Hypothesis 2.7 by the following assumptions, which allow us to obtain further information about the center manifolds in this case.

Hypothesis 3.20 (Empty unstable spectrum) *Assume that* $\sigma_+ = \varnothing$ *and that for any* $\eta \in [0, \gamma]$ *the following properties hold:*

(i) For any $f \in \mathscr{F}_\eta(\mathbb{R}, \mathscr{Y}_h)$ *the linear problem*

$$\frac{du_h}{dt} = \mathbf{L}_h u_h + f$$

has a unique solution $u_h = \mathbf{K}_h f \in \mathscr{F}_\eta(\mathbb{R}, \mathscr{Z}_h)$*. Furthermore, the linear map* $\mathbf{K}_h$ *belongs to* $\mathscr{L}(\mathscr{F}_\eta(\mathbb{R}, \mathscr{Y}_h), \mathscr{F}_\eta(\mathbb{R}, \mathscr{Z}_h))$*, and there exists a continuous map* $C : [0, \gamma] \to \mathbb{R}$ *such that*

$$\|\mathbf{K}_h\|_{\mathscr{L}(\mathscr{F}_\eta(\mathbb{R},\mathscr{Y}_h),\mathscr{F}_\eta(\mathbb{R},\mathscr{Z}_h))} \leq C(\eta).$$

(ii) The linear initial value problem

$$\frac{du_h}{dt} = \mathbf{L}_h u_h, \quad u_h|_{t=0} = u_h(0) \in \mathscr{Z}_h,$$

has a unique solution $u_h(t) \in \mathscr{C}^0(\mathbb{R}^+, \mathscr{Z}_h)$, *which satisfies*

$$\|u_h(t)\|_{\mathscr{Z}} \leq c_\eta e^{-\eta t} \text{ for all } t \geq 0$$

for some positive constant c_η.

As for Hypothesis 2.7, we have that these assumptions are satisfied, provided Hypothesis 2.15 holds (see Remark B.2 in Appendix B.2).

Exercise 3.21 *Prove that Hypothesis 3.20 is satisfied in finite dimensions when* $\mathscr{X} = \mathbb{R}^n$ *and* $\sigma_+ = \varnothing$.

Theorem 3.22 (Center manifold theorem for empty unstable spectrum) *Under the assumptions of Theorem 2.9, further assume that Hypothesis 3.20 holds. Then in addition to the properties in Theorem 2.9 the following holds.*

The local center manifold $\mathscr{M}_0$ *is locally attracting, i.e., any solution of (2.1) that stays in* $\mathscr{O}$ *for all* $t > 0$ *tends exponentially towards a solution of (2.1) on* $\mathscr{M}_0$. *More precisely, if* $u(0) \in \mathscr{O}$ *and the solution* $u(t;u(0))$ *of (2.1) satisfies* $u(t;u(0)) \in \mathscr{O}$ *for all* $t > 0$, *then there exists* $\widetilde{u} \in \mathscr{M}_0 \cap \mathscr{O}$ *and* $\gamma' > 0$ *such that*

$$u(t;u(0)) = u(t;\widetilde{u}) + O(e^{-\gamma' t}) \text{ as } t \to \infty.$$

(Here we denoted by $u(t;u(0))$ *the solution of (2.1) satisfying* $u|_{t=0} = u(0)$*).*

We prove this result in Appendix B.5. In addition, since according to the proof of Theorem 3.3, the parameter-dependent equation (3.1) can be regarded as a particular case of equation (2.1), we can extend the result above to equation (3.1).

Theorem 3.23 (Parameter-dependent center manifolds) *Assume that Hypotheses 3.1, 2.4, 2.7, and 3.20 hold. Then in addition to the properties in Theorem 3.3 the following holds.*

The local center manifold $\mathscr{M}_0(\mu)$ *is locally attracting, i.e., any solution of (3.1) that stays in* $\mathscr{O}_u$ *for all* $t > 0$ *tends exponentially towards a solution of (3.1) on* $\mathscr{M}_0(\mu)$. *More precisely, if* $u(0) \in \mathscr{O}_u$ *and the solution* $u(t;u(0))$ *of (3.1) satisfies* $u(t;u(0)) \in \mathscr{O}_u$ *for all* $t > 0$, *then there exists* $\widetilde{u} \in \mathscr{M}_0(\mu) \cap \mathscr{O}_u$ *and* $\gamma' > 0$ *such that*

$$u(t;u(0)) = u(t;\widetilde{u}) + O(e^{-\gamma' t}) \text{ as } t \to \infty.$$

(Here we denoted by $u(t;u(0))$ *the solution of (3.1) satisfying* $u|_{t=0} = u(0)$*.)*

2.4 Further Examples and Exercises

We end this chapter with some further examples in which we apply the different variants of center manifold Theorem 2.9 presented in Section 2.3. In each example we show how to check the hypotheses and discuss the reduced system. In contrast to the second example given in Section 2.2.4, here we work in Hilbert spaces, which, in particular, simplifies the checking of Hypothesis 2.7 (see Remark 2.16). In addition, these examples are such that $u = 0$ is a solution of the system for all values of the parameter(s), except for the example in Section 2.4.3, case V, and the example in Section 2.4.4. This property allows us to use the result in Exercise 3.5, and so simplify some computations.

2.4.1 A Fourth Order ODE

Consider the fourth order ODE

$$u^{(4)} - u'' - \mu u - au^2 = 0, \tag{4.1}$$

where μ is a small parameter and a a given real number. For $\mu = 0$ this is precisely equation (2.11), studied in Section 2.2.4.

Formulation as a First Order System

We start by writing equation (2.11) in the form (3.1). As in Section 2.2.4, we set $U = (u, u_1, u_2, u_3)$ with $u_1 = u'$, $u_2 = u'' - u$, $u_3 = u_2'$, and then the equation is equivalent to the system

$$\frac{dU}{dt} = \mathbf{L}U + \mathbf{R}(U, \mu), \tag{4.2}$$

in which

$$\mathbf{L} = \begin{pmatrix} 0 & 1 & 0 & 0 \\ 1 & 0 & 1 & 0 \\ 0 & 0 & 0 & 1 \\ 0 & 0 & 0 & 0 \end{pmatrix}, \qquad \mathbf{R}(U, \mu) = \begin{pmatrix} 0 \\ 0 \\ 0 \\ \mu u + au^2 \end{pmatrix}.$$

Here $\mathbf{L}$ is the same 4×4-matrix, and $\mathbf{R} : \mathbb{R}^4 \times \mathbb{R} \to \mathbb{R}^4$ is a smooth map, so that we choose again

$$\mathscr{X} = \mathscr{Y} = \mathscr{Z} = \mathbb{R}^4.$$

In addition, notice that the system (4.2) possesses a reversibility symmetry, i.e., $\mathbf{L}$ and $\mathbf{R}(\cdot, \mu)$ anticommute with

$$\mathbf{S} = \begin{pmatrix} 1 & 0 & 0 & 0 \\ 0 & -1 & 0 & 0 \\ 0 & 0 & 1 & 0 \\ 0 & 0 & 0 & -1 \end{pmatrix}.$$

This symmetry is a consequence of the fact that the equation (4.1) is invariant under the reflection $t \mapsto -t$.

Checking the Hypotheses

Clearly, Hypothesis 3.1 is satisfied for $\mathbf{L}$ and $\mathbf{R}$ as above for any $k \geq 2$, and neighborhoods $\mathscr{V}_u = \mathbb{R}^4$ and $\mathscr{V}_\mu = \mathbb{R}$. We have seen, in Section 2.2.4, that $\mathbf{L}$ satisfies Hypothesis 2.4 with

$$\sigma_+ = \{1\}, \quad \sigma_0 = \{0\}, \quad \sigma_- = \{-1\},$$

and that Hypothesis 2.7 holds because $\mathscr{X}$ is finite-dimensional. Consequently, we can apply center manifold Theorem 3.3, and conclude the existence of a local two-dimensional center manifold of class $\mathscr{C}^k$ for any arbitrary, but fixed, $k \geq 2$ for any μ sufficiently small.

In addition, since system (4.2) is reversible, Hypothesis 3.14 is also satisfied, so that according to Theorem 3.15 the reduced equation is reversible, i.e., the vector field in this equation anticommutes with the symmetry $\mathbf{S}_0$ induced by $\mathbf{S}$ on $\mathscr{E}_0$.

Reduced Equation

We compute now the Taylor expansion, up to order 2, of the vector field in the reduced equation. Clearly, for $\mu = 0$ we have the expansion found in Section 2.2.4.

Consider the basis $\{\zeta_0, \zeta_1\}$ of $\mathscr{E}_0$ computed in Section 2.2.4, and notice that $\mathbf{S}$ acts on this basis through

$$\mathbf{S}\zeta_0 = \zeta_0, \quad \mathbf{S}\zeta_1 = -\zeta_1,$$

so that

$$\mathbf{S}_0 = \begin{pmatrix} 1 & 0 \\ 0 & -1 \end{pmatrix}.$$

Then, according to Theorems 3.3 and 3.15, solutions on the center manifold are of the form

$$U(t) = U_0(t) + \Psi(U_0(t), \mu), \tag{4.3}$$

in which $\Psi(0,\mu) = 0$, $D\Psi(0,0) = 0$, $\Psi(\mathbf{S}_0 U_0, \mu) = \mathbf{S}\Psi(U_0, \mu)$, and $U_0(t) \in \mathscr{E}_0$, so that

$$U_0(t) = A(t)\zeta_0 + B(t)\zeta_1, \tag{4.4}$$

where A and B are real-valued functions. Notice that $\Psi(0,\mu) = 0$, because $\mathbf{R}(0,\mu) = 0$ (see Exercise 3.5). The reduced system is an ODE for $U_0 = (A, B)$, which now depends upon μ, and according to (3.6) it is given by

$$\frac{dU_0}{dt} = \mathbf{L}_0 U_0 + \mathbf{P}_0 \mathbf{R}(U_0 + \Psi(U_0, \mu), \mu), \tag{4.5}$$

where $\mathbf{L}_0$ and $\mathbf{P}_0$ are as in Section 2.2.4. Again, since the last component of the vector ζ_0^* vanishes, we have that

$$\mathbf{P}_0\mathbf{R}(U_0+\Psi(U_0,\mu),\mu) = \langle \mathbf{R}(U_0+\Psi(U_0,\mu),\mu), \zeta_1^*\rangle \zeta_1,$$

and since now $\Psi(U_0) = O(\|U_0\|(|\mu| + \|U_0\|))$, we conclude that

$$\mathbf{P}_0\mathbf{R}(U_0+\Psi(U_0,\mu),\mu) = \langle \mathbf{R}(U_0), \zeta_1^*\rangle \zeta_1 + O(\|U_0\|(|\mu|^2 + \|U_0\|^2)).$$

The explicit formulas for $\mathbf{P}_0$, $\mathbf{R}$, and U_0 give

$$\mathbf{R}(U_0) = \mathbf{R}(A\zeta_0 + B\zeta_1) = \begin{pmatrix} 0 \\ 0 \\ 0 \\ -\mu A + aA^2 \end{pmatrix},$$

so that

$$\mathbf{P}_0\mathbf{R}(U_0+\Psi(U_0,\mu),\mu) = \left(-\mu A + aA^2 + O((|A|+|B|)(|\mu|^2+|A|^2+|B|^2))\right)\zeta_1.$$

We conclude that the reduced system (2.15), in the basis $\{\zeta_0, \zeta_1\}$, is

$$\begin{aligned} \frac{dA}{dt} &= B \\ \frac{dB}{dt} &= -\mu A + aA^2 + O((|A|+|B|)(|\mu|^2+|A|^2+|B|^2)). \end{aligned}$$

In addition, the vector field in this system anticommutes with the matrix $\mathbf{S}_0$, which implies that the right hand side in the second equation above is even in B, so that the higher order terms in the expansion are in fact of order $O((|A|+|B|^2)(|\mu|^2+|A|^2+|B|^4))$.

Remark 4.1 *(i) For the calculation of the terms that are linear in A and B in the reduced equation, we can also use the result in Exercise 3.5. According to this result, the two eigenvalues of the 2×2-matrix obtained by linearizing the vector field in the reduced equation at $(A,B) = (0,0)$ are precisely the two eigenvalues of the matrix*

$$\mathbf{L}_\mu = \mathbf{L} + D\mathbf{R}(0,\mu) = \begin{pmatrix} 0 & 1 & 0 & 0 \\ 1 & 0 & 1 & 0 \\ 0 & 0 & 0 & 1 \\ \mu & 0 & 0 & 0 \end{pmatrix},$$

which are the continuation of the double eigenvalue 0 of $\mathbf{L}$ for small μ. A direct calculation gives the eigenvalues

$$\lambda^2 = \frac{1 \pm \sqrt{1+4\mu}}{2}.$$

Hence, the two eigenvalues close to 0 satisfy $\lambda_\pm^2 = -\mu + O(|\mu|^2)$. Next, the 2×2-matrix obtained by linearizing the vector field in the reduced equation at $(A,B) = (0,0)$ is of the form

$$\begin{pmatrix} 0 & 1 \\ \alpha(\mu) & 0 \end{pmatrix},$$

since, as we have seen above, B is the only term in the first component of the vector field, and the second component is even in B. Consequently,

$$\alpha(\mu) = -\mu + O(|\mu|^2),$$

which gives the same result as above.

(ii) For the computation of an expansion up to order 3*, or higher, one needs to compute the terms of order* 2 *in the expansion of* Ψ *(see also Remark 2.21). This can be done by substituting the Ansatz*

$$\begin{aligned}\Psi(A,B) = {} & \Psi_{101}\mu A + \Psi_{011}\mu B + \Psi_{200}A^2 + \Psi_{110}AB + \Psi_{020}B^2 \\ & + O((|A|+|B|)(|\mu|^2+|A|^2+|B|^2)) \end{aligned} \tag{4.6}$$

in the identity (3.7)*, and then the vectors* Ψ_{ijk} *are determined by identifying powers of* μ*, A, and B. Besides the fact that these vectors belong to the space* $(\mathbb{I} - \mathbf{P}_0)\mathbb{R}^4$*, so that they are orthogonal to both* ζ_0^* *and* ζ_1^**, due to the reversibility symmetry they also satisfy*

$$\mathbf{S}\Psi_{101} = \Psi_{101}, \quad \mathbf{S}\Psi_{011} = -\Psi_{011}, \quad \mathbf{S}\Psi_{200} = \Psi_{200},$$
$$\mathbf{S}\Psi_{110} = -\Psi_{110}, \quad \mathbf{S}\Psi_{020} = \Psi_{020}.$$

(iii) An alternative way of computing the reduced system is to directly substitute formulas (4.3), (4.4)*, and* (4.6) *into the first order system* (4.2) *and calculate the Taylor expansions of both sides of the resulting system. We use this approach in examples that follow, in this section.*

(iv) The terms in the expansion of the vector field that do not depend upon μ *can be computed separately, by setting* $\mu = 0$ *from the beginning. The other terms, depending upon* μ*, can be calculated afterwards by restricting to such terms in the Taylor expansions.*

2.4.2 Burgers Model

We consider the initial boundary value problem

$$\frac{\partial \phi}{\partial t} = \frac{1}{\mathscr{R}}\frac{\partial^2 \phi}{\partial x^2} + \phi - \frac{\partial(\phi^2)}{\partial x} + U\phi, \tag{4.7}$$

$$\frac{dU}{dt} = -\frac{1}{\mathscr{R}}U - \int_0^1 \phi^2(x,t)dx, \tag{4.8}$$

$$\phi(0,t) = \phi(1,t) = 0, \tag{4.9}$$

where $\phi(x,t) \in \mathbb{R}$ and $U(t) \in \mathbb{R}$ for $(x,t) \in (0,1) \times \mathbb{R}$. This model equation, introduced by J. M. Burgers [11], is a one-dimensional model used for understanding instabilities in viscous fluid flows. In this system ϕ represents a velocity fluctuation, U is the induced perturbation on the mean basic flow, and $\mathscr{R}$ is the Reynolds number, proportional to the inverse of viscosity. The product $U\phi$ represents the interaction between the mean flow and the perturbation, the derivative of ϕ^2 represents inertial terms, and the integral represents Reynolds stresses.

Formulation as a First Order Equation

We start by writing the problem (4.7)–(4.9) in the form (2.1), but now with linear part $\mathbf{L}$ depending upon the parameter $\mathscr{R}$, $\mathbf{L} = \mathbf{L}_{\mathscr{R}}$. We set

$$u = \begin{pmatrix} \phi \\ U \end{pmatrix}, \quad \mathbf{L}_{\mathscr{R}} u = \begin{pmatrix} \frac{1}{\mathscr{R}} \frac{\partial^2 \phi}{\partial x^2} + \phi \\ -\frac{1}{\mathscr{R}} U \end{pmatrix}, \quad \mathbf{R}(u) = \begin{pmatrix} -\frac{\partial(\phi^2)}{\partial x} + U\phi \\ -\int_0^1 \phi^2(x,\cdot)dx \end{pmatrix},$$

and choose the Hilbert space

$$\mathscr{X} = L^2(0,1) \times \mathbb{R}.$$

As in the example given in Section 2.2.4, we include the boundary conditions (4.9) in the domain of definition $\mathscr{Y}$ of the operator $\mathbf{L}_{\mathscr{R}}$, by taking

$$\mathscr{Z} = (H^2(0,1) \cap H_0^1(0,1)) \times \mathbb{R}.$$

Finally, we set

$$\mathscr{Y} = H_0^1(0,1) \times \mathbb{R},$$

so that $\mathbf{R}(u) \in \mathscr{Z}$ for $u \in \mathscr{Y}$. Notice that the system commutes with the symmetry $\mathbf{T}$ defined by

$$\mathbf{T} \begin{pmatrix} \phi(x) \\ U \end{pmatrix} = \begin{pmatrix} -\phi(1-x) \\ U \end{pmatrix},$$

which is an isometry in both $\mathscr{X}$ and $\mathscr{Z}$.

This formulation of the problem does not quite enter into the setting of center manifold theorems presented in the previous sections, because the linear operator depends upon the parameter $\mathscr{R}$. The next step consists in determining the spectrum of this operator in order to detect the "critical" values of the parameter $\mathscr{R}$, where its spectrum contains purely imaginary eigenvalues. These values are bifurcation points. Then we choose such a bifurcation point and apply the result in the parameter-dependent version of the center manifold theorem, Theorem 3.3, by taking $\mathbf{L}$ to be the operator $\mathbf{L}_{\mathscr{R}}$ at this bifurcation point.

Spectrum of the Linear Operator

The linear operator $\mathbf{L}_{\mathscr{R}}$ is a closed operator in $\mathscr{X}$ with domain $\mathscr{Z}$. Since the domain $\mathscr{Z}$ is compactly embedded in $\mathscr{X}$, the operator $\mathbf{L}_{\mathscr{R}}$ has compact resolvent. Consequently, its spectrum consists of isolated eigenvalues, only, which all have finite algebraic multiplicity. In order to determine the spectrum we then solve the eigenvalue problem

$$\mathbf{L}_{\mathscr{R}}u = \lambda u, \quad u \in \mathscr{Z},$$

which is equivalent to the system

$$\phi'' + \mathscr{R}(1-\lambda)\phi = 0 \quad \phi(0) = \phi(1) = 0,$$
$$\left(\lambda + \frac{1}{\mathscr{R}}\right) U = 0.$$

The two equations in this system are decoupled, so that we can determine ϕ and U separately. The second equation gives the eigenvalue $\lambda_0 = -1/\mathscr{R}$, with eigenvector $(0,1)$, whereas by solving the first equation we find the sequence of eigenvalues $\lambda_k = 1 - k^2\pi^2/\mathscr{R}$, with eigenvectors $(\sin(k\pi x), 0)$ for $k \in \mathbb{N}^*$. Upon varying the parameter $\mathscr{R}$, we find that there is a sequence $(\mathscr{R}_k)_{k\in\mathbb{N}^*}$ of critical values of $\mathscr{R}$, where the part σ_0 of the spectrum of $\mathbf{L}_{\mathscr{R}}$ is not empty:

$$\mathscr{R}_k = k^2\pi^2, \quad k \in \mathbb{N}^*.$$

At each such value, $\sigma_0 = \{0\}$ and it is easy to check that the operators $\mathbf{L}_{\mathscr{R}_k}$ satisfy spectral Hypothesis 2.4. Furthermore, in each case the kernel of the operator $\mathbf{L}_{\mathscr{R}_k}$ is one-dimensional, spanned by the vector $(\sin(k\pi x), 0)$, so that 0 has geometric multiplicity one, and by arguing as in the example in Section 2.2.4 we conclude that its algebraic multiplicity is also one.

Checking Hypotheses 3.1 and 2.7

We restrict our analysis to the first bifurcation point $\mathscr{R} = \mathscr{R}_1 = \pi^2$. We set $\mu = \mathscr{R} - \mathscr{R}_1$ and write the system in the form (3.1) by taking

$$\mathbf{L} = \mathbf{L}_{\mathscr{R}_1}, \quad \mathbf{R}(u,\mu) = \mathbf{R}(u) + (\mathbf{L}_{\mathscr{R}_1+\mu} - \mathbf{L}_{\mathscr{R}_1})u.$$

Then $\mathbf{L}$ satisfies Hypothesis 3.1, whereas we now have $\mathbf{R}(u,\mu) \in \mathscr{X}$, instead of $\mathscr{Y}$, for $u \in \mathscr{Z}$, because of the term $(\mathbf{L}_{\mathscr{R}_1+\mu} - \mathbf{L}_{\mathscr{R}_1})u$, which belongs to $\mathscr{X}$ but not to $\mathscr{Y}$. Since $\mathbf{R}(u)$ is quadratic, and

$$\|\mathbf{R}(u)\|_{\mathscr{X}} \leq C\|u\|_{\mathscr{Z}}^2 \text{ for all } u \in \mathscr{Z},$$

for some positive constant C, we have that $\mathbf{R} \in C^k(\mathscr{Z} \times \mathscr{V}_\mu, \mathscr{X})$ for any positive integer k, where $\mathscr{V}_\mu = \mathbb{R} \setminus \{\mathscr{R}_1\}$. Consequently, $\mathbf{R}$ satisfies Hypothesis 3.1 with $\mathscr{X}$

instead of $\mathscr{Y}$. We are in the presence of a "quasilinear" equation with this formulation.

Remark 4.2 *Alternatively, one could go back to the original system (4.7)–(4.9), and rescale the time t through $t = \mathscr{R}t'$, which then allows us to recover a formulation for which Hypothesis 3.1 holds with the space $\mathscr{Z}$ introduced above. With this second formulation we are in the presence of a "semilinear" equation. Since our problem is formulated in Hilbert spaces we can apply the center manifold theorem to both formulations, Theorem 2.20 to the first one and Theorem 2.17 to the second one. We choose here the first formulation above as a quasilinear equation. However, this won't be possible in Banach spaces, e.g., if the Sobolev spaces H^k are replaced by C^k, in which one has to choose this second formulation as a semilinear equation (see Section 2.2.3).*

It remains to check that Hypothesis 2.7 holds. For this we use now the result in Theorem 2.20 which shows that it is enough to check the estimate on the resolvent (2.9). For $f = (\psi, V) \in \mathscr{X}$, we have to show that the solution $u = (\phi, U) \in \mathscr{Y}$ of the system

$$(i\omega - 1)\phi - \frac{1}{\pi^2}\phi'' = \psi$$
$$\left(i\omega + \frac{1}{\pi^2}\right)U = V,$$

satisfies

$$\|u\|_{\mathscr{X}} = \left(\|\phi\|^2_{L^2(0,1)} + |U|^2\right)^{1/2} \leq \frac{c}{|\omega|}\|f\|_{\mathscr{X}} = \frac{c}{|\omega|}\left(\|\psi\|^2_{L^2(0,1)} + |V|^2\right)^{1/2},$$

for $|\omega| \geq \omega_0$ and some positive constant c. First, from the second equation we immediately find

$$|U| = \frac{\pi^2}{\sqrt{1+\pi^4\omega^2}}|V|, \tag{4.10}$$

whereas for the solution ϕ of the first equation we can proceed as in the example in Section 2.2.4 (explicitly compute the solution and then estimate its norm). Alternatively, we can make use of the fact that we know that this solution exists and belongs to $H^2(0,1) \cap H^1_0(0,1)$ for $\psi \in L^2(0,1)$, when $\omega \neq 0$, since any $i\omega \neq 0$ belongs to the resolvent set of $\mathbf{L}_{\mathscr{R}_1}$. Then multiplying the equation by $\bar{\phi}$, integrating over $(0,1)$, and integrating once by parts we obtain

$$(i\omega - 1)\|\phi\|^2_{L^2(0,1)} + \frac{1}{\pi^2}\|\phi'\|^2_{L^2(0,1)} = \int_0^1 \psi(x)\bar{\phi}(x)\,dx.$$

Upon taking the imaginary parts of both sides of this equality we find

$$\omega\|\phi\|^2_{L^2(0,1)} = \operatorname{Im}\int_0^1 \psi(x)\bar{\phi}(x)\,dx,$$

so that

$$|\omega| \|\phi\|_{L^2(0,1)}^2 \leq \int_0^1 |\psi(x)\bar{\phi}(x)|\, dx \leq \|\psi\|_{L^2(0,1)} \|\phi\|_{L^2(0,1)}.$$

Consequently,

$$\|\phi\|_{L^2(0,1)} \leq \frac{1}{|\omega|} \|\psi\|_{L^2(0,1)},$$

which together with (4.10) gives the desired estimate and proves that Hypothesis 2.7 holds.

Center Manifold

Hypotheses 3.1, 2.4, and 2.7 being satisfied, we can now apply center manifold Theorem 3.3. Since 0 is a simple eigenvalue, the space $\mathscr{E}_0$ is one-dimensional, which gives us the family of one-dimensional center manifolds $\mathscr{M}_0(\mu)$, as in (3.3), for sufficiently small μ. As in the example in Section 2.2.4, we have that $\mathbf{L}_0 u_0 = 0$, so that the linear term in the reduced system (2.7) vanishes. Further denote by ξ_0 the eigenvector

$$\xi_0 = (\sin(\pi x), 0)$$

which spans $\mathscr{E}_0$, and write

$$u_0(t) = A(t)\xi_0 \in \mathscr{E}_0, \quad A(t) \in \mathbb{R}.$$

Replacing this formula in the reduced system (3.6) we obtain a first order ODE for A,

$$\frac{dA}{dt} = f_0(A, \mu),$$

with $f_0(A,\mu) = O(|A|(|\mu| + |A|))$, as $(A,\mu) \to (0,0)$.

Now, recall that the system commutes with the symmetry $\mathbf{T}$, so that the result in Theorem 3.13 holds, as well. Then the vector field in the reduced system commutes with the induced symmetry $\mathbf{T}_0$ on $\mathscr{E}_0$. Since $\mathbf{T}\xi_0 = -\xi_0$, this symmetry acts on A through $A \mapsto -A$. In particular, this shows that the vector field f_0 is odd in A, so that we may write

$$\frac{dA}{dt} = a\mu A + bA^3 + O(|A|(|\mu|^2 + A^4)).$$

We expect to find here a pitchfork bifurcation (see Section 1.1.2, Chapter 1). In order to analyze this bifurcation we compute the coefficients a and b.

Pitchfork Bifurcation

The coefficient a can be computed with the help of the result in Exercise 3.5, which shows that $\partial f_0/\partial A(0,\mu)$ is the eigenvalue of $\mathbf{L}_{\mathscr{R}_1+\mu}$ vanishing at $\mu = 0$. This latter eigenvalue is

$$\lambda_1 = 1 - \frac{\pi^2}{\mathscr{R}_1 + \mu} = \frac{\mu}{\pi^2} - \frac{\mu^2}{\pi^4} + O(|\mu|^3),$$

so that we find

$$a = \frac{1}{\pi^2}.$$

Next, in order to compute b we write for u on the center manifold

$$u(t) = A(t)\xi_0 + \Psi(A(t), \mu), \tag{4.11}$$

in which $u_0(t) = A(t)\xi_0$ and Ψ is the reduction function. Recall that $\mathbf{R}(u,0) = \mathbf{R}(u)$ is quadratic, so that we may write

$$\mathbf{R}(u,0) = \mathbf{R}_2(u,u), \quad \mathbf{R}_2(u,v) = \begin{pmatrix} -\frac{\partial(\phi\psi)}{\partial x} + \frac{1}{2}U\psi + \frac{1}{2}V\phi \\ -\int_0^1 \phi(x,\cdot)\psi(x,\cdot)dx \end{pmatrix},$$

where $v = (\psi, V)$. We set $\mu = 0$ in the following calculations, and consider the expansion

$$\Psi(A,0) = A^2\Psi_2 + A^3\Psi_3 + O(A^4),$$

in which $\mathbf{T}\Psi_2 = \Psi_2$, and $\mathbf{T}\Psi_3 = -\Psi_3$, because Ψ commutes with the symmetry $\mathbf{T}$. Now we substitute u from (4.11) into

$$\frac{du}{dt} = \mathbf{L}u + \mathbf{R}_2(u,u), \tag{4.12}$$

and taking into account that

$$\frac{dA}{dt} = bA^3 + O(|A|^5)$$

when $\mu = 0$, we identify the powers of A in this equality. At orders $O(A^2)$ and $O(A^3)$, we find, respectively,

$$\mathbf{L}\Psi_2 = -\mathbf{R}_2(\xi_0, \xi_0),$$
$$\mathbf{L}\Psi_3 = -2\mathbf{R}_2(\xi_0, \Psi_2) + b\xi_0.$$

A necessary condition for solving these equations is that the right hand sides of both equalities lie in the range of $\mathbf{L}$, or equivalently, lie in the space orthogonal to the kernel of the adjoint of $\mathbf{L}$. A direct calculation shows that here $\mathbf{L}^* = \mathbf{L}$, i.e., $\mathbf{L}$ is self-adjoint, so that its kernel is spanned by ξ_0. Further, recall that $\Psi(A,\mu)$ belongs to $\mathscr{Z}_h$, the space defined by $\mathscr{Z}_h = (\mathbb{I} - \mathbf{P}_0)\mathscr{Z}$, where $\mathbf{P}_0$ is the spectral projection onto $\mathscr{E}_0$, associated with σ_0. It is this property which allows one to uniquely determine Ψ_2 and Ψ_3 from the equalities above. However, in this particular example we can get the desired result without explicitly computing the projection $\mathbf{P}_0$.

First,

$$\mathbf{R}_2(\xi_0, \xi_0) = \begin{pmatrix} -\pi\sin(2\pi x) \\ -\frac{1}{2} \end{pmatrix},$$

which is clearly orthogonal to ξ_0 in $\mathscr{X}$, and a direct calculation gives

$$\Psi_2 = \begin{pmatrix} -\frac{\pi}{3}\sin(2\pi x) \\ -\frac{\pi^2}{2} \end{pmatrix} + \alpha\xi_0$$

for some $\alpha \in \mathbb{R}$. Now, recall that $\mathbf{T}\Psi_2 = \Psi_2$, which together with the fact that $\mathbf{T}\xi_0 = -\xi_0$, implies that $\alpha = 0$. Next, we compute

$$2\mathbf{R}_2(\xi_0, \Psi_2) = \begin{pmatrix} \pi^2\sin(3\pi x) - \frac{5\pi^2}{6}\sin(\pi x) \\ 0 \end{pmatrix}.$$

The solvability condition for the second equation is

$$0 = \langle b\xi_0 - 2\mathbf{R}_2(\xi_0, \Psi_2), \xi_0\rangle = \frac{1}{2}b + \frac{5\pi^2}{12},$$

so that

$$b = -\frac{5\pi^2}{6}.$$

Summarizing, the reduced equation is

$$\frac{dA}{dt} = \frac{1}{\pi^2}\mu A - \frac{5\pi^2}{6}A^3 + O(|A|(|\mu|^2 + |A|^4)),$$

in which the right hand side is odd in A. According to the result in Theorem 1.9 in Chapter 1, we have here a *supercritical pitchfork bifurcation*, in which a pair of steady solutions emerges from 0 as $\mathscr{R}$ crosses $\mathscr{R}_1$. These steady solutions are stable, whereas the trivial solution $A = 0$ is stable for $\mathscr{R} < \mathscr{R}_1$ and unstable for $\mathscr{R} > \mathscr{R}_1$ (see Figure 4.1).

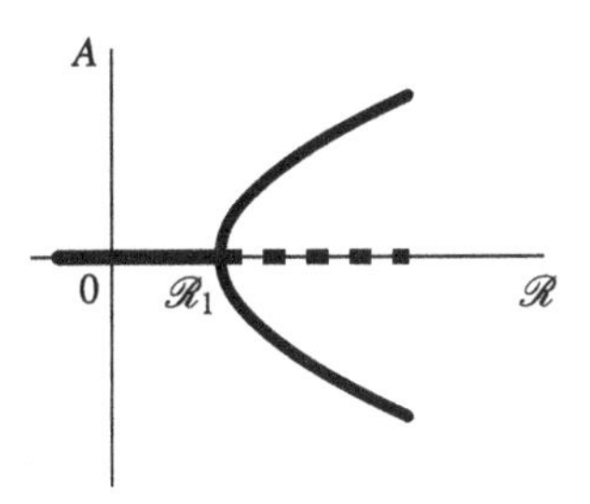

Fig. 4.1 Supercritical pitchfork bifurcation, which occurs at the first bifurcation point $\mathscr{R}_1 = \pi^2$ in the Burgers model.

Exercise 4.3 *Consider the integro-differential equation*

$$\frac{\partial u}{\partial t} = \frac{\partial^2 u}{\partial x^2} + 1 - e^{-\nu u} - K\int_0^{\pi} u(x,t)dx,$$
$$\frac{\partial u}{\partial x}\Big|_{x=0} = \frac{\partial u}{\partial x}\Big|_{x=\pi} = 0,$$

where $u(x,t) \in \mathbb{R}$ for $(x,t) \in (0,\pi)\times\mathbb{R}$, and K, ν are real parameters.

(i) Check that $u = 0$ is a solution of this problem for all K and ν. Write the system in the form (2.1) with linear operator $\mathbf{L} = \mathbf{L}_{K,\nu}$, depending upon the two parameters K and ν.
(ii) Show that the system is equivariant under the symmetry $\mathbf{T}$ defined by

$$\mathbf{T}u(x,t) = u(\pi - x,t).$$

(iii) Show that the spectrum of $\mathbf{L}_{K,\nu}$ is a discrete set, $\sigma = \{\lambda_n \in \mathbb{R}; n \in \mathbb{N}\}$, consisting of the eigenvalues

$$\lambda_0 = \nu - K\pi, \quad \lambda_n = \nu - n^2, \quad n = \mathbb{N}^*,$$

with associated eigenvectors

$$\xi_n = \cos(nx), \quad n \in \mathbb{N}.$$

Give the action of the symmetry $\mathbf{T}$ on these eigenvectors.
(iv) Assume $K\pi > 1$, and set $\nu = 1 + \mu$. Write the system in the form (3.1) and show that it possesses a center manifold of dimension 1. Show that the reduced equation takes the form

$$\frac{dA}{dt} = \mu A + bA^3 + O(|A|(|\mu|^2 + |A|^4)), \quad b = \frac{1}{6} + \frac{1}{4(K\pi - 1)} > 0.$$

(Notice that the coefficient b tends towards ∞ when $K\pi \to 1$. This is due to the invalidity of the study when $K\pi$ is close to 1, since at $K\pi = 1$ there are two "critical" eigenvalues, λ_0 and λ_1, instead of only one for $K\pi > 1$.)
(v) Consider $K\pi$ and ν close to 1, and set $\mu = \nu - 1$ and $\varepsilon = \nu - K\pi$. Write the system in the form (3.1) and show that it possesses a center manifold of dimension 2. Show that the reduced system is given by

$$\frac{dA}{dt} = \mu A - AB + \frac{1}{6}A^3 + h.o.t.$$
$$\frac{dB}{dt} = (\mu - \varepsilon)B - \frac{1}{4}A^2 - \frac{1}{2}B^2 + h.o.t.,$$

in which the first component of the vector field is odd in A, and the second component is even in A. Here and in the remainder of this book "h.o.t." denotes higher order terms.

2.4.3 Swift–Hohenberg Equation

We consider the Swift–Hohenberg equation (SHE)

$$\frac{\partial u}{\partial t} = -\left(1 + \frac{\partial^2}{\partial x^2}\right)^2 u + \mu u - u^3, \tag{4.13}$$

where $u = u(x,t) \in \mathbb{R}$ for $(x,t) \in \mathbb{R}^2$, and μ is a real parameter. The Swift–Hohenberg equation arises as a model for hydrodynamical instabilities. We refer to [18] for a detailed analysis of this equation.

Notice that $u = 0$ is a solution of (4.13) and that the equation is invariant under spatial translations $x \mapsto x + \alpha$, $\alpha \in \mathbb{R}$, and the reflections $x \mapsto -x$ and $u \mapsto -u$.

Linear Stability Analysis

We first analyze the linear stability of the trivial solution $u = 0$. We look for solutions of the form

$$u(x,t) = \widehat{u}e^{ikx+\lambda t}, \tag{4.14}$$

where k is a real wavenumber and λ and $\widehat{u}$ may be complex numbers, of the linearized SHE

$$\frac{\partial u}{\partial t} = -\left(1 + \frac{\partial^2}{\partial x^2}\right)^2 u + \mu u.$$

Inserting (4.14) into the linearized equation gives the linear dispersion relation

$$\lambda(\mu,k) = \mu - (1-k^2)^2. \tag{4.15}$$

The solution $u = 0$ is linearly stable (resp., unstable) with respect to the mode e^{ikx} if $\mathrm{Re}\,\lambda(\mu,k) < 0$ (resp., $\mathrm{Re}\,\lambda(\mu,k) > 0$).

The dispersion relation (4.15) shows that $\lambda(\mu,k)$ is real for all k and μ. For a fixed μ, the solution $u = 0$ is stable with respect to all modes e^{ikx} for which $\mu < (1-k^2)^2$, and unstable with respect to all modes for which $\mu > (1-k^2)^2$. The modes e^{ikx} such that $(1-k^2)^2 = \mu$ are the critical modes at the threshold from stability to instability. We plot in Figure 4.2 the curve $\lambda(\mu,k) = 0$. This shows that,

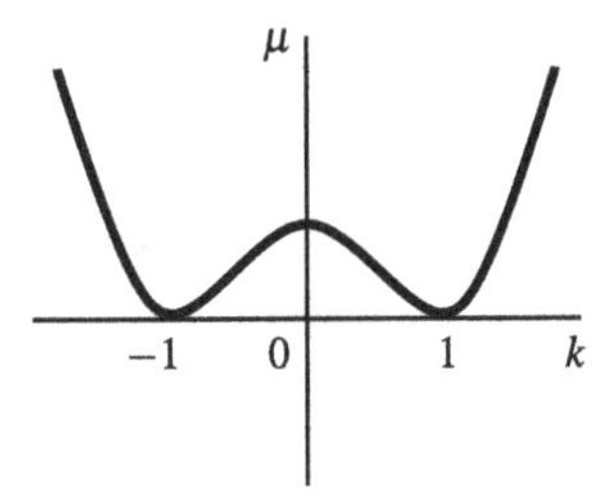

Fig. 4.2 Critical curve $\lambda(\mu,k) = 0$ for the Swift–Hohenberg equation.

upon increasing μ, the first critical modes, $k = \pm 1$, occur at $\mu = 0$. These modes correspond to 2π-periodic solutions $e^{\pm ix}$ of the linearized equation, at the threshold of linear instability. We therefore expect spatially 2π-periodic solutions to play a particular role in the dynamics of the equation, and restrict ourselves to this type of solutions in our analysis.

Center Manifolds

We write the equation in the form (2.1), with linear operator $\mathbf{L} = \mathbf{L}_\mu$ depending upon the parameter μ, by setting

$$\mathbf{L}_\mu = -\left(1 + \frac{\partial^2}{\partial x^2}\right)^2 + \mu, \quad \mathbf{R}(u) = -u^3,$$

and choosing the spaces of 2π-periodic functions

$$\mathscr{X} = L^2_{\mathrm{per}}(0, 2\pi), \quad \mathscr{Y} = \mathscr{Z} = H^4_{\mathrm{per}}(0, 2\pi).$$

Then $\mathbf{L}_\mu$ is a closed operator in $\mathscr{X}$ with domain $\mathscr{Z}$, and $\mathbf{R}$ is a cubic map in $\mathscr{Z}$, satisfying

$$\|\mathbf{R}(u)\|_{\mathscr{Z}} \leq C\|u\|^3_{\mathscr{Z}},$$

so that $\mathbf{R} \in C^k(\mathscr{Z})$ for any positive integer k.

Next, we compute the spectrum of $\mathbf{L}_\mu$. As for the operator in the previous example, Section 2.4.2, the domain $\mathscr{Y}$ of $\mathbf{L}_\mu$ is compactly embedded in $\mathscr{X}$, so that $\mathbf{L}_\mu$ has a compact resolvent. Consequently, its spectrum consists only of isolated eigenvalues with finite multiplicities. Since we work in spaces of 2π-periodic functions, we can use Fourier analysis to solve the eigenvalue problem and conclude that

$$\sigma = \{\lambda_n = \mu - (1 - n^2)^2 \; ; \; n \in \mathbb{N}\}.$$

All these eigenvalues are real, and there is a sequence $(\mu_n = (1 - n^2)^2)_{n \in \mathbb{N}}$ of values of μ for which 0 is an eigenvalue of $\mathbf{L}_\mu$. The smallest value, $\mu_1 = 0$, is the one at which the solution $u = 0$ loses its stability when increasing μ. We apply center manifold Theorem 3.3 for values of μ close to this critical value $\mu_1 = 0$.

We proceed as in the example in Section 2.4.2 and first rewrite the equation in the form (3.1), with

$$\mathbf{L} = \mathbf{L}_0, \quad \mathbf{R}(u, \mu) = \mathbf{R}(u) + (\mathbf{L}_\mu - \mathbf{L}_0)u.$$

From the arguments above it follows that $\mathbf{L}$ and $\mathbf{R}$ satisfy Hypothesis 3.1 and that Hypothesis 2.4 holds with $\sigma_0 = \{0\}$. Furthermore, 0 is an eigenvalue with geometric multiplicity two, with associated eigenvectors $e^{\pm ix}$, and by arguing as in Section 2.2.4, one can show that its algebraic multiplicity is two as well. (Alternatively, notice that $\mathbf{L}_\mu$ is self-adjoint in $\mathscr{X}$ so that its eigenvalues are all semisimple. In particular, 0 is then a double eigenvalue of $\mathbf{L}$.) Finally, Hypothesis 2.7 can be checked as in the example in Section 2.4.2. Applying Theorem 3.3, we conclude that the equation possesses a two-dimensional center manifold for μ sufficiently small.

Symmetries

An important role in this example is played by the different symmetries of the SHE mentioned above. The invariance under spatial translations $x \mapsto x+\alpha$, $\alpha \in \mathbb{R}$, and the reflections $x \mapsto -x$ and $u \mapsto -u$ imply that the equation is equivariant with respect to the isometries defined by

$$(\mathbf{T}_\alpha u)(x) = u(x+\alpha), \quad \alpha \in \mathbb{R}, \quad (\mathbf{T}u)(x) = u(-x), \quad (\mathbf{U}u)(x) = -u(x).$$

All these symmetries, $(\mathbf{T}_\alpha)_{\alpha\in\mathbb{R}}$, $\mathbf{T}$, and $\mathbf{U}$, satisfy Hypothesis 3.12. Consequently, the result in Theorem 3.13 holds with any of these symmetries. The family $(\mathbf{T}_\alpha)_{\alpha\in\mathbb{R}}$ also satisfies Hypothesis 3.17. However, we haven't in this case a nontrivial equilibrium satisfying Hypothesis 3.18, so that we cannot argue as for Theorem 3.19 in this example.

In addition, notice that

$$\mathbf{T}_\alpha = \mathbf{T}_{\alpha+2\pi}, \quad \mathbf{T}\mathbf{T}_\alpha = \mathbf{T}_{-\alpha}\mathbf{T}, \quad \mathbf{U}\mathbf{T}_\alpha = \mathbf{T}_\alpha\mathbf{U}, \;\; \alpha \in \mathbb{R}.$$

The first equality is a consequence of the fact that we restrict our analysis to 2π-periodic functions in x. In particular, the first two equalities show that (4.13) is equivariant under the representation of the group $O(2)$ by $(\mathbf{T}, (\mathbf{T}_\alpha)_{\alpha\in\mathbb{R}/2\pi\mathbb{Z}})$.

Steady $O(2)$ Bifurcation

We discuss now the reduced system given by Theorems 3.3 and 3.13. Recall that the subspace $\mathscr{E}_0$ is two-dimensional, spanned by the complex conjugated eigenvector $\zeta = e^{ix}$ and $\overline{\zeta} = e^{-ix}$, so that it is convenient in this case to write

$$u_0 = A\zeta + \bar{A}\overline{\zeta}, \quad A(t) \in \mathbb{C},$$

for real-valued $u_0(t) \in \mathscr{E}_0$. Then we set for the real-valued solutions on the center manifold

$$u = A\zeta + \bar{A}\overline{\zeta} + \Psi(A, \bar{A}, \mu), \quad A(t) \in \mathbb{C},$$

where $\Psi(A(t), \bar{A}(t), \mu) \in \mathscr{Z}_h$. The reduced equation reads

$$\frac{dA}{dt} = f(A, \bar{A}, \mu), \tag{4.16}$$

together with the complex conjugated equation for $\bar{A}$. In addition, since the original equation is equivariant under the actions of $\mathbf{T}_\alpha$ and $\mathbf{T}$, by the result in Theorem 3.13, we have that the reduced vector field $(f, \overline{f})$ is equivariant under the actions of the induced symmetries. Since

$$\mathbf{T}_\alpha\zeta = e^{i\alpha}\zeta, \quad \mathbf{T}_\alpha\overline{\zeta} = e^{-i\alpha}\overline{\zeta}, \quad \mathbf{T}\zeta = \overline{\zeta}, \quad \mathbf{T}\overline{\zeta} = \zeta,$$

the action of the induced symmetries on the pair $(A,\bar{A})$ is given by the 2×2-matrices

$$\mathbf{T}_\alpha : \begin{pmatrix} e^{i\alpha} & 0 \\ 0 & e^{-i\alpha} \end{pmatrix}, \quad \mathbf{T} : \begin{pmatrix} 0 & 1 \\ 1 & 0 \end{pmatrix}.$$

This shows that we are in the setting of the study made in Section 1.2.4, Chapter 1, on steady bifurcations with $O(2)$ symmetry. Consequently, we have that

$$f(A,\bar{A},\mu) = Ag(|A|^2,\mu),$$

where the function g is of class C^{k-1} in $(A,\bar{A},\mu)$ and real-valued. We consider the Taylor expansion of g and write

$$\frac{dA}{dt} = aA\mu + bA|A|^2 + O(|A|(|\mu|^2 + |A|^4)).$$

In polar coordinates, for $A = re^{i\phi}$, this gives the system (2.43)–(2.44) studied in Section 1.2.4.

We now compute the coefficients a and b in order to determine the nature of this bifurcation. For this we proceed as in the previous example in Section 2.4.2. First, using the result in the Exercise 3.5, we obtain

$$\frac{\partial f}{\partial A}(0,\mu) = \lambda_1 = \mu,$$

so that

$$a = 1.$$

Next, we set $\mu = 0$ in the following calculations and consider the expansion of the reduction function Ψ,

$$\Psi(A,\bar{A},0) = \sum_{p,q} \Psi_{pq} A^p \overline{A}^q.$$

Here $\Psi_{qp} \in \mathscr{Z}_h$ are such that

$$\Psi_{qp} = \overline{\Psi_{pq}}, \quad \Psi_{00} = \Psi_{10} = \Psi_{01} = 0.$$

The first equality shows that Ψ is real-valued, whereas the last equalities come from (3.2). Furthermore, from the equivariance of the equation with respect to $\mathbf{U}$, we conclude that $\Psi(-A,-\overline{A},0) = -\Psi(A,\overline{A},0)$ for all A, and thus $\Psi_{pq} = 0$ when $p+q$ is even. Summarizing, we find the expansion

$$\Psi(A,\bar{A},0) = \Psi_{30}A^3 + \Psi_{03}\bar{A}^3 + \Psi_{21}A^2\bar{A} + \Psi_{12}A\bar{A}^2 + O(|A|^5),$$

where $\Psi_{03} = \overline{\Psi_{30}}$ and $\Psi_{12} = \overline{\Psi_{21}}$.

Now by arguing as in the calculation of the coefficient b in the example in Section 2.4.2, we obtain the equalities

$$\mathbf{L}\Psi_{30} = e^{3ix},$$
$$\mathbf{L}\Psi_{21} = 3e^{ix} + be^{ix}.$$

The solvability condition for the second equation gives

$$b = -3.$$

Summarizing, the reduced equation is

$$\frac{dA}{dt} = \mu A - 3A|A|^2 + O(|A|(|\mu|^2 + |A|^4)), \tag{4.17}$$

and the reduced vector field possesses an $O(2)$ equivariance, just as in Hypothesis 2.14. According to the result in Theorem 2.18 in Chapter 1, we have here a *steady bifurcation with* $O(2)$ *symmetry*, in which a family $(A_\alpha)_{\alpha \in \mathbb{R}/2\pi\mathbb{Z}}$ of stable equilibria emerges from 0, as μ crosses 0. A direct calculation gives

$$A_\alpha = \sqrt{\frac{\mu}{3}} e^{i\alpha} + O(|\mu|^{3/2})$$

for $\mu > 0$, and the corresponding family of steady 2π-periodic solutions of SHE,

$$u_\alpha(x) = 2\sqrt{\frac{\mu}{3}} \cos(x + \alpha) + O(|\mu|^{3/2}). \tag{4.18}$$

We point out that $u_\alpha = \mathbf{T}_\alpha u_0$, so that the solutions in this family are obtained by spatially translating u_0.

Remark 4.4 *These steady* 2π*-periodic solutions of the SHE are called* roll solutions. *Actually, such solutions exist for a range of periods close to* 2π*, for any sufficiently small* μ*. One can prove the existence of all these rolls in a similar way. Looking for periodic solutions of the SHE with wavenumbers* k *close to* 1*, instead of wavenumbers* $k = 1$*, only, and normalizing the period to* 2π *in the equation, one finds an equation having an additional parameter, the wavenumber* k*. The normalization of the period allows us to use the same function spaces* $\mathscr{X}$ *and* $\mathscr{Z}$*, and this reduction procedure can be performed with two parameters,* k *close to* 1 *and* μ *small.*

Symmetry Breaking

We briefly discuss here several scenarios in which we perturb the Swift–Hohenberg equation, by adding a small term, in such a way that one, or more, of the symmetries of the SHE is broken. We are interested in the effect of the perturbation on the reduced equation (4.17).

I. First we consider the perturbed equation obtained by adding the term εu^2 in the right hand side of the SHE, with ε a small real parameter. This term breaks the equivariance of the equation with respect to the symmetry $\mathbf{U}$ but preserves the $O(2)$ equivariance with respect to $(\mathbf{T},(\mathbf{T}_\alpha)_{\alpha\in\mathbb{R}/2\pi\mathbb{Z}})$. The center manifold analysis remains the same, up to the equivariance in $\mathbf{U}$, which is lost, and to the appearance of the additional small parameter ε. However, this parameter does not play a role in checking the different hypotheses, its effect being that now the reduced vector field $(f,\bar{f})$ depends upon ε as well. Since the $O(2)$ equivariance is preserved, we still have the particular form

$$f(A,\bar{A},\mu,\varepsilon)=Ag(|A|^2,\mu,\varepsilon),$$

with g of class C^{k-1} and real-valued.

Notice that at $\varepsilon=0$ we find exactly the reduced vector field obtained for the unperturbed equation. Furthermore, we have here a new symmetry, which is the invariance of the SHE under $(u,\varepsilon)\mapsto(-u,-\varepsilon)$. It is then straightforward to check that this induces the invariance of the reduced equation under the action of $(A,\varepsilon)\mapsto(-A,-\varepsilon)$. In particular, this shows that the map g above is even in ε. This fact is useful in the computation of the Taylor expansion of g.

II. Next, we add the term $\varepsilon\partial u/\partial x$ in the right hand side of the SHE, with ε a small real parameter. This situation actually reduces to the unperturbed SHE, by the change of variables $u(x,t)=\widetilde{u}(x+\varepsilon t,t)$. It is easy to see that u is a solution of the perturbed SHE if and only if $\widetilde{u}$ is a solution of the unperturbed SHE. In particular, our previous analysis gives us in this case the family of traveling wave solutions $u_\alpha(x+\varepsilon t)$, with u_α the steady 2π-periodic solution in (4.18). These traveling waves have small speeds $-\varepsilon$, are 2π-periodic in the spatial variable x, and are periodic in time with large period $2\pi/\varepsilon$.

Our interest in considering this example is to see the effect of such a term on the different symmetries of the SHE and then on the reduced system. This term breaks the symmetry $\mathbf{T}$, but preserves the symmetries $\mathbf{T}_\alpha$ and $\mathbf{U}$. In particular, instead of an $O(2)$ equivariance we have now an $SO(2)$ equivariance. However, one can argue as in Section 1.2.4 and conclude that the map f in the reduced system is of the form

$$f(A,\bar{A},\mu,\varepsilon)=Ag(|A|^2,\mu,\varepsilon),$$

with g of class C^{k-1}, and complex-valued but not necessarily real-valued anymore.

In this situation, we have the additional invariance of the SHE under $(x,\varepsilon)\mapsto(-x,-\varepsilon)$. On the center manifold, this induces the symmetry acting by $(A,\varepsilon)\mapsto(\bar{A},-\varepsilon)$, so that g satisfies

$$g(|A|^2,\mu,\varepsilon)=\overline{g(|A|^2,\mu,-\varepsilon)}.$$

Consequently, the real part g_r of g is even in ε, whereas the imaginary part g_i of g is odd in ε. This leads to the equation

$$\frac{dA}{dt}=(\mu+c\varepsilon^2+id\varepsilon)A-3A|A|^2+h.o.t.,$$

which in polar coordinates $A = re^{i\phi}$ reads

$$\begin{aligned}\frac{dr}{dt} &= (\mu + c\varepsilon^2)r - 3r^3 + h.o.t. \\ \frac{d\phi}{dt} &= d\varepsilon + h.o.t.\end{aligned} \tag{4.19}$$

Here the real coefficients c and d can be computed explicitly, just as the coefficients a and b in (4.16), and we have used the fact that the reduced system at $\varepsilon = 0$ is the same as the reduced system found for the unperturbed equation. It is then straightforward to find the solutions

$$r_0(\mu, \varepsilon^2) = \left(\frac{\mu + c\varepsilon^2}{3}\right)^{1/2} + h.o.t., \quad \phi_0 = \omega t + \alpha, \quad \omega = d\varepsilon + h.o.t.,$$

with any $\alpha \in \mathbb{R}$. These give the solutions of the perturbed SHE equation

$$u(x,t) = 2r_0(\mu, \varepsilon^2)\cos(x + \omega t + \alpha) + h.o.t..$$

The lowest order term in this solution is clearly a traveling wave, with speed $-\omega$. A careful use of the symmetries mentioned above, together with the invariance of the equation under translations in the time t, allows us to show that these solutions are indeed traveling waves.

Exercise 4.5 *Show that $c = 0$ and $d = 1$ in the reduced system (4.19).*

III. Consider now the additional term $\varepsilon u\partial u/\partial x$ on the right hand side of the SHE. This term breaks the symmetries $\mathbf{T}$ and $\mathbf{U}$, but preserves the composed symmetry $\widetilde{\mathbf{T}} = \mathbf{T} \circ \mathbf{U}$ and the family $(\mathbf{T}_\alpha)_{\alpha \in \mathbb{R}}$. Consequently, we still have an $O(2)$ equivariance of the system, but now with $\widetilde{\mathbf{T}}$ instead of $\mathbf{T}$. The action of $\widetilde{\mathbf{T}}$ on the pair $(A, \bar{A})$ is given by the 2×2-matrix

$$\begin{pmatrix} 0 & -1 \\ -1 & 0 \end{pmatrix}.$$

However this does not change the form of the reduced equation, the map f being again of the form

$$f(A, \bar{A}, \mu, \varepsilon) = Ag(|A|^2, \mu, \varepsilon).$$

In addition, we have here the symmetry $(u, \varepsilon) \mapsto (-u, -\varepsilon)$, which implies that

$$g(|A|^2, \mu, -\varepsilon) = g(|A|^2, \mu, \varepsilon).$$

IV. We introduce now an additional term $\varepsilon_1 u\partial u/\partial x + \varepsilon_2 u^2$, in which we have two small parameters ε_1 and ε_2. This term breaks the symmetries $\mathbf{T}$, $\mathbf{U}$, and also $\widetilde{\mathbf{T}} = \mathbf{T} \circ \mathbf{U}$, but preserves the symmetries $\mathbf{T}_\alpha$, $\alpha \in \mathbb{R}$. Consequently, we still have an $SO(2)$ equivariance, just as in the case **II**, which allows us to conclude that the map f in the reduced system is of the form

$$f(A,\bar{A},\mu,\varepsilon_1,\varepsilon_2) = Ag(|A|^2,\mu,\varepsilon_1,\varepsilon_2),$$

with g of class C^{k-1} and complex-valued.

In addition, we now find the new symmetries

$$(u,\varepsilon_1,\varepsilon_2) \mapsto (-u,-\varepsilon_1,-\varepsilon_2), \quad (u(x),\varepsilon_1,\varepsilon_2) \mapsto (u(-x),-\varepsilon_1,\varepsilon_2).$$

Their action on $(A,\bar{A})$ is given by

$$(A,\bar{A},\varepsilon_1,\varepsilon_2) \mapsto (-A,-\overline{A},-\varepsilon_1,-\varepsilon_2), \quad (A,\bar{A},\varepsilon_1,\varepsilon_2) \mapsto (\bar{A},A,-\varepsilon_1,\varepsilon_2).$$

We can then conclude that the map g satisfies

$$g(|A|^2,\mu,\varepsilon_1,\varepsilon_2) = g(|A|^2,\mu,-\varepsilon_1,-\varepsilon_2), \quad g(|A|^2,\mu,\varepsilon_1,\varepsilon_2) = \overline{g(|A|^2,\mu,-\varepsilon_1,\varepsilon_2)},$$

so that the reduced equation is

$$\frac{dA}{dt} = \mu A - 3A|A|^2 + (c_1\varepsilon_1^2 + id\varepsilon_1\varepsilon_2 + c_2\varepsilon_2^2)A|A|^2 + h.o.t..$$

In polar coordinates $A = re^{i\phi}$, we find the system

$$\begin{aligned}\frac{dr}{dt} &= \mu r - 3r^3 + (c_1\varepsilon_1^2 + c_2\varepsilon_2^2)r^3 + h.o.t.\\ \frac{d\phi}{dt} &= d\varepsilon_1\varepsilon_2 r^2 + h.o.t..\end{aligned}$$

By arguing as for the system (4.19) in case **II**, one can show in this case the existence of bifurcating traveling waves with speeds of order $O(\mu\varepsilon_1\varepsilon_2)$.

Exercise 4.6 *Show that* $c_1 = -1/9$, $d = 4/3$, *and* $c_2 = 20/9$ *in the reduced system.*

V. Consider now the case of an inhomogeneous additional term $\varepsilon h(x)$, on the right hand side of the SHE, where $h : \mathbb{R} \to \mathbb{R}$ is an even 2π-periodic function and ε a small parameter, again. Notice that in this case the trivial solution $u = 0$ is no longer a solution for $\varepsilon \neq 0$.

This term now breaks the translation invariance $\mathbf{T}_\alpha$, $\alpha \in \mathbb{R}$, and the reflection $\mathbf{U}$, but preserves the symmetry $\mathbf{T}$. As in the previous cases we find a two-dimensional center manifold and a reduced equation of the form

$$\frac{dA}{dt} = f(A,\bar{A},\mu,\varepsilon)$$

for $A(t) \in \mathbb{C}$. At $\varepsilon = 0$, the map f is the one obtained for the unperturbed equation,

$$f(A,\bar{A},\mu,0) = Ag(|A|^2,\mu) = \mu A - 3A|A|^2 + h.o.t.,$$

whereas for $\varepsilon \neq 0$ the equivariance with respect to $\mathbf{T}$ implies that

$$f(A,\bar{A},\mu,\varepsilon) = \overline{f(\bar{A},A,\mu,\varepsilon)}.$$

Consequently, the reduced equation is of the form

$$\frac{dA}{dt} = c\varepsilon + \mu A - 3A|A|^2 + h.o.t.,$$

where c is a real constant. Notice that the constant term on the right hand side of this equation is real, because of the property of f above, and nonzero, since $u = 0$ is no longer a solution of the perturbed equation.

Exercise 4.7 *Show that the coefficient c in the reduced system is given by*

$$c = \frac{1}{2\pi}\int_0^{2\pi} h(x)\cos x dx.$$

Remark 4.8 (Steady solutions) *Notice that the steady solutions of this system are easy to compute. They are real, $A = A_r$, with A_r satisfying*

$$c\varepsilon + A_r(\mu - 3A_r^2) + h.o.t. = 0.$$

We plot in Figure 4.3 the bifurcation diagram for the steady solutions of this reduced equation. As for the stability of these steady solutions, it can be determined from the

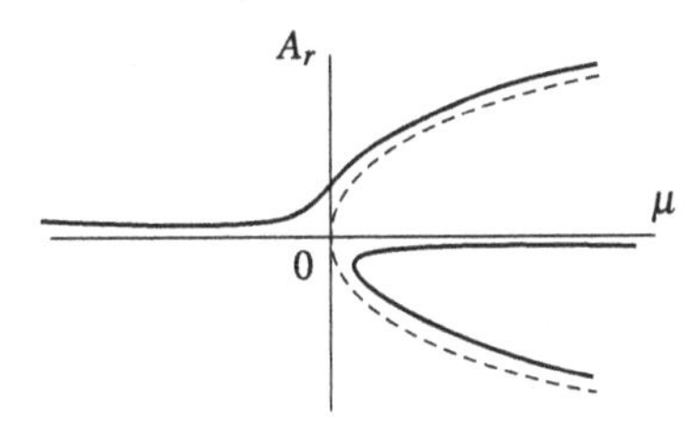

Fig. 4.3 Bifurcation diagram in the (μ, A_r)-plane for the steady solutions of the reduced system in the SHE perturbed by an inhomogeneity $\varepsilon h(x)$ in the case $c\varepsilon > 0$. The solid lines represent the branches of steady solutions for a fixed, small ε, whereas the dashed lines represent the branch of steady solutions for $\varepsilon = 0$.

eigenvalues of the linearized vector field at $A = A_r$. A direct calculation gives the two eigenvalues $\mu - 9A_r^2 + h.o.t.$ and $\mu - 3A_r^2 + h.o.t.$. In particular, in the case represented in the bifurcation diagram in Figure 4.3, the upper branch is stable (both eigenvalues are negative), while the lower branch is unstable (at least one eigenvalue is positive). We point out that this result differs from the classical result occuring in a perturbed pitchfork bifurcation. Notice that one eigenvalue is 0 at the turning point of the lower branch, but that this does not change the stability here, because of the second eigenvalue. Moreover, observe that all these steady solutions are symmetric, invariant under **T**, *since they are real.*

VI. Finally, we consider the Swift–Hohenberg equation (4.13), but instead of looking for solutions that are 2π-periodic in x, we seek solutions that satisfy the boundary conditions

$$u(\pm h,t) = \frac{\partial u}{\partial x}(\pm h,t) = 0 \tag{4.20}$$

on some interval $[-h,h]$. We assume that h is large enough, so that we regard this new problem as a "small" perturbation of the equation (4.13).

Replacing the spatial periodicity of the solutions by the boundary conditions (4.20) breaks the translational invariance $\mathbf{T}_a$, but does not break the symmetries $\mathbf{T}$ and $\mathbf{U}$, and $u = 0$ remains a solution of the new problem. As a consequence, the eigenvalues of the linear operator $\mathbf{L}_\mu$ are no longer double, and for $\mu = 0$ the former 0 eigenvalue splits into two simple, negative eigenvalues, which are close to 0, of order $O(1/h^3)$ as $h \longrightarrow \infty$. The other eigenvalues are all negative and at least of order $O(1/h^2)$. It is then convenient to rescale the variables in order to push the eigenvalues of order $O(1/h^2)$ at a distance of order $O(1)$ from the imaginary axis. Then the two eigenvalues of order $O(1/h^3)$ are changed into eigenvalues of order $O(1/h)$, which allows us to use a center manifold reduction, as described in Remark 3.6, when the critical spectrum σ_0 does not lie on the imaginary axis, but stays close to it. In addition to the original parameter μ, we now have a second small parameter $\varepsilon = O(1/h)$, so that this case is indeed a small perturbation of the original problem.

Taking into account the fact that 0 is always a solution, and that in this new problem only the translational symmetry is broken, by arguing as in the previous cases one finds that the reduced equation is now modified at main orders as follows:

$$\frac{dA}{dt} = (\mu + a\varepsilon)A + b\varepsilon\bar{A} - 3A|A|^2,$$

where a and b are real coefficients. Using polar coordinates $A = re^{i\phi}$, we find the system

$$\begin{aligned}\frac{dr}{dt} &= r(\mu + a\varepsilon + b\varepsilon\cos 2\phi - 3r^2)\\ \frac{d\phi}{dt} &= -b\varepsilon\sin(2\phi).\end{aligned}$$

Steady solutions are found for $\phi \in \{0, \pi/2, \pi, 3\pi/2\}$. Note that changing $\phi \mapsto \phi + \pi$ is equivalent to changing $r \mapsto -r$, so that we can restrict to the two cases $\phi = 0$ and $\phi = \pi/2$. The case $\phi = 0$ leads to *symmetric solutions*, i.e., invariant under $\mathbf{U}$, since $A = \bar{A}$, whereas the case $\phi = \pi/2$ leads to *antisymmetric solutions*, since $\bar{A} = -A$. It turns out that *symmetric solutions* bifurcate for $\mu = -(a+b)\varepsilon$ and have the amplitude given by $r_S^2 = 1/3(\mu + (a+b)\varepsilon)$. Their stability is determined by the sign of the two eigenvalues $-6r_S^2$, $-2b\varepsilon$. *Antisymmetric solutions* bifurcate for $\mu = (b-a)\varepsilon$, and have the amplitude given by $r_A^2 = 1/3(\mu + (a-b)\varepsilon)$. Their stability is determined by the sign of the two eigenvalues $-6r_A^2$, $2b\varepsilon$. In particular, it follows that the *stabilities of these two branches of solutions are opposite* (see Figure 4.4 for a typical bifurcation diagram).

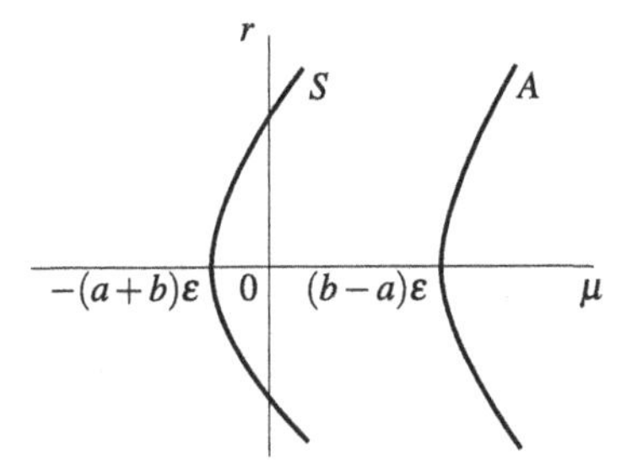

Fig. 4.4 Bifurcation diagram for the Swift–Hohenberg equation with boundary conditions (4.20), for a fixed $\varepsilon = O(1/h)$. The two curves S and A represent the branches of symmetric and antisymmetric solutions, respectively.

Remark 4.9 *This question has a major physical importance for many hydrodynamic stability problems where, for a large aspect ratio apparatus, one replaces, for mathematical convenience, the physical boundary conditions by periodic boundary conditions (large periods), as for example in Section 5.1 of Chapter 5. On the model equation SHE, a complete mathematical justification of the new amplitude equation obtained for Dirichlet–Neumann boundary conditions, as a perturbation of the periodic case, can be found in [125], while this is still a mathematically open problem for classical hydrodynamic stability problems like the ones in Section 5.1 of Chapter 5.*

2.4.4 Brusselator Model

Consider the system of PDEs

$$\begin{aligned}\frac{\partial u_1}{\partial t} &= \delta_1 \frac{\partial^2 u_1}{\partial x^2} - (\beta+1)u_1 + u_1^2 u_2 + \alpha \\ \frac{\partial u_2}{\partial t} &= \delta_2 \frac{\partial^2 u_2}{\partial x^2} + \beta u_1 - u_1^2 u_2,\end{aligned} \tag{4.21}$$

in which δ_1, δ_2, α, and β are positive constants, and $u = (u_1, u_2)$ is a function of $(x,t) \in (0,1) \times \mathbb{R}$, together with the boundary conditions

$$u_1(0,t) = u_1(1,t) = \alpha, \quad u_2(0,t) = u_2(1,t) = \frac{\beta}{\alpha}. \tag{4.22}$$

Remark 4.10 *This system is called* inhomogeneous Brusselator, *and arises in modeling an autocatalytic chemical reaction as described in Remark 2.9 of Chapter 1. In contrast to the homogeneous Brusselator considered in Section 1.2.2, in the inhomogeneous case the products are not homogeneously mixed during the reaction. In such a case, diffusion phenomena occur, so that u_1 and u_2 are now functions of a* *space variable x, besides the time t. We assume that* $x \in (0,1)$, *which is, of course, a simplification of the reality. The coefficients* δ_1 *and* δ_2 *in the system* (4.21) *represent the diffusion coefficients of the products X and Y. The Dirichlet boundary conditions* (4.22) *correspond in the chemical reaction to a control at the boundary for the concentrations of the products X and Y, which are maintained at the constant values given by the equilibrium solution of the homogeneous system. Notice that with such boundary conditions, the equilibrium* $(u_1,u_2) = (\alpha, \beta/\alpha)$ *found in Section 1.2.2 remains a solution of the PDE, but the periodic solution arising in the Hopf bifurcation for the homogeneous system is no longer a solution. It is not difficult to check that this solution does not satisfy the boundary conditions.*

First Formulation and Spectrum

We set

$$(u_1,u_2) = \left(\alpha, \frac{\beta}{\alpha}\right) + (v_1,v_2).$$

Then $v = (v_1,v_2)$ satisfies the system

$$\begin{aligned}\frac{\partial v_1}{\partial t} &= \delta_1 \frac{\partial^2 v_1}{\partial x^2} + (\beta - 1)v_1 + \alpha^2 v_2 + 2\alpha v_1 v_2 + \frac{\beta}{\alpha} v_1^2 + v_1^2 v_2 \\ \frac{\partial v_2}{\partial t} &= \delta_2 \frac{\partial^2 v_2}{\partial x^2} - \beta v_1 - \alpha^2 v_2 - 2\alpha v_1 v_2 - \frac{\beta}{\alpha} v_1^2 - v_1^2 v_2,\end{aligned} \tag{4.23}$$

and the Dirichlet boundary conditions

$$v_1(0,t) = v_1(1,t) = 0, \quad v_2(0,t) = v_2(1,t) = 0. \tag{4.24}$$

In this way we have replaced the constant solution $(u_1,u_2) = (\alpha, \beta/\alpha)$ by the trivial solution $v = 0$. This system is of the form (2.1) with

$$\mathbf{L} = \begin{pmatrix} \delta_1 \frac{d^2}{dx^2} + \beta - 1 & \alpha^2 \\ -\beta & \delta_2 \frac{d^2}{dx^2} - \alpha^2 \end{pmatrix}, \quad \mathbf{R}(v) = \begin{pmatrix} 2\alpha v_1 v_2 + \frac{\beta}{\alpha} v_1^2 + v_1^2 v_2 \\ -2\alpha v_1 v_2 - \frac{\beta}{\alpha} v_1^2 - v_1^2 v_2 \end{pmatrix},$$

where both $\mathbf{L}$ and $\mathbf{R}$ depend upon parameters.

Next, we choose the spaces

$$\mathscr{X} = (L^2(0,1))^2, \quad \mathscr{Z} = \mathscr{Y} = (H^2(0,1) \cap H_0^1(0,1))^2,$$

such that the boundary conditions are included in the definition of $\mathscr{Y}$. Then $\mathbf{L}$ is a closed operator in $\mathscr{X}$, with domain $\mathscr{Z}$, and $\mathbf{R}$ a smooth map in $\mathscr{Y}$. As in the previous examples, $\mathscr{Y}$ is compactly embedded in $\mathscr{X}$, so that $\mathbf{L}$ has compact resolvent and its spectrum consists of isolated eigenvalues with finite multiplicities.

We determine now the spectrum of $\mathbf{L}$. Since the set $\{\sin(n\pi x)\,;\, n \in \mathbb{N}\}$ forms a basis of $H_0^1(0,1)$, we can look for solutions $v = (v_1,v_2)$ of the eigenvalue problem

$$\mathbf{L}v = \lambda v, \quad \lambda \in \mathbb{C},$$

of the form

$$v_1 = \sum_{n\in\mathbb{N}} v_1^{(n)} \sin(n\pi x), \quad v_2 = \sum_{n\in\mathbb{N}} v_2^{(n)} \sin(n\pi x).$$

Then λ is an eigenvalue of $\mathbf{L}$ if there exists $n \neq 0$ such that there exists a nontrivial solution $(v_1^{(n)}, v_2^{(n)})$ of the system

$$\begin{aligned} (\delta_1 n^2\pi^2 - \beta + 1 + \lambda)v_1^{(n)} - \alpha^2 v_2^{(n)} &= 0 \\ \beta v_1^{(n)} + (\delta_2 n^2\pi^2 + \alpha^2 + \lambda)v_2^{(n)} &= 0. \end{aligned} \tag{4.25}$$

Consequently, the eigenvalues λ are roots of the characteristic polynomials

$$P_n(X) = X^2 + (\beta_n - \beta)X + \delta_2 n^2\pi^2(\gamma_n - \beta),$$

where

$$\beta_n = 1 + \alpha^2 + n^2\pi^2(\delta_1 + \delta_2), \quad \gamma_n = 1 + \alpha^2\frac{\delta_1}{\delta_2} + n^2\pi^2\delta_1 + \frac{\alpha^2}{n^2\pi^2\delta_2}.$$

The two roots of P_n have negative real parts provided

$$\beta < \beta_n, \quad \beta < \gamma_n.$$

Notice that the sequence $(\beta_n)_{n\geq 1}$ is increasing and that $\gamma_n \geq \left(1 + \alpha\sqrt{\delta_1/\delta_2}\right)^2$, so that for any β satisfying

$$\beta < \beta_1, \quad \beta < \left(1 + \alpha\sqrt{\frac{\delta_1}{\delta_2}}\right)^2,$$

the roots of all these polynomials have negative real parts. When

$$\beta = \beta_1 < \left(1 + \alpha\sqrt{\frac{\delta_1}{\delta_2}}\right)^2,$$

the polynomial P_1 has purely imaginary roots, whereas the other polynomials all have roots with negative real parts. We conclude that the eigenvalues of $\mathbf{L}$ have negative real parts (are all stable) if $\beta < \beta_1$, and that a pair of eigenvalues crosses the imaginary axis at $\beta = \beta_1$, provided

$$\beta_1 < \left(1+\alpha\sqrt{\frac{\delta_1}{\delta_2}}\right)^2.$$

This inequality is equivalent to

$$\alpha^2\left(\frac{\delta_1}{\delta_2}-1\right)+2\alpha\sqrt{\frac{\delta_1}{\delta_2}}-\pi^2(\delta_1+\delta_2)>0, \tag{4.26}$$

and we assume in the following that this condition holds, so that we have a Hopf bifurcation at $\beta=\beta_1$.

Center Manifolds

We focus on this Hopf bifurcation and set $\beta=\beta_1+\mu$. In order to apply the result in Theorem 3.3, we rewrite the system (4.23) in the form

$$\frac{dv}{dt}=\mathbf{L}v+\mathbf{R}(v,\mu) \tag{4.27}$$

in the space $\mathscr{X}$, with

$$\mathbf{L}=\begin{pmatrix}\delta_1\frac{d^2}{dx^2}+\beta_1-1 & \alpha^2\\ -\beta_1 & \delta_2\frac{d^2}{dx^2}-\alpha^2\end{pmatrix},$$

and

$$\mathbf{R}(v,\mu)=\mu\mathbf{R}_{01}v+\mathbf{R}_{20}(v,v)+\mathbf{R}_{30}(v,v,v)+\mu\mathbf{R}_{21}(v,v), \tag{4.28}$$

where

$$\mathbf{R}_{01}v=\begin{pmatrix}v_1\\ -v_1\end{pmatrix},\quad \mathbf{R}_{20}(u,v)=\begin{pmatrix}\alpha(u_1v_2+u_2v_1)+\frac{\beta_1}{\alpha}u_1v_1\\ -\alpha(u_1v_2+u_2v_1)-\frac{\beta_1}{\alpha}u_1v_1\end{pmatrix},$$

$$3\mathbf{R}_{30}(u,v,w)=\begin{pmatrix}u_1v_1w_2+u_1v_2w_1+u_2v_1w_1\\ -u_1v_1w_2-u_1v_2w_1-u_2v_1w_1\end{pmatrix},\quad \mathbf{R}_{21}(u,v)=\begin{pmatrix}\frac{1}{\alpha}u_1v_1\\ -\frac{1}{\alpha}u_1v_1\end{pmatrix}.$$

Then $\mathbf{L}$, which is closed in $\mathscr{X}$ with domain $\mathscr{Z}$, and $\mathbf{R}(v,\mu)$, which has a polynomial form and is continuous in $\mathscr{Z}$, satisfy Hypothesis 3.1. Next, the analysis above implies that the operator $\mathbf{L}$ satisfies Hypothesis 2.4, with

$$\sigma_0=\{\pm i\omega\},\quad \omega^2=\alpha^2+\alpha^2\pi^2(\delta_1-\delta_2)-\delta_2^2\pi^4.$$

The eigenvectors ζ and $\bar{\zeta}$ associated with the eigenvalues $i\omega$ and $-i\omega$, respectively, are given by

$$\zeta = \sin(\pi x)\begin{pmatrix}1\\ \gamma\end{pmatrix}, \quad \gamma = \frac{i\omega - \alpha^2 - \delta_2\pi^2}{\alpha^2} = \frac{-\beta_1}{i\omega + \alpha^2 + \delta_2\pi^2}.$$

Notice that $\omega^2 > 0$ thanks to the condition (4.26), since

$$\omega^2 = \pi^2\delta_2\left(\alpha^2\left(\frac{\delta_1}{\delta_2} - 1\right) + 2\alpha\sqrt{\frac{\delta_1}{\delta_2}} - \pi^2(\delta_1 + \delta_2)\right) + (\alpha - \pi^2\sqrt{\delta_1\delta_2})^2.$$

Finally, one can proceed as in the example in Section 2.4.2 and check the inequality (2.9), which implies that Hypothesis 2.7 holds as well.

Applying Theorem 3.3 we conclude that the system possesses a two-dimensional center manifold for sufficiently small μ. For the solutions on the center manifold we write

$$v = z\zeta + \overline{z\zeta} + \Psi(z,\bar{z},\mu), \quad z \in \mathbb{C}, \tag{4.29}$$

in which $v_0(t) = z(t)\zeta + \overline{z\zeta(t)} \in \mathscr{E}_0$ and Ψ takes values in $\mathscr{Z}_h$. The reduced equation

$$\frac{dz}{dt} = f(z,\bar{z},\mu),$$

together with the complex conjugated equation, has the linear part

$$\begin{pmatrix} \frac{\partial f}{\partial z}(0,0,0) & \frac{\partial f}{\partial \bar{z}}(0,0,0) \\ \frac{\partial \bar{f}}{\partial z}(0,0,0) & \frac{\partial \bar{f}}{\partial \bar{z}}(0,0,0) \end{pmatrix} = \begin{pmatrix} i\omega & 0 \\ 0 & -i\omega \end{pmatrix},$$

in which $\pm i\omega$ are the two eigenvalues in σ_0. In particular, it is of the form (2.5), so that we can use the results on the Hopf bifurcation in Section 1.2.1, Chapter 1, to analyze this reduced equation.

Hopf Bifurcation

According to the analysis in Section 1.2.1, there is a polynomial change of variables that transforms this reduced equation into the normal form

$$\frac{dA}{dt} = i\omega A + a\mu A + bA|A|^2 + O(|A|(|\mu| + |A|^2)^2). \tag{4.30}$$

Our goal now is to compute the coefficients a and b in this normal form. To do so it is convenient to incorporate this change of variables in the formula (4.29), and write

$$v = A\zeta + \overline{A\zeta} + \Psi(A,\bar{A},\mu), \quad A \in \mathbb{C}, \tag{4.31}$$

in which $v_0(t) = A(t)\zeta + \overline{A(t)\zeta} \in \mathscr{E}_0$, but now Ψ takes values in $\mathscr{Z}$, instead of $\mathscr{Z}_h$.

First, according to the result in the Exercise 3.5, the coefficient a can be obtained from the eigenvalue λ of $\mathbf{L} + \mu\mathbf{R}_{01}$ which is equal to $i\omega$ when $\mu = 0$. More precisely,

a is the coefficient of the $O(\mu)$ term in the expansion in μ of this eigenvalue. Going back to the eigenvalue problem (4.25) we find that this eigenvalue λ is a root of the characteristic polynomial P_1 for $\beta = \beta_1 + \mu$, i.e.,

$$\lambda^2 - \mu\lambda + \omega^2 - \mu\pi^2\delta_2 = 0.$$

Here we have used the fact that $i\omega$ is a root of P_1 when $\beta = \beta_1$. This gives the solutions

$$\lambda_+ = i\omega + \mu\left(\frac{1}{2} - i\pi^2\frac{\delta_2}{2\omega}\right) + O(\mu^2), \quad \lambda_- = \bar{\lambda}_+,$$

so that the coefficient a in (4.30) is

$$a = \frac{1}{2} - i\pi^2\frac{\delta_2}{2\omega}. \tag{4.32}$$

To compute b we proceed as in the previous examples. We set $\mu = 0$ in the following calculations. Inserting v from (4.31) into (4.27), we find the equality

$$(\zeta + \partial_A\Psi)\frac{dA}{dt} + (\bar{\zeta} + \partial_{\bar{A}}\Psi)\frac{d\bar{A}}{dt} = \mathbf{L}(A\zeta + \overline{A\zeta} + \Psi) + \mathbf{R}(A\zeta + \overline{A\zeta} + \Psi, 0). \tag{4.33}$$

Using the expansion (4.28) of $\mathbf{R}$, the expansion of Ψ

$$\Psi(A, \bar{A}, 0) = \sum_{p,q} \Psi_{pq} A^p \bar{A}^q,$$

where $\Psi_{qp} \in \mathscr{X}$ are such that

$$\Psi_{qp} = \overline{\Psi_{pq}}, \quad \Psi_{00} = \Psi_{10} = \Psi_{01} = 0,$$

and replacing dA/dt by the right hand side of (4.30), we identify the powers of $(A, \bar{A})$ in (4.33). At orders A^2, $A\bar{A}$, and $A^2\bar{A}$, we find, respectively,

$$(2i\omega - \mathbf{L})\Psi_{20} = \mathbf{R}_{20}(\zeta, \zeta), \tag{4.34}$$

$$-\mathbf{L}\Psi_{11} = 2\mathbf{R}_{20}(\zeta, \bar{\zeta}), \tag{4.35}$$

$$(i\omega - \mathbf{L})\Psi_{21} = -b\zeta + 2\mathbf{R}_{20}(\bar{\zeta}, \Psi_{20}) + 2\mathbf{R}_{20}(\zeta, \Psi_{11}) + 3\mathbf{R}_{30}(\zeta, \zeta, \bar{\zeta}). \tag{4.36}$$

Recall that $\pm i\omega$ are the only purely imaginary eigenvalues of $\mathbf{L}$. Then from the first two equations we can compute Ψ_{20} and Ψ_{11}, since $(2i\omega - \mathbf{L})$ and $\mathbf{L}$ are invertible, and from the solvability condition for the third equation we find b. We show below how these quantities can be explicitly computed. The arguments are typical for such types of bifurcation problems arising for PDEs.

First, we obtain

$$\mathbf{R}_{20}(\zeta,\zeta) = \left(\alpha\gamma + \frac{\beta_1}{2\alpha}\right)(1-\cos(2\pi x))\begin{pmatrix}1\\-1\end{pmatrix},$$
$$2\mathbf{R}_{20}(\zeta,\bar{\zeta}) = \left(2\alpha\gamma_r + \frac{\beta_1}{\alpha}\right)(1-\cos(2\pi x))\begin{pmatrix}1\\-1\end{pmatrix},$$

where γ_r in the second formula represents the real part of γ. The equation (4.34) is a linear nonhomogeneous system of two differential equations of second order. Its solution set is a four-dimensional space, and the solution Ψ_{20} is uniquely determined by the Dirichlet boundary conditions at $x = 0$ and $x = 1$, which must be satisfied by the functions in $\mathscr{X}$. We introduce the following 2×2-matrices:

$$\mathscr{M}(ni\omega, \nu^2) = \begin{pmatrix} ni\omega + 1 - \beta_1 - \delta_1\nu^2 & -\alpha^2 \\ \beta_1 & ni\omega + \alpha^2 - \delta_2\nu^2 \end{pmatrix},$$

representing the action of the operator $in\omega - \mathbf{L}$ on the exponential $e^{\nu x}v$, with $v \in \mathbb{C}^2$, so that

$$(ni\omega - \mathbf{L})e^{\nu x}v = \mathscr{M}(ni\omega, \nu^2)e^{\nu x}v.$$

Then the solutions of the homogeneous equation $(2i\omega - \mathbf{L})v = 0$ are linear combinations of the four basic solutions,

$$v_{1+}e^{\nu_1 x}, \quad v_{1-}e^{-\nu_1 x}, \quad v_{2+}e^{\nu_2 x}, \quad v_{2-}e^{-\nu_2 x},$$

in which $\pm\nu_1$, $\pm\nu_2$ are the four solutions of

$$\det(\mathscr{M}(2i\omega, \nu^2)) = 0$$

and the vectors $v_{1\pm} \in \mathbb{C}^2$ and $v_{2\pm} \in \mathbb{C}^2$ belong to the kernels of $\mathscr{M}(2i\omega, \nu_1^2)$ and $\mathscr{M}(2i\omega, \nu_2^2)$, respectively. Next, notice that the operator $2i\omega - \mathbf{L}$ preserves the linear subspaces spanned by $\cos(n\pi x)$ (and also $\sin(n\pi x)$), so that we can look for a particular solution of (4.34) in the form

$$\Psi_{20}^0 = \alpha_{20} + \beta_{20}\cos(2\pi x),$$

in which $\alpha_{20} \in \mathbb{C}^2$ and $\beta_{20} \in \mathbb{C}^2$ are the unique solutions of

$$\mathscr{M}(2i\omega, 0)\alpha_{20} = \left(\alpha\gamma + \frac{\beta_1}{2\alpha}\right)\begin{pmatrix}1\\-1\end{pmatrix},$$
$$\mathscr{M}(2i\omega, -4\pi^2)\beta_{20} = \left(\alpha\gamma + \frac{\beta_1}{2\alpha}\right)\begin{pmatrix}-1\\1\end{pmatrix}. \tag{4.37}$$

Summarizing, we have that

$$\Psi_{20} = \alpha_{20} + \beta_{20}\cos(2\pi x) + \gamma_{20}e^{\nu_1 x} + \delta_{20}e^{-\nu_1 x} + \chi_{20}e^{\nu_2 x} + \kappa_{20}e^{-\nu_2 x},$$

in which α_{20} and β_{20} are uniquely determined from (4.37), the vectors γ_{20}, δ_{20}, χ_{20}, and κ_{20} satisfy

$$\mathcal{M}(2i\omega, \nu_1^2)\gamma_{20} = 0, \quad \mathcal{M}(2i\omega, \nu_1^2)\delta_{20} = 0,$$
$$\mathcal{M}(2i\omega, \nu_2^2)\chi_{20} = 0, \quad \mathcal{M}(2i\omega, \nu_2^2)\kappa_{20} = 0,$$

and are uniquely determined from the Dirichlet boundary conditions at $x = 0$ and $x = 1$ for Ψ_{20}:

$$\alpha_{20} + \beta_{20} + \gamma_{20} + \delta_{20} + \chi_{20} + \kappa_{20} = 0,$$
$$\alpha_{20} + \beta_{20} + \gamma_{20}e^{\nu_1} + \delta_{20}e^{-\nu_1} + \chi_{20}e^{\nu_2} + \kappa_{20}e^{-\nu_2} = 0.$$

In the same way, we solve equation (4.35) and find the solution

$$\Psi_{11} = \alpha_{11} + \beta_{11}\cos(2\pi x) + \gamma_{11}e^{\mu_1 x} + \delta_{11}e^{-\mu_1 x} + \chi_{11}e^{\mu_2 x} + \kappa_{11}e^{-\mu_2 x},$$

where $\pm\mu_1 \in \mathbb{C}$ and $\pm\mu_2 \in \mathbb{C}$ are the four solutions of

$$\det(\mathcal{M}(0, \mu^2)) = 0,$$

and the vectors on the right hand side are uniquely determined from

$$\mathcal{M}(0,0)\alpha_{11} = \left(2\alpha\gamma_r + \frac{\beta_1}{\alpha}\right)\begin{pmatrix}1\\-1\end{pmatrix}, \quad \mathcal{M}(0,-4\pi^2)\beta_{11} = \left(2\alpha\gamma_r + \frac{\beta_1}{\alpha}\right)\begin{pmatrix}-1\\1\end{pmatrix},$$

$$\mathcal{M}(0,\mu_1^2)\gamma_{11} = 0, \quad \mathcal{M}(0,\mu_1^2)\delta_{11} = 0, \quad \mathcal{M}(0,\mu_2^2)\chi_{11} = 0, \quad \mathcal{M}(0,\mu_2^2)\kappa_{11} = 0,$$

and the equalities

$$\alpha_{11} + \beta_{11} + \gamma_{11} + \delta_{11} + \chi_{11} + \kappa_{11} = 0,$$
$$\alpha_{11} + \beta_{11} + \gamma_{11}e^{\mu_1} + \delta_{11}e^{-\mu_1} + \chi_{11}e^{\mu_2} + \kappa_{11}e^{-\mu_2} = 0.$$

Finally, in equation (4.36) we compute

$$2\mathbf{R}_{20}(\bar{\zeta}, \Psi_{20}) + 2\mathbf{R}_{20}(\zeta, \Psi_{11}) + 3\mathbf{R}_{30}(\zeta, \zeta, \bar{\zeta}) = \begin{pmatrix}f(x)\\-f(x)\end{pmatrix},$$

where

$$f(x) = (2\gamma + \bar{\gamma})\left(\frac{3}{4}\sin(\pi x) - \frac{1}{4}\sin(3\pi x)\right)$$
$$+\left(2\alpha(\psi_{20}^{(2)} + \psi_{11}^{(2)}) + \left(2\alpha\bar{\gamma} + \frac{2\beta_1}{\alpha}\right)\psi_{20}^{(1)} + \left(2\alpha\gamma + \frac{2\beta_1}{\alpha}\right)\psi_{11}^{(1)}\right)\sin(\pi x).$$

Here we have denoted

$$\Psi_{20} = \begin{pmatrix}\psi_{20}^{(1)}\\ \psi_{20}^{(2)}\end{pmatrix}, \quad \Psi_{11} = \begin{pmatrix}\psi_{11}^{(1)}\\ \psi_{11}^{(2)}\end{pmatrix}.$$

The solvability condition for the equation (4.36) is that its right hand side should be orthogonal to the kernel of the adjoint operator, $-i\omega + \mathbf{L}^*$. A direct computation shows that this kernel is one-dimensional and spanned by the vector

$$\zeta^* = \sin(\pi x)\begin{pmatrix} -\gamma \\ 1 \end{pmatrix},$$

from which we obtain that

$$b = \frac{2(1+\bar{\gamma})}{\bar{\gamma}-\gamma}\int_0^1 \sin(\pi x) f(x) dx.$$

According to the result in Theorem 2.6 in Chapter 1, on the Hopf bifurcation, a supercritical (resp., subcritical) Hopf bifurcation occurs at $\mu = 0$ if the real part b_r of b is negative (resp., positive). Notice that the bifurcating periodic solution corresponds to an *oscillating chemical reaction*.

2.4.5 Elliptic PDE in a Strip

Consider the elliptic problem

$$\begin{aligned} &\frac{\partial^2 v}{\partial x^2} + \frac{\partial^2 v}{\partial y^2} + \nu v + g\left(v, \frac{\partial v}{\partial x}, \frac{\partial v}{\partial y}\right) = 0, \\ &v(x,0) = v(x,\pi) = 0, \end{aligned}$$

where $v(x,y) \in \mathbb{R}$ for $(x,y) \in \mathbb{R} \times (0,\pi)$, ν is a real parameter, and we assume that $g \in C^k(\mathbb{R}^3, \mathbb{R})$, with $g(u,v,w) = O(|u|^2 + |v|^2 + |w|^2)$, as $(u,v,w) \to 0$. We further assume that g is even in its second argument.

Formulation and Symmetries

This problem enters our setting when we take as our time variable the unbounded spatial variable $x \in \mathbb{R}$, and so write the problem in the form

$$\frac{du}{dx} = \mathbf{L}_\nu u + \mathbf{R}(u). \tag{4.38}$$

This formulation of the problem is also called "spatial dynamics" formulation. The idea of spatial dynamics goes back to [80] and was used for finding bounded solutions of elliptic PDEs in cylindrical domains.

We obtain the equation (4.38) by taking $u = (u_1, u_2) = (v, \partial v/\partial x)$, and

$$\mathbf{L}_\nu = \begin{pmatrix} 0 & 1 \\ -\frac{d^2}{dy^2} - \nu & 0 \end{pmatrix}, \quad \mathbf{R}(u) = \begin{pmatrix} 0 \\ -g\left(u_1, u_2, \frac{du_1}{dy}\right) \end{pmatrix}.$$

We choose the spaces

$$\mathscr{X} = H_0^1(0,\pi) \times L^2(0,\pi), \quad \mathscr{Z} = \left(H^2(0,\pi) \cap H_0^1(0,\pi)\right) \times H_0^1(0,\pi),$$

such that $\mathbf{L}_\nu$ is a closed operator in $\mathscr{X}$ with domain $\mathscr{Z}$, which contains the boundary conditions, and

$$\mathscr{Y} = \left(H^2(0,\pi) \cap H_0^1(0,\pi)\right) \times H^1(0,\pi),$$

such that $\mathbf{R}(u) \in \mathscr{Y}$ for $u \in \mathscr{Z}$, and $\mathbf{R}$ is of class $C^k(\mathscr{Z},\mathscr{Y})$.

Notice that the elliptic equation is invariant under $(x,y) \mapsto (-x,y)$, since we assumed that g is even in its second argument. This induces a reversibility symmetry for (4.38), i.e., the vector field on the right hand side anticommutes with the symmetry $\mathbf{S}$ defined by

$$\mathbf{S}\begin{pmatrix} u_1 \\ u_2 \end{pmatrix} = \begin{pmatrix} u_1 \\ -u_2 \end{pmatrix}.$$

As in the previous examples we next look at the spectrum of $\mathbf{L}_\nu$. We have again that $\mathscr{Y}$ is compactly embedded in $\mathscr{X}$, so $\mathbf{L}_\nu$ has a compact resolvent and its spectrum consists of isolated eigenvalues with finite multiplicities. The eigenvalue problem reads

$$\begin{aligned} u_2 &= \lambda u_1 \\ -u_1'' - \nu u_1 &= \lambda u_2, \end{aligned}$$

in which u_1 satisfies the boundary conditions $u_1(0) = u_1(\pi) = 0$. Then a direct computation shows that the spectrum σ of $\mathbf{L}_\nu$ is

$$\sigma = \{\lambda_n^\pm = \pm\sqrt{n^2 - \nu}\,;\, n \in \mathbb{N}^*\}. \tag{4.39}$$

Notice that σ is symmetric with respect to both the imaginary and the real axis in the complex plane, due to the reversibility symmetry and the fact that $\mathbf{L}_\nu$ is a real operator. When $\nu \neq p^2$, for any integer p, the eigenvalues are all simple and real except for a finite number that are purely imaginary. When $\nu = p^2$ for some nonzero integer p, we find that 0 is a double eigenvalue, $\lambda_p^\pm = 0$, and the eigenvalues $\lambda_n^\pm$ with $n < p$ are purely imaginary, whereas the eigenvalues $\lambda_n^\pm$ with $n > p$ are real. Consequently, we can use the center manifold theorem, for values of ν close to $\nu_p = p^2$, for any $p \geq 1$.

Reduced System

We focus here on values of ν close to $\nu_1 = 1$ and set $\nu = 1 + \mu$. We rewrite the equation (4.38) in the form

$$\frac{du}{dx} = \mathbf{L}u + \mathbf{R}(u,\mu),$$

with

$$\mathbf{L} = \mathbf{L}_1, \quad \mathbf{R}(u,\mu) = \mathbf{R}(u) + (\mathbf{L}_{1+\mu} - \mathbf{L}_1)u.$$

From the arguments above it is now easy to check that Hypotheses 3.1 and 2.4 hold. In Hypothesis 2.4 we have $\sigma_0 = \{0\}$ with 0 a geometrically simple and algebraically double eigenvalue. The corresponding spectral subspace $\mathscr{E}_0$ is spanned by

$$\zeta_0 = \begin{pmatrix} \sin y \\ 0 \end{pmatrix}, \quad \zeta_1 = \begin{pmatrix} 0 \\ \sin y \end{pmatrix},$$

which satisfy $\mathbf{L}\zeta_0 = 0$ and $\mathbf{L}\zeta_1 = \zeta_0$, respectively. Also notice that

$$\mathbf{S}\zeta_0 = \zeta_0, \quad \mathbf{S}\zeta_1 = -\zeta_1.$$

Further, Hypothesis 2.7 can be checked as in the example in Section 2.4.2, using the result in Theorem 2.17 and showing that the estimate (2.9) holds. In addition, the reversibility symmetry $\mathbf{S}$ satisfies Hypothesis 3.14.

We can now apply the results in Theorems 3.3 and 3.15 and obtain a family of two-dimensional center manifolds for μ sufficiently small. For solutions on the center manifold we write

$$u = A\zeta_0 + B\zeta_1 + \Psi(A,B,\mu),$$

where $A(t) \in \mathbb{R}$, $B(t) \in \mathbb{R}$, and Ψ takes values in $\mathscr{Z}_h$. This leads to a reduced equation of the form

$$\begin{aligned} \frac{dA}{dx} &= f(A,B,\mu) \\ \frac{dB}{dx} &= g(A,B,\mu), \end{aligned} \tag{4.40}$$

in which the vector field (f,g) satisfies

$$(f,g)(0,0,\mu) = (0,0), \quad D(f,g)(0,0,0) = \begin{pmatrix} 0 & 1 \\ 0 & 0 \end{pmatrix}.$$

In addition, the vector field is reversible, it anticommutes with the symmetry $\mathbf{S}_0$ induced by $\mathbf{S}$ acting through

$$\mathbf{S}_0(A,B) = (A,-B),$$

and the reduction function Ψ commutes with $\mathbf{S}$,

$$\mathbf{S}\Psi(A,B,\mu) = \Psi(A,-B,\mu). \tag{4.41}$$

We shall further analyze this reduced system in Chapter 4, which is devoted to reversible systems.

Chapter 3
Normal Forms

In this chapter we present a number of results from the theory of normal forms. The idea of normal forms consists in finding a polynomial change of variable which "improves" locally a nonlinear system, in order to more easily recognize its dynamics. As we shall see, normal form transformations apply to general classes of nonlinear systems in $\mathbb{R}^n$ near a fixed point, here the origin, by just assuming a certain smoothness of the vector field. In particular, this theory applies to the reduced systems provided by the center manifold theory given in the previous chapter.

3.1 Main Theorem

We consider a differential equation in $\mathbb{R}^n$ of the form

$$\frac{du}{dt} = \mathbf{L}u + \mathbf{R}(u), \tag{1.1}$$

in which $\mathbf{L}$ and $\mathbf{R}$ represent the linear and nonlinear terms, respectively. More precisely, we assume that the following holds.

Hypothesis 1.1 *Assume that* $\mathbf{L}$ *and* $\mathbf{R}$ *in (1.1) have the following properties:*

(i) $\mathbf{L}$ *is a linear map in* $\mathbb{R}^n$*;*
(ii) for some $k \geq 2$*, there exists a neighborhood* $\mathscr{V} \subset \mathbb{R}^n$ *of* 0 *such that* $\mathbf{R} \in \mathscr{C}^k(\mathscr{V}, \mathbb{R}^n)$ *and*

$$\mathbf{R}(0) = 0, \quad D\mathbf{R}(0) = 0.$$

Our purpose is to transform this system, in a neighborhood of the origin, in such a way that the Taylor expansion of the transformed nonlinear vector field contains a *minimal number of terms at every order*. The following result shows the existence of a polynomial change of variables leading to a transformed vector field, which, as we shall see later, has this property.

M. Haragus, G. Iooss, *Local Bifurcations, Center Manifolds, and Normal Forms in Infinite-Dimensional Dynamical Systems*, Universitext,
DOI 10.1007/978-0-85729-112-7_3, © EDP Sciences 2011

Theorem 1.2 (Normal form theorem) *Consider the system (1.1) and assume that Hypothesis 1.1 holds. Then for any positive integer p, $2 \leq p \leq k$, there exists a polynomial $\Phi : \mathbb{R}^n \to \mathbb{R}^n$ of degree p, with*

$$\Phi(0) = 0, \quad D\Phi(0) = 0,$$

and such that the change of variable

$$u = v + \Phi(v) \tag{1.2}$$

defined in a neighborhood of the origin in $\mathbb{R}^n$ transforms the equation (1.1) into the "normal form"

$$\frac{dv}{dt} = \mathbf{L}v + \mathbf{N}(v) + \rho(v), \tag{1.3}$$

with the following properties:

(i) $\mathbf{N} : \mathbb{R}^n \to \mathbb{R}^n$ *is a polynomial of degree p, satisfying*

$$\mathbf{N}(0) = 0, \quad D\mathbf{N}(0) = 0.$$

(ii) The equality

$$\mathbf{N}(e^{t\mathbf{L}^*}v) = e^{t\mathbf{L}^*}\mathbf{N}(v), \tag{1.4}$$

holds for all $(t, v) \in \mathbb{R} \times \mathbb{R}^n$, where $\mathbf{L}^*$ represents the adjoint of $\mathbf{L}$.

(iii) ρ is a map of class $\mathscr{C}^k$ in a neighborhood of 0, *such that*

$$\rho(v) = o(\|v\|^p).$$

Remark 1.3 (Equivalent characterization of the normal form) *Instead of the characterization (1.4) for the polynomial* $\mathbf{N}$, *it may be advantageous to use the following equivalent characterization*

$$D\mathbf{N}(v)\mathbf{L}^*v = \mathbf{L}^*\mathbf{N}(v) \text{ for all } v \in \mathbb{R}^n. \tag{1.5}$$

Indeed, the following identity is valid for any $(t, v) \in \mathbb{R} \times \mathbb{R}^n$:

$$\frac{d}{dt}\left(e^{-t\mathbf{L}^*}\mathbf{N}(e^{t\mathbf{L}^*}v)\right) = e^{-t\mathbf{L}^*}\left(-\mathbf{L}^*\mathbf{N}(e^{t\mathbf{L}^*}v) + D\mathbf{N}(e^{t\mathbf{L}^*}v)\mathbf{L}^*e^{t\mathbf{L}^*}v\right).$$

Consequently, if (1.4) holds, then the left hand side in the above equality vanishes, and by taking $t = 0$ in the right hand side we obtain (1.5). Conversely, writing (1.5) with $e^{t\mathbf{L}^*}v$ instead of v implies that $e^{-t\mathbf{L}^*}\mathbf{N}(e^{t\mathbf{L}^*}v)$ is independent of t, which gives (1.4).

Remark 1.4 (Uniqueness of the normal form) *As we shall see from the proof of this theorem, the choice of the polynomial* $\mathbf{N}$ *is not unique. Actually, one can add to the polynomial* $\mathbf{N}$ *satisfying one of the equivalent characterizations (1.4) or (1.5) any polynomial* $\mathbf{Q}$ *which belongs to the range of the linear operator $\mathscr{A}_\mathbf{L}$ acting on the space of polynomials $\Phi : \mathbb{R}^n \to \mathbb{R}^n$ defined by*

$$(\mathscr{A}_{\mathbf{L}}\Phi)(v) = D\Phi(v)\mathbf{L}v - \mathbf{L}\Phi(v) \text{ for all } v \in \mathbb{R}^n.$$

Of course the new polynomial $\mathbf{N}+\mathbf{Q}$ *does not satisfy (1.4) and (1.5) anymore, but the change of variables* Φ *still exists. This property may sometimes allow one to further simplify the normal form (e.g., see Remark 1.10).*

Remark 1.5 *In applications we often use the characterizations (1.4) or (1.5) in a complex basis in which* $\mathbf{L}^*$ *is diagonal, or triangular (Jordan form). The formulations of (1.4) and (1.5) are valid in such a basis, as well. Indeed, denote by* $\mathbf{P}$ *the matrix for a change of basis, which may be complex, such that*

$$\mathbf{P}^{-1}\mathbf{L}^*\mathbf{P} = \mathbf{T}^*.$$

Replacing $v = \mathbf{P}w$ *into (1.5) we find*

$$D_v\mathbf{P}^{-1}\mathbf{N}(\mathbf{P}w)\mathbf{P}\mathbf{T}^*w = \mathbf{T}^*\mathbf{P}^{-1}\mathbf{N}(\mathbf{P}w).$$

Consequently, the polynomial $\widetilde{\mathbf{N}}$ *defined through*

$$\widetilde{\mathbf{N}}(w) \stackrel{def}{=} \mathbf{P}^{-1}\mathbf{N}(\mathbf{P}w)$$

satisfies

$$D_w\widetilde{\mathbf{N}}(w)\mathbf{T}^*w = \mathbf{T}^*\widetilde{\mathbf{N}}(w),$$

which is equivalent to (1.5).

Remark 1.6 *(i) Theorem 1.2 has been proved in [25] in its elementary formulation, given below in Section 3.1.1. The characterization (1.5) is in fact contained (a little hidden) in the general work of Belitskii [7], using more sophisticated methods of algebraic geometry.*

(ii) Some normal form results are also available in infinite- dimensional spaces for very specific problems, but there is no general result in this situation. The result in Theorem 1.2 suffices for the analysis of the reduced systems obtained by a center manifold reduction, since these are all finite-dimensional.

3.1.1 Proof of Theorem 1.2

We give in this section the proof of the normal form Theorem 1.2.

Consider the Taylor expansion of $\mathbf{R}$,

$$\mathbf{R}(u) = \sum_{2\leq q\leq p} \mathbf{R}_q(u^{(q)}) + o(\|u\|^p)$$

for a given p, $2 \leq p \leq k$, where $u^{(q)} = (u,\ldots,u) \in (\mathbb{R}^n)^q$, with $u \in \mathbb{R}^n$ repeated q times, and $\mathbf{R}_q$ is the q-linear symmetric map on $(\mathbb{R}^n)^q$ given through

$$\mathbf{R}_q(u^{(q)}) = \frac{1}{q!}D^q\mathbf{R}(0)(u^{(q)}).$$

Similarly, we write the polynomials $\boldsymbol{\Phi}$ and $\mathbf{N}$ in the form

$$\boldsymbol{\Phi}(v) = \sum_{2\leq q\leq p} \boldsymbol{\Phi}_q(v^{(q)}), \quad \mathbf{N}(v) = \sum_{2\leq q\leq p} \mathbf{N}_q(v^{(q)}),$$

with $\boldsymbol{\Phi}_q$ and $\mathbf{N}_q$ q-linear symmetric maps on $(\mathbb{R}^n)^q$.

Differentiating (1.2) with respect to t and replacing du/dt and dv/dt from (1.1) and (1.3), respectively, leads to the identity

$$(\mathbb{I} + D\boldsymbol{\Phi}(v))(\mathbf{L}v + \mathbf{N}(v) + \rho(v)) = \mathbf{L}(v + \boldsymbol{\Phi}(v)) + \mathbf{R}(v + \boldsymbol{\Phi}(v)), \tag{1.6}$$

which should be valid for all v in a neighborhood of 0. Our purpose it to determine $\boldsymbol{\Phi}$ and $\mathbf{N}$ from this equality. By identifying the Taylor expansions on both sides, we obtain at order 2

$$D\boldsymbol{\Phi}_2(v^{(2)})\mathbf{L}v - \mathbf{L}\boldsymbol{\Phi}_2(v^{(2)}) = \mathbf{R}_2(v^{(2)}) - \mathbf{N}_2(v^{(2)}), \tag{1.7}$$

and then at any order q, $3 \leq q \leq p$, we have

$$D\boldsymbol{\Phi}_q(v^{(q)})\mathbf{L}v - \mathbf{L}\boldsymbol{\Phi}_q(v^{(q)}) = \mathbf{Q}_q(v^{(q)}) - \mathbf{N}_q(v^{(q)}), \tag{1.8}$$

with

$$\begin{aligned}\mathbf{Q}_q(v^{(q)}) = &- \sum_{2\leq r\leq q-1} D\boldsymbol{\Phi}_r(v^{(r)})\mathbf{N}_{q-r+1}(v^{(q-r+1)}) + \\ &+ \sum_{r_1+\dots+r_\ell=q,\quad r_j\geq 1} \mathbf{R}_\ell\left(\boldsymbol{\Phi}_{r_1}(v^{(r_1)}), \boldsymbol{\Phi}_{r_2}(v^{(r_2)}), \dots, \boldsymbol{\Phi}_{r_\ell}(v^{(r_\ell)})\right),\end{aligned}$$

where we have set $\boldsymbol{\Phi}_1(v) = v$. Notice that if $\boldsymbol{\Phi}_l$ and $\mathbf{N}_l$ are known for any l, $2 \leq l \leq q-1$, then $\mathbf{Q}_q$ is known. Therefore, we can determine $\boldsymbol{\Phi}$ and $\mathbf{N}$ by successively finding $(\boldsymbol{\Phi}_2, \mathbf{N}_2)$, $(\boldsymbol{\Phi}_3, \mathbf{N}_3)$, and so on, from (1.7) and (1.8).

The equations (1.7) and (1.8) have the same structure; more precisely, they are both of the form

$$\mathscr{A}_{\mathbf{L}}\boldsymbol{\Phi}_q = \mathbf{Q}_q - \mathbf{N}_q, \tag{1.9}$$

in which $\mathscr{A}_{\mathbf{L}}$ is a linear map (also called "homological operator") acting on the space of polynomials $\boldsymbol{\Phi} : \mathbb{R}^n \to \mathbb{R}^n$ through

$$(\mathscr{A}_{\mathbf{L}}\boldsymbol{\Phi})(v) = D\boldsymbol{\Phi}(v)\mathbf{L}v - \mathbf{L}\boldsymbol{\Phi}(v). \tag{1.10}$$

A key property of $\mathscr{A}_{\mathbf{L}}$ is that it leaves invariant the subspace $\mathscr{H}_q$ of homogeneous polynomials of degree q, for any positive integer q. In the equality (1.9), $\mathbf{Q}_q$ is known, and we have to determine $\boldsymbol{\Phi}_q$ and $\mathbf{N}_q$. It is clear that if $\mathscr{A}_{\mathbf{L}}|_{\mathscr{H}_q}$ is invertible, then we can take $\mathbf{N}_q = 0$, which gives the simplest solution here. However, this is not always the case, and the condition for solving (1.9) is that $\mathbf{Q}_q - \mathbf{N}_q$ lies in the range of the operator $\mathscr{A}_{\mathbf{L}}$. We claim that this condition is achieved when (1.4), or equivalently (1.5), is satisfied by $\mathbf{N}_q$.

Indeed, we define below a scalar product in the space $\mathscr{H}$ of polynomials of degree p, such that the adjoint operator $(\mathscr{A}_{\mathbf{L}})^*$ of $\mathscr{A}_{\mathbf{L}}$ with respect to this scalar product is $\mathscr{A}_{\mathbf{L}^*}$, where $\mathbf{L}^*$ is the adjoint of $\mathbf{L}$ with respect to the canonical Euclidean scalar product in $\mathbb{R}^n$. Then $\mathbf{Q}_q - \mathbf{N}_q$ belongs to the range of $\mathscr{A}_{\mathbf{L}}$ if

$$\mathbf{Q}_q - \mathbf{N}_q \in \ker(\mathscr{A}_{\mathbf{L}^*})^{\perp} = \operatorname{im}(\mathscr{A}_{\mathbf{L}}),$$

or, equivalently,

$$\mathbf{P}_{\ker(\mathscr{A}_{\mathbf{L}^*})}(\mathbf{Q}_q - \mathbf{N}_q) = 0,$$

where $\mathbf{P}_{\ker(\mathscr{A}_{\mathbf{L}^*})}$ is the orthogonal projection on $\ker(\mathscr{A}_{\mathbf{L}^*})$ in the space $\mathscr{H}$ of polynomials of degree p. It is then natural to choose

$$\mathbf{N}_q = \mathbf{P}_{\ker(\mathscr{A}_{\mathbf{L}^*})}\mathbf{Q}_q.$$

Of course, this choice is not unique, since we can add to $\mathbf{N}_q$ any term in the range of $\mathscr{A}_{\mathbf{L}}$ (this then implies Remark 1.4). Furthermore, we shall see that the projection $\mathbf{P}_{\ker(\mathscr{A}_{\mathbf{L}^*})}$ leaves invariant the subspace $\mathscr{H}_q$, so that $\mathbf{N}_q \in \ker \mathscr{A}_{\mathbf{L}^*}|_{\mathscr{H}_q}$. In particular, this shows that (1.5) holds for $\mathbf{N}_q$. With this choice for $\mathbf{N}_q$, we can now solve (1.9) and obtain a solution $\boldsymbol{\Phi}_q$, which is determined up to an arbitrary element in the kernel of $\mathscr{A}_{\mathbf{L}}$. A possible, but not unique, choice is to choose the unique solution $\boldsymbol{\Phi}_q$ orthogonal to $\ker \mathscr{A}_{\mathbf{L}}$ in $\mathscr{H}_q$. Summarizing, this shows that (1.9) possesses a solution $(\boldsymbol{\Phi}_q, \mathbf{N}_q)$ with $\mathbf{N}_q$ satisfying (1.5). Solving successively for $q = 2, \ldots, p$, we obtain the polynomials $\boldsymbol{\Phi}$ and $\mathbf{N}$ in the theorem, with $\mathbf{N}$ satisfying (1.5).

To finish the proof, it remains to define the scalar product in the space $\mathscr{H}$ such that

$$(\mathscr{A}_{\mathbf{L}})^* = \mathscr{A}_{\mathbf{L}^*}, \tag{1.11}$$

and to check that the orthogonal projection $\mathbf{P}_{\ker(\mathscr{A}_{\mathbf{L}^*})}$ on $\ker(\mathscr{A}_{\mathbf{L}^*})$ leaves invariant the subspace $\mathscr{H}_q$.

For a pair of scalar polynomials $P, P' : \mathbb{R}^n \to \mathbb{R}$ we define

$$\langle P|P'\rangle \stackrel{\text{def}}{=} P(\partial_u)P'(u)|_{u=0}, \tag{1.12}$$

where $u = (u_1, \ldots, u_n) \in \mathbb{R}^n$ and $\partial_u = (\partial/\partial u_1, \ldots, \partial/\partial u_n)$. The equality (1.12) defines a scalar product in the linear space of scalar polynomials $P : \mathbb{R}^n \to \mathbb{R}$. To see this, it is sufficient to take the canonical basis of the space of scalar polynomials, consisting of monomials $u_1^{\alpha_1} \ldots u_n^{\alpha_n}$, and to check that

$$\langle u_1^{\alpha_1} \ldots u_n^{\alpha_n} | u_1^{\beta_1} \ldots u_n^{\beta_n}\rangle = \alpha_1! \ldots \alpha_n! \delta_{\alpha_1\beta_1} \cdots \delta_{\alpha_n\beta_n},$$

where $\delta_{\alpha_j\beta_j} = 1$ if $\alpha_j = \beta_j$, and $\delta_{\alpha_j\beta_j} = 0$ otherwise. (Notice that this scalar product can be extended to complex-valued polynomials $P : \mathbb{C}^n \to \mathbb{C}$ by taking

$$\langle P|P'\rangle \stackrel{\text{def}}{=} P(\partial_u)\overline{P}'(u)|_{u=0},$$

where $\overline{P}(u) \stackrel{\text{def}}{=} \overline{P(\overline{u})}$.)

Now we define a scalar product on $\mathscr{H}$ by taking

$$\langle \Phi | \Phi' \rangle = \sum_{j=1}^{n} \langle \Phi_j | \Phi'_j \rangle$$

for $\Phi = (\Phi_1, \ldots, \Phi_n) \in \mathscr{H}$, $\Phi' = (\Phi'_1, \ldots, \Phi'_n) \in \mathscr{H}$. An important property of this scalar product (used in theoretical physics) is that the adjoint of the multiplication by u_j is the differentiation with respect to u_j,

$$\langle u_j P | P' \rangle = \partial_{u_j} P(\partial_u) P'(u)|_{u=0} = P(\partial_u) \partial_{u_j} P'(u)|_{u=0} = \langle P | \partial_{u_j} P' \rangle.$$

For our purpose, the most interesting property is the equality

$$\langle P \circ \mathbf{T} | P' \rangle = \langle P | P' \circ \mathbf{T}^* \rangle, \tag{1.13}$$

in which $\mathbf{T}$ is any invertible linear map, and $\mathbf{T}^*$ is the adjoint of $\mathbf{T}$ with respect to the Euclidean scalar product in $\mathbb{R}^n$. To show (1.13), consider the change of variable $u = \mathbf{T}^* v$, which means

$$u_i = \sum_{j=1}^{n} T_{ji} v_j,$$

for $u = (u_1, \ldots, u_n)$, $v = (v_1, \ldots, v_n)$ and $\mathbf{T} = (T_{ij})_{1 \le i,j \le n}$. Then

$$\frac{\partial u_i}{\partial v_j} = T_{ji}, \qquad \frac{\partial}{\partial v_j} = \sum_{i=1}^{n} T_{ji} \frac{\partial}{\partial u_i},$$

so that $\partial_v = \mathbf{T} \partial_u$. Using this equality and the fact that $u = 0$ is equivalent to $v = 0$, we find

$$\langle P \circ \mathbf{T} | P' \rangle = P(\mathbf{T} \partial_u) P'(u)|_{u=0} = P(\partial_v) P'(\mathbf{T}^* v)|_{v=0} = \langle P | P' \circ \mathbf{T}^* \rangle,$$

which proves (1.13).

We use the identity (1.13) to determine the adjoint of $\mathscr{A}_{\mathbf{L}}$. We take $\mathbf{T} = e^{-t\mathbf{L}}$, for which we find $\mathbf{T}^* = e^{-t\mathbf{L}^*}$ and $\mathbf{T}^{-1} = e^{t\mathbf{L}}$. Then from (1.13) we obtain

$$\langle e^{-t\mathbf{L}} \Phi \circ e^{t\mathbf{L}} | \Phi' \rangle = \langle \Phi | e^{-t\mathbf{L}^*} \Phi' \circ e^{t\mathbf{L}^*} \rangle$$

for any $\Phi, \Phi' \in \mathscr{H}$. Differentiating this equality with respect to t at $t = 0$, leads to

$$\langle \mathscr{A}_{\mathbf{L}} \Phi | \Phi' \rangle = \langle \Phi | \mathscr{A}_{\mathbf{L}^*} \Phi' \rangle.$$

This proves the formula for the adjoint (1.11).

Finally, the identity above also holds in the subspaces $\mathscr{H}_q$ of homogeneous polynomials of degree q, which are all invariant under the actions of both $\mathscr{A}_{\mathbf{L}}$ and $\mathscr{A}_{\mathbf{L}^*}$. Consequently,

$$\ker(\mathscr{A}_{\mathbf{L}^*}|_{\mathscr{H}_p}) = \ker \mathscr{A}_{\mathbf{L}^*} \cap \mathscr{H}_p,$$

and since monomials with different degrees are orthogonal to each other, this implies the invariance of $\mathscr{H}_p$ under the orthogonal projection $\mathbf{P}_{\ker \mathscr{A}_{\mathbf{L}^*}}$. This ends the proof of Theorem 1.2.

In the next sections, we apply this theorem to different cases in dimensions 2, 3, and 4. In all these cases the linear map $\mathbf{L}$ has purely imaginary eigenvalues, only, just as the linear part has in the reduced systems obtained from the center manifold reduction.

3.1.2 Examples in Dimension 2: $i\omega$, 0^2

We discuss in this section two cases in dimension 2: $i\omega$, where $\mathbf{L}$ has a pair of simple complex eigenvalues $\pm i\omega$, and 0^2, where $\mathbf{L}$ has a double zero eigenvalue with a Jordan block of length 2.

The case $i\omega$ corresponds to a matrix

$$\mathbf{L} = \begin{pmatrix} 0 & -\omega \\ \omega & 0 \end{pmatrix},$$

where $\omega > 0$, and $\mathbf{L}$ has the simple eigenvalues $\pm i\omega$. In this situation it is more convenient to identify $\mathbb{R}^2$ with the diagonal $\{(z,\bar{z}); z \in \mathbb{C}\}$ in $\mathbb{C}^2$ and to choose a complex basis of eigenvectors $\{\zeta, \overline{\zeta}\}$ with $\zeta = (1,-i)$, such that $\mathbf{L}$ becomes

$$\mathbf{L} = \begin{pmatrix} i\omega & 0 \\ 0 & -i\omega \end{pmatrix}. \tag{1.14}$$

A vector in $\mathbb{R}^2$ is now represented as

$$u = A\zeta + \overline{A\zeta}, \quad A \in \mathbb{C}.$$

Applying Theorem 1.2, we now prove the following result.

Lemma 1.7 **($i\omega$ normal form)** *Assume that the 2×2-matrix $\mathbf{L}$ takes the form (1.14) in a complex basis $\{\zeta, \bar{\zeta}\}$, in which a vector $u \in \mathbb{R}^2$ is represented by $u = (A, \overline{A})$, with $A \in \mathbb{C}$. Then the polynomial $\mathbf{N}$ in Theorem 1.2 is of the form*

$$\mathbf{N}(u) = (AQ(|A|^2), \overline{AQ}(|A|^2)),$$

where Q is a complex-valued polynomial in its argument, satisfying $Q(0) = 0$.

Proof In order to determine the normal form in this case, it is convenient to use the identity (1.4) and Remark 1.5. We have

$$e^{t\mathbf{L}^*} = \begin{pmatrix} e^{-i\omega t} & 0 \\ 0 & e^{i\omega t} \end{pmatrix},$$

and denoting $\mathbf{N} = (P(A,\overline{A}), \overline{P}(A,\overline{A}))$, from (1.4) we obtain that

$$P(e^{-i\omega t}A, e^{i\omega t}\overline{A}) = e^{-i\omega t}P(A,\overline{A}).$$

In particular, this shows that the normal form in this case commutes with all rotations in the complex plane (with this choice of the basis). Using Lemma 2.4 in Chapter 1, we find that

$$P(A,\overline{A}) = AQ(|A|^2),$$

where Q is a complex-valued polynomial in its argument. Moreover, $Q(0) = 0$ since $D\mathbf{N}(0) = 0$, which completes the proof. □

Exercise 1.8 *Compute the terms up to order* 2 *in the normal of the system* (2.31) *in Chapter 1, with* $\mu = 0$.
Hint: *Redo the calculations in Section 1.2.2 with* $\mu = 0$.

Next we consider the case 0^2 where $\mathbf{L}$ has a double zero eigenvalue with a Jordan block of length 2.

Lemma 1.9 (**0^2 normal form**) *Assume that the matrix* $\mathbf{L}$ *is in Jordan form*

$$\mathbf{L} = \begin{pmatrix} 0 & 1 \\ 0 & 0 \end{pmatrix},$$

in a basis of $\mathbb{R}^2$ *in which a vector* $u \in \mathbb{R}^2$ *is represented by* $u = (A,B) \in \mathbb{R}^2$. *Then the polynomial* $\mathbf{N}$ *in Theorem 1.2 is of the form*

$$\mathbf{N}(u) = (AP(A), BP(A) + Q(A)),$$

where P *and* Q *are real-valued polynomials, satisfying* $P(0) = Q(0) = Q'(0) = 0$.

Proof We set

$$\mathbf{N}(u) = (\Phi_1(A,B), \Phi_2(A,B)),$$

where Φ_1 and Φ_2 are polynomials in (A,B). Then we have $\mathbf{L}^*(A,B) = (0,A)$ and using the identity (1.5) we obtain

$$A\frac{\partial \Phi_1}{\partial B} = 0, \quad A\frac{\partial \Phi_2}{\partial B} = \Phi_1.$$

Consequently, Φ_1 does not depend upon B, $\Phi_1(A,B) = \phi_1(A)$, and since the polynomial $A\partial\Phi_2/\partial B = \Phi_1$ is divisible by A, there exists a polynomial P such that

$$\Phi_1(A,B) = AP(A).$$

Then the equation for the polynomial Φ_2 leads to

$$\Phi_2(A,B) = BP(A) + Q(A),$$

with Q a polynomial. Finally, we find that $P(0) = Q(0) = Q'(0) = 0$, since $\mathbf{N}(0) = 0$ and $D\mathbf{N}(0) = 0$. □

Remark 1.10 *(i) Notice that the kernel of the operator $\mathscr{A}_{\mathbf{L}^*}$ in the proof of Theorem 1.2 in the space $\mathscr{H}_p$ of homogeneous polynomials of degree q is in this case two-dimensional, spanned by*

$$(A^q, BA^{q-1}), \quad (0, A^q).$$

Furthermore, $(-A^q, qBA^{q-1})$ is orthogonal to this two-dimensional space, so that it belongs to the range of $\mathscr{A}_{\mathbf{L}}$. As it was noticed in Remark 1.4, we can add to **N** *any term in the range of $\mathscr{A}_{\mathbf{L}}$. In particular, in this case we can then choose* **N** *such that its first component is 0, which gives a simpler normal form,*

$$\mathbf{N}(u) = (0, BP_1(A) + Q_1(A)),$$

where P_1 and Q_1 are polynomials such that $P_1(0) = Q_1(0) = Q_1'(0) = 0$.

(ii) Alternatively, we can obtain this simpler normal form starting from the result in Lemma 1.9, which gives the system

$$\begin{aligned}\frac{dA}{dt} &= B + AP(A) + \rho_0(A,B) \\ \frac{dB}{dt} &= BP(A) + Q(A) + \rho_1(A,B),\end{aligned} \tag{1.15}$$

by making the change of variables

$$\widetilde{B} = B + AP(A) + \rho_0(A,B). \tag{1.16}$$

By the implicit function theorem, this change of variables is invertible:

$$B = \widetilde{B} - AP(A) + \widetilde{\rho}_0(A, \widetilde{B}),$$

and leads to the system

$$\begin{aligned}\frac{dA}{dt} &= \widetilde{B} \\ \frac{d\widetilde{B}}{dt} &= \widetilde{B}P_1(A) + Q_1(A) + \widetilde{\rho}_1(A, \widetilde{B}),\end{aligned}$$

with

$$P_1(A) = P(A) + \frac{d}{dA}(AP(A)), \quad Q_1(A) = Q(A) - A(P(A))^2.$$

Notice that in contrast to the result in the first part of this remark, in the first equation of the system above there is no longer a remainder. In turn, when going back to the change of variables from $(A, \widetilde{B})$ to u, this transformation is now not a polynomial.

Example: Computation of a 0^2 Normal Form

Consider the following second order differential equation

$$u'' = \alpha u^2 + \beta uu' + \gamma(u')^2, \tag{1.17}$$

with α, β, and γ real numbers.

Normal Form

We set $U = (u,v)$, so that the equation takes the form

$$\frac{dU}{dt} = \mathbf{L}U + \mathbf{R}_2(U,U), \tag{1.18}$$

with

$$\mathbf{L} = \begin{pmatrix} 0 & 1 \\ 0 & 0 \end{pmatrix}, \quad \mathbf{R}_2(U,\widetilde{U}) = \begin{pmatrix} 0 \\ \alpha u\widetilde{u} + \frac{\beta}{2}(u\widetilde{v} + \widetilde{u}v) + \gamma v\widetilde{v} \end{pmatrix}.$$

We are interested in computing the normal form of this system up to terms of order 2. Therefore it is enough to use the result in the normal form Theorem 1.2 with $p = 2$, i.e., to take the polynomial Φ of the form

$$\Phi(A,B) = A^2\Phi_{20} + AB\Phi_{11} + B^2\Phi_{02}.$$

Then, according to Lemma 1.9 and Remark 1.10(ii), the change of variables

$$U = A\zeta_0 + B\zeta_1 + \Phi(A,B), \tag{1.19}$$

where

$$\zeta_0 = \begin{pmatrix} 1 \\ 0 \end{pmatrix}, \quad \zeta_1 = \begin{pmatrix} 0 \\ 1 \end{pmatrix},$$

transforms system (1.18) into the normal form

$$\begin{aligned} \frac{dA}{dt} &= B \\ \frac{dB}{dt} &= aA^2 + bAB + O(|A| + |B|)^3, \end{aligned} \tag{1.20}$$

where a and b are real constants.

Computation of the Coefficients a and b

In order to compute the coefficients a and b in this normal form we proceed as in the computation of the Hopf bifurcation in Chapter 1 (see Sections 1.2.1 and 1.2.2).

First, substituting the change of variables (1.19) into the system (1.18) we find the equation

$$\frac{dA}{dt}\zeta_0 + \frac{dB}{dt}\zeta_1 + \partial_A\Phi(A,B)\frac{dA}{dt} + \partial_B\Phi(A,B)\frac{dB}{dt}$$
$$= B\zeta_0 + \mathbf{L}\Phi + \mathbf{R}_2(A\zeta_0 + B\zeta_1 + \Phi, A\zeta_0 + B\zeta_1 + \Phi),$$

where we have used the fact that $\mathbf{L}\zeta_0 = 0$ and $\mathbf{L}\zeta_1 = \zeta_0$. Next, we substitute the expressions of dA/dt and dB/dt from (1.20) in the left hand side of the equality above. In the resulting equality we identify the monomials A^2, AB, B^2, and find that

$$a\zeta_1 = \mathbf{L}\Phi_{20} + \mathbf{R}_2(\zeta_0,\zeta_0), \tag{1.21}$$
$$b\zeta_1 + 2\Phi_{20} = \mathbf{L}\Phi_{11} + 2\mathbf{R}_2(\zeta_0,\zeta_1), \tag{1.22}$$
$$\Phi_{11} = \mathbf{L}\Phi_{02} + \mathbf{R}_2(\zeta_1,\zeta_1), \tag{1.23}$$

where

$$\mathbf{R}_2(\zeta_0,\zeta_0) = \begin{pmatrix} 0 \\ \alpha \end{pmatrix}, \quad 2\mathbf{R}_2(\zeta_0,\zeta_1) = \begin{pmatrix} 0 \\ \beta \end{pmatrix}, \quad \mathbf{R}_2(\zeta_1,\zeta_1) = \begin{pmatrix} 0 \\ \gamma \end{pmatrix}.$$

Each of equations (1.21)–(1.23) are nonhomogeneous linear systems of the form

$$\mathbf{L}\Phi = R, \quad \Phi, R \in \mathbb{R}^2,$$

which are not uniquely solvable, since $\mathbf{L}$ is not invertible. Notice that the range $\mathrm{im}\,\mathbf{L}$ of $\mathbf{L}$ is given by $\mathrm{im}\,\mathbf{L} = \{(u,0); u \in \mathbb{R}\} \subset \mathbb{R}^2$ and that the kernel $\ker\mathbf{L}$ is spanned by ζ_0. Consequently, the system $\mathbf{L}\Phi = R$ has a solution if and only if $R \in \mathrm{im}\,\mathbf{L}$ and this solution is unique up to an element in $\ker\mathbf{L}$.

For the equation (1.21) we find

$$a\zeta_1 - \mathbf{R}_2(\zeta_0,\zeta_0) = \begin{pmatrix} 0 \\ a-\alpha \end{pmatrix},$$

so that the solvability condition $a\zeta_1 - \mathbf{R}_2(\zeta_0,\zeta_0) \in \mathrm{im}\,\mathbf{L}$ is satisfied when

$$a = \alpha,$$

which determines the coefficient a in the normal form. Then the solution Φ_{20} is any element of the kernel of $\mathbf{L}$,

$$\Phi_{20} = \phi_{20}\zeta_0, \quad \phi_{20} \in \mathbb{R}.$$

Next, for equation (1.22) we have

$$b\zeta_1 + 2\Phi_{20} - 2\mathbf{R}_2(\zeta_0,\zeta_1) = \begin{pmatrix} 2\phi_{20} \\ b-\beta \end{pmatrix},$$

so that the solvability condition for this equation determines the coefficient b, namely,

$$b = \beta.$$

This completes the calculation of the coefficients a and b.

Notice that it is not necessary to compute the solution Φ_{11} of the equation (1.22) and to solve the equation (1.23), unless one needs to also compute the polynomial Φ in the change of variables. Here we find

$$\Phi_{11} = 2\phi_{20}\zeta_1 + \phi_{11}\zeta_0, \quad 2\phi_{20} = \gamma, \quad \Phi_{02} = \phi_{11}\zeta_1 + \phi_{02}\zeta_0,$$

where the second equality is the solvability condition for the equation (1.23). In particular, this uniquely determines ϕ_{20}, whereas ϕ_{11} and ϕ_{02} are arbitrary. We can choose, for instance, $\phi_{11} = \phi_{02} = 0$, which then leads to the formula for Φ:

$$\Phi(A,B) = \frac{\gamma}{2}A^2\zeta_0 + \gamma AB\zeta_1.$$

Remark 1.11 *In this example it was easy to determine the range* $\operatorname{im}\mathbf{L}$ *of* $\mathbf{L}$*, and so to obtain the solvability conditions for the equations* (1.21)–(1.23). *In general, a convenient way of finding these solvability conditions is with the help of the adjoint* $\mathbf{L}^*$*, since the kernel of the adjoint* $\mathbf{L}^*$ *is orthogonal to the range of* $\mathbf{L}$*. This means that the solvability conditions are orthogonality conditions on the kernel of the adjoint* $\mathbf{L}^*$.

3.1.3 Examples in Dimension 3: $0(i\omega)$, 0^3

We present in this section two cases in dimension 3: $0(i\omega)$, where $\mathbf{L}$ has a pair of simple complex eigenvalues $\pm i\omega$ and a simple eigenvalue at 0, and 0^3, where $\mathbf{L}$ has a triple zero eigenvalue with a Jordan block of length 3.

Lemma 1.12 ($0(i\omega)$ normal form) *Assume that the matrix* $\mathbf{L}$ *is of the form*

$$\mathbf{L} = \begin{pmatrix} 0 & 0 & 0 \\ 0 & i\omega & 0 \\ 0 & 0 & -i\omega \end{pmatrix}$$

for some $\omega > 0$*, in a basis of* $\mathbb{R}^3$ *in which a vector* $u \in \mathbb{R}^3$ *is represented by* $u = (A, B, \overline{B})$*, with* $A \in \mathbb{R}$ *and* $B \in \mathbb{C}$*. Then the polynomial* $\mathbf{N}$ *in Theorem 1.2 is of the form*

$$\mathbf{N}(u) = (P(A, |B|^2), BQ(A, |B|^2), \overline{BQ}(A, |B|^2)),$$

where P *and* Q *are polynomials in their arguments, taking values in* $\mathbb{R}$ *and* $\mathbb{C}$*, respectively, and satisfying* $P(0,0) = \partial P/\partial A(0,0) = Q(0,0) = 0$.

Proof We set

$$\mathbf{N}(u) = (P_0(A,B,\overline{B}), Q_0(A,B,\overline{B}), \overline{Q_0}(A,B,\overline{B})).$$

Then identity (1.4) leads to

$$\begin{aligned} P_0(A, e^{-i\omega t}B, e^{i\omega t}\overline{B}) &= P_0(A,B,\overline{B}), \\ Q_0(A, e^{-i\omega t}B, e^{i\omega t}\overline{B}) &= e^{-i\omega t}Q_0(A,B,\overline{B}), \end{aligned}$$

which holds for all $t \in \mathbb{R}$ and all $(A,B,\overline{B}) \in \mathbb{R} \times \mathbb{C}^2$. First, the same arguments as in the proof of Lemma 2.4 in Chapter 1, give the form of the dependency of Q_0 upon B, namely,

$$Q_0(A,B,\overline{B}) = BQ(A,|B|^2).$$

Since Q_0 is a polynomial in $(A,B,\overline{B})$ with $Q_0(0,0,0) = 0$ and $DQ_0(0,0,0) = 0$, we conclude that Q is a polynomial in its arguments with $Q(0,0) = 0$. Next, for the polynomial P_0 we take successively $\omega t = \arg B$ and $\omega t = \pi$, which give that

$$P_0(A,B,\overline{B}) = P_0(A,|B|,|B|) = P_0(A,-B,-\overline{B}).$$

Consequently, P_0 is of the form

$$P_0(A,B,\overline{B}) = P(A,|B|^2),$$

where P is a polynomial in its arguments and satisfies $P(0,0) = \partial P/\partial A(0,0) = 0$. □

In the case 0^3, we prove in Appendix C.1 that the following result holds.

Lemma 1.13 **(0^3 normal form)** *Assume that the matrix* $\mathbf{L}$ *is in Jordan form*

$$\mathbf{L} = \begin{pmatrix} 0 & 1 & 0 \\ 0 & 0 & 1 \\ 0 & 0 & 0 \end{pmatrix}$$

in a basis of $\mathbb{R}^3$ *in which a vector* $u \in \mathbb{R}^3$ *is represented by* $u = (A,B,C)$, *with* $A,B,C \in \mathbb{R}$. *Then the polynomial* $\mathbf{N}$ *in Theorem 1.2 is of the form*

$$\mathbf{N}(u) = (AP_1(A,\widetilde{B}), BP_1(A,\widetilde{B}) + AP_2(A,\widetilde{B}), CP_1(A,\widetilde{B}) + BP_2(A,\widetilde{B}) + P_3(A,\widetilde{B})),$$

where

$$\widetilde{B} = B^2 - 2AC,$$

and P_1, P_2, *and* P_3 *are real-valued polynomials such that* $P_1(0,0) = P_2(0,0) = P_3(0,0) = \partial P_3/\partial A(0,0) = 0$.

Remark 1.14 *As in the case* 0^2, *we can use here Remark 1.4 and choose* $\mathbf{N}$ *such that its two first components vanish, i.e.,*

$$\mathbf{N}(u) = (0, 0, CP_1(A,\widetilde{B}) + BP_2(A,\widetilde{B}) + P_3(A,\widetilde{B})),$$

where the polynomials P_1, P_2, and P_3 are real-valued such that $P_1(0,0) = P_2(0,0) = P_3(0,0) = \partial P_3/\partial A(0,0) = 0$.

3.1.4 Examples in Dimension 4: $(i\omega_1)(i\omega_2)$, $(i\omega)^2$, $0^2(i\omega)$, 0^20^2

In this section we consider four cases of matrices $\mathbf{L}$ in $\mathbb{R}^4$. The first case is that in which $\mathbf{L}$ has two pairs of simple purely imaginary eigenvalues, $\pm i\omega_1$ and $\pm i\omega_2$.

Lemma 1.15 (**$(i\omega_1)(i\omega_2)$ normal form**) *Assume that the matrix $\mathbf{L}$ is of the form*

$$\mathbf{L} = \begin{pmatrix} i\omega_1 & 0 & 0 & 0 \\ 0 & i\omega_2 & 0 & 0 \\ 0 & 0 & -i\omega_1 & 0 \\ 0 & 0 & 0 & -i\omega_2 \end{pmatrix},$$

where $\omega_1 \neq \omega_2$ are positive real numbers, in a basis of $\mathbb{R}^4$ in which a vector $u \in \mathbb{R}^4$ is represented by $u = (A, B, \overline{A}, \overline{B})$, with $A, B \in \mathbb{C}$.

(i) Assume that $\omega_1/\omega_2 \notin \mathbb{Q}$. Then the polynomial $\mathbf{N}$ in Theorem 1.2 is of the form

$$\mathbf{N}(u) = (AP(|A|^2,|B|^2), BQ(|A|^2,|B|^2), \overline{AP}(|A|^2,|B|^2), \overline{BQ}(|A|^2,|B|^2)),$$

where P and Q are complex-valued polynomials in their arguments such that $P(0,0) = Q(0,0) = 0$.

(ii) Assume that $\omega_1/\omega_2 = r/s \in \mathbb{Q}$. Then the polynomial $\mathbf{N}$ in Theorem 1.2 is of the form

$$\begin{aligned} \mathbf{N}(u) = \Big(& AP_1(|A|^2,|B|^2,A^s\overline{B}^r) + \overline{A}^{s-1}B^rP_2(|A|^2,|B|^2,\overline{A}^sB^r), \\ & BQ_1(|A|^2,|B|^2,\overline{A}^sB^r) + A^s\overline{B}^{r-1}Q_2(|A|^2,|B|^2,A^s\overline{B}^r), \\ & (\overline{A}\overline{P_1(|A|^2,|B|^2,A^s\overline{B}^r)} + A^{s-1}\overline{B}^r\overline{P_2(|A|^2,|B|^2,\overline{A}^sB^r)}, \\ & \overline{B}\overline{Q_1(|A|^2,|B|^2,\overline{A}^sB^r)} + \overline{A}^sB^{r-1}\overline{Q_2(|A|^2,|B|^2,A^s\overline{B}^r)}\Big), \end{aligned}$$

where P_1, P_2, Q_1, and Q_2 are complex-valued polynomials in their arguments and $P_1(0,0,0) = Q_1(0,0,0) = 0$.

Proof We set

$$\mathbf{N}(u) = (\Phi_1(A,B,\overline{A},\overline{B}), \Phi_2(A,B,\overline{A},\overline{B}), \overline{\Phi}_1(A,B,\overline{A},\overline{B}), \overline{\Phi}_2(A,B,\overline{A},\overline{B})),$$

and then from (1.4) we find

$$\begin{aligned} \Phi_1(e^{-i\omega_1 t}A, e^{-i\omega_2 t}B, e^{i\omega_1 t}\overline{A}, e^{i\omega_2 t}\overline{B}) &= e^{-i\omega_1 t}\Phi_1(A,B,\overline{A},\overline{B}) \\ \Phi_2(e^{-i\omega_1 t}A, e^{-i\omega_2 t}B, e^{i\omega_1 t}\overline{A}, e^{i\omega_2 t}\overline{B}) &= e^{-i\omega_2 t}\Phi_2(A,B,\overline{A},\overline{B}) \end{aligned} \tag{1.24}$$

for all $t \in \mathbb{R}$, and $A, B \in \mathbb{C}$.

Consider the monomials

$$\phi^{(1)}_{p_1q_1p_2q_2}A^{p_1}\overline{A}^{q_1}B^{p_2}\overline{B}^{q_2} \quad \text{and} \quad \phi^{(2)}_{p_1q_1p_2q_2}A^{p_1}\overline{A}^{q_1}B^{p_2}\overline{B}^{q_2}$$

in the polynomials Φ_1 and Φ_2, respectively. Then (1.24) implies that

$$\omega_1(p_1-q_1-1)+\omega_2(p_2-q_2)=0.$$

If $\omega_1/\omega_2 \notin \mathbb{Q}$, we then have

$$p_1=q_1+1, \quad p_2=q_2,$$

from which we conclude the result in part (i).

If $\omega_1/\omega_2=r/s\in\mathbb{Q}$, then the relation above gives

$$r(p_1-q_1-1)+s(p_2-q_2)=0,$$

and since r and s have no common divisor, we obtain

$$p_1-q_1-1=ls, \quad p_2-q_2=-lr$$

for some $l\in\mathbb{Z}$. For $l\geq 0$, this gives

$$p_1=q_1+1+ls, \quad q_2=p_2+lr,$$

which corresponds to a polynomial of the form $AP_1(|A|^2,|B|^2,A^s\overline{B}^r)$, where P_1 is a polynomial in its arguments. For $l=-l'<0$, we find

$$q_1=p_1+s-1+(l'-1)s, \quad p_2=q_2+r+(l'-1)r,$$

which gives a polynomial of the form $\overline{A}^{s-1}B^rP_2(|A|^2,|B|^2,\overline{A}^sB^r)$, where P_2 is a polynomial in its arguments. The same arguments work for the polynomial Φ_2. Notice that the lowest order terms in these polynomials, which are not of the standard form found in the irrational case, are of degree $r+s-1\geq 2$ (we assumed $\omega_1\neq\omega_2$, which implies that r and s are different positive integers). This ends the proof of the lemma. □

Exercise 1.16 (Generalization) *Consider the matrix* $\mathbf{L}$ *in* $\mathbb{R}^{2n}$ *with the pairs of simple eigenvalues* $\pm i\omega_1,\ldots,\pm i\omega_n$.

(i) Assume that $\langle\alpha,\omega\rangle\neq 0$ *for any* $\alpha\in\mathbb{Z}^n\setminus\{0\}$, *where* $\langle\cdot,\cdot\rangle$ *denotes the scalar product in* $\mathbb{R}^n$, *and* $\omega=(\omega_1,\ldots,\omega_n)$. *Show that the polynomial* $\mathbf{N}$ *in Theorem 1.2 is of form*

$$\begin{aligned}\mathbf{N}(u) = {} & (A_1P_1(|A_1|^2,\ldots,|A_n|^2),\ldots,A_nP_n(|A_1|^2,\ldots,|A_n|^2),\\ & \overline{A_1P_1}(|A_1|^2,\ldots,|A_n|^2),\ldots,\overline{A_nP_n}(|A_1|^2,\ldots,|A_n|^2)),\end{aligned}$$

where the P_j, $j=1,\ldots,n$, *are complex-valued polynomials in their arguments such that* $P_j(0,\ldots,0)=0$.

(ii) *Set* $|\alpha_0| = \min\{|\alpha|\,;\,\langle\alpha,\omega\rangle = 0, \alpha\in\mathbb{Z}^n\setminus\{0\}\} < \infty$, *where* $|\alpha| = \sum_{j=1}^{n}|\alpha_j|$, *for* $\alpha = (\alpha_1,\ldots,\alpha_n)\in\mathbb{Z}^n$. *Show that the lowest order terms in the polynomial* $\mathbf{N}$ *in Theorem 1.2, which are not of the "standard form" obtained in the case (i), are of degree* $|\alpha_0|-1$.

In the remainder of this section, we give the normal forms in the cases $(i\omega)^2$, $0^2(i\omega)$, and 0^20^2. The proofs of the following results are given in Appendices C.2, C.3, and C.4. The first two proofs can be also found in [25], while the latter one can be found in [59]. We also refer to [20] for different proofs of the results in the cases $(i\omega)^2$ and 0^20^2.

Lemma 1.17 (**$(i\omega)^2$ normal form**) *Assume that the matrix* $\mathbf{L}$ *is of the form*

$$\mathbf{L} = \begin{pmatrix} i\omega & 1 & 0 & 0\\ 0 & i\omega & 0 & 0\\ 0 & 0 & -i\omega & 1\\ 0 & 0 & 0 & -i\omega \end{pmatrix},$$

where $\omega>0$, *in a basis of* $\mathbb{R}^4$ *in which a vector* $u\in\mathbb{R}^4$ *is represented by* $u = (A,B,\overline{A},\overline{B})$, *with* $A,B\in\mathbb{C}$. *Then the polynomial* $\mathbf{N}$ *in Theorem 1.2 is of the form*

$$\begin{aligned}\mathbf{N}(u) = {} & (AP(|A|^2, i(A\overline{B}-\overline{A}B)),\, BP(|A|^2, i(A\overline{B}-\overline{A}B)) + AQ(|A|^2, i(A\overline{B}-\overline{A}B)),\\ & \overline{AP}(|A|^2, i(A\overline{B}-\overline{A}B)),\, \overline{BP}(|A|^2, i(A\overline{B}-\overline{A}B)) + \overline{AQ}(|A|^2, i(A\overline{B}-\overline{A}B))),\end{aligned}$$

where P *and* Q *are complex-valued polynomials in their arguments, satisfying* $P(0,0) = Q(0,0) = 0$.

Lemma 1.18 (**$0^2(i\omega)$ normal form**) *Assume that the matrix* $\mathbf{L}$ *is of the form*

$$\mathbf{L} = \begin{pmatrix} 0 & 1 & 0 & 0\\ 0 & 0 & 0 & 0\\ 0 & 0 & i\omega & 0\\ 0 & 0 & 0 & -i\omega \end{pmatrix},$$

where $\omega>0$, *in a basis of* $\mathbb{R}^4$ *in which a vector* $u\in\mathbb{R}^4$ *is represented by* $u = (A,B,C,\overline{C})$, *with* $A,B\in\mathbb{R}$ *and* $C\in\mathbb{C}$. *Then the polynomial* $\mathbf{N}$ *in Theorem 1.2 is of the form*

$$\begin{aligned}N(u) = {} & (AP_0(A,|C|^2),\, BP_0(A,|C|^2) + P_1(A,|C|^2),\\ & CP_2(A,|C|^2),\, \overline{CP_2}(A,|C|^2)),\end{aligned}$$

where P_0 *and* P_1 *are real-valued polynomials, and* P_2 *is a complex-valued polynomial, satisfying*

$$P_0(0,0) = P_1(0,0) = P_2(0,0) = \frac{\partial P_1}{\partial A}(0,0) = 0.$$

Lemma 1.19 (0^20^2 **normal form)** *Assume that the matrix* $\mathbf{L}$ *is in Jordan form*

$$\mathbf{L} = \begin{pmatrix} 0 & 1 & 0 & 0 \\ 0 & 0 & 0 & 0 \\ 0 & 0 & 0 & 1 \\ 0 & 0 & 0 & 0 \end{pmatrix}$$

in a basis of $\mathbb{R}^4$ *in which a vector* $u \in \mathbb{R}^4$ *is represented by* $u = (A,B,C,D)$, *with* $A,B,C,D \in \mathbb{R}$. *Then the polynomial* $\mathbf{N}$ *in Theorem 1.2 is of the form*

$$\begin{aligned}\mathbf{N}(u) = (&AP_1(A,C,\widetilde{B}) + CP_2(A,C,\widetilde{B}), BP_1(A,C,\widetilde{B}) + DP_2(A,C,\widetilde{B}) + P_3(A,C),\\ &AP_4(A,C,\widetilde{B}) + CP_5(A,C,\widetilde{B}), BP_4(A,C,\widetilde{B}) + DP_5(A,C,\widetilde{B}) + P_6(A,C)),\end{aligned}$$

where $\widetilde{B} = BC - AD$, *and* P_1, P_2, P_3, P_4, P_5, *and* P_6 *are real-valued polynomials satisfying*

$$\begin{aligned}&P_1(0,0,0) = P_2(0,0,0) = P_4(0,0,0) = P_5(0,0,0) = P_3(0,0) = P_6(0,0) = 0,\\ &\frac{\partial}{\partial A}P_3(0,0) = \frac{\partial}{\partial C}P_3(0,0) = \frac{\partial}{\partial A}P_6(0,0) = \frac{\partial}{\partial C}P_6(0,0) = 0.\end{aligned}$$

Remark 1.20 *We shall discuss the normal form in the case* 0^4 *in Section 4.3.5, in Chapter 4, in the particular case when the system possesses a reversibility symmetry. The interested reader may find other normal forms in literature, as for example* 0^20^3 *in [25],* $(i\omega_1)^2(i\omega_2)$ *in [60, 93],* $(i\omega)^5$ *with spherical symmetry* $O(3)$ *in [70].*

3.2 Parameter-Dependent Normal Forms

3.2.1 *Main Result*

In the same framework as above, we are interested now in parameter-dependent equations of the form

$$\frac{du}{dt} = \mathbf{L}u + \mathbf{R}(u,\mu), \tag{2.1}$$

in which we assume that $\mathbf{L}$ and $\mathbf{R}$ satisfy the following hypothesis.

Hypothesis 2.1 *Assume that* $\mathbf{L}$ *and* $\mathbf{R}$ *in (2.1) have the following properties:*

(i) $\mathbf{L}$ *is a linear map in* $\mathbb{R}^n$;
(ii) for some $k \geq 2$, *there exist neighborhoods* $\mathscr{V}_u \subset \mathbb{R}^n$ *and* $\mathscr{V}_\mu \subset \mathbb{R}^m$ *of* 0 *such that* $\mathbf{R} \in \mathscr{C}^k(\mathscr{V}_u \times \mathscr{V}_\mu, \mathbb{R}^n)$ *and*

$$\mathbf{R}(0,0) = 0, \quad D_u\mathbf{R}(0,0) = 0.$$

In this situation we have the following result.

Theorem 2.2 (Normal form for perturbed vector fields) *Assume that Hypothesis 2.1 holds. Then for any positive integer p,* $2 \leq p \leq k$*, there exist neighborhoods* $\mathcal{V}_1$ *and* $\mathcal{V}_2$ *of 0 in* $\mathbb{R}^n$ *and* $\mathbb{R}^m$*, respectively, such that for any* $\mu \in \mathcal{V}_2$*, there is a polynomial* $\mathbf{\Phi}_\mu : \mathbb{R}^n \to \mathbb{R}^n$ *of degree p with the following properties:*

(i) The coefficients of the monomials of degree q in $\mathbf{\Phi}_\mu$ *are functions of* μ *of class* $\mathscr{C}^{k-q}$*, and*

$$\Phi_0(0) = 0, \quad D_u\Phi_0(0) = 0.$$

(ii) For $v \in \mathcal{V}_1$*, the polynomial change of variable*

$$u = v + \mathbf{\Phi}_\mu(v), \tag{2.2}$$

transforms equation (2.1) into the "normal form"

$$\frac{dv}{dt} = \mathbf{L}v + \mathbf{N}_\mu(v) + \rho(v,\mu), \tag{2.3}$$

and the following properties hold:

a. For any $\mu \in \mathcal{V}_2$*,* $\mathbf{N}_\mu$ *is a polynomial* $\mathbb{R}^n \to \mathbb{R}^n$ *of degree p, with coefficients depending upon* μ*, such that the coefficients of the monomials of degree q are of class* $\mathscr{C}^{k-q}$*, and*

$$\mathbf{N}_0(0) = 0, \quad D_v\mathbf{N}_0(0) = 0.$$

b. The equality

$$\mathbf{N}_\mu(e^{t\mathbf{L}^*}v) = e^{t\mathbf{L}^*}\mathbf{N}_\mu(v) \tag{2.4}$$

holds for all $(t,v) \in \mathbb{R} \times \mathbb{R}^n$ *and* $\mu \in \mathcal{V}_2$*.*

c. The map ρ *belongs to* $\mathscr{C}^k(\mathcal{V}_1 \times \mathcal{V}_2, \mathbb{R}^n)$*, and*

$$\rho(v,\mu) = o(\|v\|^p)$$

for all $\mu \in \mathcal{V}_2$*.*

We give the proof of this theorem in Appendix C.5. We point out that in most results on normal forms in the literature the normal form $\mathbf{N}_\mu$ is a polynomial in both v and μ, whereas here it is only a polynomial in v. To our knowledge, a proof of this latter type of result is not available in the literature.

Remark 2.3 *(i) As for Theorem 1.2, identity (2.4) is equivalent to the identity*

$$D_v\mathbf{N}_\mu(v)\mathbf{L}^*v = \mathbf{L}^*\mathbf{N}_\mu(v) \text{ for all } v \in \mathbb{R}^n, \ \mu \in \mathcal{V}_2.$$

(ii) Notice that the origin is not necessarily an equilibrium of (2.1) when $\mu \neq 0$*. Then* $\mathbf{N}_\mu(0)$ *is, in general, not 0, and the equality above shows that in this case*

$$\mathbf{N}_\mu(0) \in \ker \mathbf{L}^*.$$

(iii) In Theorem 2.2, the polynomials $\mathbf{\Phi}_\mu$ *and* $\mathbf{N}_\mu$ *have coefficients depending upon* μ*. The regularity with respect to* μ *of these coefficients decreases as the de-*

gree of the corresponding monomial increases. In applications, we actually compute the Taylor expansions of the coefficients of the polynomials Φ_μ *and* $\mathbf{N}_\mu$ *up to a needed degree in* μ *(see Section 3.2.3 below). Also notice, that the remainder* ρ *in (2.3) is uniformly estimated for* $\mu \in \mathscr{V}_2$. *This is sometimes useful when one is looking for the optimal behavior of certain solutions as* $t \to \pm\infty$.

(iv) *We can consider again the examples in Sections 3.1.2–3.1.4, now in the context of the parameter-dependent equation (2.1). In each case, we find that the parameter-dependent normal form polynomial* $\mathbf{N}_\mu$ *has the same structure as the unperturbed polynomial* $\mathbf{N}$*, but now with coefficients depending upon the parameter* μ.

3.2.2 Linear Normal Forms

An interesting particular case occurs when the map $\mathbf{R}(u,\mu)$ is linear in u, so that we have a linear equation

$$\frac{du}{dt} = \mathbf{L}u + \mathbf{R}_\mu u.$$

Assuming that $\mathbf{R}_0 = 0$, Hypothesis 2.1 is satisfied and the result in Theorem 2.2 holds. According to Remark C.1 in the Appendix C.5, the polynomial Φ_μ is of degree 1 in this case, so that we have a linear change of variables. The normal form is also linear,

$$\frac{dv}{dt} = (\mathbf{L} + \mathbf{N}_\mu)v,$$

in which the map $\mu \mapsto \mathbf{N}_\mu$ is of class $\mathscr{C}^{k-1}$ in a neighborhood of 0, and now

$$\mathbf{N}_\mu \mathbf{L}^* = \mathbf{L}^* \mathbf{N}_\mu. \tag{2.5}$$

This result was proved in [4] and is of particular interest since it gives a *smooth* unfolding of a linear map $\mathbf{L}$, which is, in general, not the case when one uses the classical transformation into Jordan form. For example, assume that $\mathbf{L}$ is not diagonalizable, but $\mathbf{L} + \mathbf{R}_\mu$ is diagonalizable for $\mu \neq 0$. Then the linear change of variables, which transforms $\mathbf{L} + \mathbf{R}_\mu$ into a diagonal matrix, is singular in $\mu = 0$.

Exercise 2.4 *Consider the* 3×3*-matrix*

$$\mathbf{L} = \begin{pmatrix} 0 & 1 & 0 \\ 0 & 0 & 0 \\ 0 & 0 & \lambda \end{pmatrix},$$

in which λ *is a real parameter, and consider a linear perturbation* $\mathbf{R}_\mu$ *depending smoothly upon* $\mu \in \mathbb{R}^m$*, such that* $\mathbf{R}_0 = 0$.

(i) *Assume that* $\lambda \neq 0$. *Show that there is a linear change of variables in* $\mathbb{R}^3$*, which is smooth in* μ *in a neighborhood of 0, such that the transformed matrix is of the form*

$$\begin{pmatrix} \alpha_\mu & 1 & 0 \\ \beta_\mu & \alpha_\mu & 0 \\ 0 & 0 & \lambda+\gamma_\mu \end{pmatrix},$$

where α_μ, β_μ, and γ_μ depend smoothly upon μ. Compute the first two leading order terms in the Taylor expansions in μ of the vectors in the basis $\{\zeta_1(\mu),\zeta_2(\mu),\zeta_3(\mu)\}$ of $\mathbb{R}^3$ consisting of generalized eigenvectors of the new matrix, which is the smooth continuation of the basis $\{\xi_1,\xi_2,\xi_3\}$ such that

$$\mathbf{L}\xi_1 = 0, \quad \mathbf{L}\xi_2 = \xi_1, \quad \mathbf{L}\xi_3 = 0.$$

Hint: *Use (2.5) to prove the first part. For the second part, use the dual basis $\{\xi_1^*,\xi_2^*,\xi_3^*\}$ such that*

$$\mathbf{L}^*\xi_1^* = \xi_2^*, \quad \mathbf{L}^*\xi_2^* = 0, \quad \mathbf{L}^*\xi_3^* = \lambda\xi_3^*, \quad \langle \xi_j,\xi_l^*\rangle = \delta_{kl},$$

and identify the different powers of μ in the identities

$$\begin{aligned}(\mathbf{L}+\mathbf{R}_\mu)\zeta_1(\mu) &= \alpha_\mu\zeta_1(\mu)+\beta_\mu\zeta_2(\mu),\\ (\mathbf{L}+\mathbf{R}_\mu)\zeta_2(\mu) &= \zeta_1(\mu)+\alpha_\mu\zeta_2(\mu),\\ (\mathbf{L}+\mathbf{R}_\mu)\zeta_3(\mu) &= (\lambda+\gamma(\mu))\zeta_3(\mu).\end{aligned}$$

(ii) Assume that $\lambda = 0$. Show that there is a linear change of variables in $\mathbb{R}^3$, which is smooth in μ in a neighborhood of 0, such that the transformed matrix is of the form

$$\begin{pmatrix} \alpha_\mu & 1 & 0 \\ \beta_\mu & \alpha_\mu & \varepsilon_\mu \\ \delta_\mu & 0 & \gamma_\mu \end{pmatrix},$$

where α_μ, β_μ, γ_μ, δ_μ, and ε_μ depend smoothly upon μ. Describe a method for computing the Taylor expansions in μ of the vectors in the basis $\{\zeta_1(\mu),\zeta_2(\mu),\zeta_3(\mu)\}$ of $\mathbb{R}^3$ consisting of generalized eigenvectors of the new matrix, which is the smooth continuation of the basis $\{\xi_1,\xi_2,\xi_3\}$ above. Show that in general the eigenvalues of the transformed matrix do not depend smoothly upon μ, even for a single parameter $\mu\in\mathbb{R}$.

3.2.3 Derivation of the Parameter-Dependent Normal Form

In this section we give a method of computing the Taylor expansions of the polynomials $\boldsymbol{\Phi}_\mu$ and $\mathbf{N}_\mu$ given by Theorem 2.2. We have already used this method in the particular case of the Hopf bifurcation in Section 1.2.1, Chapter 1, and without parameters in the example of a 0^2 normal form in Section 3.1.2.

We write the Taylor expansion of $\mathbf{R}$ and rewrite polynomials $\boldsymbol{\Phi}_\mu$ and $\mathbf{N}_\mu$ as follows:

$$\begin{aligned}\mathbf{R}(u,\mu) &= \sum_{1\le q+l\le p} \mathbf{R}_{ql}(u^{(q)},\mu^{(l)}) + o((\|u\|+\|\mu\|)^p), \quad \mathbf{R}_{10}=0,\\ \boldsymbol{\Phi}_\mu(v) &= \sum_{1\le q+l\le p} \boldsymbol{\Phi}_{ql}(v^{(q)},\mu^{(l)}) + o((\|v\|+\|\mu\|)^p), \quad \boldsymbol{\Phi}_{10}=0,\\ \mathbf{N}_\mu(v) &= \sum_{1\le q+l\le p} \mathbf{N}_{ql}(v^{(q)},\mu^{(l)}) + o((\|v\|+\|\mu\|)^p), \quad \mathbf{N}_{10}=0,\end{aligned}$$

where $\mathbf{R}_{ql}$, Φ_{ql}, and $\mathbf{N}_{ql}$ are $(q+l)$-linear maps on $(\mathbb{R}^n)^q \times (\mathbb{R}^m)^l$, $u^{(q)} = (u,\dots,u) \in (\mathbb{R}^n)^q$, and $\mu^{(l)} = (\mu,\dots,\mu) \in (\mathbb{R}^m)^l$. Furthermore, $\mathbf{R}_{ql}(\cdot,\mu^{(l)})$ and $\mathbf{R}_{ql}(u^{(q)},\cdot)$ are q-linear symmetric and l-linear symmetric, respectively, and similar properties hold for Φ_{ql}, and $\mathbf{N}_{ql}$. Notice that the terms $o((\|v\|+\|\mu\|)^p)$ in the expansions of Φ and $\mathbf{N}$ come from the fact that these are polynomials in v with coefficients that are functions of μ, of class $\mathscr{C}^{k-q}$ for the monomials of degree q.

Now we proceed as in the proof of Theorem 1.2. Differentiating (2.2) with respect to t and replacing du/dt and dv/dt from (2.1) and (2.3), respectively, we obtain the identity

$$\mathscr{A}_{\mathbf{L}}\Phi_\mu(v) + \mathbf{N}_\mu(v) = \Pi_p\left(\mathbf{R}(v+\Phi_\mu(v),\mu) - D_v\Phi_\mu(v)\mathbf{N}_\mu(v)\right). \tag{2.6}$$

Here $\mathscr{A}_{\mathbf{L}}$ is the homological operator given by (1.10), and Π_p represents the linear map which associates to a map of class $\mathscr{C}^p$ the polynomial of degree p in its Taylor expansion. Identifying the coefficients of the monomials of degree q in u and of degree 0 in μ leads to

$$\begin{aligned}
\mathscr{A}_{\mathbf{L}}\Phi_{20} + \mathbf{N}_{20} &= \mathbf{R}_{20},\\
\mathscr{A}_{\mathbf{L}}\Phi_{30} + \mathbf{N}_{30} &= \mathbf{Q}_{30},
\end{aligned}$$

with

$$\mathbf{Q}_{30}(v^{(3)}) = \mathbf{R}_{30}(v^{(3)}) + 2\mathbf{R}_{20}(v,\Phi_{20}(v)) - 2\Phi_{20}(v,\mathbf{N}_{20}(v^{(2)}))$$

for $q=2$ and $q=3$, respectively, and similar equalities hold for $q \geq 4$, just as in (1.8). Then the equation for $\mathscr{A}_{\mathbf{L}}\Phi_{q0} + \mathbf{N}_{q0}$ only contains in the right hand side terms involving $\Phi_{q'0}$ and $\mathbf{N}_{q'0}$, with $q' \leq q-1$, so that we can successively determine Φ_{q0} and $\mathbf{N}_{q0}$.

Next, we consider the monomials of degree q in u and of degree 1 in μ, and obtain

$$\begin{aligned}
\mathscr{A}_{\mathbf{L}}\Phi_{01} + \mathbf{N}_{01} &= \mathbf{R}_{01},\\
\mathscr{A}_{\mathbf{L}}\Phi_{11}(v,\mu) + \mathbf{N}_{11}(v,\mu) &= \mathbf{R}_{11}(v,\mu) - 2\Phi_{20}(v,\mathbf{N}_{01}(\mu))
\end{aligned}$$

for $q=0$ and $q=2$, respectively, and

$$\mathscr{A}_{\mathbf{L}}\Phi_{q1}(v^{(q)},\mu) + \mathbf{N}_{q1}(v^{(q)},\mu) = \mathbf{Q}_{q1}(v^{(q)},\mu)$$

for $q \geq 2$, where $\mathbf{Q}_{q1}$ depends upon $\Phi_{q'1}, \mathbf{N}_{q'1}, \Phi_{q''0}, \mathbf{N}_{q''0}$ such that $q' \leq q-1$ and $q'' \leq q+1$. Consequently, once we have found $(\Phi_{q0},\mathbf{N}_{q0})$, $q=2,\dots,p$, we can determine $(\Phi_{q1},\mathbf{N}_{q1})$ by successively solving the equations above for $q = 0,1,\dots,p-1$. More generally, we obtain

$$\mathscr{A}_{\mathbf{L}}\Phi_{ql}(v^{(q)},\mu^{(l)}) + \mathbf{N}_{ql}(v^{(q)},\mu^{(l)}) = \mathbf{Q}_{ql}(v^{(q)},\mu^{(l)}),$$

which is of the same form as above, with $\mathbf{Q}_{ql}$ depending upon $\Phi_{q'l'}$ and $\mathbf{N}_{q'l'}$ either such that $q'+l' \leq q+l-1$ with $l' \leq l$, or such that $q'+l' = q+l$ with $l' \leq l-1$. This shows that once we found $(\Phi_{qj},\mathbf{N}_{qj})$, for $q+j \leq p$, $j=0,1,\dots,l$, then we can determine $(\Phi_{q'l'}, \mathbf{N}_{q'l'})$ for $l' = l+1$ and $q' \leq p-l-1$. We indicate in Figure 2.1 the way in which $(\Phi_{ql}, \mathbf{N}_{ql})$ depend upon $(\Phi_{q'l'}, \mathbf{N}_{q'l'})$.

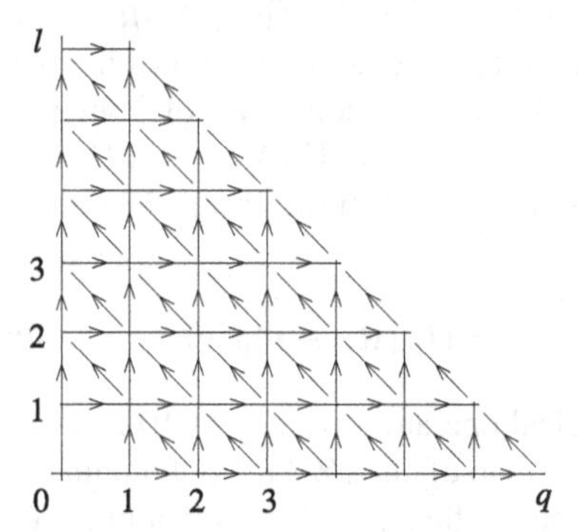

Fig. 2.1 Plot of the indices (q,l) of $(\Phi_{ql}, \mathbf{N}_{ql})$. The arrows indicate the dependence of $(\Phi_{ql}, \mathbf{N}_{ql})$ at the position (q,l) upon $(\Phi_{q'l'}, \mathbf{N}_{q'l'})$ at the position (q',l').

3.2.4 Example: 0^2 *Normal Form with Parameters*

Consider the second order differential equation

$$u'' = \mu_0 + \mu_1 u + \mu_2 u' + \alpha u^2 + \beta u u' + \gamma (u')^2,$$

where α, β, γ are real constants, and μ_0, μ_1, μ_2 small parameters. Notice that for $\mu_0 = \mu_1 = \mu_2 = 0$ this is precisely the equation (1.17) for which the normal form has been computed in Section 3.1.2. Therefore, it remains to compute the terms in the normal form involving the three small parameters μ_0, μ_1, and μ_2.

Normal Form

We set $U = (u,v)$ and $\mu = (\mu_0, \mu_1, \mu_2) \in \mathbb{R}^3$, so that the equation becomes

$$\frac{dU}{dt} = \mathbf{L}U + \mathbf{R}(U,\mu), \qquad \mathbf{R}(U,\mu) = \mathbf{R}_{01}(\mu) + \mathbf{R}_2(U,U) + \mathbf{R}_{11}(U,\mu), \quad (2.7)$$

where $\mathbf{L}$ and $\mathbf{R}_2$ are as in (1.18), and

$$\mathbf{R}_{01}(\mu) = \begin{pmatrix} 0 \\ \mu_0 \end{pmatrix}, \quad \mathbf{R}_{11}(U,\mu) = \begin{pmatrix} 0 \\ \mu_1 u + \mu_2 v \end{pmatrix}.$$

We are interested in computing the normal form of this system up to terms of order 2, so that it is enough to consider a polynomial Φ_μ of degree 2,

$$\Phi_\mu(A,B) = \Phi_{001}(\mu) + A\Phi_{101}(\mu) + B\Phi_{011}(\mu) + A^2\Phi_{200} + AB\Phi_{110} + B^2\Phi_{020},$$

where $\Phi_{ij1} : \mathbb{R}^3 \to \mathbb{R}^2$ are linear maps. Since for $\mu = 0$ the result is the same as the one found for the equation (1.17) in Section 3.1.2, it is clear that here

$$\Phi_{200} = \Phi_{20}, \quad \Phi_{110} = \Phi_{11}, \quad \Phi_{020} = \Phi_{02},$$

where

$$\Phi_{20} = \frac{\gamma}{2}\zeta_0, \quad \Phi_{11} = \gamma\zeta_1, \quad \Phi_{02} = 0$$

have been computed in Section 3.1.2. According to Lemma 1.9 and Remark 1.10(ii), and taking into account the result found for $\mu = 0$ in Section 3.1.2, it follows that the change of variables

$$U = A\zeta_0 + B\zeta_1 + \Phi_\mu(A,B), \tag{2.8}$$

where

$$\zeta_0 = \begin{pmatrix} 1 \\ 0 \end{pmatrix}, \quad \zeta_1 = \begin{pmatrix} 0 \\ 1 \end{pmatrix},$$

transforms the system (2.7) into the normal form

$$\begin{aligned} \frac{dA}{dt} &= B \\ \frac{dB}{dt} &= \alpha_1(\mu) + \alpha_2(\mu)A + \alpha_3(\mu)B + \alpha A^2 + \beta AB \\ &\quad + O(|\mu|^2 + |\mu|(|A|+|B|)^2 + (|A|+|B|)^3), \end{aligned} \tag{2.9}$$

in which $\alpha_j : \mathbb{R}^3 \to \mathbb{R}$, $j = 0,1,2$, are linear maps.

Computation of α_0, α_1, *and* α_2

We proceed as indicated in Section 3.2.3, and also as in the previous computations. We substitute the change of variables (2.8) into the system (2.7), and then replace the derivatives dA/dt and dB/dt from (2.9). In the resulting equality we now identify the terms of orders $O(\mu)$, $O(\mu A)$, and $O(\mu B)$, which gives the equations

$$\alpha_1(\mu)\zeta_1 = \mathbf{L}\Phi_{001}(\mu) + \mathbf{R}_{01}(\mu), \tag{2.10}$$

$$\alpha_2(\mu)\zeta_1 + \alpha_1(\mu)\Phi_{110} = \mathbf{L}\Phi_{101}(\mu) + \mathbf{R}_{11}(\zeta_0,\mu) + 2\mathbf{R}_2(\zeta_0, \Phi_{001}(\mu)), \tag{2.11}$$

$$\alpha_3(\mu)\zeta_1 + 2\alpha_1(\mu)\Phi_{020} + \Phi_{101}(\mu) = \mathbf{L}\Phi_{011}(\mu) + \mathbf{R}_{11}(\zeta_1,\mu) + 2\mathbf{R}_2(\zeta_1, \Phi_{001}(\mu)). \tag{2.12}$$

Using the fact that the range $\operatorname{im}\mathbf{L}$ of $\mathbf{L}$ is given by $\operatorname{im}\mathbf{L} = \{(u,0); u \in \mathbb{R}\} \subset \mathbb{R}^2$ and that the kernel $\ker\mathbf{L}$ is spanned by ζ_0, we can solve these equations and determine α_j from the corresponding solvability conditions.

Solving these three equations we find, successively,

$$\alpha_1(\mu) = \mu_0, \quad \Phi_{001}(\mu) = \phi_{001}(\mu)\zeta_0,$$

$$\alpha_2(\mu) = -\gamma\mu_0 + \mu_1 + 2\alpha\phi_{001}(\mu), \quad \Phi_{101}(\mu) = \phi_{101}(\mu)\zeta_0,$$

and

$$\alpha_3(\mu) = \mu_2 + \beta\phi_{001}(\mu), \quad \Phi_{011}(\mu) = \phi_{101}(\mu)\zeta_1 + \phi_{011}(\mu)\zeta_0,$$

in which $\phi_{001}, \phi_{101}, \phi_{011} : \mathbb{R}^3 \to \mathbb{R}$ are arbitrary linear maps. A simple choice is of course $\phi_{001} = \phi_{101} = \phi_{011} = 0$, which then gives

$$\alpha_1(\mu) = \mu_0, \quad \alpha_2(\mu) = -\gamma\mu_0 + \mu_1, \quad \alpha_3(\mu) = \mu_2.$$

Alternatively, if $\beta \neq 0$ we may choose $\phi_{001}(\mu)$ such that $\alpha_3(\mu) = 0$, i.e.,

$$\phi_{001}(\mu) = -\frac{\mu_2}{\beta},$$

which gives

$$\alpha_1(\mu) = \mu_0, \quad \alpha_2(\mu) = -\gamma\mu_0 + \mu_1 - \frac{2\alpha}{\beta}\mu_2, \quad \alpha_3(\mu) = 0,$$

whereas if $\alpha \neq 0$ we may choose $\phi_{001}(\mu)$ such that $\alpha_2(\mu) = 0$, i.e.,

$$\phi_{001}(\mu) = \frac{1}{2\alpha}(\gamma\mu_0 - \mu_1),$$

which gives

$$\alpha_1(\mu) = \mu_0, \quad \alpha_2(\mu) = 0, \quad \alpha_3(\mu) = \mu_2 + \frac{\beta}{2\alpha}(\gamma\mu_0 - \mu_1).$$

Actually, these choices can be made in general for a Takens–Bogdanov bifurcation (see Section 3.4.4).

3.3 Symmetries and Reversibility

In this section, we consider the particular cases where the equation is equivariant under the action of a symmetry and where it possesses a reversibility symmetry. In both cases we show that the symmetry is inherited by the normal form. We state our results for equation (1.1), but the same results also hold for the parameter-dependent equation (2.1).

3.3.1 Equivariant Vector Fields

We start with the case of an equation that is equivariant under the action of a linear symmetry. More precisely, we make the following assumption.

Hypothesis 3.1 (Equivariant vector field) *Assume that there exists an isometry* $\mathbf{T} \in \mathscr{L}(\mathbb{R}^n)$ *which commutes with the vector field in the equation (1.1),*

$$\mathbf{TL}u = \mathbf{LT}u, \quad \mathbf{TR}(u) = \mathbf{R}(\mathbf{T}u) \text{ for all } u \in \mathbb{R}^n.$$

In this situation, we prove the following result.

Theorem 3.2 (Equivariant normal forms) *Under the assumptions of Theorem 1.2, further assume that Hypothesis 3.1 holds. Then the polynomials* $\boldsymbol{\Phi}$ *and* $\mathbf{N}$ *in Theorem 1.2 commute with* $\mathbf{T}$.

Proof We consider the linear operator $\mathbf{J}$ in the space $\mathscr{H}$ of polynomials $\boldsymbol{\Phi} : \mathbb{R}^n \to \mathbb{R}^n$ of degree p defined through

$$\mathbf{J}\boldsymbol{\Phi} = \mathbf{T}^{-1}\boldsymbol{\Phi} \circ \mathbf{T} \text{ for all } \boldsymbol{\Phi} \in \mathscr{H}.$$

Then notice that the equality $\mathbf{J}\boldsymbol{\Phi} = \boldsymbol{\Phi}$ is equivalent to the fact that $\boldsymbol{\Phi}$ commutes with $\mathbf{T}$ and that $\mathbf{J}$ leaves invariant the subspace $\mathbf{H}_q$ of homogeneous polynomials of degree q.

Now, we go back to equation (1.9) in the proof of Theorem 1.2, which determines the homogeneous parts $\boldsymbol{\Phi}_q$ and $\mathbf{N}_q$ of degree q in the polynomials $\boldsymbol{\Phi}$ and $\mathbf{N}$. At each step q, we have to solve an equation of the form

$$\mathscr{A}_{\mathbf{L}}\boldsymbol{\Phi} = \mathbf{Q} - \mathbf{N}, \tag{3.1}$$

in which $\mathbf{Q}$ is known, and we have dropped the subscripts q for notational simplicity. We prove by induction in q that there is a solution $(\boldsymbol{\Phi}, \mathbf{N})$ of (3.1) satisfying

$$\mathbf{J}\boldsymbol{\Phi} = \boldsymbol{\Phi}, \quad \mathbf{JN} = \mathbf{N}, \tag{3.2}$$

when $\mathbf{Q}$ satisfies

$$\mathbf{JQ} = \mathbf{Q}. \tag{3.3}$$

Assuming that Hypothesis 3.1 holds, it is straightforward to check that (3.3) holds for $q = 2$. We assume now that (3.3) is satisfied at degree q and show that there is a solution $(\boldsymbol{\Phi}, \mathbf{N})$ of (3.1) satisfying (3.2), and then, that in the equation at degree $q+1$ the term $\mathbf{Q}$ also satisfies (3.3). According to the proof of Theorem 1.2, the equation (3.1) has a unique solution with the property

$$(\boldsymbol{\Phi}, \mathbf{N}) \in (\ker \mathscr{A}_{\mathbf{L}})^{\perp} \times \ker \mathscr{A}_{\mathbf{L}^*}.$$

We show that this solution has the required property.

Recall that $\mathbf{T}$ is an isometry, so that $\mathbf{T}^{-1} = \mathbf{T}^*$. Since $\mathbf{T}$ commutes with $\mathbf{L}$ we then find that $\mathbf{T}^{-1}$ commutes with $\mathbf{L}$, and both $\mathbf{T}$ and $\mathbf{T}^{-1}$ commute with the adjoint $\mathbf{L}^*$. This further implies that $\mathbf{J}$ commutes with $\mathscr{A}_{\mathbf{L}}$ and $\mathscr{A}_{\mathbf{L}^*}$,

$$\begin{aligned}(\mathbf{J}\mathscr{A}_{\mathbf{L}}\Phi)(u) &= \mathbf{T}^{-1}D\Phi(\mathbf{T}u)\mathbf{L}\mathbf{T}u - \mathbf{T}^{-1}\mathbf{L}\Phi(\mathbf{T}u)\\ &= D\mathbf{T}^{-1}\Phi(\mathbf{T}u)\mathbf{T}\mathbf{L}u - \mathbf{L}\mathbf{T}^{-1}\Phi(\mathbf{T}u) = (\mathscr{A}_{\mathbf{L}}\mathbf{J}\Phi)(u)\end{aligned}$$

and

$$\begin{aligned}(\mathbf{J}\mathscr{A}_{\mathbf{L}^*}\Phi)(u) &= \mathbf{T}^{-1}D\Phi(\mathbf{T}u)\mathbf{L}^*\mathbf{T}u - \mathbf{T}^{-1}\mathbf{L}^*\Phi(\mathbf{T}u)\\ &= D\mathbf{T}^{-1}\Phi(\mathbf{T}u)\mathbf{T}\mathbf{L}^*u - \mathbf{L}^*\mathbf{T}^{-1}\Phi(\mathbf{T}u) = (\mathscr{A}_{\mathbf{L}^*}\mathbf{J}\Phi)(u).\end{aligned}$$

In particular, this shows that the subspaces $\mathrm{im}\,(\mathscr{A}_{\mathbf{L}^*}) = (\ker\mathscr{A}_{\mathbf{L}})^{\perp}$ and $\ker\mathscr{A}_{\mathbf{L}^*}$ are invariant under the action of $\mathbf{J}$.

Next, consider the unique solution $(\Phi,\mathbf{N})$ of (3.1), constructed in the proof of Theorem 1.2, satisfying

$$(\Phi,\mathbf{N}) \in (\ker\mathscr{A}_{\mathbf{L}})^{\perp} \times \ker\mathscr{A}_{\mathbf{L}^*}.$$

Applying $\mathbf{J}$ to the equation (3.1) and taking into account the fact that $\mathbf{J}$ commutes with $\mathscr{A}_{\mathbf{L}}$, and that $\mathbf{J}\mathbf{Q} = \mathbf{Q}$, we find that $(\mathbf{J}\Phi,\mathbf{J}\mathbf{N})$ is also a solution of (3.1). Furthermore, since $(\ker\mathscr{A}_{\mathbf{L}})^{\perp}$ and $\ker\mathscr{A}_{\mathbf{L}^*}$ are invariant under the action of $\mathbf{J}$, we have

$$(\mathbf{J}\Phi,\mathbf{J}\mathbf{N}) \in (\ker\mathscr{A}_{\mathbf{L}})^{\perp} \times \ker\mathscr{A}_{\mathbf{L}^*}.$$

The uniqueness of the solution $(\Phi,\mathbf{N})$, now implies that $\mathbf{J}\Phi = \Phi$ and $\mathbf{J}\mathbf{N} = \mathbf{N}$. Furthermore, from the formula for $\mathbf{Q}$ in the proof of Theorem 1.2 it is now easy to check that the term $\mathbf{Q}$ in the equation at degree $q+1$ satisfies $\mathbf{J}\mathbf{Q} = \mathbf{Q}$, which completes the proof. □

3.3.2 Reversible Vector Fields

Next, we consider the case of reversible equations for which we assume that the following assumptions are satisfied.

Hypothesis 3.3 (Reversible vector field) *Assume that there exists an isometry* $\mathbf{S} \in \mathscr{L}(\mathbb{R}^n)$, *with*

$$\mathbf{S}^2 = \mathbb{I}, \quad \mathbf{S} \neq \mathbb{I},$$

and which anticommutes with the vector field in (1.1),

$$\mathbf{S}\mathbf{L}u = -\mathbf{L}\mathbf{S}u, \quad \mathbf{S}\mathbf{R}(u) = -\mathbf{R}(\mathbf{S}u) \textit{ for all } u \in \mathbb{R}^n.$$

In this case we prove the following result.

Theorem 3.4 (Reversible normal forms) *Under the assumptions of Theorem 1.2, further assume that Hypothesis (3.3) holds. Then the polynomial $\boldsymbol{\Phi}$ in Theorem 1.2 commutes with* $\mathbf{S}$, *whereas the polynomial* $\mathbf{N}$ *anticommutes with* $\mathbf{S}$.

Proof As in the proof of Theorem 3.2, we consider the linear operator $\mathbf{J}$ in the space $\mathscr{H}$ of polynomials $\boldsymbol{\Phi} : \mathbb{R}^n \to \mathbb{R}^n$ of degree p, defined through

$$\mathbf{J}\boldsymbol{\Phi} = \mathbf{S}\boldsymbol{\Phi} \circ \mathbf{S} \text{ for all } \boldsymbol{\Phi} \in \mathscr{H}.$$

Recall that here $\mathbf{S} = \mathbf{S}^{-1}$, so that $\boldsymbol{\Phi}$ commutes (resp., anticommutes) with $\mathbf{S}$ if $\mathbf{J}\boldsymbol{\Phi} = \boldsymbol{\Phi}$ (resp., $\mathbf{J}\boldsymbol{\Phi} = -\boldsymbol{\Phi}$). In addition, we have that $\mathbf{J}^2 = \mathbb{I}$. By arguing as in the proof of Theorem 3.2, from the fact that $\mathbf{S}$ anticommutes with $\mathbf{L}$, we obtain here that

$$\mathbf{J}\mathscr{A}_{\mathbf{L}} = -\mathscr{A}_{\mathbf{L}}\mathbf{J}, \quad \mathbf{J}\mathscr{A}_{\mathbf{L}^*} = -\mathscr{A}_{\mathbf{L}^*}\mathbf{J}, \tag{3.4}$$

which further implies that the subspaces $\operatorname{im}\mathscr{A}_{\mathbf{L}^*} = (\ker\mathscr{A}_{\mathbf{L}})^{\perp}$ and $\ker\mathscr{A}_{\mathbf{L}^*}$ are invariant under the action of $\mathbf{J}$.

Now, we consider again the equation

$$\mathscr{A}_{\mathbf{L}}\boldsymbol{\Phi} = \mathbf{Q} - \mathbf{N}, \tag{3.5}$$

which determines the homogeneous parts $\boldsymbol{\Phi}_q$ and $\mathbf{N}_q$ of degree q in the polynomials $\boldsymbol{\Phi}$ and $\mathbf{N}$, as in the proof of Theorem 1.2. We proceed now as in the proof of Theorem 3.2, and solve this equation by induction in q. Assuming that Hypothesis 3.3 holds, we have that

$$\mathbf{J}\mathbf{Q} = -\mathbf{Q} \tag{3.6}$$

holds for $q = 2$. We assume that this equality holds for $\mathbf{Q}$ at degree q, and now show that the unique solution $(\boldsymbol{\Phi}, \mathbf{N})$ of (3.5) satisfying

$$(\boldsymbol{\Phi}, \mathbf{N}) \in (\ker\mathscr{A}_{\mathbf{L}})^{\perp} \times \ker\mathscr{A}_{\mathbf{L}^*},$$

from the proof of Theorem 1.2, now also satisfies

$$\mathbf{J}\boldsymbol{\Phi} = \boldsymbol{\Phi}, \quad \mathbf{J}\mathbf{N} = -\mathbf{N},$$

which further implies that (3.6) holds for $\mathbf{Q}$ at degree $q+1$. Indeed, applying $\mathbf{J}$ to the equation (3.5) and taking into account the fact that $\mathbf{J}$ anticommutes with $\mathscr{A}_{\mathbf{L}}$, and that $\mathbf{J}\mathbf{Q} = -\mathbf{Q}$, we find that $(\mathbf{J}\boldsymbol{\Phi}, -\mathbf{J}\mathbf{N})$ is also a solution of (3.5). Furthermore, since $(\ker\mathscr{A}_{\mathbf{L}})^{\perp}$ and $\ker\mathscr{A}_{\mathbf{L}^*}$ are invariant under the action of $\mathbf{J}$, we have

$$(\mathbf{J}\boldsymbol{\Phi}, -\mathbf{J}\mathbf{N}) \in (\ker\mathscr{A}_{\mathbf{L}})^{\perp} \times \ker\mathscr{A}_{\mathbf{L}^*}.$$

The uniqueness of the solution $(\boldsymbol{\Phi}, \mathbf{N})$ now implies that $\mathbf{J}\boldsymbol{\Phi} = \boldsymbol{\Phi}$ and $\mathbf{J}\mathbf{N} = -\mathbf{N}$, which proves the result in the theorem. □

3.3.3 Example: van der Pol System

Consider the van der Pol system [104],

$$\begin{aligned} u_1' &= \mu u_1 - u_2 - u_1^3 \\ u_2' &= u_1, \end{aligned}$$

in which μ is a small parameter. (This system models an electrical circuit with a triode vacuum tube, nowadays replaced by a transistor.) Notice that the system is invariant under the reflection $(u_1,u_2) \mapsto -(u_1,u_2)$.

Normal Form

We set $U = (u_1,u_2)$, so that the system is of the form

$$\frac{dU}{dt} = \mathbf{L}U + \mathbf{R}(U,\mu), \qquad \mathbf{R}(U,\mu) = \mu\mathbf{R}_{11}(U) + \mathbf{R}_{30}(U,U,U), \tag{3.7}$$

where

$$\mathbf{L} = \begin{pmatrix} 0 & -1 \\ 1 & 0 \end{pmatrix}, \quad \mathbf{R}_{11}(U) = \begin{pmatrix} u_1 \\ 0 \end{pmatrix}, \quad \mathbf{R}_{30}(U,V,W) = \begin{pmatrix} -u_1v_1w_1 \\ 0 \end{pmatrix}.$$

Due to the reflection invariance mentioned above, the system (3.7) is equivariant under the action of

$$\mathbf{T} = -\mathbb{I}.$$

The linear map $\mathbf{L}$ has a pair of complex conjugated eigenvalues $\pm i$, with associated eigenvectors

$$\zeta = \begin{pmatrix} 1 \\ -i \end{pmatrix}, \quad \overline{\zeta} = \begin{pmatrix} 1 \\ i \end{pmatrix}.$$

This implies that $\mu = 0$ is a bifurcation point, at which we expect a Hopf bifurcation to occur. We are interested in computing the normal form of this system up to terms of order 3, taking into account the equivariance of the system under the action of $\mathbf{T}$.

We consider the change of variables

$$U = A\zeta + \overline{A\zeta} + \Phi_\mu(A,\overline{A}),$$

with $A(t) \in \mathbb{C}$ and Φ_μ a polynomial of degree 3, since we are interested in the normal form up to terms of order 3. According to the result in Lemma 1.7 and Theorem 3.2 there exists a polynomial Φ_μ which commutes with $\mathbf{T}$ and such that the system is transformed into the normal form

$$\frac{dA}{dt} = a\mu A + bA|A|^2 + O(\mu^2|A| + |\mu||A|^3| + |A|^5).$$

Since $\boldsymbol{\Phi}_\mu$ commutes with $\mathbf{T}$, it follows that $\boldsymbol{\Phi}_\mu$ is an odd polynomial, hence

$$\boldsymbol{\Phi}_\mu(A,\overline{A}) = \mu A\boldsymbol{\Phi}_{101} + \mu\overline{A}\boldsymbol{\Phi}_{011} + A^3\boldsymbol{\Phi}_{300} + A^2\overline{A}\boldsymbol{\Phi}_{210} + A\overline{A}^2\boldsymbol{\Phi}_{120} + \overline{A}^3\boldsymbol{\Phi}_{030}.$$

Computation of the Coefficients a and b

We proceed as in the computation of the Hopf bifurcation in Section 1.2.1, which leads for a general Hopf bifurcation to the system (2.22)–(2.27) in Chapter 1. Here, due to the equivariance under $\mathbf{T}$, implying in particular that $\boldsymbol{\Phi}_\mu$ is an odd polynomial, several terms in this calculation vanish, so that we find the system

$$a\zeta + (i-\mathbf{L})\boldsymbol{\Phi}_{101} = \mathbf{R}_{11}(\zeta) \tag{3.8}$$

$$(3i-\mathbf{L})\boldsymbol{\Phi}_{300} = \mathbf{R}_{30}(\zeta,\zeta,\zeta) \tag{3.9}$$

$$b\zeta + (i-\mathbf{L})\boldsymbol{\Phi}_{210} = 3\mathbf{R}_{30}(\zeta,\zeta,\overline{\zeta}), \tag{3.10}$$

instead of the general system (2.22)–(2.27) in Chapter 1. Now, the coefficients a and b are easily computed from the solvability conditions for the equations (3.8) and (3.10). Recall that these conditions are orthogonality conditions on the kernel of the adjoint matrix, namely,

$$(i-\mathbf{L})^* = -i-\mathbf{L}^* = -i+\mathbf{L},$$

e.g., see Section 1.2.1, which is here one-dimensional and spanned by

$$\zeta^* = \frac{1}{2}\begin{pmatrix} 1 \\ -i \end{pmatrix}.$$

This vector is chosen such that $\langle\zeta,\zeta_*\rangle = 1$, and then the solvability conditions lead to

$$a = \langle\mathbf{R}_{11}(\zeta),\zeta^*\rangle = \frac{1}{2}, \quad b = \langle 3\mathbf{R}_{30}(\zeta,\zeta,\overline{\zeta}),\zeta^*\rangle = -\frac{3}{2}.$$

Notice that $b < 0$, which implies that we have a *supercritical Hopf bifurcation*. Since $a > 0$ the branch of stable periodic solutions bifurcates for $\mu > 0$.

Exercise 3.5 *Compute the higher orders terms and show that*

$$\begin{aligned}\boldsymbol{\Phi}_\mu(A,\overline{A}) = {}& \mu A\begin{pmatrix} 0 \\ 1/2 \end{pmatrix} + A^3\begin{pmatrix} 3i/8 \\ 1/8 \end{pmatrix} + A^2\overline{A}\begin{pmatrix} 0 \\ -3/2 \end{pmatrix} \\ & + \overline{A}^3\begin{pmatrix} -3i/8 \\ 1/8 \end{pmatrix} + A\overline{A}^2\begin{pmatrix} 0 \\ -3/2 \end{pmatrix} + O(|A|^5 + |\mu||A|^3 + |\mu|^2|A|),\end{aligned}$$

and that the normal form is

$$\frac{dA}{dt} = \left(i + \frac{1}{2}\mu - \frac{i}{8}\mu^2\right)A - \frac{3}{2}(1-i\mu)A|A|^2 - \frac{63i}{16}A|A|^4 + h.o.t..$$

Remark 3.6 *In Chapter 4 we give examples of reversible bifurcations for which we apply the result in Theorem 3.4.*

3.4 Normal Forms for Reduced Systems on Center Manifolds

Consider an infinite-dimensional system of the form

$$\frac{du}{dt} = \mathbf{L}u + \mathbf{R}(u,\mu), \tag{4.1}$$

which satisfies the assumptions in center manifold Theorem 3.3 in Chapter 2. Then the reduced system is of the form (2.1) and satisfies Hypothesis 2.1, so that we can apply normal form Theorem 2.2. We show in this section how to compute the normal form of the reduced system directly from the infinite-dimensional, *without computing the reduced system*, as this was already done in the example in Section 2.4.4 in Chapter 2. Of course, this is the most efficient way of computation in applications.

3.4.1 Computation of Center Manifolds and Normal Forms

Recall that the center manifold theorem gives solutions of the form

$$u = u_0 + \Psi(u_0,\mu),$$

with $u_0 \in \mathscr{E}_0$ and $\Psi(u_0,\mu) \in \mathscr{Z}_h$. Then the normal form theorem applied to the reduced system for u_0 in the finite-dimensional subspace $\mathscr{E}_0$ shows that

$$u_0 = v_0 + \Phi_\mu(v_0),$$

which leads to the normal form

$$\frac{dv_0}{dt} = \mathbf{L}_0 v_0 + \mathbf{N}_\mu(v_0) + \rho(v_0,\mu). \tag{4.2}$$

Consequently, we can write

$$u = v_0 + \widetilde{\Psi}(v_0,\mu), \tag{4.3}$$

with

$$\widetilde{\Psi}(v_0,\mu) = \Phi_\mu(v_0) + \Psi(v_0 + \Phi_\mu(v_0),\mu) \in \mathscr{Z}.$$

Notice that here $\widetilde{\Psi}(v_0,\mu)$ belongs to the entire space $\mathscr{Z}$, and not to $\mathscr{Z}_h$ as $\Psi(u_0,\mu)$. To obtain the normal form, we can now use the Ansatz (4.3), and proceed as for the algorithmic derivation in Section 3.2.3.

First, differentiating (4.3) with respect to t and replacing du/dt and dv_0/dt from (4.1) and (4.2), respectively, gives the identity

$$D_{v_0}\widetilde{\Psi}(v_0,\mu)\mathbf{L}_0 v_0 - \mathbf{L}\widetilde{\Psi}(v_0,\mu) + \mathbf{N}_\mu(v_0) = \mathbf{Q}(v_0,\mu), \tag{4.4}$$

where

$$\mathbf{Q}(v_0,\mu) = \Pi_p\left(\mathbf{R}(v_0+\widetilde{\Psi}(v_0,\mu),\mu) - D_{v_0}\widetilde{\Psi}(v_0,\mu)\mathbf{N}_\mu(v_0)\right).$$

Here Π_p represents the linear map that associates to a map of class $\mathscr{C}^p$ the polynomial of degree p in its Taylor expansion. Next, we set

$$\widetilde{\Psi}(v_0,\mu) = \widetilde{\Psi}_0(v_0,\mu) + \widetilde{\Psi}_h(v_0,\mu),$$

where $\widetilde{\Psi}_0 = \mathbf{P}_0\widetilde{\Psi}$ and $\widetilde{\Psi}_h = \mathbf{P}_h\widetilde{\Psi}$ take values in $\mathscr{E}_0$ and $\mathscr{Z}_h$, respectively, according to the decomposition $\mathscr{Z} = \mathscr{E}_0 + \mathscr{Z}_h$. Projecting the identity (4.4) successively on $\mathscr{E}_0$ and $\mathscr{Z}_h$ with the projectors $\mathbf{P}_0$ and $\mathbf{P}_h$, respectively, gives the following system:

$$\mathscr{A}_{\mathbf{L}_0}\widetilde{\Psi}_0(v_0,\mu) + \mathbf{N}_\mu(v_0) = \mathbf{Q}_0(v_0,\mu) \tag{4.5}$$

$$D_{v_0}\widetilde{\Psi}_h(v_0,\mu)\mathbf{L}_0 v_0 - \mathbf{L}_h\widetilde{\Psi}_h(v_0,\mu) = \mathbf{Q}_h(v_0,\mu), \tag{4.6}$$

where

$$\mathbf{Q}_0(v_0,\mu) = \mathbf{P}_0\mathbf{Q}(v_0,\mu), \quad \mathbf{Q}_h(v_0,\mu) = \mathbf{P}_h\mathbf{Q}.$$

We can solve both equations in this system using again the Taylor expansions of $\mathbf{R}$, $\widetilde{\Psi}_0$, $\widetilde{\Psi}_h$, and $\mathbf{N}_\mu$. Then equation (4.5) leads to an equation of the form (2.6), with $\Phi_\mu(v)$ replaced by $\widetilde{\Psi}_0(v_0,\mu)$ and can be solved as described in Section 3.2.3. Parallel to this, we have to solve the second equation, which determines $\widetilde{\Psi}_h(v_0,\mu)$. This is also done with the help of the Taylor expansions, which lead at every order to an equation of the form

$$D_{v_0}\widetilde{\Psi}_h(v_0)\mathbf{L}_0 v_0 - L_h\widetilde{\Psi}_h(v_0) = \mathbf{Q}_h(v_0),$$

in which the right hand side is known. At this point we have to make sure that this equation has a solution $\widetilde{\Psi}_h(v_0) \in \mathscr{Z}_h$. For this, notice that the equation above is obtained from the equation

$$\frac{d}{dt}\widetilde{\Psi}_h(e^{\mathbf{L}_0 t}v_0) = \mathbf{L}_h\widetilde{\Psi}_h(e^{\mathbf{L}_0 t}v_0) + \mathbf{Q}_h(e^{\mathbf{L}_0 t}v_0)$$

by taking $t=0$. Here the map $t \mapsto \mathbf{Q}_h(e^{\mathbf{L}_0 t}v_0)$ belongs to $\mathscr{C}_\eta(\mathbb{R},\mathscr{Y}_h)$ for any $\eta > 0$, so that by Hypothesis (2.7) of Chapter 2 this equation possesses a unique solution $\mathbf{K}_h\mathbf{Q}_h(e^{\mathbf{L}_0\cdot}v_0)$. Consequently, we may take

$$\widetilde{\Psi}_h(v_0) = \left(\mathbf{K}_h\mathbf{Q}_h(e^{\mathbf{L}_0\cdot}v_0)\right)(0),$$

which then shows that (4.6) can be solved at every order.

We show in the following sections, how to simultaneously compute the center manifold and the normal form for three different bifurcations in infinite-dimensional equations.

3.4.2 Example 1: Hopf Bifurcation

Consider an equation of the form (4.1), with a single parameter $\mu \in \mathbb{R}$, and satisfying the hypotheses in the center manifold Theorem 3.3, Chapter 2. Further assume that the spectrum of the linear operator $\mathbf{L}$ contains precisely two purely imaginary eigenvalues $\pm i\omega$, which are simple.

Normal Form

Under these assumptions, we have that $\sigma_0 = \{\pm i\omega\}$ and that the associated spectral subspace $\mathscr{E}_0$ is two-dimensional spanned by the eigenvectors ζ and $\overline{\zeta}$ associated with $i\omega$ and $-i\omega$, respectively. The center manifold Theorem 3.3, Chapter 2, gives

$$u = u_0 + \Psi(u_0,\mu), \quad u_0 \in \mathscr{E}_0, \quad \Psi(u_0,\mu) \in \mathscr{Y}_h,$$

and applying the normal form Theorem 2.2 to the reduced system we find

$$u_0 = v_0 + \Phi_\mu(v_0),$$

which gives the equality (4.3),

$$u = v_0 + \widetilde{\Psi}(v_0,\mu), \quad v_0 \in \mathscr{E}_0, \quad \widetilde{\Psi}(u_0,\mu) \in \mathscr{Y}.$$

For $v_0(t) \in \mathscr{E}_0$, it is convenient to write

$$v_0(t) = A(t)\zeta + \overline{A(t)\zeta}, \quad A(t) \in \mathbb{C},$$

and according to the Lemma 1.7 (see Remark 2.3(iv)), the polynomial $\mathbf{N}_\mu(A,\overline{A})$ in the normal form is of the form

$$\mathbf{N}_\mu(A,\overline{A}) = (AQ(|A|^2,\mu), \overline{AQ}(|A|^2,\mu)),$$

with Q a complex-valued polynomial in its first argument satisfying $Q(0,0) = 0$.

Computation of the Normal Form

Our purpose it to show how to compute the two leading order coefficients in the expression of $\mathbf{N}_\mu$, i.e., the coefficients a and b in the expression

$$Q(|A|^2,\mu) = a\mu + b|A|^2 + O((|\mu| + |A|^2)^2).$$

(An example of such a computation is given in the example in Section 2.4.4, Chapter 2.) For this calculation we proceed as indicated in Section 3.4.

We start from the identity (4.4) in which we replace the Taylor expansions of $\mathbf{R}$ and $\widetilde{\Psi}$. With the notations from Section 3.2.3, we set

$$\widetilde{\Psi}_{ql}(v_0^{(q)},\mu^{(l)}) = \mu^l \sum_{q_1+q_2=q} A^{q_1}\overline{A}^{q_2}\Psi_{q_1q_2l}, \quad \Psi_{q_1q_2l} \in \mathscr{Y}.$$

By identifying in (4.4) the terms of order $O(\mu)$, $O(A^2)$, and $O(A\overline{A})$, we obtain

$$\begin{aligned} -\mathbf{L}\Psi_{001} &= \mathbf{R}_{01}, \\ (2i\omega - \mathbf{L})\Psi_{200} &= \mathbf{R}_{20}(\zeta,\zeta), \\ -\mathbf{L}\Psi_{110} &= 2\mathbf{R}_{20}(\zeta,\overline{\zeta}). \end{aligned}$$

Here the operators $\mathbf{L}$ and $(2i\omega - \mathbf{L})$ on the left hand sides are invertible, so that Ψ_{001}, Ψ_{200}, and Ψ_{110} are uniquely determined from these equalities. Next, we identify the terms of order $O(\mu A)$ and $O(A^2\overline{A})$ and find

$$\begin{aligned} (i\omega - \mathbf{L})\Psi_{101} &= -a\zeta + \mathbf{R}_{11}(\zeta) + 2\mathbf{R}_{20}(\zeta,\Psi_{001}), \\ (i\omega - \mathbf{L})\Psi_{210} &= -b\zeta + 2\mathbf{R}_{20}(\zeta,\Psi_{110}) + 2\mathbf{R}_{20}(\overline{\zeta},\Psi_{200}) + 3\mathbf{R}_{30}(\zeta,\zeta,\overline{\zeta}). \end{aligned}$$

Since $i\omega$ is a simple isolated eigenvalue of $\mathbf{L}$, the range of $(i\omega - \mathbf{L})$ is of codimension 1, so that we can solve these equations and determine Ψ_{101} and Ψ_{200}, provided the right hand sides satisfy one solvability condition. It is this solvability condition which allows us to compute the coefficients a and b, just in the finite-dimensional case. In the case where $\mathbf{L}$ has an adjoint $\mathbf{L}^*$ acting in the dual space $\mathscr{X}^*$, the solvability condition is that the right hand sides be orthogonal to the kernel of the adjoint $(-i\omega - \mathbf{L}^*)$ of $(i\omega - \mathbf{L})$. The kernel of $(-i\omega - \mathbf{L}^*)$ is one-dimensional, just as the kernel of $(i\omega - \mathbf{L})$, spanned by $\zeta^* \in \mathscr{X}^*$ that we choose such that $\langle\zeta,\zeta^*\rangle = 1$. Here $\langle\cdot,\cdot\rangle$ denotes the duality product between $\mathscr{X}$ and $\mathscr{X}^*$, where it is semilinear with respect to the second argument. Then in this situation we find

$$\begin{aligned} a &= \langle\mathbf{R}_{11}(\zeta) + 2\mathbf{R}_{20}(\zeta,\Psi_{001}),\zeta^*\rangle, \\ b &= \langle 2\mathbf{R}_{20}(\zeta,\Psi_{110}) + 2\mathbf{R}_{20}(\overline{\zeta},\Psi_{200}) + 3\mathbf{R}_{30}(\zeta,\zeta,\overline{\zeta}),\zeta^*\rangle. \end{aligned}$$

Notice that here it is not necessary to further solve the equations and compute Ψ_{101} and Ψ_{210}.

Now, if the adjoint $\mathbf{L}^*$ does not exist, we still have a Fredholm alternative for the equations above. Indeed, both equations are of the form

$$(i\omega - \mathbf{L})\Psi = \mathbf{R}, \tag{4.7}$$

with $\mathbf{R} \in \mathscr{X}$. Projecting with $\mathbf{P}_0$ and $\mathbf{P}_h$ on the subspaces $\mathscr{E}_0$ and $\mathscr{X}_h$, respectively, we obtain

$$\begin{aligned}(i\omega - \mathbf{L}_0)\mathbf{P}_0\Psi &= \mathbf{P}_0\mathbf{R},\\ (i\omega - \mathbf{L}_h)\mathbf{P}_h\Psi &= \mathbf{P}_h\mathbf{R}.\end{aligned}$$

The operator on the left hand side of the second equation is invertible, since the spectrum of $\mathbf{L}_h$ is $\sigma_- \cup \sigma_+$, which is bounded away from the imaginary axis (see Hypothesis 2.4 in Chapter 2). Then the second equation has a unique solution

$$\mathbf{P}_h\Psi = (i\omega - \mathbf{L}_h)^{-1}\mathbf{P}_h\mathbf{R}, \quad (i\omega - \mathbf{L}_h)^{-1} : \mathscr{X}_h \to \mathscr{Z}_h.$$

The first equation is two-dimensional, so that there is a solution Ψ_0, provided the following solvability condition holds

$$\langle \mathbf{R}_0, \zeta_0^* \rangle = 0,$$

where $\zeta_0^* \in \mathscr{E}_0$ is the eigenvector in the kernel of the adjoint $(-i\omega - \mathbf{L}_0^*)$ in $\mathscr{E}_0$ chosen such that $\langle \zeta, \zeta_0^* \rangle = 1$. We rewrite this solvability condition as

$$\langle \mathbf{R}_0, \zeta_0^* \rangle = \langle \mathbf{P}_0\mathbf{R}, \zeta_0^* \rangle = \langle \mathbf{R}, \mathbf{P}_0^*\zeta_0^* \rangle = 0, \tag{4.8}$$

in which $\mathbf{P}_0^*$ is the adjoint of the projector $\mathbf{P}_0$, and the last bracket represents the duality product between $\mathscr{X}$ and $\mathscr{X}^*$. Upon setting

$$\zeta^* = \mathbf{P}_0^*\zeta_0^* \in \mathscr{X}^*,$$

the solvability condition becomes $\langle \mathbf{R}, \zeta^* \rangle = 0$, which then leads to formulas for the coefficients a and b as above.

We point out that the range of $i\omega - \mathbf{L}$ is orthogonal to the vector ζ^* constructed above, with respect to the duality product between $\mathscr{X}$ and $\mathscr{X}^*$, and actually, its range is precisely the space orthogonal to ζ^*. Indeed, since $i\omega$ is an isolated simple eigenvalue of $\mathbf{L}$, the operator is Fredholm with index zero, so its range is closed and has a codimension equal to the dimension of the kernel, which is 1.

Reduced Dynamics

The dynamics of the reduced equation, which is two-dimensional, is as described in Theorem 2.6 in Chapter 1, so that we are here in the presence of a Hopf bifurcation. We then have a branch of equilibria for small μ and a family of periodic solutions of size $O(|\mu|^{1/2})$, which bifurcate at $\mu = 0$ for μ such that $a_r b_r \mu < 0$. Here a_r and b_r denote the real parts of a and b, respectively.

We point out that such a Hopf bifurcation typically occurs for equations for which the unstable spectrum σ_+ in Hypothesis 2.4, Chapter 2, is empty, $\sigma_+ = \varnothing$. In this situation, the stability of both equilibria and periodic solutions is the same in the reduced system and in the full equation. Indeed, for all these solutions, one has a strong stable manifold of codimension 2 corresponding to perturbations of the stable spectrum σ_- of $\mathbf{L}$, and the remaining dynamics are found on the center manifold. For example, assume that $a_r > 0$. Then the family of equilibria is stable for $\mu < 0$ and loses its stability when μ crosses 0 (see Theorem 2.6 of Chapter 1). In the supercritical case, when $b_r < 0$, we have an *attracting periodic solution on the center manifold for* $\mu > 0$, for which we can compute the Floquet exponents. The most unstable exponents correspond to the flow on the center manifold, which give here 0, due to the invariance under translations in time t of (4.1), and a real negative exponent, close to 0. The other exponents correspond to perturbations of the stable eigenvalues in σ_- of $\mathbf{L}$, and give a strong stable manifold of codimension 2, transverse to the weakly stable mode obtained from the dynamics on the center manifold. It results that in the supercritical case the bifurcating periodic solution is also stable in $\mathscr{Y}$. In the subcritical case, when $b_r > 0$, the periodic solution occurs for $\mu < 0$ and is unstable, since it is already unstable on the center manifold.

3.4.3 Example 2: Hopf Bifurcations with Symmetries

We discuss in this section two examples of Hopf bifurcations, with symmetries $SO(2)$ and $O(2)$. While in the first case the symmetry implies that the reduced system is always in normal form, in the second case we apply the result in Theorem 3.2 to determine the normal form of the reduced system.

Hopf Bifurcation with $SO(2)$ Symmetry

Consider the situation in Section 3.4.2 of an equation of the form (4.1), with a single parameter $\mu \in \mathbb{R}$, satisfying the hypotheses in center manifold Theorem 3.3, Chapter 2, and such that the spectrum of the linear operator $\mathbf{L}$ contains precisely two purely imaginary eigenvalues $\pm i\omega$, which are simple. We now further assume that there is a one-parameter continuous family of linear maps $\mathbf{R}_\varphi \in \mathscr{L}(\mathscr{X}) \cap \mathscr{L}(\mathscr{Z})$ for $\varphi \in \mathbb{R}/2\pi\mathbb{Z}$, with the following properties:

(i) $\mathbf{R}_\varphi \circ \mathbf{R}_\psi = \mathbf{R}_{\varphi+\psi}$ for all $\varphi,\ \psi \in \mathbb{R}/2\pi\mathbb{Z}$;
(ii) $\mathbf{R}_0 = \mathbb{I}$;
(iii) $\mathbf{R}_\varphi \mathbf{L} = \mathbf{L}\mathbf{R}_\varphi$ and $\mathbf{R}(\mathbf{R}_\varphi u, \mu) = \mathbf{R}_\varphi \mathbf{R}(u, \mu)$ for all $\varphi \in \mathbb{R}/2\pi\mathbb{Z}$, $u \in \mathscr{Z}$, and $\mu \in \mathbb{R}$.

In particular, the group $\{\mathbf{R}_\varphi; \varphi \in \mathbb{R}/2\pi\mathbb{Z}\}$ is a representation of an $SO(2)$ symmetry in $\mathscr{X}$ and $\mathscr{Z}$. As in the two-dimensional case discussed in Section 1.2.1 in Chapter 1, these properties allow us to simplify the analysis of the reduced equation, and induce some symmetry properties for the bifurcating periodic solutions.

Reduced System

Consider the eigenvector ζ associated to the simple eigenvalue $i\omega$ of $\mathbf{L}$. Then, by arguing as in Section 1.2.1 from the fact that $\mathbf{R}_\varphi$ commutes with $\mathbf{L}$, we find that

$$\mathbf{R}_\varphi \zeta = e^{im\varphi}\zeta,$$

for some $m \in \mathbb{Z}$. In the case $m = 0$, which means that the action of all $\mathbf{R}_\varphi$ on the subspace $\mathscr{E}_0$ is trivial, the results in Section 3.4.2 hold with the additional property that the periodic solution is pointwise invariant under the "rotations" $\mathbf{R}_\varphi$.

Assume that $m \neq 0$. Then we choose a norm on $\mathscr{E}_0$ such that $\mathbf{R}_\varphi$ is an isometry, and applying the result in Theorem 3.13 in Chapter 2, we find that the reduction function Ψ satisfies

$$\mathbf{R}_\varphi \Psi(u_0, \mu) = \Psi(\mathbf{R}_\varphi u_0, \mu) \text{ for all } u_0 \in \mathscr{E}_0,\ \mu \in \mathbb{R}.$$

We set again

$$u_0(t) = A(t)\zeta + \overline{A(t)\zeta}$$

for $u_0(t) \in \mathscr{E}_0$, with A a complex-valued function, and then the reduced system is

$$\frac{dA}{dt} = i\omega A + f(A, \overline{A}, \mu),$$

together with the complex conjugated equation. In addition, the vector field commutes with $\mathbf{R}_\varphi|_{\mathscr{E}_0}$, which together with the equality $\mathbf{R}_\varphi \zeta = e^{im\varphi}\zeta$ implies that

$$f(e^{im\varphi}A, e^{-im\varphi}\overline{A}, \mu) = e^{im\varphi} f(A, \overline{A}, \mu).$$

According to Lemma 2.4 in Chapter 1, we then have that

$$f(A, \overline{A}, \mu) = Ag(|A|^2, \mu),$$

with g of class C^{k-1}, so that in this case the reduced equation in already in normal form.

Reduced Dynamics

This situation was discussed in Section 1.2.1, in Chapter 1. First, $A = 0$ is always an equilibrium of the reduced system, which gives the equilibria $u = \Psi(0, \mu)$. These equilibria are invariant under the action of $\mathbf{R}_\varphi$. Next, according to the results in Corollary 2.13 in Chapter 1, the reduced equation possesses a family of periodic solutions

$$A(t,\mu) = r(\mu)e^{i\omega_*(\mu)t}, \quad r(\mu) = O(|\mu|^{1/2}),$$

which are rotating waves, with

$$A(t,\mu) = \mathbf{R}_{\frac{\omega_*(\mu)t}{m}} A(0,\mu)$$

satisfying

$$\mathbf{R}_\varphi A(t,\mu)\zeta = A(t+\frac{m\varphi}{\omega_*},\mu)\zeta.$$

Using the fact that $\mathbf{R}_\varphi$ commutes with the reduction function Ψ, we find that the corresponding solutions $u(\cdot,\mu)$ of the full equation satisfy

$$\begin{aligned}\mathbf{R}_\varphi u(t,\mu) &= \mathbf{R}_\varphi(u_0(t,\mu)+\Psi(u_0(t,\mu),\mu)) \\ &= u_0(t+\frac{m\varphi}{\omega_*},\mu)+\Psi(u_0(t+\frac{m\varphi}{\omega_*},\mu),\mu) = u(t+\frac{m\varphi}{\omega_*},\mu).\end{aligned}$$

By arguing as for (2.38) in Chapter 1, this implies that $u(\cdot,\mu)$ is also *a rotating wave*, i.e.,

$$u(t,\mu) = \mathbf{R}_{\frac{\omega_*(\mu)t}{m}} u(0,\mu). \tag{4.9}$$

Hopf Bifurcation with $O(2)$ Symmetry

In the same setting as above, we now further assume that there exists a symmetry $\mathbf{S}$, with $\mathbf{S}^2 = \mathbb{I}$, such that the vector field is equivariant under the action of $\mathbf{S}$,

$$\mathbf{SL} = \mathbf{LS}, \quad \mathbf{R}(\mathbf{S}u,\mu) = \mathbf{SR}(u,\mu) \text{ for all } \mu \in \mathbb{R} \tag{4.10}$$

and that

$$\mathbf{R}_\varphi \mathbf{S} = \mathbf{SR}_{-\varphi} \text{ for all } \varphi \in \mathbb{R}/2\pi\mathbb{Z}. \tag{4.11}$$

Then the group $\{\mathbf{R}_\varphi, \mathbf{S}; \varphi \in \mathbb{R}/2\pi\mathbb{Z}\}$ is a representation of an $O(2)$ symmetry in $\mathscr{X}$ and $\mathscr{Z}$. We already met this type of symmetry in Section 1.2.4 in Chapter 1 and in the example in Section 2.4.3 in Chapter 2.

A key property here is that generically the eigenvalues of the linear operator $\mathbf{L}$ are at least geometrically double. Indeed, by arguing as in Section 1.2.4, one concludes that any eigenvalue λ of $\mathbf{L}$ that has an eigenvector ζ which is not invariant under the action of $\mathbf{R}_\varphi$ (i.e., $\mathbf{R}_\varphi\zeta \neq \zeta$ for some $\varphi \in \mathbb{R}/2\pi\mathbb{Z}$) is at least geometrically double. We shall therefore assume in this example that $\sigma_0 = \{\pm i\omega\}$, where $\pm i\omega$ are algebraically and geometrically double eigenvalues, with associated eigenvectors that are not invariant under the action of $\mathbf{R}_\varphi$. Then the restriction of the action of $\mathbf{R}_\varphi$ to the eigenspaces associated with the eigenvalues $\pm i\omega$ is not trivial, and the result in (2.39) in Chapter 1 shows that we can choose the eigenvectors $\{\zeta_0, \zeta_1\}$ associated with $i\omega$ such that

$$\mathbf{R}_\varphi \zeta_0 = e^{im\varphi}\zeta_0, \quad \mathbf{R}_\varphi \zeta_1 = e^{-im\varphi}\zeta_1, \quad \mathbf{S}\zeta_0 = \zeta_1, \quad \mathbf{S}\zeta_1 = \zeta_0. \tag{4.12}$$

Clearly, $\{\overline{\zeta}_0, \overline{\zeta}_1\}$ are the eigenvectors associated with $-i\omega$.

Normal Form

We can now choose a norm on $\mathscr{E}_0$ such that $\mathbf{R}_\varphi$ and $\mathbf{S}$ are isometries, and applying the result in Theorem 3.13 in Chapter 2, we find that the reduction function Ψ satisfies

$$\Psi(\mathbf{R}_\varphi u_0, \mu) = \mathbf{R}_\varphi \Psi(u_0, \mu), \quad \Psi(\mathbf{S}u_0, \mu) = \mathbf{S}\Psi(u_0, \mu) \text{ for all } u_0 \in \mathscr{E}_0,\ \mu \in \mathbb{R}.$$

Further applying Theorems 2.2 and 3.2 to the reduced equation, we write

$$u = v_0 + \widetilde{\Psi}(v_0, \mu), \quad v_0 \in \mathscr{E}_0, \quad \widetilde{\Psi}(v_0, \mu) \in \mathscr{Z},$$

and set

$$v_0(t) = A(t)\zeta_0 + B(t)\zeta_1 + \overline{A(t)}\overline{\zeta}_0 + \overline{B(t)}\overline{\zeta}_1.$$

Here A and B are complex-valued functions, and $\widetilde{\Psi}(\cdot, \mu)$ commutes with $\mathbf{R}_\varphi$ and $\mathbf{S}$.

The polynomial $\mathbf{N}_\mu$ in the resulting normal form satisfies the characterization (2.4) and also commutes with $\mathbf{R}_\varphi$ and $\mathbf{S}$. We write

$$\mathbf{N}_\mu = (\Phi_0, \Phi_1, \overline{\Phi}_0 \overline{\Phi}_1)$$

where Φ_j, $j = 0, 1$, are polynomials of $(A, B, \overline{A}, \overline{B})$ with coefficients depending upon μ. Using successively the characterization (2.4) and the fact that $\mathbf{N}_\mu$ commutes with $\mathbf{R}_\varphi$ and $\mathbf{S}$, we find that

$$\begin{aligned}
\Phi_0(e^{-i\omega t}A, e^{-i\omega t}B, e^{i\omega t}\overline{A}, e^{i\omega t}\overline{B}) &= e^{-i\omega t}\Phi_0(A, B, \overline{A}, \overline{B}),\\
\Phi_1(e^{-i\omega t}A, e^{-i\omega t}B, e^{i\omega t}\overline{A}, e^{i\omega t}\overline{B}) &= e^{-i\omega t}\Phi_1(A, B, \overline{A}, \overline{B}),\\
\Phi_0(e^{im\varphi}A, e^{-im\varphi}B, e^{-im\varphi}\overline{A}, e^{im\varphi}\overline{B}) &= e^{im\varphi}\Phi_0(A, B, \overline{A}, \overline{B}),\\
\Phi_1(e^{im\varphi}A, e^{-im\varphi}B, e^{-im\varphi}\overline{A}, e^{im\varphi}\overline{B}) &= e^{-im\varphi}\Phi_1(A, B, \overline{A}, \overline{B}),\\
\Phi_0(B, A, \overline{B}, \overline{A}) &= \Phi_1(A, B, \overline{A}, \overline{B})
\end{aligned} \tag{4.13}$$

for all $t \in \mathbb{R}$ and $\varphi \in \mathbb{R}/2\pi\mathbb{Z}$.

To exploit these identities we proceed as follows. The first and third identities lead to

$$\Phi_0(e^{i(m\varphi-\omega t)}A, e^{-i(m\varphi+\omega t)}B, e^{i(\omega t-m\varphi)}\overline{A}, e^{i(m\varphi+\omega t)}\overline{B}) = e^{i(m\varphi-\omega t)}\Phi_0(A, B, \overline{A}, \overline{B})$$

for any $t \in \mathbb{R}$ and $\varphi \in \mathbb{R}/2\pi\mathbb{Z}$. We choose (t, φ) such that

$$m\varphi - \omega t = -\arg A, \quad m\varphi + \omega t = \arg B,$$

which implies that

$$\Phi_0(A,B,\overline{A},\overline{B}) = e^{i\arg A}\Phi_0(|A|,|B|,|A|,|B|).$$

Then we choose (t,φ) such that

$$m\varphi - \omega t = \pi, \quad m\varphi + \omega t = 0,$$

which gives

$$\Phi_0(-A,B,-\overline{A},\overline{B}) = -\Phi_0(A,B,\overline{A},\overline{B}),$$

and finally we choose (t,φ) such that

$$m\varphi - \omega t = 0, \quad m\varphi + \omega t = \pi,$$

which shows that

$$\Phi_0(A,-B,\overline{A},-\overline{B}) = \Phi_0(A,B,\overline{A},\overline{B}).$$

Since Φ_0 is a polynomial, it follows now that there is a polynomial P_0 such that

$$\Phi_0(A,B,\overline{A},\overline{B}) = AP_0(|A|^2,|B|^2),$$

and similarly we obtain that there is a polynomial P_1 such that

$$\Phi_1(A,B,\overline{A},\overline{B}) = BP_1(|A|^2,|B|^2).$$

In addition, from the last identity in (4.13) we conclude that

$$P_1(|A|^2,|B|^2) = P_0(|B|^2,|A|^2).$$

Summarizing, we have the normal form

$$\begin{aligned}\frac{dA}{dt} &= i\omega A + AP(|A|^2,|B|^2,\mu) + \rho(A,B,\overline{A},\overline{B},\mu)\\ \frac{dB}{dt} &= i\omega B + BP(|B|^2,|A|^2,\mu) + \rho(B,A,\overline{B},\overline{A},\mu),\end{aligned} \tag{4.14}$$

in which P is a polynomial of degree p in its first two arguments with coefficients depending upon μ, as given in Theorem 2.2, and $\rho(A,B,\overline{A},\overline{B},\mu) = O((|A|+|B|)^{2p+3})$. Furthermore, notice here the particular form of the remainder ρ, which is due to the fact that the vector field in this system commutes with $\mathbf{S}$, whereas from the fact that the vector field commutes with $\mathbf{R}_\varphi$ we have in addition that

$$\rho(e^{im\varphi}A, e^{im\varphi}B, e^{-im\varphi}\overline{A}, e^{-im\varphi}\overline{B},\mu) = e^{im\varphi}\rho(A,B,\overline{A},\overline{B},\mu).$$

Exercise 4.1 (Computation of the normal form) *Consider the normal form truncated at order 3,*

$$\frac{dA}{dt} = i\omega A + A(a\mu + b|A|^2 + c|B|^2),$$
$$\frac{dB}{dt} = i\omega B + B(a\mu + b|B|^2 + c|A|^2),$$

with complex coefficients a, b, and c, and the Taylor expansion of $\widetilde{\Psi}$,

$$\widetilde{\Psi}(A,B,\overline{A},\overline{B},\mu) = \sum_{p+q+r+s+l\geq 1} \Psi_{pqrsl} A^p \overline{A}^q B^r \overline{B}^s \mu^l,$$

in which $\Psi_{10000} = \Psi_{01000} = \Psi_{00100} = \Psi_{00010} = 0$. *Show that*

$$\Psi_{00001} = -\mathbf{L}^{-1}\mathbf{R}_{01}, \quad \Psi_{20000} = (2i\omega - \mathbf{L})^{-1}\mathbf{R}_{20}(\zeta_0,\zeta_0),$$
$$\Psi_{11000} = -2\mathbf{L}^{-1}\mathbf{R}_{20}(\zeta_0,\overline{\zeta}_0), \quad \Psi_{00110} = \mathbf{S}\Psi_{11000},$$
$$\Psi_{10100} = 2(2i\omega - \mathbf{L})^{-1}\mathbf{R}_{20}(\zeta_0,\zeta_1), \quad \Psi_{10010} = -2\mathbf{L}^{-1}\mathbf{R}_{20}(\zeta_0,\overline{\zeta}_1),$$

and that the coefficients a,b,c *are given by*

$$a = \langle \mathbf{R}_{11}(\zeta_0) + 2\mathbf{R}_{20}(\zeta_0,\Psi_{00001}), \zeta_0^* \rangle,$$
$$b = \langle 2\mathbf{R}_{20}(\zeta_0,\Psi_{11000}) + 2\mathbf{R}_{20}(\overline{\zeta}_0,\Psi_{20000}) + 3\mathbf{R}_{30}(\zeta_0,\zeta_0,\overline{\zeta}_0), \zeta_0^* \rangle,$$
$$c = \langle 2\mathbf{R}_{20}(\zeta_0,\Psi_{00110}) + 2\mathbf{R}_{20}(\zeta_1,\Psi_{10010}) + 2\mathbf{R}_{20}(\overline{\zeta}_1,\Psi_{10100}) + 6\mathbf{R}_{30}(\zeta_0,\zeta_1,\overline{\zeta}_1), \zeta_0^* \rangle,$$

where $\zeta_0^* \in \mathscr{X}^*$ *is constructed as* ζ^* *in Section 3.4.2.*

Reduced Dynamics

The study of the dynamics of the system (4.14) strongly relies upon the study of the normal form truncated at order 3. In polar coordinates

$$A = r_0 e^{i\theta_0}, \quad B = r_1 e^{i\theta_1},$$

the truncated normal form becomes

$$\begin{aligned}
\frac{dr_0}{dt} &= r_0(a_r\mu + b_r r_0^2 + c_r r_1^2),\\
\frac{dr_1}{dt} &= r_1(a_r\mu + b_r r_1^2 + c_r r_0^2),\\
\frac{d\theta_0}{dt} &= \omega + a_i\mu + b_i r_0^2 + c_i r_1^2,\\
\frac{d\theta_1}{dt} &= \omega + a_i\mu + b_i r_1^2 + c_i r_0^2,
\end{aligned} \tag{4.15}$$

where the subscripts r and i indicate the real and the imaginary parts, respectively, of a complex number. Here the two first equations for (r_0, r_1) decouple from the last two equations for the phases (θ_0, θ_1), so that we can solve them separately.

The dynamics of these two equations are rather simple and are summarized in the case $a_r\mu > 0$ in Figure 4.1. (Similar phase portraits can be found in the other cases.)

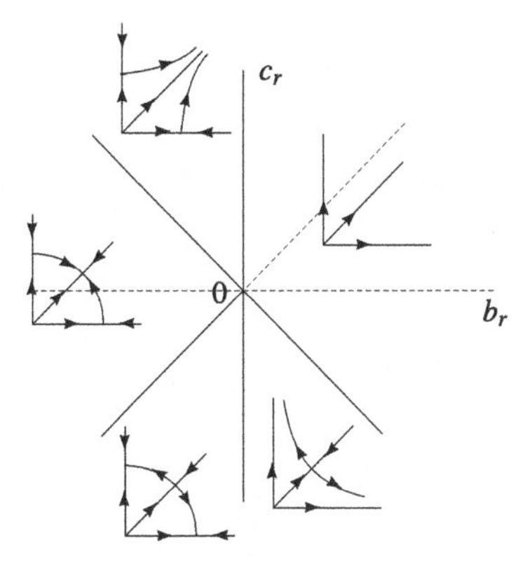

Fig. 4.1 Phase portraits in the (r_0, r_1)-plane of the equations for (r_0, r_1), depending upon (b_r, c_r) in the case $a_r\mu > 0$.

In particular, for $b_r < 0$ in this case, one finds two pair of equilibria $(\pm r_*(\mu), 0)$ and $(0, \pm r_*(\mu))$ on the r_0- and r_1-axis, respectively. These equilibria correspond to *rotating waves*, just as for the Hopf bifurcation in the presence of $SO(2)$ symmetry. Here, the symmetry $\mathbf{S}$ exchanges the two axes, so that it exchanges the rotating waves corresponding to $r_0 = 0$ into the rotating waves corresponding to $r_1 = 0$. Their stability is indicated in Figure 4.1, and we refer for instance to [56] for a proof of the persistence of these rotating waves for the full system (4.14). Next, for $b_r + c_r < 0$ in this case, there is another pair of equilibria with $r_0 = r_1$, which correspond to *standing waves*, another class of bifurcating periodic solutions (e.g., see [56] for a proof of the persistence of these solutions for (4.14)). These correspond to a torus of solutions of the normal form

$$\begin{aligned} v_0(t,\mu,\delta_0,\delta_1) &= r_0(\mu)\left(e^{i(\omega_*(\mu)t+\delta_0)}\zeta_0 + e^{i(\omega_*(\mu)t+\delta_1)}\zeta_1\right) \\ &+ r_0(\mu)\left(e^{-i(\omega_*(\mu)t+\delta_0)}\overline{\zeta}_0 + e^{-i(\omega_*(\mu)t+\delta_1)}\overline{\zeta}_1\right) \end{aligned}$$

for any $(\delta_0, \delta_1) \in \mathbb{R}^2$, which induces a torus of solutions $u(t,\mu,\delta_0,\delta_1)$ in $\mathscr{Y}$ of the full system (4.1). Notice that these standing waves possess the following symmetry properties:

$$\begin{aligned} &\mathbf{R}_{\frac{\delta_1-\delta_0}{m}}\mathbf{S}u(t,\mu,\delta_0,\delta_1) = u(t,\mu,\delta_0,\delta_1), \quad \mathbf{R}_{\frac{2\pi}{m}}u(t,\mu,\delta_0,\delta_1) = u(t,\mu,\delta_0,\delta_1), \\ &\mathbf{R}_{\frac{\pi}{m}}u(t,\mu,\delta_0,\delta_1) = u(t+\frac{\pi}{\omega_*(\mu)},\mu,\delta_0,\delta_1), \quad \mathbf{S}u(t,\mu,\delta_0,\delta_0) = u(t,\mu,\delta_0,\delta_0). \end{aligned}$$

Exercise 4.2 *Consider a system of the form (4.1) with $\mu \in \mathbb{R}^2$ satisfying the hypotheses of Theorem 3.3 of Chapter 2. Further assume that*

(i) *the linear operator $\mathbf{L}$ has precisely three eigenvalues on the imaginary axis, $\sigma_0 = \{\pm i\omega, 0\}$, which are all simple;*

(ii) $\mathbf{L}$ *and* $\mathbf{R}(\cdot,\mu)$ *commute with a symmetry* $\mathbf{S}$*, with* $\mathbf{S}^2 = \mathbb{I}$*;*
(iii) the eigenvector ζ *associated with the eigenvalue* 0 *is antisymmetric,* $\mathbf{S}\zeta = -\zeta$*.*

Using the result in Lemma 1.12, derive the normal form for the three-dimensional reduced system, and give formulas for the coefficients of the linear and cubic terms. (The study of the dynamics of the reduced vector field in this situation can be found in [88].)

3.4.4 Example 3: Takens–Bogdanov Bifurcation

Consider now an equation of the form (4.1), with a parameter $\mu \in \mathbb{R}^m$, and satisfying the hypotheses in the center manifold Theorem 3.3, Chapter 2. Further assume that 0 is the only eigenvalue of $\mathbf{L}$ on the imaginary axis and that this eigenvalue is geometrically simple and algebraically double.

Normal Form

With these assumptions we have $\sigma_0 = \{0\}$, and the associated spectral subspace $\mathscr{E}_0$ is two-dimensional. We choose a basis $\{\zeta_0, \zeta_1\}$ in $\mathscr{E}_0$ such that

$$\mathbf{L}\zeta_0 = 0, \quad \mathbf{L}\zeta_1 = \zeta_0.$$

As in the previous example, center manifold Theorem 3.3, Chapter 2, gives

$$u = u_0 + \Psi(u_0,\mu), \quad u_0 \in \mathscr{E}_0, \quad \Psi(u_0,\mu) \in \mathscr{Z}_h,$$

and applying normal form Theorem 2.2 to the reduced system we find

$$u_0 = v_0 + \Phi_\mu(v_0),$$

which gives the equality (4.3),

$$u = v_0 + \widetilde{\Psi}(v_0,\mu), \quad v_0 \in \mathscr{E}_0, \quad \widetilde{\Psi}(u_0,\mu) \in \mathscr{Z}.$$

For $v_0(t) \in \mathscr{E}_0$, we now write

$$v_0(t) = A(t)\zeta_0 + B(t)\zeta_1,$$

in which A and B are real-valued. According to the result in Lemma 1.9 and Remark 1.10, we find here the normal form

$$\begin{aligned}\frac{dA}{dt} &= B\\ \frac{dB}{dt} &= BP(A,\mu) + Q(A,\mu) + \rho(A,B,\mu),\end{aligned} \tag{4.16}$$

where $P(\cdot,\mu)$ and $Q(\cdot,\mu)$ are polynomials of degree p such that

$$P(0,0) = Q(0,0) = \frac{\partial Q}{\partial A}(0,0) = 0 \tag{4.17}$$

and

$$\rho(A,B,\mu) = o((|A|+|B|)^p).$$

Computation of the Normal Form

We compute now the leading order terms in the expansion of the vector field. We set

$$\begin{aligned} \frac{dA}{dt} &= B \\ \frac{dB}{dt} &= \alpha_1(\mu) + \alpha_2(\mu)A + \alpha_3(\mu)B + \beta_1(\mu)AB + \beta_2(\mu)A^2 + \widetilde{\rho}(A,B,\mu), \end{aligned} \tag{4.18}$$

where the coefficients $\alpha_j(\mu)$ and $\beta_j(\mu)$ are such that

$$\alpha_j(\mu) = \alpha_j^{(1)}(\mu) + O(\mu^2), \quad \beta_j(\mu) = \beta_j^{(0)} + O(|\mu|),$$

with $\alpha_j^{(1)} : \mathbb{R}^m \to \mathbb{R}$ linear maps, according to (4.17), and $\widetilde{\rho}(A,B,\mu) = O(|A^2B| + |A|^3) + o((|A|+|B|)^p)$.

We proceed as for the previous example and start from identity (4.4), in which we replace the Taylor expansions of $\mathbf{R}$ and $\widetilde{\Psi}$. With the notations from Section 3.2.3, we set here

$$\widetilde{\Psi}(v_0,\mu) = \sum_{q+l+r\geq 1} A^q B^l \Psi_{qlr}(\mu^{(r)}), \quad \Psi_{100} = \Psi_{010} = 0,$$

where $\Psi_{ql0} \in \mathscr{Z}$, and Ψ_{qlr}, $r \geq 1$, is r-linear symmetric in $\mu \in \mathbb{R}^m$ with values in $\mathscr{Z}$. By identifying in (4.4) the terms of order $O(A^2)$, $O(AB)$, and $O(B^2)$, we obtain

$$\beta_2^{(0)}\zeta_1 = \mathbf{L}\Psi_{200} + \mathbf{R}_{20}(\zeta_0,\zeta_0), \tag{4.19}$$

$$\beta_1^{(0)}\zeta_1 + 2\Psi_{200} = \mathbf{L}\Psi_{110} + 2\mathbf{R}_{20}(\zeta_0,\zeta_1), \tag{4.20}$$

$$\Psi_{110} = \mathbf{L}\Psi_{020} + \mathbf{R}_{20}(\zeta_1,\zeta_1), \tag{4.21}$$

and similarly, for the terms of order $O(\mu)$, $O(\mu A)$, and $O(\mu B)$, we find

$$\alpha_1^{(1)}\zeta_1 = \mathbf{L}\Psi_{001} + \mathbf{R}_{01}, \tag{4.22}$$

$$\alpha_2^{(1)}\zeta_1 + \alpha_1^{(1)}\Psi_{110} = \mathbf{L}\Psi_{101} + \mathbf{R}_{11}(\zeta_0,\cdot) + 2\mathbf{R}_{20}(\zeta_0,\Psi_{001}), \tag{4.23}$$

$$\alpha_3^{(1)}\zeta_1 + 2\alpha_1^{(1)}\Psi_{020} + \Psi_{101} = \mathbf{L}\Psi_{011} + \mathbf{R}_{11}(\zeta_1,\cdot) + 2\mathbf{R}_{20}(\zeta_1,\Psi_{001}). \tag{4.24}$$

Notice that each term in these three equalities is a linear map of $\mu \in \mathbb{R}^m$ with values in $\mathscr{X}$, so that the equalities hold in $\mathscr{X}$ for any $\mu \in \mathbb{R}^m$.

Next, we claim that for an equation of the form

$$\mathbf{L}\Psi = \mathbf{R}, \tag{4.25}$$

with $\mathbf{R} \in \mathscr{X}$ and $\Psi \in \mathscr{Z}$, a Fredholm alternative applies, just as in the previous example. Indeed, projecting again with the spectral projections $\mathbf{P}_0$ and $\mathbf{P}_h$, the equation decomposes as

$$\begin{aligned}\mathbf{L}_0\mathbf{P}_0\Psi &= \mathbf{P}_0\mathbf{R},\\ \mathbf{L}_h\mathbf{P}_h\Psi &= \mathbf{P}_h\mathbf{R}.\end{aligned}$$

Since $\mathbf{L}_h : \mathscr{X}_h \to \mathscr{Z}_h$ is invertible, the second equation has the unique solution

$$\mathbf{P}_h\Psi = \mathbf{L}_h{}^{-1}\mathbf{P}_h\mathbf{R}, \quad \mathbf{L}_h{}^{-1} : \mathscr{Z}_h \to \mathscr{X}_h.$$

The first equation is two-dimensional, and the linear operator $\mathbf{L}_0$ has a one-dimensional kernel spanned by ζ_0 and a two-dimensional generalized kernel spanned by ζ_0 and ζ_1. Then we can choose a dual basis $\{\zeta_{00}^*, \zeta_{01}^*\}$ for the generalized kernel of the adjoint $\mathbf{L}_0^*$, with the properties

$$\mathbf{L}_0^*\zeta_{01}^* = 0, \quad \mathbf{L}_0^*\zeta_{00}^* = \zeta_{01}^*,$$

and

$$\langle \zeta_0, \zeta_{00}^* \rangle = 1, \quad \langle \zeta_1, \zeta_{00}^* \rangle = 0, \quad \langle \zeta_0, \zeta_{01}^* \rangle = 0, \quad \langle \zeta_1, \zeta_{01}^* \rangle = 1.$$

The solvability condition is now

$$\langle \mathbf{P}_0\mathbf{R}, \zeta_{01}^* \rangle = 0,$$

and a solution $\mathbf{P}_0\Psi$ is determined up an element in the kernel of $\mathbf{L}_0$. Among these solutions there is precisely one solution, $\mathbf{P}_0\widetilde{\Psi}$, which is orthogonal to ζ_{00}^*, and summarizing we have that the solutions are of the form

$$\mathbf{P}_0\Psi = \mathbf{P}_0\widetilde{\Psi} + \alpha\zeta_0, \quad \langle \mathbf{P}_0\widetilde{\Psi}, \zeta_{00}^* \rangle = 0, \quad \alpha \in \mathbb{R}.$$

Taking now the adjoint $\mathbf{P}_0^*$ of $\mathbf{P}_0$, we can rewrite the solvability condition

$$\langle \mathbf{R}, \zeta_1^* \rangle = 0, \quad \zeta_1^* = \mathbf{P}_0^*\zeta_{01}^*,$$

and the solutions

$$\mathbf{P}_0\Psi = \mathbf{P}_0\widetilde{\Psi} + \alpha\zeta_0, \quad \langle \widetilde{\Psi}, \zeta_0^* \rangle = 0, \quad \zeta_0^* = \mathbf{P}_0^*\zeta_{00}^*, \quad \alpha \in \mathbb{R},$$

with $\widetilde{\Psi}$ uniquely determined by the condition $\langle \widetilde{\Psi}, \zeta_0^* \rangle = 0$, and α an arbitrary real number. In the case when the operator $\mathbf{L}$ has an adjoint $\mathbf{L}^*$ in $\mathscr{X}^*$, then ζ_0^* and ζ_1^* above are the vectors in the dual basis of the generalized kernel of the $\mathbf{L}^*$, with the properties

$$\mathbf{L}^*\zeta_1^* = 0, \quad \mathbf{L}^*\zeta_0^* = \zeta_1^*,$$

and

$$\langle\zeta_0,\zeta_0^*\rangle = 1, \quad \langle\zeta_1,\zeta_0^*\rangle = 0, \quad \langle\zeta_0,\zeta_1^*\rangle = 0, \quad \langle\zeta_1,\zeta_1^*\rangle = 1.$$

Notice that in this case again we have that the range of $\mathbf{L}$ is the space orthogonal to ζ_1^*.

Going back to the equalities (4.19)–(4.24), we can now determine the different coefficients in (4.18) from the solvability conditions, which give,

$$\begin{aligned}
\beta_2^{(0)} &= \langle\mathbf{R}_{20}(\zeta_0,\zeta_0),\zeta_1^*\rangle,\\
\beta_1^{(0)} &= \langle 2\mathbf{R}_{20}(\zeta_0,\zeta_1) - 2\Psi_{200},\zeta_1^*\rangle,\\
\alpha_1^{(1)}(\mu) &= \langle\mathbf{R}_{01}(\mu),\zeta_1^*\rangle,\\
\alpha_2^{(1)}(\mu) &= \langle -\alpha_1^{(1)}(\mu)\Psi_{110} + \mathbf{R}_{11}(\zeta_0,\mu) + 2\mathbf{R}_{20}(\zeta_0,\Psi_{001}(\mu)),\zeta_1^*\rangle,\\
\alpha_3^{(1)}(\mu) &= \langle -2\alpha_1^{(1)}(\mu)\Psi_{020} - \Psi_{101}(\mu) + \mathbf{R}_{11}(\zeta_1,\mu) + 2\mathbf{R}_{20}(\zeta_1,\Psi_{001}(\mu)),\zeta_1^*\rangle.
\end{aligned}$$

Here, the terms Ψ_{200}, Ψ_{110}, $\Psi_{001}(\mu)$, Ψ_{020} and $\Psi_{101}(\mu)$ are obtained by solving successively the equations (4.19), (4.20), (4.22), (4.21), and (4.23), using the procedure explained above. First, from (4.19) we find

$$\Psi_{200} = \widetilde{\Psi}_{200} + \psi_{200}\zeta_0, \quad \langle\widetilde{\Psi}_{200},\zeta_0^*\rangle = 0, \quad \psi_{200}\in\mathbb{R},$$

and then from (4.20) we obtain

$$\Psi_{110} = \widetilde{\Psi}_{110} + 2\psi_{200}\zeta_1 + \psi_{110}\zeta_0, \quad \langle\widetilde{\Psi}_{110},\zeta_0^*\rangle = 0, \quad \psi_{110}\in\mathbb{R}.$$

The solvability condition for equation (4.21) determines the coefficient ψ_{200},

$$2\psi_{200} = \langle\mathbf{R}_{20}(\zeta_1,\zeta_1) - \widetilde{\Psi}_{110},\zeta_1^*\rangle,$$

and then solving (4.21) we find

$$\Psi_{020} = \widetilde{\Psi}_{020} + \psi_{110}\zeta_1 + \psi_{020}\zeta_0, \quad \langle\widetilde{\Psi}_{020},\zeta_0^*\rangle = 0, \quad \psi_{020}\in\mathbb{R}.$$

Next, from (4.22) we obtain

$$\Psi_{001}(\mu) = \widetilde{\Psi}_{001}(\mu) + \psi_{001}(\mu)\zeta_0, \quad \langle\widetilde{\Psi}_{001},\zeta_0^*\rangle = 0, \quad \psi_{001}(\mu)\in\mathbb{R},$$

and solving (4.23) we find

$$\begin{aligned}
&\Psi_{101}(\mu) = \widetilde{\Psi}_{101}(\mu) + \alpha_1^{(1)}(\mu)\psi_{110}\zeta_1 + \psi_{101}(\mu)\zeta_0 + 2\psi_{001}(\mu)\widetilde{\Psi}_{200},\\
&\langle\widetilde{\Psi}_{101},\zeta_0^*\rangle = 0, \quad \psi_{101}(\mu)\in\mathbb{R},
\end{aligned}$$

where we have used in (4.23) the equality (4.19) which gives

$$2\mathbf{R}_{20}(\zeta_0,\Psi_{001}(\mu)) = 2\mathbf{R}_{20}(\zeta_0,\widetilde{\Psi}_{001}(\mu)) + 2\psi_{001}(\mu)(\beta_2^{(0)}\zeta_1 - \mathbf{L}\widetilde{\Psi}_{200}).$$

Notice that we do not need to solve (4.24) and determine $\Psi_{011}(\mu)$.

In the formulas above we have determined Ψ_{110}, Ψ_{020}, $\Psi_{001}(\mu)$, and $\Psi_{101}(\mu)$, up to an element $\psi_{110}\zeta_0$, $\psi_{020}\zeta_0$, $\psi_{001}(\mu)\zeta_0$, and $\psi_{101}(\mu)\zeta_0$, respectively, which belongs to the kernel of $\mathbf{L}$ and is arbitrary. The simplest choice is to take

$$\psi_{110} = \psi_{020} = \psi_{001}(\mu) = \psi_{101}(\mu) = 0.$$

However, notice that the coefficients $\beta_2^{(0)}$, $\beta_1^{(0)}$, and $\alpha_1^{(1)}(\mu)$ are uniquely determined, whereas $\alpha_2^{(1)}(\mu)$ and $\alpha_3^{(1)}(\mu)$ depend upon the choice of ψ_{110} and $\psi_{001}(\mu)$. We can then make use of the fact that ψ_{110} and $\psi_{001}(\mu)$ are arbitrary, in order to further simplify the normal form.

Further Transformation

Consider the coefficient $\alpha_2^{(1)}(\mu)$ that we rewrite as

$$\begin{aligned}\alpha_2^{(1)}(\mu) = \langle -\alpha_1^{(1)}(\mu)\widetilde{\Psi}_{110} - 2\psi_{200}\mathbf{R}_{01}(\mu) + \mathbf{R}_{11}(\zeta_0,\mu) + 2\mathbf{R}_{20}(\zeta_0,\widetilde{\Psi}_{001}(\mu)),\zeta_1^*\rangle \\ +2\beta_2^{(0)}\psi_{001}(\mu).\end{aligned}$$

If the coefficient $\beta_2^{(0)} = 0$, then $\alpha_2^{(1)}(\mu)$ is uniquely determined. If $\beta_2^{(0)} \neq 0$, then we can choose the arbitrary coefficient $\psi_{001}(\mu)$ such that $\alpha_2^{(1)}(\mu) = 0$. Indeed, this is achieved by taking

$$\begin{aligned}\psi_{001}(\mu) = \frac{1}{2\beta_2^{(0)}}\langle \alpha_1^{(1)}(\mu)\widetilde{\Psi}_{110} + 2\psi_{200}\mathbf{R}_{01}(\mu) \\ - \mathbf{R}_{11}(\zeta_0,\mu) - 2\mathbf{R}_{20}(\zeta_0,\widetilde{\Psi}_{001}(\mu)),\zeta_1^*\rangle.\end{aligned}$$

Similarly, for $\alpha_3^{(1)}(\mu)$ we write

$$\begin{aligned}\alpha_3^{(1)}(\mu) = \langle -2\alpha_1^{(1)}(\mu)\widetilde{\Psi}_{020} - \widetilde{\Psi}_{101}(\mu) + \mathbf{R}_{11}(\zeta_1,\mu) + 2\mathbf{R}_{20}(\zeta_1,\widetilde{\Psi}_{001}(\mu)),\zeta_1^*\rangle \\ -3\alpha_1^{(1)}(\mu)\psi_{110} + \beta_1^{(0)}\psi_{001}(\mu).\end{aligned}$$

Then if $\beta_1^{(0)} \neq 0$ we can take $\psi_{110} = 0$ and

$$\psi_{001}(\mu) = \frac{1}{\beta_1^{(0)}}\langle 2\alpha_1^{(1)}(\mu)\widetilde{\Psi}_{020} + \widetilde{\Psi}_{101}(\mu) - \mathbf{R}_{11}(\zeta_1,\mu) - 2\mathbf{R}_{20}(\zeta_1,\widetilde{\Psi}_{001}(\mu)),\zeta_1^*\rangle,$$

and then $\alpha_3^{(1)}(\mu) = 0$.

Remark 4.3 *(i) Alternatively, we can obtain that either $\alpha_2^{(1)}(\mu) = 0$ or $\alpha_3^{(1)}(\mu) = 0$ by making a change of variables of the form*

$$\widetilde{A} = A - A_*(\mu),$$

with $A_*(\mu)$ suitably chosen, provided $\beta_2^{(0)} \neq 0$ or $\beta_1^{(0)} \neq 0$, respectively. Indeed, we have that $\alpha_2^{(1)}(\mu) = 0$ after the change of variables above, provided $A_*(\mu)$ satisfies

$$\frac{\partial Q(A_*(\mu),\mu)}{\partial A} = 0.$$

The existence of $A_*(\mu)$ with this property is obtained by solving equation

$$\frac{\partial Q(A,\mu)}{\partial A} = 0.$$

Since $\partial Q/\partial A(0,0) = 0$ and $\partial^2 Q/\partial A^2(0,0) = 2\beta_2^{(0)}$, the implicit function theorem gives a unique solution $A_*(\mu)$ of this equation for μ sufficiently small, provided $\beta_2^{(0)} \neq 0$. In a similar way, by solving $P(A_*(\mu),\mu) = 0$, for which $\partial P/\partial A(0,0) = \beta_1^{(0)}$, one finds $\alpha_3^{(1)}(\mu) = 0$ when $\beta_1^{(0)} \neq 0$.

(ii) An example of a second order ODE which has a normal form as described here is given in Section 3.2.4.

Reduced Dynamics

The dynamics of systems of the form (4.18) have been extensively studied in the literature. In particular, we refer the reader to [38] for an analysis of the Takens–Bogdanov bifurcation, which is generically of codimension 2, arising for two small parameters, the coefficients α_1 and α_3 in (4.18). Varying these two coefficients, one finds here saddle-node, Hopf, and homoclinic bifurcations. In Sections 4.1.1 and 4.1.2, Chapter 4, we analyze this system under the additional assumption that it possesses a reversibility symmetry.

3.4.5 Example 4: $(i\omega_1)(i\omega_2)$ bifurcation

Consider again an equation of the form (4.1), with a parameter $\mu \in \mathbb{R}^m$ and satisfying the hypotheses in center manifold Theorem 3.3, Chapter 2. We assume now that the spectrum of the linear operator $\mathbf{L}$ contains precisely two pairs of eigenvalues on the imaginary axis, $\pm i\omega_1$ and $\pm i\omega_2$, with $0 < \omega_1 < \omega_2$. Furthermore, we assume that these eigenvalues are simple, and that $\omega_1/\omega_2 = r/s \in \mathbb{Q}$, where r and s are positive integers, $r < s$, and the fraction is irreducible.

We point out that, since we can use as many parameters as needed, in practical situations when ω_1/ω_2 is irrational, or rational $\omega_1/\omega_2 = r/s$ with large r and s, then it is more convenient to consider these cases as perturbations of the case with

$\omega_1/\omega_2 = r/s$, where r/s is a rational number, with smallest r and s, sufficiently close to ω_1/ω_2 (see also Remark 4.5).

Normal Form

With these assumptions we have $\sigma_0 = \{\pm i\omega_1, \pm i\omega_2\}$, and the associated spectral subspace $\mathscr{E}_0$ is four-dimensional. We choose a basis $\{\zeta_1, \zeta_2, \overline{\zeta}_1, \overline{\zeta}_2\}$ in $\mathscr{E}_0$ consisting of the eigenvectors associated to the eigenvalues $i\omega_1$, $i\omega_2$, $-i\omega_1$, and $-i\omega_2$, respectively. As in the previous examples, center manifold Theorem 3.3, Chapter 2, gives

$$u = u_0 + \Psi(u_0,\mu), \quad u_0 \in \mathscr{E}_0, \quad \Psi(u_0,\mu) \in \mathscr{Z}_h,$$

and applying normal form Theorem 2.2 to the reduced system we find

$$u_0 = v_0 + \Phi_\mu(v_0),$$

which gives the equality (4.3),

$$u = v_0 + \widetilde{\Psi}(v_0,\mu), \quad v_0 \in \mathscr{E}_0, \quad \widetilde{\Psi}(u_0,\mu) \in \mathscr{Z}.$$

For $v_0(t) \in \mathscr{E}_0$, we now write

$$v_0(t) = A(t)\zeta_1 + B(t)\zeta_2 + \overline{A(t)\zeta_1} + \overline{B(t)\zeta_2},$$

in which A and B are complex-valued. According to the result in Lemma 1.15, we find here the normal form

$$\begin{aligned}
\frac{dA}{dt} &= i\omega_1 A + AP_1(|A|^2,|B|^2,A^s\overline{B}^r,\mu) + \overline{A}^{s-1}B^r P_2(|A|^2,|B|^2,\overline{A}^s B^r,\mu) \\
&\quad + \rho_1(A,B,\overline{A},\overline{B},\mu) \\
\frac{dB}{dt} &= i\omega_2 B + BQ_1(|A|^2,|B|^2,\overline{A}^s B^r,\mu) + A^s\overline{B}^{r-1} Q_2(|A|^2,|B|^2,A^s\overline{B}^r,\mu) \\
&\quad + \rho_2(A,B,\overline{A},\overline{B},\mu),
\end{aligned} \tag{4.26}$$

with P_j and Q_j polynomials in their first three arguments satisfying $P_1(0,0,0,0) = Q_1(0,0,0,0) = 0$, and $\rho_j(A,B,\overline{A},\overline{B},\mu) = O((|A|+|B|)^{2p+2})$, $j = 1,2$.

Computation of the Normal Form

We proceed now as in the previous examples and compute the leading order terms in this normal form. We write

$$\begin{aligned}\frac{dA}{dt} &= (i\omega_1 + \alpha_1(\mu))A + A(a|A|^2 + b|B|^2) + \beta_1 \overline{A}^{s-1} B^r + \widetilde{\rho}_1(A,B,\overline{A},\overline{B},\mu)\\ \frac{dB}{dt} &= (i\omega_2 + \alpha_2(\mu))B + B(c|A|^2 + d|B|^2) + \beta_2 A^s \overline{B}^{r-1} + \widetilde{\rho}_2(A,B,\overline{A},\overline{B},\mu), \quad (4.27)\end{aligned}$$

where

$$\alpha_j(\mu) = \alpha_j^{(1)}(\mu) + O(|\mu|^2), \quad j = 1,2,$$

with $\alpha_j^{(1)}$, $j = 1,2$, linear maps in μ, the coefficients a, b, c, d, β_1, and β_2 complex numbers, and

$$\widetilde{\rho}_j(A,B,\overline{A},\overline{B},\mu) = O(|\mu|(|A|+|B|)^3 + (|A|+|B|)^4 + |\mu|(|A|+|B|)^{r+s-1}).$$

Here $r+s \geq 3$, so that the coefficients β_1 and β_2 are relevant in this expansion only in the cases $(r,s) = (1,2)$ and $(r,s) = (1,3)$, which correspond to $\omega_2 = 2\omega_1$ and $\omega_2 = 3\omega_1$, respectively. Therefore the cases $(r,s) = (1,2)$ and $(r,s) = (1,3)$ are also called *strongly resonant cases*, whereas the cases when $r+s \geq 5$ are called *weakly resonant cases.*

The computation of these coefficients can be done exactly as in the previous two examples. We shall therefore only give the results here. First, by looking at the terms of orders $O(\mu A)$ and $O(\mu B)$ we obtain

$$\begin{aligned}\alpha_1^{(1)} &= \langle \mathbf{R}_{11}(\zeta_1) + 2\mathbf{R}_{20}(\zeta_1, \Psi_{00001}), \zeta_1^* \rangle,\\ \alpha_2^{(1)} &= \langle \mathbf{R}_{11}(\zeta_2) + 2\mathbf{R}_{20}(\zeta_2, \Psi_{00001}), \zeta_2^* \rangle,\end{aligned}$$

where

$$\Psi_{00001} = -\mathbf{L}^{-1}\mathbf{R}_{01}.$$

Here ζ_1^* and ζ_2^* belong to $\mathscr{X}^*$, and span the orthogonal to the range of $i\omega_1 - \mathbf{L}$ and $i\omega_2 - \mathbf{L}$, respectively, just as the vector ζ^* constructed in Section 3.4.2. Next, by considering the terms of order 2 in $(A,\overline{A},B,\overline{B})$, in the case $\omega_2 \neq 2\omega_1$, we find

$$\begin{aligned}\Psi_{20000} &= (2i\omega_1 - \mathbf{L})^{-1}\mathbf{R}_{20}(\zeta_1, \zeta_1),\\ \Psi_{10100} &= 2(i(\omega_1 + \omega_2) - \mathbf{L})^{-1}\mathbf{R}_{20}(\zeta_1, \zeta_2),\\ \Psi_{10010} &= 2(i(\omega_1 - \omega_2) - \mathbf{L})^{-1}\mathbf{R}_{20}(\zeta_1, \overline{\zeta_2}),\\ \Psi_{11000} &= -2\mathbf{L}^{-1}\mathbf{R}_{20}(\zeta_1, \overline{\zeta_1}),\\ \Psi_{00200} &= (2i\omega_2 - \mathbf{L})^{-1}\mathbf{R}_{20}(\zeta_2, \zeta_2),\\ \Psi_{00110} &= -2\mathbf{L}^{-1}\mathbf{R}_{20}(\zeta_2, \overline{\zeta_2}),\end{aligned}$$

whereas if $\omega_2 = 2\omega_1$ we need to solve the equations

$$\begin{aligned}\beta_1\zeta_1 + (i\omega_1 - \mathbf{L})\overline{\Psi_{10010}} &= 2\mathbf{R}_{2,0}(\zeta_2, \overline{\zeta_1}),\\ \beta_2\zeta_2 + (i\omega_2 - \mathbf{L})\Psi_{20000} &= \mathbf{R}_{20}(\zeta_1, \zeta_1).\end{aligned}$$

The solvability conditions for these two equations give the formulas for the coefficients β_1 and β_2, in this case,

$$\beta_1 = \langle 2\mathbf{R}_{20}(\zeta_2, \overline{\zeta_1}), \zeta_1^* \rangle,$$
$$\beta_2 = \langle \mathbf{R}_{20}(\zeta_1, \zeta_1), \zeta_2^* \rangle.$$

Finally, by considering the terms of order 3, we find in the case $\omega_2 \neq 3\omega_1$ that

$$\begin{aligned}
a &= \langle 2\mathbf{R}_{20}(\zeta_1, \Psi_{11000}) + 2\mathbf{R}_{20}(\overline{\zeta_1}, \Psi_{20000}) + 3\mathbf{R}_{30}(\zeta_1, \zeta_1, \overline{\zeta_1}), \zeta_1^* \rangle, \\
b &= \langle 2\mathbf{R}_{20}(\zeta_1, \Psi_{00110}) + 2\mathbf{R}_{20}(\zeta_2, \Psi_{10010}) + 2\mathbf{R}_{20}(\overline{\zeta_2}, \Psi_{10100}) \\
&\quad + 6\mathbf{R}_{30}(\zeta_1, \zeta_2, \overline{\zeta_2}), \zeta_1^* \rangle, \\
c &= \langle 2\mathbf{R}_{20}(\zeta_1, \overline{\Psi_{10010}}) + 2\mathbf{R}_{20}(\zeta_2, \Psi_{11000}) + 2\mathbf{R}_{20}(\overline{\zeta_1}, \Psi_{10100}) \\
&\quad + 6\mathbf{R}_{30}(\zeta_1, \zeta_2, \overline{\zeta_1}), \zeta_2^* \rangle, \\
d &= \langle 2\mathbf{R}_{20}(\zeta_2, \Psi_{00110}) + 2\mathbf{R}_{20}(\overline{\zeta_2}, \Psi_{00200}) + 3\mathbf{R}_{30}(\zeta_2, \zeta_2, \overline{\zeta_2}), \zeta_2^* \rangle.
\end{aligned}$$

Exercise 4.4 *Compute the coefficients a, b, c, d, β_1, and β_2 in the case $\omega_2 = 3\omega_1$.*

Reduced Dynamics

Finding the full bifurcation diagram of a parameter-dependent dynamical system in high dimensions is beyond the scope of this book. This is also the case for the system (4.26), which is four-dimensional. Instead, the analysis is often restricted to the questions of finding bounded orbits, such as equilibria, periodic orbits, invariant tori, homoclinic or heteroclinic orbits, and determining their stability properties. In particular, one way of treating the existence question is to first show the existence of some bounded orbit for the truncated normal form, obtained by removing the small remainder ρ, e.g., by removing $\widetilde{\rho}$ in (4.27), and then show the persistence of this orbit for the full system. For equilibria and periodic orbits the persistence question can be often solved by an adapted implicit function theorem, but this question is much more delicate for invariant tori, homoclinics, and heteroclinics, and may be wrong. We discuss this type of difficulty in more detail in Chapter 4 in the case of reversible systems.

We do not attempt to discuss here these issues for the system (4.26), for which we refer for instance to [38]. Instead, we only mention some basic facts for the generic situation in which all the coefficients in (4.27) are nonzero. For the μ-dependent coefficients $\alpha_1(\mu)$ and $\alpha_2(\mu)$, which are small, since $\alpha_1(0) = \alpha_2(0) = 0$, we write

$$\alpha_j(\mu) = \nu_j + i\chi_j, \quad j = 1, 2,$$

and assume that the small real parts ν_j are nonzero. A convenient way of studying system (4.27) is in polar coordinates, by setting

$$A = r_1 e^{i\theta_1}, \quad B = r_2 e^{i\theta_2}.$$

Restricting ourselves to the leading order system obtained by removing the terms $\widetilde{\rho}_j$, $j = 1,2$, in (4.27), we find three equations which decouple:

$$\begin{aligned} \frac{dr_1}{dt} &= \nu_1 r_1 + r_1(a_r r_1^2 + b_r r_2^2) + r_1^{s-1} r_2^r \operatorname{Re}(\beta_1 e^{-i\Theta}) \\ \frac{dr_2}{dt} &= \nu_2 r_2 + r_2(c_r r_1^2 + d_r r_2^2) + r_1^s r_2^{r-1} \operatorname{Re}(\beta_2 e^{i\Theta}) \\ \frac{d\Theta}{dt} &= \gamma + (sa_i - rc_i) r_1^2 + (sb_i - rd_i) r_2^2 + r_1^{s-2} r_2^{r-2} \operatorname{Im}(r_2^2 s\beta_1 e^{-i\Theta} - r_1^2 r\beta_2 e^{i\Theta}), \end{aligned} \tag{4.28}$$

in which

$$\Theta = s\theta_1 - r\theta_2,$$

together with an equation for θ_1. Here $\gamma = s\chi_1 - r\chi_2$ is a *detuning parameter*, and the subscripts r and i indicate the real and the imaginary parts, respectively, of a complex number.

In particular, the equilibria (r_1, r_2, Θ) of the three equations which decouple depend upon the values of the coefficients a, b, c, d, β_1, and β_2, and upon the three small parameters ν_1, ν_2, and γ. These equilibria correspond to periodic solutions for the four-dimensional truncated system (4.27), because of the additional phase θ_1, and, provided they persist, also for the full system (4.26).

Looking at (4.28) we notice again the fundamental difference between the weakly resonant cases where $r+s \geq 5$, and the strongly resonant cases where $r+s \leq 4$. Indeed, in the weakly resonant cases the Θ-dependent terms in the equations for r_1 and r_2 are of an order higher than 3, so that these two equations decouple in the truncation at order 3. One can first solve these two equations, for which we are in the presence of a bifurcation of codimension 2, with two small parameters ν_1 and ν_2. We refer to [38] for a detailed analysis of this situation. However, when including the higher order terms, we observe that the two first equations give the equilibria r_1 and r_2 as functions of Θ, which are generically of size $O((|\nu_1| + |\nu_2|)^{1/2})$. Then the equation for Θ leads to a condition between the small parameters γ and ν_1, ν_2, represented in the three-dimensional parameter space by a "resonance tongue."

Remark 4.5 *The case when ω_1/ω_2 is irrational is similar to the weakly resonant cases discussed above and can be analyzed in the same way. However, we point out that this irrationality condition is physically hard to check, so that in practical situations it is more convenient to regard this situation as a perturbation of a weakly resonant case by considering the closest rational number r/s to ω_1/ω_2 which has the smallest sum $r+s$, and then taking a detuning parameter $\delta = s\omega_1 - r\omega_2$, which is added to the detuning γ in the system (4.28). This allows us to regard this situation as a small perturbation of the case $\omega_1/\omega_2 = r/s$. On the contrary, in the strongly resonant cases the terms in $r_1^{s-1} r_2^r$ and $r_1^s r_2^{r-1}$ are of order 2 or 3, i.e., they are larger or comparable to the cubic terms. This introduces a number of difficulties in the bifurcation study. We refer to [89], and the references therein, for a discussion of the case $\omega_1/\omega_2 = 1/2$.*

Remark 4.6 (**$(i\omega)^2$ bifurcation (1:1 resonance)**) *In the same context as above, one can consider the case $\omega_1 = \omega_2$. The most interesting situation arises when these eigenvalues are double, non-semisimple. In this case the center manifold is four-dimensional and the normal form is given by Lemma 1.17. We refer to [27] for an analysis of the generic cases, in which one finds a bifurcation of codimension 3, i.e., involving three small parameters. In Chapter 4, Section 4.3.3, we shall discuss this situation in the case of reversible systems, where it turns out that the bifurcation is of codimension 1, only.*

3.5 Further Normal Forms

3.5.1 Time-Periodic Normal Forms

A situation which arises quite often in applications is that of a periodically forced system. Here we consider the cases where the system is nonautonomous, as in Section 2.3.2, in Chapter 2, with $\mathbf{R}$ being periodic in t. In particular, this means that the time-dependency occurs as a small perturbation near the origin. This is not the general case of systems with time-periodic coefficients, and also not the case of autonomous systems near a closed orbit, for which normal forms may be found for general cases in [53, 56].

We consider a differential equation in $\mathbb{R}^n$ of the form

$$\frac{du}{dt} = \mathbf{L}u + \mathbf{R}(u,\mu,t), \tag{5.1}$$

for which we assume that the following hypothesis holds.

Hypothesis 5.1 *Assume that* $\mathbf{L}$ *and* $\mathbf{R}$ *in (5.1) have the following properties:*

(i) $\mathbf{L}$ *is a linear map in* $\mathbb{R}^n$*;*
(ii) for some $k \geq 2$ *and* $l \geq 1$*, there exists a neighborhood* $\mathscr{V}$ *of the origin in* $\mathbb{R}^n \times \mathbb{R}^m$ *such that the map* $t \mapsto \mathbf{R}(\cdot,\cdot,t)$ *belongs to* $H^l(\mathbb{R}, \mathscr{C}^k(\mathscr{V}, \mathbb{R}^n))$*;*
(iii) $\mathbf{R}(0,0,t) = 0$ *and* $D_u\mathbf{R}(0,0,t) = 0$ *for all* $t \in \mathbb{R}$*;*
(iv) there exists $\tau > 0$*, such that*

$$\mathbf{R}(u,\mu,t+\tau) = \mathbf{R}(u,\mu,t) \textit{ for all } t \in \mathbb{R}, (u,\mu) \in \mathscr{V}.$$

Notice that the time dependency is taken in the Sobolev space H^l with $l \geq 1$. This is to insure that we can multiply two such functions, since $H^1(\mathbb{R}/\tau\mathbb{Z})$ is an algebra. We could use continuous functions instead, but H^l is really useful when we are looking at infinite-dimensional problems.

Theorem 5.2 (Periodically forced normal form) *Consider the system (5.1) and assume that Hypothesis 5.1 holds. Then for any positive integer* $p \leq k$ *there exist* *neighborhoods* $\mathscr{V}_1$ *and* $\mathscr{V}_2$ *of* 0 *in* $\mathbb{R}^n$ *and* $\mathbb{R}^m$*, respectively, and a* τ*-periodic function* $t \mapsto \Phi(\cdot,\cdot,t)$*, which belongs to* $H^l(\mathbb{R}/\tau\mathbb{Z}, \mathscr{C}^k(\mathbb{R}^n \times \mathscr{V}_2, \mathbb{R}^n))$*, with the following properties:*

(i) Φ *is a polynomial of degree* p *in its first argument, and the coefficients of the monomials of degree* q *belong to* $H^l(\mathbb{R}/\tau\mathbb{Z}, \mathscr{C}^{k-q}(\mathscr{V}_2, \mathbb{R}^n))$*. Furthermore,*

$$\Phi(0,0,t) = 0, \quad D_u\Phi(0,0,t) = 0 \textit{ for all } t \in \mathbb{R}.$$

(ii) *For* $v \in \mathscr{V}_1$*, the polynomial change of variable*

$$u = v + \Phi(v,\mu,t),$$

transforms system (2.1) into the "normal form"

$$\frac{dv}{dt} = \mathbf{L}v + \mathbf{N}(v,\mu,t) + \rho(v,\mu,t), \tag{5.2}$$

with the following properties:

a. *The map* $t \mapsto \mathbf{N}(\cdot,\cdot,t)$ *is* τ*-periodic and satisfies*

$$\mathbf{N}(0,0,t) = 0, \quad D_v\mathbf{N}(0,0,t) = 0 \textit{ for all } t \in \mathbb{R}.$$

Furthermore, $\mathbf{N}$ *is a polynomial of degree* p *in its first argument and the coefficients of the monomials of degree* q *belong to* $H^l(\mathbb{R}/\tau\mathbb{Z}, \mathscr{C}^{k-q}(\mathscr{V}_2, \mathbb{R}^n))$*.*

b. *The equality*

$$e^{t\mathbf{L}^*}\mathbf{N}(e^{-t\mathbf{L}^*}v,\mu,t) = \mathbf{N}(v,\mu,0) \tag{5.3}$$

holds for all $(t,v) \in \mathbb{R} \times \mathbb{R}^n$ *and* $\mu \in \mathscr{V}_2$*.*

c. *The map* ρ *belongs to* $H^l(\mathbb{R}/\tau\mathbb{Z}, \mathscr{C}^k(\mathscr{V}_1 \times \mathscr{V}_2, \mathbb{R}^n))$ *and*

$$\rho(v,\mu,t) = o(\|v\|^p) \textit{ for all } (t,v) \in \mathbb{R} \times \mathscr{V}_1,\ \mu \in \mathscr{V}_2.$$

A preliminary version of this theorem appeared in [24].

Remark 5.3 *As in Theorem 1.2 we can replace (5.3) by*

$$\frac{\partial \mathbf{N}(v,\mu,t)}{\partial t} = D_v\mathbf{N}(v,\mu,t)\mathbf{L}^*v - \mathbf{L}^*\mathbf{N}(v,\mu,t) \textit{ for all } (t,v) \in \mathbb{R} \times \mathbb{R}^n,\ \mu \in \mathscr{V}_2. \tag{5.4}$$

Via Fourier analysis this equation is, for every Fourier mode, of the same form as (1.5).

Proof (of Theorem 5.2) Proceeding as in the proof of Theorems 1.2 in Section 3.1.1, we are lead to solve the equation

$$\frac{\partial \Phi}{\partial t} + \mathscr{A}_{\mathbf{L}}\Phi + \mathbf{N} = \Pi_p(\mathbf{R}(\cdot + \Phi, \mu) - D_v\Phi \cdot \mathbf{N}) \tag{5.5}$$

with respect to $(\Phi,\mathbf{N})$, which are unknown functions of (v,μ,t). This equation is the analogue for this situation of the equality (C.10) in the proof of Theorem 2.2, and the notations $\mathscr{A}_\mathbf{L}$ and Π_p below have the same meaning as in this proof (see also equality (2.6)).

We start by solving the equation at $\mu = 0$. Then, at each degree q in v, we have to solve a linear equation of the form

$$\frac{\partial \Phi}{\partial t} + \mathscr{A}_\mathbf{L}\Phi = \mathbf{Q} - \mathbf{N}, \tag{5.6}$$

in which $\mathbf{Q} \in H^l(\mathbb{R}/\tau\mathbb{Z}, \mathscr{H}_q)$, where $\mathscr{H}_q$ is the space of homogeneous polynomials of degree q, as in Section 3.1.1. Taking the Fourier expansion with respect to t of (5.6), we find for the kth Fourier coefficient,

$$\left(\frac{2ik\pi}{\tau} + \mathscr{A}_\mathbf{L}\right)\Phi^{(k)} = \mathbf{Q}^{(k)} - \mathbf{N}^{(k)}.$$

This equation is now solved using the scalar product introduced in Section 3.1.1. It follows that we may choose $\mathbf{N}^{(k)}$ as the orthogonal projection of $\mathbf{Q}^{(k)}$ on the kernel of the adjoint of $(2ik\pi/\tau + \mathscr{A}_\mathbf{L})$, which is $(-2ik\pi/\tau + \mathscr{A}_{\mathbf{L}^*})$, and $\Phi^{(k)}$ orthogonal to the kernel of $(2ik\pi/\tau + \mathscr{A}_\mathbf{L})$. In fact, this is equivalent to considering the scalar product in $L^2(\mathbb{R}/\tau\mathbb{Z}, \mathscr{H}_q)$ defined through

$$\langle \Phi, \Psi \rangle_\tau = \frac{1}{\tau}\int_0^\tau \langle \Phi(\cdot,t), \overline{\Psi}(\cdot,t)\rangle dt, \tag{5.7}$$

and then directly solving (5.6) with the help of the formal adjoint $-\partial/\partial t + \mathscr{A}_{\mathbf{L}^*}$ of the linear operator $\partial/\partial t + \mathscr{A}_\mathbf{L}$ in $L^2(\mathbb{R}/\tau\mathbb{Z}, \mathscr{H})$.

The Fourier analysis above shows that there is a unique solution $(\Phi,\mathbf{N})$ of (5.6) satisfying

$$\Phi \in H^{l+1}(\mathbb{R}/\tau\mathbb{Z}, \mathscr{H}_q), \quad \Phi \in \left(\ker\left(\frac{\partial}{\partial t} + \mathscr{A}_\mathbf{L}\right)\right)^\perp,$$

and

$$\mathbf{N} \in H^l(\mathbb{R}/\tau\mathbb{Z}, \mathscr{H}_q), \quad \mathbf{N} \in \ker\left(-\frac{\partial}{\partial t} + \mathscr{A}_{\mathbf{L}^*}\right),$$

for any $\mathbf{Q} \in H^l(\mathbb{R}/\tau\mathbb{Z}, \mathscr{H}_q)$. Furthermore, the linear mapping $\mathbf{Q} \mapsto (\Phi,\mathbf{N})$ is bounded from $H^l(\mathbb{R}/\tau\mathbb{Z}, \mathscr{H}_q)$ to $(H^l(\mathbb{R}/\tau\mathbb{Z}, \mathscr{H}_q))^2$.

Finally, we solve the equation (5.5) for small μ. The proof is done in the same way as the proof of Theorem 2.2 given in Appendix C.5, and we omit the details here. □

Remark 5.4 *Consider an infinite-dimensional system, as in Chapter 2, and assume that the center manifold theorems in Chapter 2, Theorem 3.9 with periodic time-dependence, and Theorem 3.3 for perturbed vector fields apply (e.g., see the example of periodically forced Hopf bifurcation below). We then obtain a reduced* *finite-dimensional system in $\mathscr{E}_0$ which is of the form (5.1). Hence, we can apply Theorem 5.2 to this reduced system. For the computation of the normal form, we can make it directly on the infinite-dimensional system, as it is not necessary to split the computation into the computation of the center manifold and the computation of the normal form, just as in the computation made in Section 3.4.*

3.5.2 Example: Periodically Forced Hopf Bifurcation

Consider an infinite-dimensional system of the form

$$\frac{du}{dt} = \mathbf{L}u + \mathbf{R}(u, \mu, t), \tag{5.8}$$

where, with the notations from Chapter 2,

$$\mathbf{L} \in \mathscr{L}(\mathscr{Z}, \mathscr{X}) \text{ and } \mathbf{R} \in H^l(\mathbb{R}, \mathscr{C}^k(\mathscr{V}, \mathscr{Y})),$$

for $k \geq 2$, $l \geq 1$, and $\mathscr{V}$ a neighborhood of the origin in $\mathscr{Z} \times \mathbb{R}^m$. We assume that $\mathbf{R}$ is τ-periodic in t,

$$\mathbf{R}(u, \mu, t + \tau) = \mathbf{R}(u, \mu, t) \text{ for all } (u, \mu) \in \mathscr{V},\ t \in \mathbb{R},$$

and

$$\mathbf{R}(0, 0, t) = 0, \quad D_u\mathbf{R}(0, 0, t) = 0.$$

We further assume that the hypotheses of Theorems 3.9 and 3.3 in Chapter 2 are satisfied, and that the Hypothesis 2.4 of Chapter 2 on $\mathbf{L}$ holds with $\sigma_0 = \{\pm i\omega\}$, in which $\pm i\omega$ are simple eigenvalues.

Normal Form

Under the above hypotheses, we find a 2-dimensional reduced system to which we can apply the Theorem 5.2. We choose an eigenvector ζ associated with the eigenvalue $i\omega$, so that $\{\zeta, \overline{\zeta}\}$ is a basis of $\mathscr{E}_0$. As in the previous examples, we then have

$$u(t) = v_0(t) + \widetilde{\Psi}(v_0(t), \mu, t),$$

with

$$v_0(t) = A(t)\zeta + \overline{A(t)\zeta} \in \mathscr{E}_0,$$

and $\widetilde{\Psi}(v_0, \mu, t) \in \mathscr{Z}$, for (v_0, μ) in a neighborhood of 0 in $\mathscr{E}_0 \times \mathbb{R}^m$. Furthermore, Ψ is τ-periodic in t, and

$$\widetilde{\Psi}(0, 0, t) = 0, \quad D_{v_0}\widetilde{\Psi}(0, 0, t) = 0.$$

The normal form of the reduced equation is

$$\frac{dA}{dt} = i\omega A + N(A,\overline{A},\mu,t) + \rho(A,\overline{A},\mu,t),$$

with N polynomial of degree p in $(A,\overline{A})$, with coefficients depending upon μ and t, as in Theorem 5.2, and

$$\rho(A,\overline{A},\mu,t) = O(|A|^{p+1}).$$

Moreover,

$$N(0,0,0,t) = 0, \partial_A N(0,0,0,t) = \partial_{\overline{A}} N(0,0,0,t) = 0,$$
$$N(A,\overline{A},\mu,t+\tau) = N(A,\overline{A},\mu,t),$$

and the identity (5.3) gives in this case

$$e^{-i\omega t} N(e^{i\omega t}A, e^{-i\omega t}\overline{A},\mu,t) = N(A,\overline{A},\mu,0) \tag{5.9}$$

for all $A \in \mathbb{C}$ and $t \in \mathbb{R}$.

We set $\omega_f = 2\pi/\tau$, and consider the monomials of the nth Fourier mode of $N(A,\overline{A},\mu,\cdot)$ of the form

$$\alpha_{pq}^{(n)}(\mu) A^p \overline{A}^q e^{in\omega_f t}.$$

According to (5.9), these monomials should satisfy

$$\alpha_{pq}^{(n)}(\mu)\left(e^{i((p-q-1)\omega + n\omega_f)t} - 1\right) = 0,$$

so that

$$(p-q-1)\omega + n\omega_f = 0. \tag{5.10}$$

Assume now that

$$\frac{\omega_f}{\omega} = \frac{r}{s} \in \mathbb{Q}.$$

Then the equality (5.10) leads to

$$p-q-1 = lr, \quad n = -ls, \quad l \in \mathbb{Z},$$

and we conclude that

$$N(A,\overline{A},\mu,t) = AN_0(|A|^2,(Ae^{-i\omega t})^r,\mu) + \overline{A}^{r-1} e^{ri\omega t} N_1(|A|^2,(\overline{A}e^{i\omega t})^r,\mu), \tag{5.11}$$

where N_0 and N_1 are polynomials in their first two arguments.

The leading order terms in the normal form now strongly depend upon the value of r. For $r = 1$ we find the truncated equation

$$\begin{aligned}\frac{dA}{dt} &= i\omega A + a(\mu)A + c(\mu)e^{i\omega t} + d(\mu)\overline{A}e^{2i\omega t} + e(\mu)A^2 e^{-i\omega t} + f(\mu)\overline{A}^2 e^{3i\omega t} \\ &\quad + b(\mu)A|A|^2 + g(\mu)A^3 e^{-2i\omega t} + h(\mu)\overline{A}^3 e^{4i\omega t} + j(\mu)\overline{A}|A|^2 e^{2i\omega t},\end{aligned} \tag{5.12}$$

where $a(0) = c(0) = d(0) = 0$. For $r = 2$ we obtain the equation

$$\frac{dA}{dt} = i\omega A + a(\mu)A + c(\mu)\overline{A}e^{2i\omega t} + b(\mu)A|A|^2 + d(\mu)A^3 e^{-2i\omega t} + g(\mu)\overline{A}^3 e^{4i\omega t} + f(\mu)\overline{A}|A|^2 e^{2i\omega t}, \tag{5.13}$$

with $a(0) = c(0) = 0$, whereas for $r \geq 3$ we find

$$\frac{dA}{dt} = i\omega A + a(\mu)A + b(\mu)A|A|^2 + c(\mu)\overline{A}^{r-1} e^{ri\omega t}, \tag{5.14}$$

in which $a(0) = 0$. The cases $r = 1, 2, 3$ are strongly resonant cases, leading to very rich dynamics, the "worse" being $r = 1$.

Remark 5.5 *(i) In the case of a small periodic forcing, i.e., if*

$$\partial_t \mathbf{R}(u, 0, t) = 0,$$

all the coefficients of the time-dependent terms in the above equations vanish at $\mu = 0$. We refer to [26] for an analysis of the dynamics in the cases $r = 1$ and $r = 2$.

(ii) The case when ω_f/ω is irrational is quite academic, since it is physically hard to check. Instead, it is more convenient to consider this case as a small perturbation of the case $\omega_f/\omega = r/s \in \mathbb{Q}$, by choosing a rational number r/s with minimal r close enough to ω_f/ω.

Computation of the Normal Form

We briefly describe below how to compute the terms of order $O(\mu)$ of the coefficients $a(\mu)$ and $c(\mu)$, and the coefficients $b(0)$, $d(0)$, $g(0)$, and $f(0)$ in the case $\omega_f = 2\omega$, i.e., $r = 2$ and $s = 1$.

We set

$$a(\mu) = a^{(1)}(\mu) + O(|\mu|^2), \quad c(\mu) = c^{(1)}(\mu) + O(|\mu|^2),$$

where $a^{(1)}$ and $c^{(1)}$ are linear maps in $\mu \in \mathbb{R}^m$. We proceed as in the previous examples by taking the Taylor expansions of $\mathbf{R}$ and Ψ. With similar notations, we first find at order $O(\mu)$

$$\frac{d\Psi_{001}}{dt} - \mathbf{L}\Psi_{001} = \mathbf{R}_{01}(t).$$

Here $\mathbf{R}_{01}(t)$ is τ-periodic, and after taking its Fourier expansion

$$\mathbf{R}_{01}(t) = \sum_{n \in \mathbb{Z}} \mathbf{R}_{01}^{(n)} e^{in\omega_f t}, \quad \mathbf{R}_{01}^{(n)} \in \mathscr{L}(\mathbb{R}^m, \mathscr{Y}),$$

we then have to solve the equations

$$(in\omega_f - \mathbf{L})\,\Psi_{001}^{(n)} = \mathbf{R}_{01}^{(n)}$$

for any $n \in \mathbb{Z}$. Since $\omega_f = 2\omega$, the operators $(in\omega_f - \mathbf{L})$ are invertible, so that we obtain a unique solution $\Psi_{001} \in H^l(\mathbb{R}/\tau\mathbb{Z}, \mathscr{L}(\mathbb{R}^m, \mathscr{X})$.

Next, we consider the terms of order $O(\mu A)$ and find

$$\frac{d\Psi_{101}}{dt} + (i\omega - \mathbf{L})\Psi_{101} + a_1\zeta + \overline{c}_1 e^{-2i\omega t}\overline{\zeta} = \mathbf{R}_{11}(\zeta)(t) + 2\mathbf{R}_{20}(\zeta, \Psi_{001}(t))(t).$$

Using again Fourier series, we obtain a system of equations for $n \in \mathbb{Z}$ as above. These equations are invertible for $n \notin \{0, -1\}$ and the solvability conditions for $n = 0$ and $n = -1$ determine the coefficients

$$\begin{aligned} a^{(1)} &= \langle \mathbf{R}_{11}(\zeta)(\cdot) + 2\mathbf{R}_{20}(\zeta, \Psi_{001}(\cdot))(\cdot), \zeta^* \rangle_\tau \\ \overline{c}^{(1)} &= \langle \mathbf{R}_{11}(\zeta)(\cdot) + 2\mathbf{R}_{20}(\zeta, \Psi_{001}(\cdot))(\cdot), e^{2i\omega t}\overline{\zeta}^* \rangle_\tau. \end{aligned}$$

Here $\langle \cdot, \cdot \rangle_\tau$ is the scalar product defined through (5.7), and ζ^* is taken such that $\{\zeta^*, \overline{\zeta^*}\}$ is a dual basis of $\{\zeta, \overline{\zeta}\}$ in $\mathscr{E}_0$.

Finally, we compute the coefficients $b(0)$, $d(0)$, $g(0)$, and $f(0)$ by considering successively the terms of orders $O(A^2)$, $O(A\overline{A})$, $O(A^3)$, and $O(A^2\overline{A})$. At orders $O(A^2)$ and $O(A\overline{A})$, we find

$$\begin{aligned} \frac{d\Psi_{200}}{dt} + (2i\omega - \mathbf{L})\Psi_{200} &= \mathbf{R}_{20}(\zeta, \zeta)(t), \\ \frac{d\Psi_{110}}{dt} - \mathbf{L}\Psi_{110} &= 2\mathbf{R}_{20}(\zeta, \overline{\zeta})(t), \end{aligned}$$

and $\Psi_{200}(t)$ and $\Psi_{110}(t)$ are determined just as Ψ_{001} above. At orders $O(A^3)$, and $O(A^2\overline{A})$ we obtain

$$\begin{aligned} &\frac{d\Psi_{300}}{dt} + (3i\omega - \mathbf{L})\Psi_{300} + de^{-2i\omega t}\zeta + \overline{g}e^{-4i\omega t}\overline{\zeta} \\ &\quad = 2\mathbf{R}_{20}(\zeta, \Psi_{200}(t))(t) + \mathbf{R}_{30}(\zeta, \zeta, \zeta)(t), \\ &\frac{d\Psi_{210}}{dt} + (i\omega - \mathbf{L})\Psi_{210} + b\zeta + \overline{f}e^{-2i\omega t}\overline{\zeta} \\ &\quad = 2\mathbf{R}_{20}(\overline{\zeta}, \Psi_{200}(t))(t) + 3\mathbf{R}_{30}(\zeta, \zeta, \overline{\zeta})(t) + 2\mathbf{R}_{20}(\zeta, \Psi_{110}(t))(t), \end{aligned}$$

and the coefficients are obtained from the solvability conditions for these equations:

$$\begin{aligned} b(0) &= \langle 2\mathbf{R}_{20}(\overline{\zeta}, \Psi_{200}(\cdot))(\cdot) + 3\mathbf{R}_{30}(\zeta, \zeta, \overline{\zeta})(\cdot) + 2\mathbf{R}_{20}(\zeta, \Psi_{110}(\cdot))(\cdot), \zeta^* \rangle_\tau, \\ \overline{f(0)} &= \langle 2\mathbf{R}_{20}(\overline{\zeta}, \Psi_{200}(\cdot))(\cdot) + 3\mathbf{R}_{30}(\zeta, \zeta, \overline{\zeta})(\cdot) + 2\mathbf{R}_{20}(\zeta, \Psi_{110}(\cdot))(\cdot), e^{2i\omega t}\overline{\zeta}^* \rangle_\tau, \\ d(0) &= \langle 2\mathbf{R}_{20}(\zeta, \Psi_{200}(\cdot))(\cdot) + \mathbf{R}_{30}(\zeta, \zeta, \zeta)(\cdot), e^{2i\omega t}\zeta^* \rangle_\tau, \\ \overline{g(0)} &= \langle 2\mathbf{R}_{20}(\zeta, \Psi_{200}(\cdot))(\cdot) + \mathbf{R}_{30}(\zeta, \zeta, \zeta)(\cdot), e^{4i\omega t}\overline{\zeta}^* \rangle_\tau. \end{aligned}$$

Exercise 5.6 (Periodically forced vibrating structure) *Consider a system in $\mathbb{R}^n$, $n = 2m$, of the form*

$$\frac{du}{dt} = \mathbf{L}u + \mathbf{R}(u,t),$$

in which $\mathbf{L}$ *and* $\mathbf{R}$ *have the following properties:*

(i) *the linear map* $\mathbf{L}$ *has* $2m$ *simple, purely imaginary eigenvalues* $\pm i\omega_j$, $j = 1,2,\ldots,m$;
(ii) *the map* $\mathbf{R}$ *is smooth and* τ*-periodic in* t;
(iii) $\mathbf{R}(0,t) = D_u\mathbf{R}(0,t) = 0$ *for all* $t \in \mathbb{R}$.

Further consider the change of variables in the normal form Theorem 5.2,

$$u = v + \Phi(v,t),$$

with Φ *polynomial in* v, τ*-periodic in* t, *satisfying* $\Phi(0,t) = D_v\Phi(0,t) = 0$, *and with*

$$v = \sum_{j=1}^{m} A_j\zeta_j + \sum_{j=1}^{m} \overline{A}_j\overline{\zeta}_j,$$

where ζ_j *are the eigenvectors associated with the eigenvalues* $i\omega_j$.

Set $\omega_f = 2\pi/\tau$, *and take* r_0 *and* r_j, $j = 1,\ldots,m$ *a set of integers, such that*

$$r_0\omega_f + \sum_{j=1}^{m} r_j\omega_j = 0, \quad r_0 \neq 0, \tag{5.15}$$

with a minimal total degree $|r|$ *defined by*

$$|r| = \sum_{j=1}^{m} |r_j|.$$

Assuming the nonresonance condition

$$\sum_{j=1}^{m} \alpha_j\omega_j \neq 0, \quad \alpha_j \in \mathbb{Z}, \quad |\alpha| \leq p+1$$

for some $p \geq 3$, *show that*

(i) *the normal form at order* p *reads*

$$\frac{dA_j}{dt} = i\omega_j A_j + A_j P_j(|A_1|^2 + \cdots + |A_m|^2) + Q_j(A_1,\ldots,A_m,\overline{A}_1,\ldots,\overline{A}_m,t),$$

where P_j *are polynomials, and* Q_j *are polynomials in* $(A_1,\ldots,A_m,\overline{A}_1,\ldots,\overline{A}_m)$ *with* τ*-periodic coefficients in* t;

(ii) *the lowest order monomials in the normal form that have time-dependent coefficients are of degree* $|r| - 1$, *and their coefficients are proportional to either* $e^{ir_0\omega_f t}$ *or* $e^{-ir_0\omega_f t}$.

Application: *Take* $m = 3$ *and assume that the eigenvalues* $\pm i\omega_1, \pm i\omega_2, \pm i\omega_3$ *of* $\mathbf{L}$ *satisfy*

$$\sum_{j=1}^{3} \alpha_j\omega_j \neq 0,$$

for any $\alpha = (\alpha_1,\alpha_2,\alpha_3) \in \mathbb{Z}^3$ *with* $|\alpha| \leq 4$. *Further assume that the frequency* ω_f *of the periodic forcing satisfies*

$$4\omega_f + \omega_1 - 3\omega_2 = 0,$$

and that no other integer combination corresponding to the minimal degree $|r| = 4$ *exists.*

(i) *Show that the normal form at order 3 contains* only *the following time-dependent terms:* $c_1 e^{-4i\omega_f t} A_2^3$ *in the equation for* A_1*, and* $c_2 e^{4i\omega_f t} A_1 \overline{A}_2^2$ *in the equation for* A_2*, with complex coefficients* c_1 *and* c_2*.*

(ii) *Consider polar coordinates* $\theta_j = \arg A_j$ *and set* $\Theta = \theta_1 - 3\theta_2 + 4\omega_f t$*. Show that the normal form at order 3 written in these polar coordinates leads to a four-dimensional autonomous system for* (r_1, r_2, r_3, Θ)*, which decouples from the two equations for the phases* θ_1 *and* θ_3*.*

3.5.3 Normal Forms for Analytic Vector Fields

An interesting issue about normal forms arises when the vector field in (2.1) is analytic in (u, μ). The polynomials $\boldsymbol{\Phi}$ and $\mathbf{N}$ exist for any order $p \in \mathbb{N}$, and a natural question is then the convergence of the series resulting as $p \to \infty$. In general this series does not converge, but under suitable conditions, it is possible to determine an optimal degree for the normal form polynomial that minimizes the remainder term ρ (in the sense that the remainder is exponentially small). We present in this section two recent results by Iooss and Lombardi [62, 63] which show the existence of this optimal degree.

Definition 5.7 *Consider a linear map* $\mathbf{L}$ *on* $\mathbb{C}^n$ *with eigenvalues* $\lambda_1, \ldots, \lambda_n \in \mathbb{C}$*. Set* $\lambda = (\lambda_1, \ldots, \lambda_n) \in \mathbb{C}^n$*, and consider* $\gamma > 0$ *and* $\tau > n - 1$*. The linear map* $\mathbf{L}$ *is called* (γ, τ)-homologically diophantine *if for every* $\alpha = (\alpha_1, \ldots, \alpha_n) \in \mathbb{N}^n$*, with* $|\alpha| \geq 2$*, where* $|\alpha| = \sum_{j=1}^n \alpha_j$*, the following inequality holds:*

$$|\langle \lambda, \alpha \rangle - \lambda_j| \geq \frac{\gamma}{|\alpha|^\tau},$$

whenever $\langle \lambda, \alpha \rangle - \lambda_j \neq 0$*.*

The following result is proved in [62].

Theorem 5.8 (Optimal normal form) *Consider the system (2.1) with* $\mathbf{R}$ *an analytic map in a neighborhood of the origin in* $\mathbb{R}^n \times \mathbb{R}^m$ *such that* $\mathbf{R}(0, \mu) = 0$ *for all* μ*. Assume that there exist positive constants c and r such that in the expansion*

$$\mathbf{R}(u, \mu) = \sum_{k+l \geq 2, k \geq 1} \mathbf{R}_{kl}(u^{(k)}, \mu^{(l)})$$

of $\mathbf{R}$*, the* $(k+l)$*-linear maps* $\mathbf{R}_{kl}$ *on* $(\mathbb{R}^n)^k \times (\mathbb{R}^m)^l$ *satisfy*

$$\|\mathbf{R}_{kl}(u_1, \ldots, u_k, \mu_1, \ldots, \mu_l)\| \leq c \frac{\|u_1\| \ldots \|u_k\| \, \|\mu_1\| \ldots \|\mu_l\|}{r^{k+l}}.$$

Then for any $p \geq 2$*, the result in Theorem 2.2 holds, with* $\boldsymbol{\Phi}$ *and* $\mathbf{N}$ *polynomials of degree p in* (u, μ)*. Furthermore, the following properties hold:*

(i) If the linear operator $\mathbf{L}$ *is diagonalizable and* (γ, τ)*-homologically diophantine, then there is a degree* p_{opt} *for the polynomials* Φ *and* $\mathbf{N}$ *such that the remainder* ρ *satisfies*

$$\sup_{\|v\|+\|\mu\|\leq\delta} \|\rho(v,\mu)\| \leq M(\tau)e^{-C/\delta^b},$$

where C *depends upon* (c,r,γ,n,m)*,* $M(\tau)$ *depends upon* τ *and* (c,r,γ,n,m)*,* $b=(1+\tau)^{-1}$*, and* $p_{opt} = O(\delta^{-b})$*.*

(ii) If 0 *is the only eigenvalue of* $\mathbf{L}$*, with at most one* 2×2 *or* 3×3 *Jordan block, then the above estimate for* ρ *holds with* $b=1$*.*

Remark 5.9 *(i) Notice that the optimal degree* p_{opt} *of the normal form depends upon the radius of the ball where the remainder* ρ *is estimated. In applications, this is not really a restriction, since one may choose* δ *to be of order* $O(|\mu|^\beta)$ *with* $\beta > 0$ *small enough, such that the "interesting dynamics" take place in a smaller ball. In particular, the bifurcating solutions lie inside this ball. The result shows that, under the above hypotheses, the remainder* ρ *is exponentially small with respect to the relevant terms in the bifurcation analysis of the normal form.*

(ii) The restriction (i) on the linear map $\mathbf{L}$ *in Theorem 5.8 may sometimes be overcome by using a suitable decomposition of the problem (see [61] where the case* $0^{2+}i\omega$*, with* $\mathbf{L}$ *not diagonalizable, is studied).*

A key ingredient in Theorem 5.8 is the analyticity of $\mathbf{R}$. However, this condition is not satisfied by the reduced systems given by the center manifold theorem, of interest here, in which the vector field is not analytic, even when the original vector field is analytic (see Remark 2.14 in Chapter 2). In this situation, the idea is to first use a normal form transformation on a suitably decomposed system, taking advantage of the analyticity of the vector field, and then use the center manifold reduction, taking into account the exponentially small estimate given by the normal form. In this context, the following result has been proved in [63].

Theorem 5.10 *Consider the system (2.1) with* $\mathbf{R}$ *as in Theorem 5.8. Further assume that* $\mathbf{L}$ *is the direct sum of two linear maps* $\mathbf{L}_0$ *on* $\mathbb{R}^{n_0}$ *and* $\mathbf{L}_1$ *on* $\mathbb{R}^{n_1}$*, with* $n_0+n_1 = n$*, such that* $\mathbf{L}_0$ *is diagonalizable with eigenvalues* $\lambda_1^{(0)},\dots,\lambda_{n_0}^{(0)}$*, and that there exist positive constants* γ *and* τ *such that*

$$|\langle\alpha,\lambda^{(0)}\rangle - \lambda_j^{(1)}| \geq \frac{\gamma}{|\alpha|^\tau}, \quad j=1,\dots,n_1, \tag{5.16}$$

for any $\alpha \in \mathbb{N}^{n_0}$*,* $\alpha \neq 0$*, where* $\lambda^{(0)} = (\lambda_1^{(0)},\dots,\lambda_{n_0}^{(0)})$ *and* $\lambda_1^{(1)},\dots,\lambda_{n_1}^{(1)}$ *are the eigenvalues of* $\mathbf{L}_1$*. Then, there exists a polynomial* $\Phi : \mathbb{R}^{n_0}\times\mathbb{R}^m \to \mathbb{R}^{n_1}$ *of optimal degree* p_{opt} *such that the change of variables*

$$u_1 = \widetilde{u}_1 + \Phi(u_0,\mu),$$

transforms the system (2.1) into the following system in $\mathbb{R}^{n_0}\times\mathbb{R}^{n_1}$

$$\frac{du_0}{dt} = \mathbf{L}_0 u_0 + \widetilde{\mathbf{R}}_0(u_0, \widetilde{u}_1, \mu), \tag{5.17}$$
$$\frac{d\widetilde{u}_1}{dt} = \mathbf{L}_1 \widetilde{u}_1 + \widetilde{\mathbf{R}}_1(u_0, \widetilde{u}_1, \mu) + \rho_1(u_0, \mu),$$

in which $\widetilde{\mathbf{R}}_0$, $\widetilde{\mathbf{R}}_1$, *and* ρ_1 *are analytic in their arguments,*

$$\widetilde{\mathbf{R}}_0(u_0, \widetilde{u}_1, \mu) = \mathbf{P}_0 \mathbf{R}(u_0 + \widetilde{u}_1 + \Phi(u_0, \mu), \mu),$$

where $\mathbf{P}_0$ *is the projection on the subspace* $\mathbb{R}^{n_0}$,

$$\widetilde{\mathbf{R}}_1(u_0, \widetilde{u}_1, \mu) = O(\|\widetilde{u}_1\|(\|u_0\| + \|\widetilde{u}_1\| + \|\mu\|)),$$

and, with the notations from Theorem 5.8, $p_{opt} = O(\delta^{-b})$ *where* $b = (1+\nu\tau)^{-1}$, ν *being the maximal algebraic multiplicity of eigenvalues of* $\mathbf{L}_1$, *and*

$$\sup_{\|u_0\|+\|\mu\| \leq \delta} \|\rho_1(u_0, \mu)\| \leq M(\tau) e^{-C/\delta^b}.$$

Remark 5.11 *The polynomial* Φ *in the above theorem satisfies the identity*

$$\begin{aligned} D_{u_0}\Phi(u_0, \mu)\mathbf{L}_0 u_0 - \mathbf{L}_1 \Phi(u_0, \mu) = & -D_{u_0}\Phi(u_0, \mu)\mathbf{P}_0 \mathbf{R}(u_0 + \Phi(u_0, \mu), \mu) \\ & + \mathbf{P}_1 \mathbf{R}(u_0 + \Phi(u_0, \mu), \mu) - \rho_1(u_0, \mu). \end{aligned}$$

From this identity one can compute the coefficients of the polynomial Φ *by identifying the powers of* (u_0, μ) *in the Taylor expansions of both sides (like in the computation described in Section 3.4; see also Figure 2.1). The Theorem 5.10 asserts that there is an optimal degree for the polynomial* Φ *for which the remainder* ρ_1 *is exponentially small.*

A particularly interesting situation arises when in Theorem 5.10 the spectrum of $\mathbf{L}_0$ lies on the imaginary axis, whereas the spectrum of $\mathbf{L}_1$ is hyperbolic, i.e., it has no point on the imaginary axis. In this case the condition (5.16) is always satisfied, so that the result in Theorem 5.10 holds. Notice that if ρ_1 would be identically 0, then the manifold $\widetilde{u}_1 = 0$, i.e.,

$$\{u = u_0 + u_1 = u_0 + \Phi(u_0, \mu) \; ; \; u_0 \in \mathbb{R}^{n_0}\},$$

would be an invariant center manifold for the system (2.1). This means that we have found in Theorem 5.10 an approximated center manifold, with an exponentially small error, but with the property of keeping the analyticity of the vector field. Applying now the center manifold Theorem 3.3 of Chapter 2 to the system (5.17) one finds a reduced system for $u_0 \in \mathbb{R}^{n_0}$ in which the vector field is the sum of an analytic vector field with an exponentially small remainder, and it is possible to adapt the Theorem 5.8 for this reduced system. This result can be generalized to the infinite dimensional situation treated in Section 2.3.1 in Chapter 2. More precisely, we have the following result.

Theorem 5.12 *Consider equation (4.1), under the hypotheses of the center manifold Theorem 3.3, Chapter 2. With the notations from Section 2.3.1, further assume that* $\mathbf{R}$ *is analytic on* $\mathscr{V}_u \times \mathscr{V}_\mu$ *and that* $\mathbf{L}_0$ *is diagonalizable. Then, there exists a polynomial* $\Phi : \mathscr{E}_0 \times \mathbb{R}^m \to \mathscr{Z}_h$ *of optimal degree* p_{opt}*, such that the change of variable*

$$u_h = \widetilde{u}_h + \Phi(u_0, \mu)$$

transforms equation (4.1) into a system of the form (5.17) for $u_0 \in \mathscr{E}_0$ *and* $\widetilde{u}_h \in \mathscr{Z}_h$*, with the same properties as in Theorem 5.10 where the subscript* 1 *is replaced by h.*

Remark 5.13 *As in the finite-dimensional case, one can apply center manifold Theorem 3.3 of Chapter 2 to the system given by the theorem above, and find a center manifold of the form* $\{u = u_0 + \Phi(u_0, \mu) + O(e^{-C/\delta})\ ;\ u_0 \in \mathscr{E}_0\}$ *in a ball of radius* δ *in* $\mathscr{X}$*. Again, it is possible to adapt Theorem 5.8 for the reduced system.*

Remark 5.14 *Another interesting situation in Theorem 5.10 arises when the eigenvalues of* $\mathbf{L}_0$ *and* $\mathbf{L}_1$ *are all purely imaginary. Provided they satisfy the condition (5.16), the result of the theorem allows us to give a bound for the solutions of the initial value problem, for initial values lying on the manifold* $\{u = u_0 + \Phi(u_0, \mu); u_0 \in \mathscr{E}_0\}$*. One expects that* $\widetilde{u}_1$ *stays exponentially close to* 0 *for a very long time, i.e., we don't see the eigenmodes of* $\mathbf{L}_1$ *for a very long time of order* $O(\delta^{-(b+1/\nu)})$*, where* ν *is the maximal index of the eigenvalues of* $\mathbf{L}_1$ *(see [63]). This situation occurs for instance in the theory of nonlinear vibrations of structures, where in some circumstances many modes are not excited, this being true for all times due to the existence of a small dissipation in the structure. A precise statement of this last assertion would be an interesting application of these results.*

Exercise 5.15 *Consider a system of the form (2.1) with* $\mu \in \mathbb{R}$ *and such that* $\mathbf{R}(0, \mu) = 0$ *for all* μ*. Further assume that the eigenvalues of* $\mathbf{L}$ *are all purely imaginary* $\{\pm i\omega_j; j = 0, 1, \ldots, r\}$*, with* $\pm i\omega_0$ *simple eigenvalues and such that this nonresonance condition is satisfied:*

$$n\omega_0 \neq \omega_j, \quad j = 1, \ldots, r, \quad n \in \mathbb{Z}.$$

Set

$$u = A\zeta_0 + \overline{A\zeta_0} + \Phi(A, \overline{A}, \mu) + v, \quad A \in \mathbb{C},\ v \in E_1,$$

where ζ_0 *is an eigenvector of* $\mathbf{L}$ *associated to the eigenvalue* $i\omega_0$*,* E_1 *is the spectral subspace associated to the eigenvalues* $\{\pm i\omega_j; j = 1, \ldots, r\}$*, and* Φ *is a polynomial in its arguments taking values in* $\mathbb{R}^n$*.*

(i) Check that the hypotheses of Theorem 5.10 are satisfied, with $\mathbf{L}_0$ *being the restriction of* $\mathbf{L}$ *to the spectral space associated to the eigenvalues* $\pm i\omega_0$ *and* $\mathbf{L}_1$ *the restriction of* $\mathbf{L}$ *to* E_1*.*
(ii) Show that there is a polynomial Φ *such that the system satisfied by* $(A, \overline{A}, v)$ *becomes*

$$\begin{aligned}
\frac{dA}{dt} &= Ag(|A|^2, \mu) + R_0(A, \overline{A}, v, \mu) + \rho_0(A, \overline{A}, \mu) \\
\frac{dv}{dt} &= \mathbf{L}_1 v + \mathbf{R}_1(A, \overline{A}, v, \mu) + \rho_1(A, \overline{A}, \mu),
\end{aligned}$$

with the properties:

$$g(|A|^2,\mu) = i\omega_0 + a\mu + b|A|^2 + h.o.t.,$$
$$|R_0(A,\overline{A},v,\mu)| + \|\mathbf{R}_1(A,\overline{A},v,\mu)\| = O(\|v\|(|A| + \|v\| + |\mu|)),$$
$$\sup_{\|u_0\|+\|\mu\|\leq\delta} (|\rho_0(u_0,\mu)| + \|\rho_1(u_0,\mu)\|) \leq Me^{-C/\delta^b}.$$

(iii) Determine the first order terms of the polynomial Φ. *(One finds the same formulas as for the Hopf bifurcation in Section 3.4.2.)*

(iv) Notice that if $b_r < 0$, *and if at time* $t = 0$ *the* v *component is* 0, *or exponentially small, then it stays exponentially small for a very long time.*

Chapter 4
Reversible Bifurcations

In this chapter we present a number of typical bifurcations that occur for reversible systems in dimensions 2, 3, and 4. We focus on bifurcations of codimension 1, which involve only one bifurcation parameter. Reversible systems are first order systems in which the vector field anticommutes with a linear symmetry. We already met reversible systems in Section 2.3.3 of Chapter 2, and in Section 3.3.2 of Chapter 3, where we have seen that the reversibility property is preserved by both the center manifold reduction and the normal form transformation (Theorem 3.15 in Chapter 2 and Theorem 3.4 in Chapter 3, respectively). We discuss in this chapter reversible systems for which the linearization at the origin has a spectrum lying on the imaginary axis, including in this way the reduced systems provided by the center manifold theorem. In all cases, the analysis relies upon the normal form transformation for reversible systems in Theorem 3.4 in Chapter 3.

4.1 Dimension 2

In this section, we consider reversible systems in $\mathbb{R}^2$ of the form

$$\frac{du}{dt} = \mathbf{F}(u,\mu), \tag{1.1}$$

for which we assume that the vector field $\mathbf{F}$ satisfies the following hypothesis.

Hypothesis 1.1 *Assume that the vector field* $\mathbf{F}$ *is of class* $\mathscr{C}^k$, $k \geq 3$, *in a neighborhood* $\mathscr{V}$ *of* $(0,0) \in \mathbb{R}^2 \times \mathbb{R}$, *satisfying*

$$\mathbf{F}(0,0) = 0. \tag{1.2}$$

We further assume that there is a symmetry $\mathbf{S}$ *with*

$$\mathbf{S}^2 = \mathbb{I}, \tag{1.3}$$

M. Haragus, G. Iooss, *Local Bifurcations, Center Manifolds, and Normal Forms in Infinite-Dimensional Dynamical Systems*, Universitext,
DOI 10.1007/978-0-85729-112-7_4, © EDP Sciences 2011

which anticommutes *with* $\mathbf{F}$,

$$\mathbf{F}(\mathbf{S}u,\mu) = -\mathbf{SF}(u,\mu) \textit{ for all } (u,\mu) \in \mathscr{V}. \tag{1.4}$$

Remark 1.2 *(i) Condition (1.2) shows that $u = 0$ is an equilibrium of (1.1) for $\mu = 0$. Our analysis concerns the dynamics of (1.1) close to this equilibrium.*

(ii) The property (1.4) gives by definition a reversible system (see also Section 2.3.3 for a definition in infinite dimensions). We point out that this situation occurs for instance for second order differential equations modeling mechanical systems without dissipation. In Section 2.4.5 of Chapter 2, we discussed an elliptic PDE in a strip possessing the spatial symmetry $x \mapsto -x$. As we have seen, this reflection symmetry induces a reversibility symmetry when the equation is written as a first order system with the coordinate x playing the role of the evolutionary variable (denoted by t in (1.1)). We come back to this example at the end of Section 4.1.1 since the reduced system provided by the center manifold theorem is two-dimensional and enters into the present discussion.

(iii) A key property of reversible systems is that if $t \mapsto u(t)$ is a solution of (1.1), then $t \mapsto \mathbf{S}u(-t)$ is also a solution of (1.1).

We consider the linearized operator $\mathbf{L}$ at the equilibrium $u = 0$ for $\mu = 0$,

$$\mathbf{L} = D_u\mathbf{F}(0,0).$$

An immediate consequence of reversibility property (1.4) is that $\mathbf{L}$ anticommutes with $\mathbf{S}$. This property implies that the two eigenvalues of $\mathbf{L}$ are pairs $\{\lambda, -\lambda\}$, since if λ is an eigenvalue with associated eigenvector ζ, then $-\lambda$ is an eigenvalue with associated eigenvector $\mathbf{S}\zeta$. Consequently, the spectrum $\sigma(\mathbf{L})$ of $\mathbf{L}$ is symmetric with respect to the origin in the complex plane, and since $\mathbf{L}$ is real we further conclude that it is symmetric with respect to both the real and imaginary axis. We then have the following three cases:

(i) $\sigma(\mathbf{L}) = \{\pm\lambda\}$ for some nonzero real number λ;
(ii) $\sigma(\mathbf{L}) = \{\pm i\omega\}$ for some nonzero real number ω;
(iii) $\sigma(\mathbf{L}) = \{0\}$ with 0 an algebraically double eigenvalue.

In the first two cases the linear operator $\mathbf{L}$ is invertible, and by arguing with the implicit function theorem, we find in each of these two cases a unique *family of equilibria* $u = u(\mu)$ for μ close to 0, which are solutions of

$$\mathbf{F}(u(\mu),\mu) = 0.$$

Here $u(0) = 0$ and the map $\mu \mapsto u(\mu)$ is of class $\mathscr{C}^k$ for small μ. Furthermore, since $\mathbf{S}u(\mu)$ is also an equilibrium of (1.1), due to reversibility, from the uniqueness of the equilibria $u(\mu)$ we have that

$$\mathbf{S}u(\mu) = u(\mu).$$

The linearized operators at these equilibria,

$$\mathbf{L}_\mu = D_u\mathbf{F}(u(\mu),\mu),$$

are functions of μ of class $\mathscr{C}^{k-1}$, and the same regularity is true for their eigenvalues. Indeed, the two eigenvalues of $\mathbf{L}_\mu$ are small perturbations of the two simple eigenvalues of $\mathbf{L}$, so that they are also simple for μ sufficiently small and functions of μ of class $\mathscr{C}^{k-1}$ (e.g., see [76, Chapter 2]). Moreover, since $\mathbf{S}u(\mu) = u(\mu)$, it is easy to check that $\mathbf{L}_\mu$ anticommutes with $\mathbf{S}$, just as $\mathbf{L}$ does. In particular, this shows that the set of eigenvalues of $\mathbf{L}_\mu$ is also symmetric with respect to both the real and imaginary axis.

In case (i) the situation is hyperbolic, and, as already mentioned in Section 1.2 of Chapter 1, no bifurcation occurs in this case, the Hartman–Grobman theorem showing that the phase portraits of (1.1) are qualitatively the same when varying μ in a neighborhood of 0. The dynamics in a neighborhood of 0 are the same as those of the linear equation $du/dt = \mathbf{L}u$, and are completely understood.

In case (ii), the symmetry of the spectrum of $\mathbf{L}_\mu$ observed above implies that the two purely imaginary eigenvalues $\pm i\omega$ of $\mathbf{L}$ stay purely imaginary, and nonzero, for sufficiently small μ. In particular, they do not cross the imaginary axis when varying the parameter μ, in contrast to the case of the Hopf bifurcation discussed in Section 1.2.1 of Chapter 1. It turns out that in this case the dynamics of the nonlinear system (1.1) are qualitatively the same as those of the linearized problem $du/dt = \mathbf{L}_\mu u$, i.e., the equilibrium $u(\mu)$ is surrounded by a one-parameter family of periodic orbits. This result strongly relies upon the reversibility property of the system and can be proved using the classical Lyapunov center theorem in a rather straightforward manner (e.g., see [77]). We point out that such a situation also arises for Hamiltonian systems.

The most interesting case is the last one, when 0 is an algebraically double eigenvalue. We focus on the generic case, when 0 is geometrically simple, so that $\mathbf{L}$ has a one-dimensional kernel. If 0 is geometrically double, then $\mathbf{L}$ is diagonalizable, which means here that $\mathbf{L} = 0$. This is a very special case, and might result in applications, for instance, from the occurrence of a second symmetry, or from the special value of an additional parameter, coming from a codimension 2 problem. Assuming that 0 is geometrically simple, we can choose a basis $\{\zeta_0, \zeta_1\}$ of $\mathbb{R}^2$ such that

$$\mathbf{L}\zeta_0 = 0, \quad \mathbf{L}\zeta_1 = \zeta_0. \tag{1.5}$$

Then we have

$$\mathbf{S}\mathbf{L}\zeta_0 = \mathbf{L}\mathbf{S}\zeta_0 = 0,$$

and since the kernel of $\mathbf{L}$ is one-dimensional,

$$\mathbf{S}\zeta_0 = k\zeta_0,$$

in which $k = \pm 1$, due to (1.3). Following the notations used in Chapter 3, we refer to the case $k = +1$ as an 0^{2+} bifurcation and to the case $k = -1$ as an 0^{2-} bifurcation. We discuss these cases in the following two sections. We refer for example to the recent paper [124], and the references therein, for a detailed analysis of these bifurcations.

4.1.1 Reversible Takens–Bogdanov Bifurcation 0^{2+}

We consider here the 0^{2+} bifurcation so that we assume the following holds.

Hypothesis 1.3 *Assume that* $\sigma(\mathbf{L}) = \{0\}$, *with* 0 *an algebraically double and geometrically simple eigenvalue. Further assume that the eigenvector* ζ_0 *associated with this eigenvalue satisfies*

$$\mathbf{S}\zeta_0 = \zeta_0,$$

in which $\mathbf{S}$ *is the symmetry anticommuting with* $\mathbf{L}$ *in Hypothesis 1.1.*

Normal Form

We start by constructing a convenient basis of $\mathbf{R}^2$.

Lemma 1.4 (0^{2+} **basis of** $\mathbb{R}^2$) *Assume that Hypothesis 1.3 holds. Then there exists a basis* $\{\zeta_0, \zeta_1\}$ *of* $\mathbb{R}^2$ *consisting of generalized eigenvectors of* $\mathbf{L}$ *such that*

$$\mathbf{L}\zeta_0 = 0, \quad \mathbf{L}\zeta_1 = \zeta_0, \tag{1.6}$$

$$\mathbf{S}\zeta_0 = \zeta_0, \quad \mathbf{S}\zeta_1 = -\zeta_1. \tag{1.7}$$

Proof Consider a basis $\{\zeta_0, \zeta_1'\}$ of $\mathbb{R}^2$ such that $\mathbf{L}\zeta_0 = 0$ and $\mathbf{L}\zeta_1' = \zeta_0$. Then

$$\mathbf{LS}\zeta_1' = -\zeta_0,$$

and there is $\alpha \in \mathbb{R}$ such that

$$\mathbf{S}\zeta_1' = -\zeta_1' + \alpha\zeta_0.$$

Define the vector ζ_1 by

$$\zeta_1 = \zeta_1' - \frac{\alpha}{2}\zeta_0.$$

Then

$$\mathbf{S}\zeta_1 = -\zeta_1,$$

and the lemma is proved. □

We use now the results in Chapter 3 to determine the normal form of the equation (1.1) in this case.

Lemma 1.5 (0^{2+} **normal form**) *Assume that Hypotheses 1.1 and 1.3 hold, and consider a basis* $\{\zeta_0, \zeta_1\}$ *of* $\mathbb{R}^2$ *satisfying (1.6) and (1.7). Then for any positive* *integer p, $2 \le p \le k$, there exist neighborhoods $\mathscr{V}_1$ and $\mathscr{V}_2$ of 0 in $\mathbb{R}^2$ and $\mathbb{R}$, respectively, and a map $\Phi : \mathscr{V}_1 \times \mathscr{V}_2 \to \mathbb{R}^2$ with the following properties:*

(i) Φ is of class $\mathscr{C}^k$, satisfying

$$\Phi(0,0,0) = 0, \quad \partial_{(A,B)}\Phi(0,0,0) = 0,$$

and

$$\Phi(A,-B,\mu) = \mathbf{S}\Phi(A,B,\mu).$$

(ii) For $(A,B) \in \mathscr{V}_1$, the change of variables

$$u = A\zeta_0 + B\zeta_1 + \Phi(A,B,\mu), \tag{1.8}$$

transforms the equation (1.1) into the normal form

$$\begin{aligned} \frac{dA}{dt} &= B \\ \frac{dB}{dt} &= Q(A,\mu) + \rho(A,B,\mu), \end{aligned} \tag{1.9}$$

where $Q(\cdot,\mu) : \mathbb{R} \to \mathbb{R}$ is a polynomial of degree p such that

$$Q(0,0) = 0, \quad \partial_A Q(0,0) = 0,$$

and the remainder $\rho \in \mathscr{C}^k(\mathscr{V}_1 \times \mathscr{V}_2, \mathbb{R})$ is an even function in B satisfying

$$\rho(A,B,\mu) = o((|A| + |B|)^p).$$

Proof Applying the results in Theorem 2.2 and Lemma 1.9, and taking into account Remark 1.10(ii), in Chapter 3, we obtain the existence of the map Φ, which transforms (1.1) into the normal form

$$\begin{aligned} \frac{dA}{dt} &= B \\ \frac{dB}{dt} &= BP(A,\mu) + Q(A,\mu) + \rho(A,B,\mu), \end{aligned} \tag{1.10}$$

with $P(\cdot,\mu)$ and $Q(\cdot,\mu)$ polynomials on $\mathbb{R}$.

Next, we use the reversibility symmetry of the equation. First notice that the action of $\mathbf{S}$ in the basis $\{\zeta_0, \zeta_1\}$ is given by the matrix

$$\mathbf{S} = \begin{pmatrix} 1 & 0 \\ 0 & -1 \end{pmatrix}.$$

Applying now the result in Theorem 3.4 in Chapter 3 and taking into account that the change of variables (1.16) in Remark 1.10(ii) of Chapter 3, preserves the symmetry $\mathbf{S}$, we find that the map Φ commutes with $\mathbf{S}$, i.e.,

$$\Phi(A,-B,\mu)=\mathbf{S}\Phi(A,B,\mu),$$

and the system (1.10) is reversible, i.e., it anticommutes with $\mathbf{S}$. This property implies that

$$-BP(A,\mu)+Q(A,\mu)+\rho(A,-B,\mu)=BP(A,\mu)+Q(A,\mu)+\rho(A,B,\mu)$$

for all $(A,B)\in\mathbb{R}^2$ and $\mu\in\mathscr{V}_2$. It is now straightforward to check that $P=0$ and that ρ is even in B, which completes the proof. □

Solutions of the Normal Form System

We analyze now the dynamics of the normal form system (1.9). As usual, we start by analyzing the truncated system obtained by removing the remainder ρ and then study the full system. We consider the generic situation in which the following assumption on the expansion of the vector field holds.

Hypothesis 1.6 *Assume that the expansion of Q in (1.9),*

$$Q(A,\mu)=a\mu+bA^2+O(|\mu|^2+|\mu||A|+|A|^3),$$

is such that $a\neq 0$ and $b\neq 0$.

Remark 1.7 *The coefficients a and b in the expansion above can be computed as explained in Chapter 3. We show below how to compute them when starting from a reversible infinite-dimensional system for which the center manifold theorem leads to a two-dimensional reduced system satisfying Hypotheses 1.1 and 1.3.*

Truncated System

We start with the truncated system

$$\begin{aligned}\frac{dA}{dt}&=B\\ \frac{dB}{dt}&=Q(A,\mu)\end{aligned}\tag{1.11}$$

obtained from (1.9) by removing the remainder ρ. A key property of this system it that it is *integrable*, with first integral

$$B^2=2\int_0^A Q(z,\mu)dz+H,\tag{1.12}$$

with constant $H\in\mathbb{R}$. Varying the constant H, we find the full phase portrait of the truncated system.

Assuming that Hypothesis 1.6 holds, the first integral becomes

$$B^2 = f_H(A,\mu),$$

with

$$f_H(A,\mu) = \frac{2}{3}bA^3 + 2a\mu A + H + O(|\mu|^2|A| + |\mu||A|^2 + |A|^4), \quad H \in \mathbb{R}.$$

We show in Figure 1.1 the graphs, in a neighborhood of 0, of f_0 in the case $b<0$ for $a\mu<0$, $\mu=0$, and $a\mu>0$. For $H\neq 0$, the graph of f_H in each case is obtained by a vertical translation from the graph of f_0. It is now straightforward to check that the phase portraits, in a neighborhood of the origin, of (1.11) are as indicated in Figure 1.2.

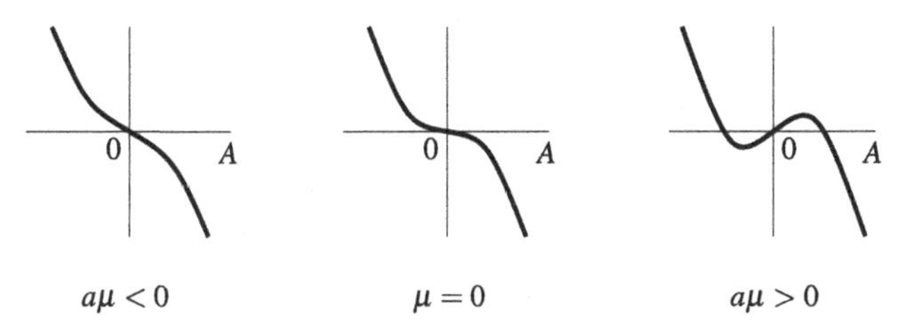

Fig. 1.1 Schematic plot of the graph of f_0 in a neighborhood of the origin in the case $b<0$.

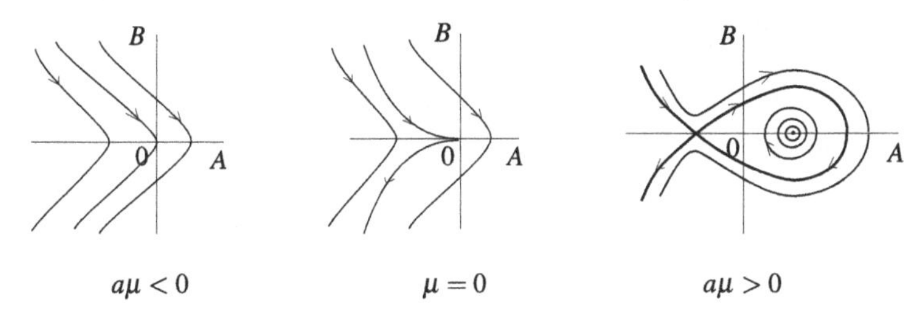

Fig. 1.2 Dynamics of the 0^{2+} bifurcation in the case $b<0$. Phase portraits in a neighborhood of the origin in the (A,B)-plane of the truncated system (1.11). The phase portraits in a neighborhood of the origin of the full normal form (1.9) are qualitatively the same.

The phase portraits are symmetric with respect to the A-axis, since the first integral is even in B. This is actually a consequence of the symmetry **S**. For $ab\mu>0$, the function f_H is monotone in a neighborhood of 0, and therefore there are no bounded orbits of (1.11) in a neighborhood of the origin in this case. For $ab\mu<0$, the function f_H has two critical points, $A_\pm^{(0)} = \pm\sqrt{-a\mu/b} + O(|\mu|)$, which correspond to two equilibria $(A_\pm^{(0)},0)$ for the truncated system (1.11). When $b<0$ the equilibrium $(A_-^{(0)},0)$ is a saddle, while $(A_+^{(0)},0)$ is a center, and the situation is reversed in the case $b>0$. The center equilibrium is surrounded by a *one-parameter family of periodic orbits,* which tend to the center equilibrium as their amplitude is decreased.

Increasing the amplitude, the periods of these periodic orbits tend to infinity, and in the limit we find the *homoclinic orbit*, which connects the saddle equilibrium to itself. This homoclinic orbit constitutes at the same time, the unstable and stable manifolds of the saddle equilibrium.

Full System

We consider now the full system (1.9). While we do not give detailed proofs, we discuss some geometrical properties which indicate that the dynamics of the full system is qualitatively the same as the one of the truncated system.

For $ab\mu > 0$ we observe that the system has no equilibria in a neighborhood of the origin, since $Q(A,\mu)$ does not vanish for A sufficiently small. Moreover, dB/dt has a definite sign for (A,B) sufficiently small, given by the sign of $Q(A,\mu)$. Hence B is monotonous in the neighborhood of the origin, and since $dA/dt = B$, the A-component reaches an extremum when $B = 0$. This allows us to show that in this case the dynamics are qualitatively the same as that of the truncated system.

For $ab\mu < 0$, by arguing with the implicit function theorem it is not difficult to check that the equilibria $(A_{\pm}^{(0)},0)$ of the truncated system persist for the full system, for which we have two *equilibria* $(A_{\pm},0)$, which are $O(\mu)$-close to $(A_{\pm}^{(0)},0)$. Furthermore, these equilibria are of the same type, one is a saddle and the other a center.

We focus here on the persistence of the *homoclinic* orbit connecting the saddle equilibrium to itself. Assume, for fixing ideas, that $b < 0$, so that $(A_{-},0)$ is a saddle and $(A_{+},0)$ is a center. The key idea is to show that the unstable manifold of $(A_{-},0)$ intersects the A-axis in some point $(A_0,0)$. This follows from the fact that the unstable manifold of $(A_{-},0)$ is $O(\mu)$-close to the unstable manifold of $(A_{-}^{(0)},0)$ of the truncated system, and that the unstable manifold of the truncated system intersects the A-axis transversely. Actually, at the intersection point the tangent to the unstable manifold is parallel to the B-axis, in both cases, due to symmetry $\mathbf{S}$. Then, by taking this point as an initial data at $t = 0$ for the system (1.9), we find one solution $t \mapsto (A_u(t), B_u(t))$, which corresponds to the unstable manifold, and tends to $(A_{-},0)$ as $t \to -\infty$. Due to reversibility we have that $t \mapsto (A_u(-t), -B_u(-t))$ is also a solution, which satisfies the same initial data, and tends to $(A_{-},0)$ as $t \to +\infty$. By the uniqueness of solutions of the differential equation, these two solutions coincide, which in particular shows that the unstable manifold tends to $(A_{-},0)$ as $t \to \infty$ and coincides with the stable manifold of the equilibrium $(A_{-},0)$.

These arguments allow one to show in this case the persistence of the homoclinic orbit for the full system (1.9). Finally, using reversibility, again, one can show that inside this homoclinic orbit there is one-parameter family of *periodic orbits*, surrounding the center equilibrium, just as for the truncated system.

Summarizing, we have the following result.

Theorem 1.8 (0^{2+} **bifurcation**) *Assume that Hypotheses 1.1, 1.3, and 1.6 hold. Then for the differential equation (1.1) a reversible Takens–Bogdanov bifurcation* *occurs at $\mu = 0$. More precisely, the following properties hold in a neighborhood of* 0 *in* $\mathbb{R}^2$ *for sufficiently small* μ*:*

(i) *For* $ab\mu > 0$ *there is no small bounded solution of (1.1).*
(ii) *For* $ab\mu < 0$*, the differential equation (1.1) possesses two equilibria of order* $O(|\mu|^{1/2})$*, one equilibrium is a saddle and the other one a center. The center equilibrium is surrounded by a one-parameter family of periodic orbits, which tend to a homoclinic orbit connecting the saddle equilibrium to itself, as the period tends to infinity.*

Remark 1.9 (i) *In the case when the coefficients a or (and) b in Hypothesis 1.6 vanish, one has to consider the next nonzero higher order terms in the expansion of Q.*
(ii) *A frequent case is when* 0 *is a solution of (1.1) for all* μ*, i.e.,* $F(0,\mu) = 0$ *for all* μ*. Then* $a = 0$ *and*

$$Q(A,\mu) = c\mu A + bA^2 + h.o.t.$$

If $b \neq 0$ *and* $c \neq 0$*, the phase portraits of the system are qualitatively the same as in the case* $ab\mu < 0$ *(see Figure 1.2). This situation appears for the reduced system in the example considered in Section 2.4.5 of Chapter 2, which is discussed at the end of this section.*
(iii) *Another important particular case is when the system (1.1) possesses an additional symmetry* $\mathbf{T}$*, which commutes with* $\mathbf{L}$ *and* $\mathbf{R}(\cdot,\mu)$*, and for which the eigenvector* ζ_0 *is antisymmetric,* $\mathbf{T}\zeta_0 = -\zeta_0$*. In this situation the normal form has an additional symmetry, i.e., the vector field is odd in* (A,B)*, and then the phase portraits are symmetric with respect to both axes in the* (A,B)*-plane (see the Exercise 1.10 below, and the example in Section 4.3.1).*

Exercise 1.10 (i) *Determine the phase portraits of the truncated normal form (1.11) with*

$$Q(A,\mu) = c\mu A + dA^3,$$

where $c \neq 0$ *and* $d \neq 0$ *(see Figure 1.3).*
(ii) *Show that the phase portraits are qualitatively the same for the full normal form (1.9), where the vector field is odd in A.*

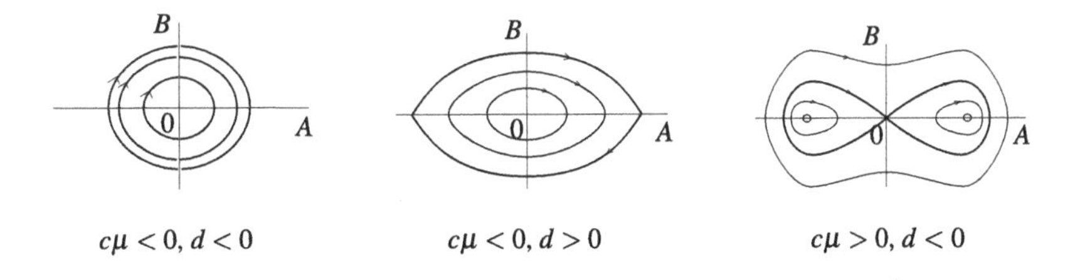

Fig. 1.3 Bounded orbits of (1.9) in the case where the vector field is odd in A.

Computation of the 0^{2+} Normal Form in Infinite Dimensions

We show below how to compute the principal coefficients in the normal form (1.9) when starting from an infinite-dimensional system

$$\frac{du}{dt} = \mathbf{L}u + \mathbf{R}(u,\mu), \tag{1.13}$$

just like the system (4.1) in Chapter 3. We assume that the parameter μ is real and that system (1.13) possesses a reversibility symmetry $\mathbf{S}$ and satisfies the hypotheses of Theorems 3.3 and 3.15 in Chapter 2. We further assume that the spectrum of the linear operator $\mathbf{L}$ is such that $\sigma_0 = \{0\}$, where 0 is an algebraically double and geometrically simple eigenvalue, with a symmetric eigenvector ζ_0 such that

$$\mathbf{S}\zeta_0 = \zeta_0.$$

Then the two-dimensional reduced system satisfies the hypotheses of Lemma 1.4, so that its normal form is given by (1.9).

For the computation of the coefficients a and b in the expansion of Q we proceed as in Section 3.4 and in the examples given in Chapter 3. We start from the equality (4.3) in Chapter 3, in which we take $v_0 = A\zeta_0 + B\zeta_1$, and then write

$$u = A\zeta_0 + B\zeta_1 + \widetilde{\Psi}(A,B,\mu), \tag{1.14}$$

where $\widetilde{\Psi}$ takes values in $\mathscr{Z}$. With the notations from Section 3.2.3, we consider the Taylor expansions of $\mathbf{R}$ and $\widetilde{\Psi}$,

$$\mathbf{R}(u,\mu) = \sum_{1\leq r+q\leq p} \mu^q \mathbf{R}_{r,q}(u^{(r)}) + o((\|u\|_{\mathscr{Y}} + |\mu|)^p), \quad \mathbf{R}_{1,0} = 0, \tag{1.15}$$

and

$$\begin{aligned}\widetilde{\Psi}(A,B,\mu) = & \sum_{1\leq r+s+q\leq p} \Psi_{rsq} A^r B^s \mu^q + o((|A|+|B|+|\mu|)^p), \\ & \Psi_{100} = \Psi_{010} = 0.\end{aligned} \tag{1.16}$$

Using the reversibility symmetry we find that

$$\mathbf{R}_{r,q}((\mathbf{S}u)^{(r)}) = -\mathbf{S}\mathbf{R}_{r,q}(u^{(r)}) \tag{1.17}$$

and

$$\mathbf{S}\Psi_{rsq} = (-1)^s \Psi_{rsq}. \tag{1.18}$$

Now we identify the different powers of (A,B,μ) in identity (4.4) in Chapter 3, obtained here by differentiating (1.14) with respect to t and replacing the derivatives du/dt and dA/dt, dB/dt from (1.13) and (1.9), respectively. This gives here the identity

$$(\partial_A \widetilde{\Psi})B + (\zeta_1 + \partial_B \widetilde{\Psi})Q(A,\mu) = \mathbf{L}\widetilde{\Psi} + \mathbf{R}(A\zeta_0 + B\zeta_1 + \widetilde{\Psi},\mu). \tag{1.19}$$

Using the expansions of $\mathbf{R}$, Ψ, and Q, we find at orders $O(\mu)$ and $O(A^2)$ the equalities

$$\begin{aligned} a\zeta_1 &= \mathbf{L}\Psi_{001} + \mathbf{R}_{0,1}, \\ b\zeta_1 &= \mathbf{L}\Psi_{200} + \mathbf{R}_{2,0}(\zeta_0,\zeta_0). \end{aligned}$$

The coefficients a and b are now found from the solvability conditions for these two equations, just as in Section 3.4.4 of Chapter 3,

$$a = \langle \mathbf{R}_{0,1}, \zeta_1^* \rangle, \tag{1.20}$$
$$b = \langle \mathbf{R}_{2,0}(\zeta_0,\zeta_0), \zeta_1^* \rangle, \tag{1.21}$$

in which ζ_1^* is the vector orthogonal to the range of $\mathbf{L}$ constructed in Section 3.4.4.

Remark 1.11 *(i) We point out that the vector ζ_1^* above can be taken such that*

$$\mathbf{S}^*\zeta_1^* = -\zeta_1^*,$$

where $\mathbf{S}^*$ is the adjoint of the (bounded) symmetry $\mathbf{S}$ in $\mathscr{X}$. Recall that $\zeta_1^* = \mathbf{P}_0^*\zeta_{01}^* \in \mathscr{X}^*$, where $\mathbf{P}_0^*$ is the adjoint of the (bounded) projection $\mathbf{P}_0$, and $\{\zeta_{00}^*, \zeta_{01}^*\}$ is a dual basis for the generalized kernel of the adjoint $\mathbf{L}_0^*$, such that

$$\mathbf{L}_0^*\zeta_{01}^* = 0, \quad \mathbf{L}_0^*\zeta_{00}^* = \zeta_{01}^*, \quad \langle \zeta_k, \zeta_{0j}^* \rangle = \delta_{kj}, \quad k,j = 0,1.$$

For any $\xi \in \mathscr{Z}$ we have

$$\langle \mathbf{L}\xi, \mathbf{S}^*\zeta_1^* \rangle = \langle \mathbf{SL}\xi, \zeta_1^* \rangle = -\langle \mathbf{LS}\xi, \zeta_1^* \rangle = 0,$$

and since $(\mathbf{S}^*)^2 = \mathbb{I}$, we deduce that

$$\mathbf{S}^*\zeta_1^* = \pm\zeta_1^*. \tag{1.22}$$

Using the identity

$$\langle \mathbf{S}\zeta_1, \mathbf{S}^*\zeta_1^* \rangle = \langle \zeta_1, \zeta_1^* \rangle = 1,$$

and $\mathbf{S}\zeta_1 = -\zeta_1$, we conclude that $\mathbf{S}^*\zeta_1^* = -\zeta_1^*$.

(ii) By taking the higher orders in (μ, A, B) in the expansion (1.19) we can compute the coefficients in the expansion of $\widetilde{\Psi}$ and the coefficients of the terms of next orders in the expansion of Q. The computation is similar to the calculation given in Section 3.4.4 of Chapter 3.

Exercise 1.12 *Consider the second order differential equation*

$$x'' - \mu x - \alpha x^2 - \beta (x')^2 - \gamma x^3 - \delta x (x')^2 = 0,$$

in which α, β, γ, δ are real constants and μ is a small real parameter.

(i) Set $u = (x,x')$ and write the equation in the form (1.1). Show that this equation satisfies the hypotheses in Lemma 1.5, with the reversibility symmetry $\mathbf{S}$ defined by

$$S\begin{pmatrix} u_1 \\ u_2 \end{pmatrix} = \begin{pmatrix} u_1 \\ -u_2 \end{pmatrix}$$

and $\zeta_0 = (1,0)$, $\zeta_1 = (0,1)$.

(ii) *With the notations from Lemma 1.5, show that*

$$\Phi(A,B,0) = \frac{\beta}{2}A^2\zeta_0 + \beta AB\zeta_1 + O((|A|+|B|)^3)$$

and

$$Q(A,\mu) = \mu A + \alpha A^2 + \gamma A^3 + O(|\mu|A^2 + A^4).$$

(iii) *Draw the phase portraits in a neighborhood of the origin in the* (A,B)*-plane of the normal form system in the cases* $\alpha \neq 0$ *and* $\alpha = 0$, $\gamma \neq 0$.

Example: Elliptic PDE in a Strip

We consider the elliptic PDE in a strip in the example in Section 2.4.5 of Chapter 2, in which we take

$$g(u_1,u_2,u_3) = -\alpha u_1^2 - \beta u_1 u_3 - \gamma u_2^2 - \delta u_3^2.$$

We saw in Section 2.4.5 that the problem can be written in the form (1.13), to which we can apply Theorems 3.3 and 3.15 in Chapter 2. The resulting two-dimensional reduced system (4.40) in Chapter 2 is reversible and satisfies the hypotheses in Lemma 1.5, so that we can choose coordinates (A,B) in $\mathbb{R}^2$ such that the reduced system is in normal form

$$\begin{aligned} \frac{dA}{dx} &= B \\ \frac{dB}{dx} &= Q(A,\mu) + \rho(A,B^2,\mu), \end{aligned} \tag{1.23}$$

with $\rho(A,B^2,\mu) = O((|A|+|B|)^p)$. Notice that here 0 is a solution of (1.13) for all μ, so that $Q(0,\mu) = 0$, which then leads to the expansion

$$Q(A,\mu) = c\mu A + bA^2 + O(|\mu|^2|A| + |A|^3).$$

Furthermore the coefficients in the expansion (1.16) of $\widetilde{\Psi}$ now satisfy

$$\Psi_{00q} = 0, \quad \mathbf{S}\Psi_{rsq} = (-1)^s \Psi_{rsq}.$$

In order to compute the coefficients c and b we proceed as in the previous section, by identifying powers of (A,B,μ) in (1.19). At orders μA and A^2 we find

$$c\zeta_1 = \mathbf{L}\Psi_{101} + \mathbf{R}_{1,1}\zeta_0 \tag{1.24}$$

$$b\zeta_1 = \mathbf{L}\Psi_{200} + \mathbf{R}_{2,0}(\zeta_0,\zeta_0), \tag{1.25}$$

so that c and b can be found from the compatibility conditions for these equations:

$$c = \langle \mathbf{R}_{1,1}\zeta_0, \zeta_1^* \rangle$$
$$b = \langle \mathbf{R}_{2,0}(\zeta_0, \zeta_0), \zeta_1^* \rangle.$$

In this case the operator $\mathbf{L}$ has a well-defined adjoint $\mathbf{L}^*$ so that ζ_1^* is the vector in the kernel of $\mathbf{L}^*$ satisfying $\langle \zeta_1, \zeta_1^* \rangle = 1$. Using the explicit formulas of the different terms in these equalities, a direct calculation gives

$$c = -1, \quad b = \frac{2}{3\pi}(4\alpha + 2\delta).$$

Assuming that $b \neq 0$ and $\mu \neq 0$, the phase portraits of this system are as the phase portrait to the right of Figure 1.2 (the case $a\mu > 0$). In particular, there is a homoclinic orbit to the saddle equilibrium, and the center equilibrium is surrounded by a family of periodic orbits. In the full elliptic PDE, the homoclinic orbit corresponds to an asymptotically homogeneous solution $v(x,y)$, which tends as $x \to \pm\infty$ towards the same x-independent solution $v_*(y)$. The periodic orbits correspond to a family of solutions of the PDE, which are periodic in x and which tend to the asymptotically homogeneous solution, as their period tends to infinity.

Remark 1.13 *(i) Alternatively, for the computation of the coefficient c we can use the result in the Exercise 3.5 in Chapter 2. This implies that the two eigenvalues of the linearization of the normal form (1.23) at 0 for small μ are precisely the two eigenvalues of the linearization $\mathbf{L} + D\mathbf{R}(0,\mu)$ that vanish at $\mu = 0$. According to the formula (4.39) in Section 2.4.5, these two eigenvalues are $\pm\sqrt{-\mu}$, which then gives $c = -1$.*

(ii) The equality (1.25) leads to the second order differential equation

$$u_{200}'' + u_{200} + b\sin y = \frac{\alpha+\delta}{2} - \frac{\alpha-\delta}{2}\cos(2y) + \frac{\beta}{2}\sin(2y),$$

in which u_{200} represents the first component of the vector Ψ_{200} and satisfies $u_{200}|_{y=0} = u_{200}|_{y=\pi} = 0$. Taking the scalar product of this equality with $\sin y$ and integrating by parts gives the formula for b above.

Exercise 1.14 *(i) Identifying the terms of orders μB, AB, and B^2 in (1.19) show that*

$$\Psi_{101} = \mathbf{L}\Psi_{011} + \mathbf{R}_{1,1}\zeta_1,$$
$$2\Psi_{200} = \mathbf{L}\Psi_{110} + 2\mathbf{R}_{2,0}(\zeta_0, \zeta_1),$$
$$\Psi_{110} = \mathbf{L}\Psi_{020} + \mathbf{R}_{2,0}(\zeta_1, \zeta_1).$$

(ii) Show that the solvability conditions for these equations are always satisfied.

(iii) Compute Ψ_{101}, Ψ_{011}, Ψ_{200}, Ψ_{110}, and Ψ_{020}.

Exercise 1.15 *Consider the example in Section 2.4.5, Chapter 2, with*

$$g(u_1,u_2,u_3) = -\beta u_1 u_3 - \chi u_1 u_2^2.$$

(i) Write the equation in the form (1.13) and show that it is reversible and that the vector field commutes with the symmetry $\mathbf{S}_1$ defined by

$$\mathbf{S}_1 \begin{pmatrix} u_1 \\ u_2 \end{pmatrix} (y) = \begin{pmatrix} -u_1 \\ -u_2 \end{pmatrix} (\pi - y).$$

(ii) Applying Lemma 1.5 to the reduced system, show that $Q(A,\mu)$ and $\rho(A,B^2,\mu)$ are odd in A and that Ψ satisfies (4.41) and

$$\Psi(-A,-B,\mu) = \mathbf{S}_1\Psi(A,B,\mu).$$

(iii) Show that

$$Q(A,\mu) = c\mu A + dA^3 + O(|\mu|^2|A| + |A|^5),$$

with $c = -1$.

(iv) Consider the expansion (1.16) of $\widetilde{\Psi}$. Show that

$$\Psi_{200} = \left(-\frac{\beta}{6}\sin(2y), 0\right), \quad \Psi_{110} = \left(0, -\frac{\beta}{3}\sin(2y)\right), \quad \Psi_{200} = \left(-\frac{\beta}{9}\sin(2y), 0\right)$$

and

$$d = \frac{\beta^2}{12}.$$

*(v) Draw the phase portraits of the reduced system for $\mu > 0$ and $\mu < 0$. In particular, show that for $\mu > 0$ there are two heteroclinic orbits, symmetric under the action of **S** and connecting two equilibria symmetric under the action of $\mathbf{S}_1$ and a one-parameter family of periodic solutions surrounding the origin. A heteroclinic orbit corresponds to a front solution $v(x,y)$ of the elliptic PDE, which tends as $x \to \pm\infty$ towards two homogeneous solutions, which are independent of x and symmetric under the action of $\mathbf{S}_1$.*

Exercise 1.16 *Consider the example in 2.4.5 of Chapter 2, with*

$$g(u_1,u_2,u_3) = -\alpha u_1^2 - \beta u_1 u_3 - \gamma u_2^2 - \delta u_3^2 + \varepsilon u_2,$$

where ε is a small parameter.

*(i) Show that the reversibility symmetry given by **S** is broken.*

(ii) Show that the vector field in (1.13) anticommutes with the symmetry $\widetilde{\mathbf{S}}$ defined by

$$\widetilde{\mathbf{S}}(u,\varepsilon) = (\mathbf{S}u, -\varepsilon).$$

(iii) Using the characterization of the normal form in Lemma 1.9, Chapter 2, and Theorem 2.2 in Chapter 2, show that the normal form of the reduced system is

$$\frac{dA}{dx} = B + \varepsilon dA + h.o.t.$$
$$\frac{dB}{dx} = c\mu A + bA^2 + \varepsilon dB + h.o.t.$$

and show that $c = -1$ and $d = -1/2$.

(iv) Assuming that $\mu < 0$, $\varepsilon > 0$, and $\varepsilon^2 \ll |\mu|$, show that the phase portrait of the reduced system is as in Figure 1.4. In particular, show that the reduced system possesses a heteroclinic orbit connecting the origin to an equilibrium close to $(\mu/b, 0)$. Give an interpretation of the solution $v(x,y)$ of the elliptic PDE corresponding to this orbit.

Exercise 1.17 *Consider the following PDE, due to Ostrowsky [99],*

$$\frac{\partial}{\partial x}\left(\frac{\partial v}{\partial t} - \beta\frac{\partial^3 v}{\partial x^3} + \frac{\partial}{\partial x}(v^2)\right) = \gamma v.$$

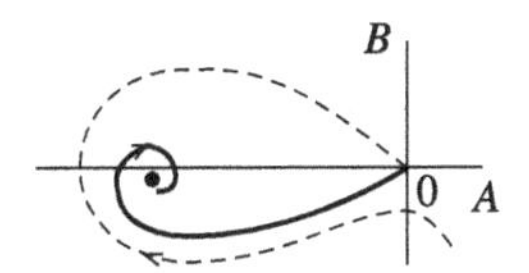

Fig. 1.4 Phase portrait in the (A,B)-plane of the reduced system in Exercise 1.16 for $\mu < 0$, $\varepsilon > 0$, and $\varepsilon^2 \ll |\mu|$.

(This equation models the unidirectional propagation of weakly nonlinear long surface and internal waves of small amplitude in a rotating fluid.) We are interested in traveling waves, i.e., solutions of the form

$$v(x,t) = \phi(x-ct).$$

(i) *Set $z = x - ct$ and show that ϕ satisfies the following fourth-order ODE:*

$$\phi^{(4)} - q\phi'' + \mu\phi + (\phi^2)'' = 0, \tag{1.26}$$

where

$$\mu = \frac{\gamma}{\beta}, \quad q = -\frac{c}{\beta}, \quad v = -\beta\phi.$$

(ii) *Set $u = (u_1,u_2,u_3,u_4) = (\phi,\phi',\phi'',\phi''')$ and rewrite (1.26) as a first order system of the form*

$$\frac{du}{dz} = \mathbf{L}_{q,\mu}u + \mathbf{R}(u,u), \tag{1.27}$$

with

$$\mathbf{L}_{q,\mu} = \begin{pmatrix} 0 & 1 & 0 & 0 \\ 0 & 0 & 1 & 0 \\ 0 & 0 & 0 & 1 \\ -\mu & 0 & q & 0 \end{pmatrix}, \quad \mathbf{R}(u,u) = \begin{pmatrix} 0 \\ 0 \\ 0 \\ -2u_1u_3 - 2u_2^2 \end{pmatrix}.$$

(iii) *Show that system (1.27) is reversible, with the reversibility symmetry $\mathbf{S}$ defined by*

$$\mathbf{S}u = (u_1, -u_2, u_3, -u_4).$$

(iv) *Show that the eigenvalues λ of $\mathbf{L}_{q,\mu}$ satisfy*

$$\lambda^4 - q\lambda^2 + \mu = 0.$$

(v) *Assume that $q > 0$ and that μ is close to 0. Show that the spectrum of $\mathbf{L}_{q,0}$ consists of a pair of simple eigenvalues $\pm\sqrt{q}$ and of 0, which is algebraically double and geometrically simple. Compute a basis $\{\zeta_0, \zeta_1\}$ of the generalized kernel of $\mathbf{L}_{q,0}$ satisfying*

$$\mathbf{S}\zeta_0 = \zeta_0, \quad \mathbf{S}\zeta_1 = -\zeta_1.$$

(vi) *Show that the system satisfies the hypotheses of center manifold Theorems 3.3 and 3.15 in Chapter 2 and of Lemma 1.5. Conclude that for solutions u of the form*

$$u = A\zeta_0 + B\zeta_1 + \widetilde{\Psi}(A,B,\mu),$$

where $\widetilde{\Psi}(A,B,\mu) \in \mathbb{R}^4$ and

$$\mathbf{S}\widetilde{\Psi}(A,B,\mu) = \widetilde{\Psi}(A,-B,\mu),$$

the functions A and B satisfy

$$\frac{dA}{dz} = B$$
$$\frac{dB}{dz} = Q(A,\mu) + \rho(A,B^2,\mu),$$

with

$$Q(A,\mu) = a\mu A + bA^2 + cA^3, \quad \rho(A,B^2,\mu) = O(|\mu|^2|A| + |\mu|A^2 + A^4).$$

(vii) Show that

$$a = \frac{1}{q}, \quad b = c = 0,$$

and, with the notations from (1.16), that

$$\Psi_{200} = \frac{1}{q}\zeta_0, \quad \Psi_{110} = \frac{2}{q}\zeta_1, \quad \Psi_{020} = \left(0, 0, \frac{2}{q}, 0\right).$$

(viii) Consider the normal form of the reduced system at order p. Show that the coefficients of the monomials A^p of the polynomial $Q(\cdot,\mu)$ vanish when $\mu = 0$.

(ix) Conclude that this method does not allow *one to prove the existence of small traveling waves for the Ostrowsky equation (1.26). (We point out that the conclusion is different if the nonlinear term $(\phi^2)''$ in (1.26) is replaced by ϕ^2; see [56, Problem 8, Question 2].)*

4.1.2 Reversible Takens–Bogdanov Bifurcation 0^{2-}

In this section, we treat in the same way the case 0^{2-}, where the following hypothesis holds.

Hypothesis 1.18 *Assume that $\sigma(\mathbf{L}) = \{0\}$, with* 0 *an algebraically double and geometrically simple eigenvalue. Further assume that the eigenvector ζ_0 associated with this eigenvalue satisfies*

$$\mathbf{S}\zeta_0 = -\zeta_0,$$

in which **S** *is the symmetry anticommuting with* **L** *in Hypothesis 1.1.*

Normal Form

As in the case 0^{2+}, we start by constructing a convenient basis of $\mathbf{R}^2$.

Lemma 1.19 (0^{2-} **basis of** $\mathbb{R}^2$) *Assume that Hypothesis 1.18 holds. Then there exists a basis $\{\zeta_0, \zeta_1\}$ of $\mathbb{R}^2$ consisting of generalized eigenvectors of* **L** *such that*

$$\mathbf{L}\zeta_0 = 0, \quad \mathbf{L}\zeta_1 = \zeta_0, \tag{1.28}$$
$$\mathbf{S}\zeta_0 = -\zeta_0, \quad \mathbf{S}\zeta_1 = \zeta_1. \tag{1.29}$$

Proof Except for minor changes, the proof is the same as that of Lemma 1.4, so we omit the details here. □

We use now the results of Chapter 3 to determine the normal form of the equation (1.1) in this case.

Lemma 1.20 (0^{2-} normal form) *Assume that Hypotheses 1.1 and 1.18 hold, and consider a basis $\{\zeta_0, \zeta_1\}$ of $\mathbb{R}^2$ satisfying (1.28) and (1.29). Then, for any positive integer p, $2 \leq p \leq k$, there exist neighborhoods $\mathscr{V}_1$ and $\mathscr{V}_2$ of 0 in $\mathbb{R}^2$ and $\mathbb{R}$, respectively, and a map $\Phi : \mathscr{V}_1 \times \mathscr{V}_2 \to \mathbb{R}^2$ with the following properties:*

(i) Φ is of class $\mathscr{C}^k$, satisfying

$$\Phi(0,0,0) = 0, \quad \partial_{(A,B)}\Phi(0,0,0) = 0,$$

and

$$\Phi(-A,B,\mu) = \mathbf{S}\Phi(A,B,\mu).$$

(ii) For $(A,B) \in \mathscr{V}_1$, the change of variables

$$u = A\zeta_0 + B\zeta_1 + \Phi(A,B,\mu), \tag{1.30}$$

transforms the equation (1.1) into the normal form

$$\begin{aligned}\frac{dA}{dt} &= B\\ \frac{dB}{dt} &= ABP(A^2,\mu) + AQ(A^2,\mu) + A\rho(A,B,\mu),\end{aligned} \tag{1.31}$$

where $P(\cdot,\mu) : \mathbb{R} \to \mathbb{R}$ and $Q(\cdot,\mu) : \mathbb{R} \to \mathbb{R}$ are polynomials of degrees $p-2$ and $p-1$, respectively, such that $Q(0,0) = 0$, and the remainder $\rho \in \mathscr{C}^{k-1}(\mathscr{V}_1 \times \mathscr{V}_2, \mathbb{R})$ is an even function in A satisfying

$$\rho(A,B,\mu) = O((|A| + |B|)^p).$$

Proof As in the proof of Lemma 1.5, using the results in Theorem 2.2, Lemma 1.9, and Remark 1.10(ii) of Chapter 3, we obtain the existence of the map Φ which transforms (1.1) into the normal form

$$\begin{aligned}\frac{dA}{dt} &= B\\ \frac{dB}{dt} &= BP(A,\mu) + Q(A,\mu) + \rho(A,B,\mu),\end{aligned}$$

with $P(\cdot,\mu)$ and $Q(\cdot,\mu)$ polynomials on $\mathbb{R}$.

Now, in the basis $\{\zeta_0, \zeta_1\}$ of $\mathbb{R}^2$ in Lemma 1.19 the reversibility symmetry is represented by the matrix

$$\mathbf{S} = \begin{pmatrix} -1 & 0 \\ 0 & 1 \end{pmatrix}.$$

Applying the result in Theorem 3.4 in Chapter 3, and taking into account that the change of variables (1.16) in Remark 1.10(ii) of Chapter 3 preserves the symmetry $\mathbf{S}$, we find that the vector field in the normal form anticommutes with $\mathbf{S}$, so that

$$BP(-A,\mu)+Q(-A,\mu)+\rho(-A,B,\mu)=-BP(A,\mu)-Q(A,\mu)-\rho(A,B,\mu)$$

for any A, B, and μ in a neighborhood of 0. This implies that P, Q, and ρ are all odd in A, so that we can write

$$BP(A,\mu)+Q(A,\mu)+\rho_2(A,B,\mu)=AB\widetilde{P}(A^2,\mu)+A\widetilde{Q}(A^2,\mu)+A\widetilde{\rho}_2(A^2,B,\mu),$$

from which we conclude that the normal form is as in (1.31). □

Remark 1.21 *Notice that in this case 0 is always a solution of the normal form (1.31), which implies that the full equation (1.1) has the family of symmetric equilibria $u=\Phi(0,0,\mu)$, for μ close to 0.*

Solutions of the Normal Form System

We discuss now the normal form (1.31). We assume that the polynomials P and Q satisfy the following hypothesis.

Hypothesis 1.22 *Assume that*

$$\begin{aligned}P(A^2,\mu)&=c+O(|\mu|+A^2),\\ Q(A^2,\mu)&=a\mu+bA^2+O(|\mu|^2+|\mu|A^2+A^4),\end{aligned}$$

with coefficients $a\neq 0$, $b\neq 0$, and $c\neq 0$.

Remark 1.23 *(i) In the case $c=0$, the phase portrait of the truncated system is as in Figure 1.3, with c and d replaced by a and b, respectively. Often this situation arises when the system possesses an additional symmetry, the case in which one can prove the persistence of these phase portraits for the full system.*
(ii) The coefficients a, b, and c in the expansion above can be computed as explained in Chapter 3. We show below how to compute them when starting from a reversible infinite-dimensional system for which the center manifold theorem leads to a two-dimensional reduced system satisfying the Hypotheses 1.1 and 1.18.

Truncated System

We start by studying the dynamics of the truncated system

$$\begin{aligned}\frac{dA}{dt}&=B\\ \frac{dB}{dt}&=a\mu A+cAB+bA^3.\end{aligned}\tag{1.32}$$

For $ab\mu>0$ there is only one equilibrium in 0, which is a saddle when $a\mu>0$ and a center when $a\mu<0$. It is then not difficult to conclude that the system has no bounded solutions when $a\mu > 0$, and that 0 is surrounded by a family of periodic orbits when $a\mu < 0$ (see Figure 1.5).

For $ab\mu < 0$, there are two additional equilibria,

$$A_+ = \left(\sqrt{\frac{-a\mu}{b}}, 0\right), \quad A_- = \left(-\sqrt{\frac{-a\mu}{b}}, 0\right),$$

which are exchanged by the symmetry $\mathbf{S}$, $\mathbf{S}A_+ = A_-$. These equilibria are both saddles if $b > 0$, both nodes, with opposite stabilities if $b < 0$ and $c^2 + 8b > 0$, and both foci, with opposite stabilities if $b < 0$ and $c^2 + 8b < 0$. Moreover, if we set

$$B = u(v), \quad A^2 = v,$$

then $u(v)$ satisfies the first order differential equation

$$2u'u = a\mu + bv + cu.$$

This equation possesses two explicit solutions,

$$u_\pm = \alpha_\pm + \beta_\pm v,$$

with

$$\beta_\pm = \frac{1}{4}\left(c \pm \sqrt{c^2 + 8b}\right), \quad \alpha_\pm = -\frac{a\mu}{2\beta_\mp}.$$

When $c^2 + 8b > 0$, these two solutions give two parabolas in the (A, B)-plane, which are invariant manifolds of the equilibria A_+ and A_-, and connect these equilibria. In particular, they correspond to *heteroclinic orbits* (see Figure 1.5(ii)–(iii)).

Summarizing, the truncated system possesses no bounded solutions (except for the equilibrium at the origin) in the case $a\mu > 0$ and $b > 0$ and has several bounded solutions in the other cases. We give in Figure 1.5 the phase portraits of the truncated system in these cases. We point out that these phase portraits are all symmetric with respect to the B-axis, due to the reversibility symmetry. In particular, we find:

(i) for $a\mu < 0$ and $b < 0$, a one-parameter family of periodic orbits surrounding the origin;
(ii) for $a\mu < 0$ and $b > 0$, two heteroclinic orbits connecting the nontrivial equilibria A_+ and A_-, together with a one-parameter family of periodic orbits, lying inside the region bounded by the heteroclinic orbits and surrounding the origin;
(iii) for $a\mu > 0$, $b < 0$, and $c^2 + 8b > 0$, a family of heteroclinic orbits connecting the equilibria A_+ and A_-, and two heteroclinic orbits connecting the origin with A_-, and A_+ with the origin;
(iv) for $a\mu > 0$, $b < 0$, and $c^2 + 8b < 0$, a homoclinic orbit to the origin, a family of periodic orbits outside this homoclinic orbit, two heteroclinics connecting the origin with A_-, and A_+ with the origin, and a family of heteroclinics connecting A_+ with A_-.

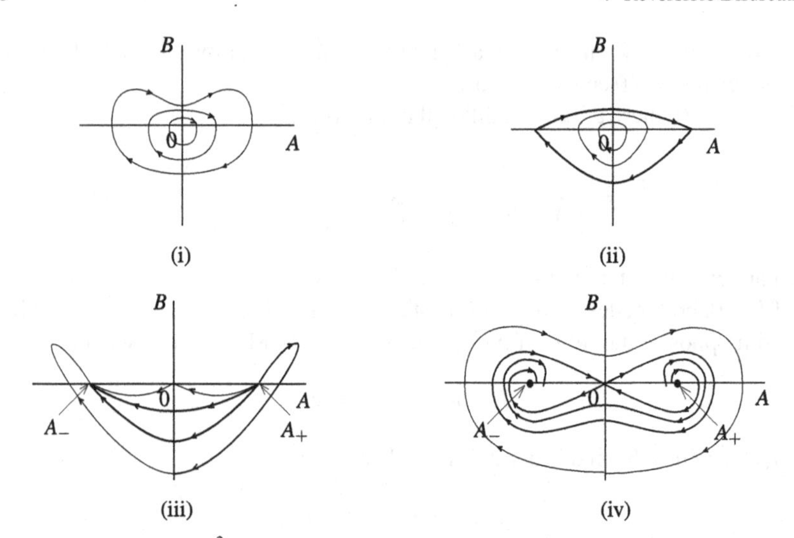

Fig. 1.5 Dynamics of the 0^{2-} bifurcation. Phase portraits in the (A,B)-plane of the truncated system (1.32) in the cases when the system has nontrivial bounded orbits: (i) $a\mu < 0, b < 0$, (ii) $a\mu < 0, b > 0$, (iii) $a\mu > 0$, $b < 0$, $c^2 + 8b > 0$, (iv) $a\mu > 0$, $b < 0$, $c^2 + 8b < 0$. These dynamics persist for the full system.

Notice a striking feature in case (iv), where the phase portrait in the domain outside the homoclinic orbit to the origin resembles a conservative system, while inside the homoclinic orbit it resembles a dissipative system.

Full System

The dynamics of the truncated system persist qualitatively for the full vector field. As for the 0^{2+} bifurcation we only discuss some properties which indicate that these dynamics persist. A detailed analytic proof may be found in [124].

First, the implicit function theorem allows one to show that the fixed points of the truncated system persist for the full system (1.31), and that they have the same stability properties, i.e., saddles, nodes, centers and foci of the truncated system stay saddles, nodes, centers and foci, respectively, for the system (1.31). Furthermore, in a neighborhood of a saddle, the stable and unstable manifolds are close to the corresponding manifolds of the truncated system.

Next, the phase portrait is symmetric with respect to the B-axis, because of the reversibility symmetry $\mathbf{S}$. As a consequence, the right half of a closed orbit for (1.32), which intersects twice the B-axis orthogonally, is perturbed in an integral curve for (1.31), which still intersects twice the B-axis and by symmetry it gives again a closed orbit. This allows one to show the persistence of the periodic orbits. For a homoclinic or a heteroclinic orbit of (1.32), which intersects the B-axis, in the full system (1.31) one has an unstable manifold, starting at $t = -\infty$, which intersects still orthogonally the B-axis; and the reversibility symmetry gives the other half of the curve. This implies the persistence of the heteroclinic orbits connecting $A_\pm$ in the case (ii), of the homoclinic orbit in the case (iv), and the same idea allows one to prove the persistence of the heteroclinic orbits connecting $A_\pm$ in the cases (iii) and (iv). Finally, the persistence of heteroclinic orbits connecting an invariant manifold of the origin with a node in the case (iii), or a focus in the case (iv), is generic.

We summarize these dynamics in the next theorem.

Theorem 1.24 (0^{2-} **bifurcation)** *Assume that Hypotheses 1.1, 1.18, and 1.22 hold. Then for the differential equation (1.1) a reversible Takens–Bogdanov bifurcation occurs at $\mu = 0$. More precisely, the following properties hold in a neighborhood of 0 in $\mathbb{R}^2$, for sufficiently small μ:*

(i) For $a\mu < 0$, $b < 0$, the origin is a center surrounded by a one-parameter family of periodic orbits.
(ii) For $a\mu < 0$, $b > 0$, the differential equation has three equilibria, a center at the origin and two saddles that are exchanged by the reversibility symmetry **S**. *In addition, there is a pair of heteroclinic orbits connecting the two saddles, and a one-parameter family of periodic orbits lying inside the closed domain bounded by the two heteroclinics, and surrounding the origin.*
(iii) For $a\mu > 0$, $b < 0$, $c^2 + 8b > 0$, the differential equation has three equilibria, a saddle at the origin, and two nodes with opposite stabilities and exchanged by the reversibility symmetry **S**. *In addition, there is one family of heteroclinic orbits connecting the two nodes, and two heteroclinics connecting the origin with the stable node, and the unstable node with the origin.*
(iv) For $a\mu > 0$, $b < 0$, $c^2 + 8b < 0$, the differential equation has three equilibria, a saddle at the origin, and two foci with opposite stabilities and exchanged by the reversibility symmetry **S**. *In addition, there is a homoclinic orbit to the saddle at the origin, a family of periodic orbits lying outside this homoclinic orbit, two heteroclinics lying inside this homoclinic orbit, one connecting the origin with the stable focus, and the other one connecting the unstable focus with the origin, and a family of heteroclinics connecting the unstable focus to the stable focus.*
(v) For $a\mu > 0$, $b > 0$, there is no small nontrivial bounded solution of (1.1).

Computation of the 0^{2-} Normal Form in Infinite Dimensions

We briefly show below how to compute the coefficients in the truncated system (1.32), when starting from an infinite-dimensional system of the form (1.13). We assume that this system satisfies the same hypotheses as in Section 4.1.1, except that now we suppose that the eigenvector ζ_0 is antisymmetric,

$$\mathbf{S}\zeta_0 = -\zeta_0.$$

Following the arguments in Section 4.1.1, we start from decomposition (1.14) and the expansions of $\mathbf{R}(u,\mu)$ and $\widetilde{\Psi}(A,B,\mu)$ in (1.15) and (1.16). Property (1.18) is now replaced by

$$\mathbf{S}\Psi_{rsq} = (-1)^r \Psi_{rsq}, \tag{1.33}$$

and the identity (1.19) becomes

$$(\partial_A \widetilde{\Psi})B + (\zeta_1 + \partial_B \widetilde{\Psi})\left(ABP(A^2,\mu) + AQ(A^2,\mu)\right) = \mathbf{L}\widetilde{\Psi} + \mathbf{R}(A\zeta_0 + B\zeta_1 + \widetilde{\Psi},\mu).$$

Then, at orders $O(\mu)$, $O(\mu A)$, $O(A^2)$, $O(AB)$, and $O(A^3)$ we find, successively,

$$0 = \mathbf{L}\Psi_{001} + \mathbf{R}_{0,1}, \tag{1.34}$$

$$a\zeta_1 = \mathbf{L}\Psi_{101} + \mathbf{R}_{1,1}\zeta_0 + 2\mathbf{R}_{2,0}(\zeta_0, \Psi_{001}), \tag{1.35}$$

$$0 = \mathbf{L}\Psi_{200} + \mathbf{R}_{2,0}(\zeta_0, \zeta_0), \tag{1.36}$$

$$c\zeta_1 + 2\Psi_{200} = \mathbf{L}\Psi_{110} + 2\mathbf{R}_{2,0}(\zeta_0, \zeta_1), \tag{1.37}$$

$$b\zeta_1 = \mathbf{L}\Psi_{300} + 2\mathbf{R}_{2,0}(\zeta_0, \Psi_{200}) + \mathbf{R}_{3,0}(\zeta_0, \zeta_0, \zeta_0). \tag{1.38}$$

The solvability conditions for the equations (1.35), (1.37), and (1.38) give the formulas for the coefficients a, c, and b,

$$a = \langle \mathbf{R}_{1,1}\zeta_0 + 2\mathbf{R}_{2,0}(\zeta_0, \Psi_{001}), \zeta_1^* \rangle, \tag{1.39}$$

$$c = 2\langle \mathbf{R}_{2,0}(\zeta_0, \zeta_1) - \Psi_{200}, \zeta_1^* \rangle, \tag{1.40}$$

$$b = \langle 2\mathbf{R}_{2,0}(\zeta_0, \Psi_{200}) + \mathbf{R}_{3,0}(\zeta_0, \zeta_0, \zeta_0), \zeta_1^* \rangle, \tag{1.41}$$

in which ζ_1^* is the vector orthogonal to the range of $\mathbf{L}$ in Section 4.1.1, and Ψ_{001} and Ψ_{200} are found by solving (1.34) and (1.36), respectively.

In order to solve (1.34) and (1.36) we have to check that the solvability conditions for these two equations are satisfied. This is a consequence of the reversibility symmetry. Indeed, following the arguments in Remark 1.11(i), we find here that

$$\mathbf{S}^* \zeta_1^* = \zeta_1^*.$$

Since

$$\mathbf{S}\mathbf{R}_{0,1} = -\mathbf{R}_{0,1},$$

$$\mathbf{S}\mathbf{R}_{2,0}(\zeta_0, \zeta_0) = -\mathbf{R}_{2,0}(\mathbf{S}\zeta_0, \mathbf{S}\zeta_0) = -\mathbf{R}_{2,0}(\zeta_0, \zeta_0),$$

we conclude that

$$\langle \mathbf{R}_{0,1}, \zeta_1^* \rangle = 0, \quad \langle \mathbf{R}_{2,0}(\zeta_0, \zeta_0), \zeta_1^* \rangle = 0,$$

which shows that the solvability conditions for equations (1.34) and (1.36) hold.

Exercise 1.25 *Compute the terms of orders $O(\mu B)$ and $O(B^2)$ and find*

$$\Psi_{101} = \mathbf{L}\Psi_{011} + \mathbf{R}_{1,1}\zeta_1 + 2\mathbf{R}_{2,0}(\zeta_1, \Psi_{001}),$$

$$\Psi_{110} = \mathbf{L}\Psi_{020} + \mathbf{R}_{2,0}(\zeta_1, \zeta_1).$$

Show that the solvability conditions for these equations hold.

4.2 Dimension 3

In this section, we consider reversible systems in $\mathbb{R}^3$ of the form

$$\frac{du}{dt} = \mathbf{F}(u, \mu), \tag{2.1}$$

for which we assume that the vector field satisfies the following hypothesis.

Hypothesis 2.1 *Assume that the vector field* $\mathbf{F}$ *is of class* $\mathscr{C}^k$, $k \geq 3$, *in a neighborhood* $\mathscr{V}$ *of* $(0,0) \in \mathbb{R}^3 \times \mathbb{R}$, *satisfying*

$$\mathbf{F}(0,0) = 0. \tag{2.2}$$

We further assume that there is a symmetry $\mathbf{S}$ *with*

$$\mathbf{S}^2 = \mathbb{I},$$

which anticommutes *with* $\mathbf{F}$,

$$\mathbf{F}(\mathbf{S}u, \mu) = -\mathbf{S}\mathbf{F}(u, \mu) \text{ for all } (u, \mu) \in \mathscr{V}. \tag{2.3}$$

We set

$$\mathbf{L} = D_u\mathbf{F}(0,0), \tag{2.4}$$

and as in the previous section we have that

$$\mathbf{SL} = -\mathbf{LS}.$$

Consequently, the spectrum of $\mathbf{L}$ is symmetric with respect to both the real and imaginary axes, and we distinguish the following cases:

(i) $\sigma(\mathbf{L}) = \{0, \pm\lambda\}$ for some nonzero real number λ;
(ii) $\sigma(\mathbf{L}) = \{0, \pm i\omega\}$ for some nonzero real number ω;
(iii) $\sigma(\mathbf{L}) = \{0\}$, with 0 an algebraically triple eigenvalue.

In case (i), the operator $\mathbf{L}$ has only one eigenvalue on the imaginary axis, and in this case we can apply the center manifold theorem in Chapter 2. This gives a one-dimensional manifold on which the dynamics are described by a scalar, reversible equation. Depending upon the action of $\mathbf{S}$ on the eigenvector ζ_0 in the kernel of $\mathbf{L}$, we distinguish here two bifurcations, 0^+ when $\mathbf{S}\zeta_0 = \zeta_0$, and 0^- when $\mathbf{S}\zeta_0 = -\zeta_0$. It is then not difficult to see that in the case 0^+ the reduced vector field vanishes, so that the dynamic is trivial. In the case 0^-, the vector field is even, and this bifurcation can be analyzed in the same way as the bifurcations in Section 1.1 of Chapter 1. Generically, one finds in this case a saddle node bifurcation in which the two bifurcating equilibria are exchanged by the symmetry $\mathbf{S}$.

The most interesting situations arise in case (iii). In this case, the type of bifurcations further depends upon the geometric multiplicity of the eigenvalue 0 and the action of $\mathbf{S}$ on the kernel of $\mathbf{L}$. The generic situation occurs when 0 is a geometrically simple eigenvalue, 0^3 bifurcation, for which we distinguish between 0^{3+}, when (for ζ_0 in the kernel of $\mathbf{L}$) we have $\mathbf{S}\zeta_0 = \zeta_0$, and 0^{3-}, when $\mathbf{S}\zeta_0 = -\zeta_0$. The next case is 00^2 when 0 is a geometrically double eigenvalue, and finally we have the case 000, when 0 is geometrically triple. The latter case is in fact a codimension 3 bifurcation, and it arises, typically, when the system possesses an additional property, due to either an additional parameter taking a special value, or an additional symmetry, for example a continuous group invariance. We shall not discuss this particular case here.

We focus in this section on the case 0^{3+}, which is the codimension 1 bifurcation arising most frequently in concrete examples. The results in the cases 0^{3-}, 00^2, and $0^{\pm}(i\omega)$ are less complete, and we only present their normal forms, and collect some known results on their dynamics.

4.2.1 Reversible 0^{3+} Bifurcation

We consider in this section the 0^{3+} bifurcation (e.g., see [55] for a detailed analysis of this bifurcation).

Hypothesis 2.2 *Assume that $\sigma(\mathbf{L}) = \{0\}$, with 0 an algebraically triple and geometrically simple eigenvalue. Further assume that the eigenvector ζ_0 associated with this eigenvalue satisfies*

$$\mathbf{S}\zeta_0 = \zeta_0,$$

in which $\mathbf{S}$ is the symmetry anticommuting with $\mathbf{L}$ in Hypothesis 1.1.

Normal Form

As in the previous cases, we start by constructing a suitable basis of $\mathbb{R}^3$.

Lemma 2.3 (0^{3+} **basis of** $\mathbb{R}^3$) *Assume that Hypothesis 2.2 holds. Then there exists a basis $\{\zeta_0, \zeta_1, \zeta_2\}$ of $\mathbb{R}^3$ consisting of generalized eigenvectors of $\mathbf{L}$ such that*

$$\mathbf{L}\zeta_0 = 0, \quad \mathbf{L}\zeta_1 = \zeta_0, \quad \mathbf{L}\zeta_2 = \zeta_1, \tag{2.5}$$

$$\mathbf{S}\zeta_0 = \zeta_0, \quad \mathbf{S}\zeta_1 = -\zeta_1, \quad \mathbf{S}\zeta_2 = \zeta_2. \tag{2.6}$$

Proof Consider a basis $\{\zeta_0, \zeta_1', \zeta_2'\}$, such that

$$\mathbf{L}\zeta_0 = 0, \quad \mathbf{L}\zeta_1' = \zeta_0, \quad \mathbf{L}\zeta_2' = \zeta_1'.$$

By Hypothesis 2.2 we have $\mathbf{S}\zeta_0 = \zeta_0$, and then

$$\mathbf{L}\mathbf{S}\zeta_1' = -\mathbf{S}\mathbf{L}\zeta_1' = -\mathbf{S}\zeta_0 = -\zeta_0.$$

Consequently, there exists $\alpha \in \mathbb{R}$ such that

$$\mathbf{S}\zeta_1' = -\zeta_1' + \alpha\zeta_0,$$

and by taking

$$\zeta_1 = \zeta_1' - \frac{\alpha}{2}\zeta_0,$$

we obtain

$$\mathbf{L}\zeta_1 = \zeta_0, \quad \mathbf{S}\zeta_1 = -\zeta_1.$$

Next, we have

$$\mathbf{LS}\zeta_2' = \zeta_1' - \alpha\zeta_0$$

so that

$$\mathbf{S}\zeta_2' = \zeta_2' - \alpha\zeta_1 + \beta\zeta_0,$$

for some $\beta \in \mathbb{R}$. Since

$$\mathbf{S}^2\zeta_2' = \zeta_2' + 2\beta\zeta_0 = \zeta_2',$$

we have that $\beta = 0$. Now setting

$$\zeta_2 = \zeta_2' - \frac{\alpha}{2}\zeta_1,$$

we find

$$\mathbf{L}\zeta_2 = \zeta_1, \quad \mathbf{S}\zeta_2 = \zeta_2,$$

which proves the lemma. □

Lemma 2.4 (0^{3+} **normal form)** *Assume that Hypotheses 2.1 and 2.2 hold, and consider a basis* $\{\zeta_0, \zeta_1, \zeta_2\}$ *of* $\mathbb{R}^3$ *satisfying (2.5) and (2.6). Then for any positive integer p,* $2 \leq p \leq k$*, there exist neighborhoods* $\mathscr{V}_1$ *and* $\mathscr{V}_2$ *of* 0 *in* $\mathbb{R}^3$ *and* $\mathbb{R}$*, respectively, a map* $\boldsymbol{\Phi} : \mathscr{V}_1 \times \mathscr{V}_2 \to \mathbb{R}^3$ *with the following properties:*

(i) $\boldsymbol{\Phi}$ *is of class* $\mathscr{C}^k$*, satisfying*

$$\boldsymbol{\Phi}(0,0,0,0) = 0, \quad \partial_{(A,B,C)}\boldsymbol{\Phi}(0,0,0,0) = 0, \tag{2.7}$$

and

$$\boldsymbol{\Phi}(A, -B, C, \mu) = \mathbf{S}\boldsymbol{\Phi}(A, B, C, \mu).$$

(ii) For $(A,B,C) \in \mathscr{V}_1$*, the change of variables*

$$u = A\zeta_0 + B\zeta_1 + C\zeta_2 + \boldsymbol{\Phi}(A,B,C,\mu), \tag{2.8}$$

transforms the equation (2.1) into the normal form

$$\begin{aligned}
\frac{dA}{dt} &= B \\
\frac{dB}{dt} &= C + AP(A, \widetilde{B}, \mu) + \rho_B(A,B,C,\mu) \\
\frac{dC}{dt} &= BP(A, \widetilde{B}, \mu) + B\rho_C(A,B,C,\mu),
\end{aligned} \tag{2.9}$$

where $\widetilde{B} = B^2 - 2AC$, $P(\cdot,\mu) : \mathbb{R}^2 \to \mathbb{R}$ *is a polynomial of degree* $p-1$ *in* (A,B,C)*, such that*

$$P(0,0,0) = 0,$$

and the remainders ρ_B *and* ρ_C *are of class* $\mathscr{C}^k$ *and* $\mathscr{C}^{k-1}$*, respectively, are both even in* B*, and satisfy*

$$\rho_B(A,B,C,\mu) = o((|A|+|B|+|C|)^p), \tag{2.10}$$

$$\rho_C(A,B,C,\mu) = o((|A|+B|+|C|)^{p-1}). \tag{2.11}$$

Proof According to Theorem 2.2 and Lemma 1.13 in Chapter 3, there is a polynomial $\Phi_1(\cdot,\mu)$ of degree p in (A,B,C), with coefficients depending upon μ, which satisfies (2.7), and such that the change of variables

$$u = A\zeta_0 + B\zeta_1 + C\zeta_2 + \Phi_1(A,B,C,\mu)$$

transforms (2.1) into

$$\frac{dA}{dt} = B + AP_1(A,\widetilde{B},\mu) + \rho_1(A,B,C,\mu)$$
$$\frac{dB}{dt} = C + BP_1(A,\widetilde{B},\mu) + AP_2(A,\widetilde{B},\mu) + \rho_2(A,B,C,\mu)$$
$$\frac{dC}{dt} = CP_1(A,\widetilde{B},\mu) + BP_2(A,\widetilde{B},\mu) + P_3(A,\widetilde{B},\mu) + \rho_3(A,B,C,\mu),$$

in which $\widetilde{B} = B^2 - 2AC$, P_1 and P_2 are polynomials of degree $p-1$ in (A,B,C), P_3 is a polynomial of degree p in (A,B,C), and ρ_j are functions of class $\mathscr{C}^k$ both satisfying (2.10).

Now, using the reversibility symmetry $\mathbf{S}$, which acts on the basis $\{\zeta_0,\zeta_1,\zeta_2\}$ through

$$\mathbf{S} = \begin{pmatrix} 1 & 0 & 0 \\ 0 & -1 & 0 \\ 0 & 0 & 1 \end{pmatrix},$$

by Theorem 3.4 in Chapter 3, we have that

$$\Phi_1(A,-B,C,\mu) = \mathbf{S}\Phi_1(A,B,C,\mu)$$

and that the vector field in the normal form anticommutes with $\mathbf{S}$. For the first component of the vector field this leads to

$$AP_1(A,\widetilde{B},\mu) + \rho_1(A,-B,C,\mu) = -AP_1(A,\widetilde{B},\mu) - \rho_1(A,B,C,\mu),$$

which implies that

$$P_1(A,\widetilde{B},\mu) = 0 \quad \text{and} \quad \rho_1 \text{ is odd in } B.$$

For the second and third components of the vector field we find

$$AP_2(A,\widetilde{B},\mu)+\rho_2(A,-B,C,\mu)=AP_2(A,\widetilde{B},\mu)+\rho_2(A,B,C,\mu),$$
$$BP_2(A,\widetilde{B},\mu)-P_3(A,\widetilde{B},\mu)-\rho_3(A,-B,C,\mu)$$
$$=BP_2(A,\widetilde{B},\mu)+P_3(A,\widetilde{B},\mu)+\rho_3(A,B,C,\mu),$$

which lead to

$$P_3(A,\widetilde{B},\mu)=0,\quad \rho_2 \text{ is even in } B \quad\text{and}\quad \rho_3 \text{ is odd in } B.$$

Consequently, we have the normal form

$$\frac{dA}{dt}=B+B\widetilde{\rho}_1(A,B,C,\mu)$$
$$\frac{dB}{dt}=C+AP_2(A,\widetilde{B},\mu)+\widetilde{\rho}_2(A,B,C,\mu)$$
$$\frac{dC}{dt}=BP_2(A,\widetilde{B},\mu)+B\widetilde{\rho}_3(A,B,C,\mu),$$

with $\widetilde{\rho}_j$ even in B, $\widetilde{\rho}_1$ and $\widetilde{\rho}_3$ satisfying (2.11), and $\widetilde{\rho}_2$ satisfying (2.10).

Finally, we obtain the desired normal form by making a change of variables as in Remark 1.10(ii), in Chapter 3. We set

$$B'=B+B\widetilde{\rho}_1(A,B,C,\mu), \tag{2.12}$$

which is invertible by the implicit function theorem,

$$B=B'+B'\widehat{\rho}_1(A,B',C,\mu),$$

where $\widehat{\rho}_1$ is of class $\mathscr{C}^{k-1}$, and even in B'. This transforms the first equation into

$$\frac{dA}{dt}=B'.$$

Furthermore, we have $B\widetilde{\rho}_1=-B'\widehat{\rho}_1$, and

$$\widetilde{B'}=B'^2-2AC=\widetilde{B}+2\widetilde{\rho}_1B^2+\widetilde{\rho}_1^2B^2,$$
$$\widetilde{B}=\widetilde{B'}+2B'^2\widehat{\rho}_1+B'^2\widehat{\rho}_1^2.$$

In addition,

$$\frac{d\widetilde{B}}{dt}=2B(\widetilde{\rho}_2-C\widetilde{\rho}_1-A\widetilde{\rho}_3),$$

so that

$$\frac{dB'}{dt}=(1+\widetilde{\rho}_1)(C+AP_2+\widetilde{\rho}_2)+B^2(1+\widetilde{\rho}_1)\partial_A\widetilde{\rho}_1$$
$$+\,2B^2(\widetilde{\rho}_2-C\widetilde{\rho}_1-A\widetilde{\rho}_3)\partial_{\widetilde{B}}\widetilde{\rho}_1+B^2(P_2+\widetilde{\rho}_3)\partial_C\widetilde{\rho}_1.$$

Then this equation can be written in the form

$$\frac{dB'}{dt} = C + AP_2(A, \widetilde{B}', \mu) + \rho_{B'}(A, B', C, \mu),$$

with $\rho_{B'}$ even in B', and a similar calculation gives the equation for dC/dt. With this transformation for B, we have the normal form in the lemma, but now with a map Φ, instead of the polynomial Φ_1. Finally, observe that Φ has the same symmetry property as Φ_1, which completes the proof. □

Solutions of the Normal Form System

We analyze now the dynamics of the normal form system (2.9). We make the following assumption on the polynomial P.

Hypothesis 2.5 *Assume that the expansion of P in (2.9),*

$$P(A, \widetilde{B}, \mu) = a\mu + bA + c\widetilde{B} + dA^2 + O(\mu^2 + |A||\mu| + (|\mu| + |A| + |B| + C|)^3),$$

is such that $a \neq 0$, and either $b \neq 0$ or $b = 0$ and $d \neq 0$.

Truncated System

Consider first the truncated system

$$\begin{aligned} \frac{dA}{dt} &= B \\ \frac{dB}{dt} &= C + AP(A, \widetilde{B}, \mu) \\ \frac{dC}{dt} &= BP(A, \widetilde{B}, \mu), \end{aligned} \tag{2.13}$$

where $\widetilde{B} = (B^2 - 2AC)$. A key property of this system is that it possesses two first integrals,

$$K = \widetilde{B} = B^2 - 2AC, \quad H = C - g_K(A, \mu),$$

where

$$g_K(A, \mu) = \int_0^A P(s, K, \mu) ds.$$

Consequently, the system is integrable and its phase portrait is obtained by varying $(H, K) \in \mathbb{R}^2$. The orbits in the (A, B, C)-space are curves given by

$$B^2 = f_{H,K}(A, \mu), \quad C = g_K(A, \mu) + H, \tag{2.14}$$

where

$$f_{H,K}(A, \mu) = 2Ag_K(A, \mu) + 2HA + K.$$

Remark 2.6 (Equilibria) *The equilibria of the truncated system satisfy*

$$B=0, \quad C+AP(A,-2AC,\mu)=0.$$

The implicit function theorem allows one to solve the second equation, and gives a one-parameter family of symmetric equilibria $(A,0,C_*(A,\mu))$ for sufficiently small μ, with

$$C_*(A,\mu)=-a\mu A-bA^2-dA^3+O(|\mu||A|^3+A^4), \quad A=O(\mu).$$

The linearization of system (2.1) at these equilibria gives a family of linear operators, which anticommute with the symmetry **S**, *because the equilibria are symmetric. In particular, the spectrum of these operators is symmetric with respect to both the real and imaginary axes, just as the spectrum of* **L**. *It is not difficult to check that 0 is an eigenvalue of each operator, with symmetric eigenvector*

$$\zeta_{A,\mu}=(1,0,\partial_A C_*(A,\mu)).$$

Notice that the vector

$$\zeta'_{A,\mu}=(0,0,\partial_\mu C(A,\mu))$$

is not an eigenvector, since the vector field depends on μ. The two other eigenvalues are symmetric with respect to the origin and are either real or purely imaginary. These eigenvalues vanish if

$$\partial_A C_*(A,\mu)+\frac{1}{A}C_*(A,\mu)=0.$$

Then, if $b\neq 0$, or if $b=0$ and $a\mu d<0$, one can show that there is an equilibrium for some $A=A_0(\mu)$ such that 0 is a triple eigenvalue of the linearized operator, with a 3×3 Jordan block.

For the study of the phase portraits, we consider the cases $b\neq 0$, and $b=0$, $d\neq 0$. In both cases we look at the projections on the (A,B)-plane of the orbits, which are given by the equation

$$B^2=f_{H,K}(A,\mu). \tag{2.15}$$

We first consider only the leading order terms in the expansion of P, which, as we shall see, give a correct description of the dynamics in a neighborhood of the origin.

Assume $b\neq 0$. Then we have at leading orders

$$B^2=f_{H,K}(A,\mu)=bA^3+2\widetilde{\mu}A^2+2HA+K, \quad \widetilde{\mu}=a\mu+cK. \tag{2.16}$$

We plot in Figure 2.1 the graphs of $f_{H,K}(\cdot,\mu)$, varying $(K,H)\in\mathbb{R}^2$ in the case $b>0$ and $\widetilde{\mu}>0$. (In the other cases the result is similar.) We find a curve in the (K,H)-plane along which $f_{H,K}(\cdot,\mu)$ has a double root, obtained by solving

$$f_{H,K}(A,\mu)=\partial_A f_{H,K}(A,\mu)=0. \tag{2.17}$$

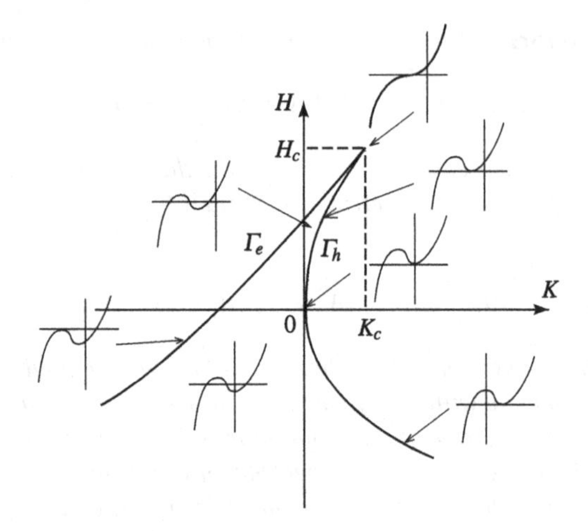

Fig. 2.1 Graphs of $f_{H,K}(\cdot,\mu)$, depending on (K,H), for $\widetilde{\mu}>0$, $b>0$.

These equalities give the parametric equations of the curve,

$$K=2bA^3+2\widetilde{\mu}A^2,\quad H=-\frac{3b}{2}A^2-2\widetilde{\mu}A.$$

At the cusp point

$$(K_c,H_c)=\left(\frac{8\widetilde{\mu}^3}{27b^2},\frac{2\widetilde{\mu}^2}{3}\right),\quad A=-\frac{2\widetilde{\mu}}{3b},$$

the polynomial $f_{K,H}(\cdot,\mu)$ has a triple root.

The double roots of $f_{H,K}(\cdot,\mu)$ correspond to equilibria of the truncated system, and their stability, in the (A,B)-plane, is determined by the sign of the second derivative $\partial_A^2 f_{K,H}(\cdot,\mu)$: the equilibrium is a saddle for positive sign, and a center for negative sign. Then, for (H,K) along the curve Γ_e we have a center equilibrium, and along Γ_h a saddle equilibrium. These equilibria are precisely the ones found in Remark 2.6, and in particular, the equilibrium given by the cusp point is the equilibrium found for $A=A_0(\mu)$, at which the linearization has a triple zero eigenvalue. Furthermore, there is no other bounded solution for (K,H) on Γ_e, whereas for (K,H) on Γ_h there is a homoclinic orbit connecting the saddle equilibrium to itself. We point out that for $H=0$ this orbit is homoclinic to the origin. For values of (K,H) lying outside the curves Γ_e and Γ_h there is no bounded orbit, while for (K,H) lying between the curves Γ_e and Γ_h there is a periodic orbit surrounding the center equilibrium. We plot in Figure 2.2 the projections in the (A,B)-plane of the orbits obtained by varying K, for a fixed $H<H_c$. As $H\to H_c$ the homoclinic orbit shrinks to the equilibrium found at the cusp point (K_c,H_c), and there are no bounded solutions for $H>H_c$.

Finally, notice that the bounded orbits found in this case are such that

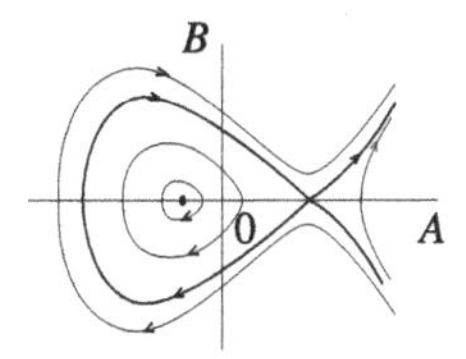

Fig. 2.2 Dynamics of the 0^{3+} bifurcation. Plot of the projections in the (A,B)-plane of the orbits of the truncated system (2.13) obtained by varying K for a fixed $H < H_c$, in the case $\widetilde{\mu} > 0$, $b > 0$. For $H > H_c$ there are no bounded orbits, and the dynamics are similar for other values of $\widetilde{\mu}$ and b.

$$K = O(|\widetilde{\mu}|^3), \quad H = O(|\widetilde{\mu}|^2),$$
$$A = O(|\widetilde{\mu}|), \quad B = O(|\widetilde{\mu}|^{3/2}), \quad C = O(|\widetilde{\mu}|^2).$$

In particular, this shows that $\widetilde{\mu} = O(\mu)$. Furthermore, the terms in the polynomial $f_{H,K}(\cdot,\mu)$ that were neglected by the restriction to the leading order terms in (2.16) are of order $O(A^2(|\mu|+|A|)^2)$, so adding these terms does not change the shape of $f_{H,K}(\cdot,\mu)$ for A in a neighborhood of size $O(\mu)$. This implies that the same phase portraits can be found for the truncated normal form (2.13) in a sufficiently small neighborhood of the origin.

Consider now the case $b = 0$ and $d \neq 0$. Then at leading orders we find

$$f_{H,K}(A,\mu) = \frac{2}{3}dA^4 + 2\widetilde{\mu}A^2 + 2HA + K, \quad \widetilde{\mu} = a\mu + cK. \tag{2.18}$$

We summarize the behavior of $f_{H,K}(\cdot,\mu)$ for different values of $\widetilde{\mu}$ and d in Figure 2.3. In particular, solving (2.17) we find that $f_{H,K}(\cdot,\mu)$ has double roots for (K,H) along the curve parameterized by

$$K = 2dA^4 + 2\widetilde{\mu}A^2, \quad H = -\frac{4d}{3}A^3 - 2\widetilde{\mu}A.$$

The situation is slightly more complicated in the case $\widetilde{\mu}d < 0$, where there is a pair of cusp points

$$(K_c, \pm H_c) = \left(-\frac{\widetilde{\mu}^2}{2d}, \pm\frac{4}{3}\sqrt{-\frac{\widetilde{\mu}^3}{2d}}\right), \quad A = \pm\sqrt{-\frac{\widetilde{\mu}}{2d}},$$

and a particular point

$$(K_0, 0) = \left(\frac{3\widetilde{\mu}^2}{2d}, 0\right), \quad A = \sqrt{-\frac{3\widetilde{\mu}}{2d}}.$$

The projection of the orbits on the (A,B)-plane obtained varying K for fixed H, in the cases of bounded orbits, is sketched in Figure 2.4. Summarizing, we have:

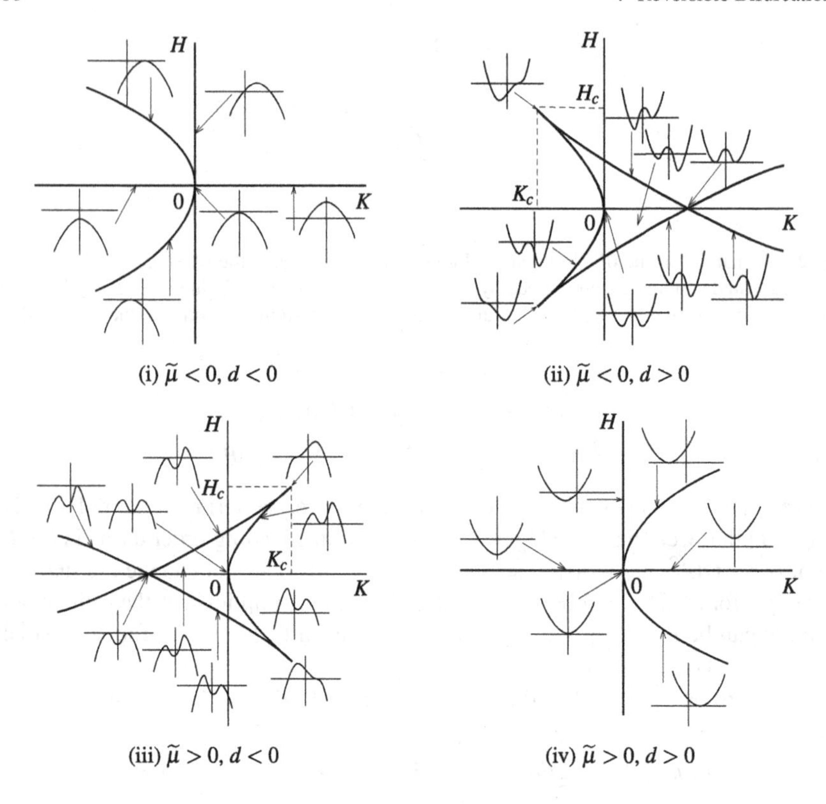

Fig. 2.3 Graphs of $f_{H,K}(\cdot,\mu)$ for $b = 0$ depending on (K,H).

(i) For $\widetilde{\mu} < 0$, $d < 0$, and any $H \in \mathbb{R}$, there is precisely a center equilibrium surrounded by a one-parameter family of periodic orbits.
(ii) For $\widetilde{\mu} < 0$, $d > 0$, and $|H| < H_c$, there are three equilibria, a center and two saddles. For $H = 0$, the origin is a center equilibrium surrounded by a one-parameter family of periodic orbits, which tend to a pair of heteroclinic orbits, exchanged by the reversibility symmetry, connecting the two saddle equilibria. For $0 < |H| < H_c$, the center equilibrium is surrounded by a one-parameter family of periodic orbits, which tend to a homoclinic orbit to one of the saddle equilibria. As $|H| \to H_c$, the three equilibria collide in a single equilibrium, which is the only bounded orbit, for $|H| > H_c$.
(iii) For $\widetilde{\mu} > 0$, $d < 0$, there are three equilibria, two centers and a saddle, when $|H| < H_c$. Each center is surrounded by a one-parameter family of periodic orbits, which tends to a homoclinic orbit to the saddle equilibrium. As $|H| \to H_c$, the saddle equilibrium and one center collide, and for $|H| = H_c$ there is a one-parameter family of equilibria surrounding the remaining center and a homoclinic orbit to the other equilibrium. For $|H| > H_c$, there is only one center equilibrium surrounded by a one-parameter family of periodic orbits.

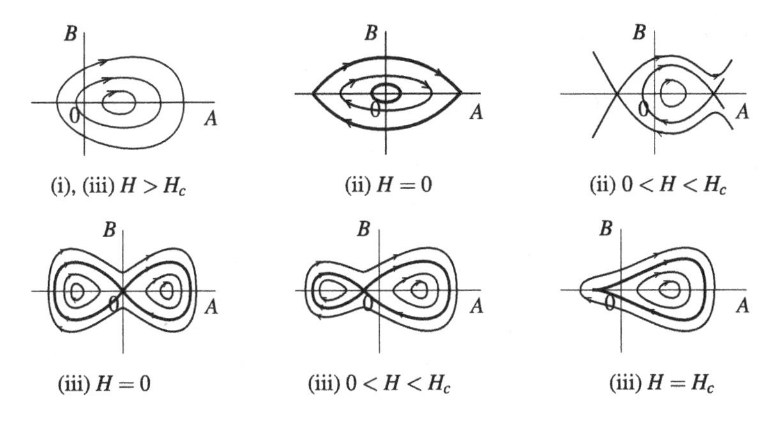

Fig. 2.4 Dynamics of the 0^{3+} bifurcation in the case $b = 0$. Plot of the projections in the (A,B)-plane of the orbits of the truncated system (2.13) obtained by varying K for a fixed H in the cases (i) $\widetilde{\mu} < 0, d < 0$, (ii) $\widetilde{\mu} < 0, d > 0$, (iii) $\widetilde{\mu} > 0, d < 0$, when the system has nonequilibrium bounded orbits.

(iv) For $\widetilde{\mu} > 0$, $d > 0$, and any $H \in \mathbb{R}$ there is a saddle equilibrium and no other bounded solutions.

Finally, we have that the small bounded solutions are such that

$$K = O(|\widetilde{\mu}|^2), \quad H = O(|\widetilde{\mu}|^{3/2}),$$

and

$$A = O(|\widetilde{\mu}|^{1/2}), \quad B = O(|\widetilde{\mu}|), \quad C = O(|\widetilde{\mu}|^{3/2}),$$

hence $\widetilde{\mu} = O(\mu)$, and the neglected terms in $f_{H,K}(A,\mu)$ are $O(A^2(|\widetilde{\mu}|^2 + |\widetilde{\mu}||A| + |A|^3))$. This implies that the leading order terms describe correctly the shape of $f_{H,K}(\cdot,\mu)$ in the neighborhood of 0 of size $O(|\mu|^{1/2})$, so that the phase portraits can be found for the truncated system (2.13).

Full System

For the persistence of the phase portraits obtained above for the full system (2.9) we proceed as in the previous sections. First, the implicit function theorem allows us to show that the equilibria persist. Next, the reversibility symmetry acting through

$$(A,B,C) \mapsto (A,-B,C)$$

implies that the phase portrait is symmetric with respect to the plane $B = 0$, which is invariant under the action of this symmetry. Then, by arguing as in the previous sections one can show that the orbits of the truncated system which intersect the plane $B=0$ persist. This allows us to prove the persistence of the periodic orbits and the homoclinic orbits that exist in the different cases.

It remains to discuss the persistence of the pair of heteroclinic orbits found in the case $\widetilde{\mu}<0$, $d>0$, for $H=0$, $K=K_0$. Recall that in this case the two equilibria belong to the one-parameter family of equilibria $(A,0,C_*(A,\mu))$ found in Remark 2.6, and that the linearization at these points possesses a zero eigenvalue and a pair of real eigenvalues, one negative and one positive. They are found for $A=A_\pm=\pm\sqrt{-3\widetilde{\mu}/2d}$, and the heteroclinic orbits connect $(A_-,0,C_*(A_-,\mu))$ with $(A_+,0,C_*(A_+,\mu))$. Each heteroclinic orbit intersects transversally the plane $A=0$ in some point $(0,B_0,C_0)$, which then implies that these orbits persist under small perturbations. Indeed, the equilibria $(A,0,C_*(A,\mu))$, for A close to $A_\pm$, have the same stability properties, and in particular a one-dimensional stable and a one-dimensional unstable manifold. Then the family of one-dimensional unstable manifolds of $(A,0,C(A,\mu))$ for A close to A_+, intersects transversally the plane $A=0$ along a curve γ_+ lying in a neighborhood of $(0,B_0,C_0)$. In the same way, the family of one-dimensional stable manifolds of $(A,0,C(A,\mu))$ for A close to A_-, intersects transversally the plane $A=0$ along a curve γ_- lying in a neighborhood of $(0,B_0,C_0)$. One shows then that γ_- intersects transversally γ_+ (e.g., see [55]). This implies the persistence of the heteroclinic orbits.

Summarizing we have the following result.

Theorem 2.7 (**0^{3+} bifurcation**) *Assume that the Hypotheses 2.1, 2.2, and 2.5 hold. Then for the differential equation (2.1) a reversible 0^{3+} bifurcation occurs at $\mu=0$. More precisely, the following properties hold in a neighborhood of 0 in $\mathbb{R}^3$ for sufficiently small μ.*

For $b\neq 0$, there are two curves of equilibria $(A,0,C_e(A,\mu))$ and $(A,0,C_h(A,\mu))$ which meet at a point $(A_0,0,C_*(A_0,\mu))$ having an algebraically and geometrically triple eigenvalue. Each equilibrium $(A,0,C_e(A,\mu))$ is surrounded by a one-parameter family of periodic orbits, which tend to a homoclinic orbit to an equilibrium $(A,0,C_h(A,\mu))$.

For $b=0$ and $d\neq 0$, we have the following cases:

(i) For $a\mu<0$ and $d<0$, there is a one-parameter family of equilibria, and each equilibrium is surrounded by a one-parameter family of periodic orbits.

(ii) For $a\mu<0$ and $d>0$, there are three curves of equilibria $(A,0,C_e(A,\mu))$ and $(A,0,C_\pm(A,\mu))$ for A close to 0. Each equilibrium $(A,0,C_e(A,\mu))$ with $A\neq 0$, is surrounded by a one-parameter family of periodic orbits which tend to a homoclinic orbit to an equilibrium $(A,0,C_\pm(A,\mu))$. The equilibrium $(0,0,C_e(0,\mu))$ is surrounded by a one-parameter family of periodic orbits which tend to a pair of heteroclinic orbits, exchanged by the reversibility symmetry, connecting two equilibria $(A,0,C_\pm(A,\mu))$.

(iii) For $a\mu>0$ and $d<0$, there are three curves of equilibria $(A,0,C_h(A,\mu))$ and $(A,0,C_\pm(A,\mu))$ for A close to 0. Each equilibrium $(A,0,C_\pm(A,\mu))$ is surrounded by a one-parameter family of periodic orbits, which tend to a homoclinic orbit to an equilibrium $(A,0,C_h(A,\mu))$.

(iv) For $a\mu > 0$ and $d > 0$, there is a one-parameter family of equilibria and no other bounded solutions.

Remark 2.8 *We show in Appendix D.1 how to compute the principal coefficients in the normal form (2.9) when starting from an infinite-dimensional system of the form (1.13).*

Exercise 2.9 *Consider a system in $\mathbb{R}^4$ of the form*

$$\frac{du}{dt} = \mathbf{F}(u,\mu).$$

Assume that the vector field $\mathbf{F}$ is of class $\mathscr{C}^k$, $k \geq 3$, in a neighborhood $\mathscr{V}$ of $(0,0) \in \mathbb{R}^4 \times \mathbb{R}$, satisfying $\mathbf{F}(0,0) = 0$, and that 0 is an algebraically quadruple and geometrically simple eigenvalue of $\mathbf{L} = D_u\mathbf{F}(0,0)$. Further assume that there is a symmetry $\mathbf{S}$ with $\mathbf{S}^2 = \mathbb{I}$, which anticommutes with $\mathbf{F}$,

$$\mathbf{F}(\mathbf{S}u,\mu) = -\mathbf{S}\mathbf{F}(u,\mu) \text{ for all } (u,\mu) \in \mathscr{V},$$

and that there is a continuous symmetry such that

$$\mathbf{F}(u + a\zeta_0,\mu) = \mathbf{F}(u,\mu) \text{ for all } a \in \mathbb{R},$$

where ζ_0 satisfies

$$\mathbf{L}\zeta_0 = 0, \quad \mathbf{S}\zeta_0 = -\zeta_0.$$

(i) Show that there is a basis $\{\zeta_0,\zeta_1,\zeta_2,\zeta_3\}$ in $\mathbb{R}^4$ such that

$$\mathbf{L}\zeta_0 = 0, \quad \mathbf{L}\zeta_1 = \zeta_0, \quad \mathbf{L}\zeta_2 = \zeta_1, \quad \mathbf{L}\zeta_3 = \zeta_2,$$
$$\mathbf{S}\zeta_0 = -\zeta_0, \quad \mathbf{S}\zeta_1 = \zeta_1, \quad \mathbf{S}\zeta_2 = -\zeta_2, \quad \mathbf{S}\zeta_3 = \zeta_3,$$

and a dual basis $\{\zeta_0^*,\zeta_1^*,\zeta_2^*,\zeta_3^*\}$ such that

$$\mathbf{L}^*\zeta_3^* = 0, \quad \mathbf{L}^*\zeta_2^* = \zeta_3^*, \quad \mathbf{L}^*\zeta_1^* = \zeta_2^*, \quad \mathbf{L}^*\zeta_0^* = \zeta_1^*, \quad \langle \zeta_j,\zeta_k^* \rangle = \delta_{jk},$$
$$\mathbf{S}^*\zeta_0^* = -\zeta_0^*, \quad \mathbf{S}^*\zeta_1^* = \zeta_1^*, \quad \mathbf{S}^*\zeta_2^* = -\zeta_2^*, \quad \mathbf{S}^*\zeta_3^* = \zeta_3^*.$$

(ii) For $u \in \mathbb{R}^4$, set

$$u = A\zeta_0 + v, \quad \langle v,\zeta_0^* \rangle = 0, \quad A \in \mathbb{R}.$$

Show that the vector field of the system satisfied by (A,v) is independent of A. (Compare with the proof of Theorem 3.19 in Chapter 2.)

(iii) Show that v satisfies a first order equation in $\mathbb{R}^3$ of the form (2.1) satisfying Hypotheses 2.1 and 2.2.

4.2.2 Reversible 0^{3-} Bifurcation (Elements)

We consider in this section the 0^{3-} bifurcation, when 0 is an algebraically triple and geometrically simple eigenvalue of $\mathbf{L}$, with associated eigenvector ζ_0 satisfying

$$\mathbf{S}\zeta_0 = -\zeta_0.$$

This type of bifurcation is still an open problem, so we shall only briefly outline the normal form and some elementary facts about its dynamics. We point out that we are not aware of any physical example leading to a bifurcation of this type.

0^{3-} *Normal Form*

First, by arguing as in the case of the 0^{3+} bifurcation, we can prove here that *there is a basis* $\{\zeta_0, \zeta_1, \zeta_2\}$ *of* $\mathbb{R}^3$ *such that*

$$\mathbf{L}\zeta_0 = 0, \quad \mathbf{L}\zeta_1 = \zeta_0, \quad \mathbf{L}\zeta_2 = \zeta_1,$$
$$\mathbf{S}\zeta_0 = -\zeta_0, \quad \mathbf{S}\zeta_1 = \zeta_1, \quad \mathbf{S}\zeta_2 = -\zeta_2.$$

Using this basis, in the same way as in the proof of Lemma 2.4, we find in this case that *for any* $p < k$ *there exists a polynomial* $\Phi(\cdot, \mu) : \mathbb{R}^3 \to \mathbb{R}^3$ *of degree* p, *with*

$$\Phi(0,0,0,0) = 0, \quad \partial_{(A,B,C)}\Phi(0,0,0,0) = 0, \quad \Phi(-A,B,-C,\mu) = \mathbf{S}\Phi(A,B,C,\mu),$$

such that the change of variables

$$u = A\zeta_0 + B\zeta_1 + C\zeta_2 + \Phi(A,B,C,\mu), \tag{2.19}$$

transforms the differential equation into the normal form

$$\begin{aligned}
\frac{dA}{dt} &= B + A^2P(A^2,\widetilde{B},\mu) + \rho_A(A,B,C,\mu)\\
\frac{dB}{dt} &= C + ABP(A^2,\widetilde{B},\mu) + AQ(A^2,\widetilde{B},\mu) + \rho_B(A,B,C,\mu)\\
\frac{dC}{dt} &= ACP(A^2,\widetilde{B},\mu) + BQ(A^2,\widetilde{B},\mu) + R(A^2,\widetilde{B},\mu) + \rho_C(A,B,C,\mu).
\end{aligned} \tag{2.20}$$

Here $\widetilde{B} = B^2 - 2AC$, *and* P, Q, *and* R *are polynomials of degree* $p-2$, $p-1$, *and* p, *respectively, in* (A,B,C). *The remainders* ρ_A *and* ρ_C *are even in* (A,C), *while* ρ_B *is odd in* (A,C), *and they are all of class* $\mathscr{C}^k$ *satisfying*

$$|\rho_A| + |\rho_B| + |\rho_C| = o((|A| + B| + |C|)^p).$$

Exercise 2.10 *Consider the leading order terms in the polynomials* P, Q, *and* R,

$$\begin{aligned}
P(A^2,\widetilde{B},\mu) &= d + h.o.t.,\\
Q(A^2,\widetilde{B},\mu) &= e\mu + h.o.t.,\\
R(A^2,\widetilde{B},\mu) &= a\mu + bA^2 + c(B^2 - 2AC) + h.o.t,
\end{aligned}$$

with real coefficients a, b, c, d, *and* e. *Using the procedure in Appendix D.1, compute the coefficients* a, b, c, d, *and* e.

Solutions of the Normal Form

The complete description of the dynamics of system (2.20) is an open problem. We only mention here that equilibria can be found as usual, and that assuming that $ab \neq 0$ we have:

(i) for $ab\mu > 0$ there is no equilibrium in a neighborhood of 0;
(ii) for $ab\mu < 0$ there is a pair of equilibria, exchanged by the symmetry $\mathbf{S}$, with

$$A = O(|\mu|^{1/2}), \quad B = O(\mu), \quad C = O(|\mu|^{3/2}).$$

The linearized operator at these equilibria has one real and two complex eigenvalues of order $O(|\mu|^{1/6})$.

For the study of the dynamics, a natural way to start is by analyzing the truncated system, which reduces in this case to a third order differential equation

$$A''' = a\mu + bA^2 + (c+3d)(A')^2 + (4d-2c)AA''. \tag{2.21}$$

4.2.3 *Reversible* 00^2 *Bifurcation (Elements)*

Consider now the 00^2 bifurcation, when 0 is an algebraically triple and geometrically double eigenvalue of $\mathbf{L}$. We briefly describe the normal form and collect some known facts about its dynamics.

00^2 *Normal Form*

Consider a basis $\{\xi_0, \zeta_0, \zeta_1\}$ of $\mathbb{R}^3$ such that

$$\mathbf{L}\xi_0 = 0, \quad \mathbf{L}\zeta_0 = 0, \quad \mathbf{L}\zeta_1 = \zeta_0. \tag{2.22}$$

Then, depending upon the action of the symmetry $\mathbf{S}$ on this basis, *we distinguish the following situations:*

(i) 0^+0^{2+} *when* $\mathbf{S}\xi_0 = \xi_0$, $\mathbf{S}\zeta_0 = \zeta_0$, $\mathbf{S}\zeta_1 = -\zeta_1$;
(ii) 0^+0^{2-} *when* $\mathbf{S}\xi_0 = \xi_0$, $\mathbf{S}\zeta_0 = -\zeta_0$, $\mathbf{S}\zeta_1 = \zeta_1$;
(iii) 0^-0^{2+} *when* $\mathbf{S}\xi_0 = -\xi_0$, $\mathbf{S}\zeta_0 = \zeta_0$, $\mathbf{S}\zeta_1 = -\zeta_1$;
(iv) 0^-0^{2-} *when* $\mathbf{S}\xi_0 = -\xi_0$, $\mathbf{S}\zeta_0 = -\zeta_0$, $\mathbf{S}\zeta_1 = \zeta_1$.

By arguing as in the previous sections, *we obtain the normal forms below. As before,* P, Q, Q_1, Q_2, R, R_1, *and* R_2 *represent polynomials, and* ρ_A, ρ_B, *and* ρ_C *denote the higher order terms.*

(i) In the case (0^+0^{2+}),

$$\begin{aligned}\frac{dA}{dt} &= C\rho_A\\ \frac{dB}{dt} &= C + C\rho_B\\ \frac{dC}{dt} &= R(A,B,\mu) + \rho_C,\end{aligned} \tag{2.23}$$

with ρ_A, ρ_B, and ρ_C even in C.

(ii) In the case (0^+0^{2-}),

$$\begin{aligned}\frac{dA}{dt} &= BP(A,B^2,\mu) + B\rho_A\\ \frac{dB}{dt} &= C + B^2Q(A,B^2,\mu) + \rho_B\\ \frac{dC}{dt} &= CBQ(A,B^2,\mu) + BR(A,B^2,\mu) + B\rho_C,\end{aligned} \tag{2.24}$$

with ρ_A, ρ_B, and ρ_C even in B.

(iii) In the case (0^-0^{2+}),

$$\begin{aligned}\frac{dA}{dt} &= P(A^2,B,\mu) + \rho_A\\ \frac{dB}{dt} &= C + ABQ(A^2,B,\mu) + \rho_B\\ \frac{dC}{dt} &= ACQ(A^2,B,\mu) + R(A^2,B,\mu) + \rho_C,\end{aligned} \tag{2.25}$$

with ρ_A, ρ_C even, and ρ_B odd in (A,C).

(iv) In the case (0^-0^{2-}),

$$\begin{aligned}\frac{dA}{dt} &= P(A^2,AB,B^2,\mu) + \rho_A\\ \frac{dB}{dt} &= C + ABQ_1(A^2,AB,B^2,\mu) + B^2Q_2(A^2,AB,B^2,\mu) + \rho_B\\ \frac{dC}{dt} &= ACQ_1(A^2,AB,B^2,\mu) + BCQ_2(A^2,AB,B^2,\mu)\\ &\quad + AR_1(A^2,AB,B^2,\mu) + BR_2(A^2,AB,B^2,\mu) + \rho_C,\end{aligned} \tag{2.26}$$

with ρ_A, ρ_B even, and ρ_C odd in (A,B).

Solutions of the Normal Form

The dynamics in the case 0^+0^{2+} are similar to that of the 0^{2+} bifurcation, with an additional coordinate that can take the role of a parameter. The invariant set for the symmetry $\mathbf{S}$ is the plane $C = 0$, so it is two-dimensional, which allows us to prove the persistence of the different bounded orbits. Similarly, the analysis in the case

0^+0^{2-} is analogue to that of the 0^{2-} bifurcation, with an additional coordinate that can take the role of a parameter. A new feature here is the existence of a family of symmetric equilibria for which the stability changes at some point. In this case again, the invariant set for the symmetry $\mathbf{S}$ is two-dimensional, the plane $B=0$, which allows us to show the persistence results.

The cases 0^-0^{2+} and 0^-0^{2-} are less understood. In contrast to the previous cases, here the invariant set for the symmetry $\mathbf{S}$ is only one-dimensional, which creates a serious difficulty for the proofs of persistence results. We mention that in the case when the vector field is analytic, one could use the optimal normal form in Theorem 5.8 in Chapter 2. The exponentially small size of the remainder in this theorem may be helpful to show the persistence of bounded orbits (e.g., see the $0^{2+}(i\omega)$ bifurcation in Section 4.3.1).

4.2.4 Reversible $0(i\omega)$ Bifurcation (Elements)

Consider now the case $0(i\omega)$, when $\mathbf{L}$ has a simple 0 eigenvalue and a pair of purely imaginary eigenvalues $i\omega$.

$0(i\omega)$ *Normal Form*

Consider a basis $\{\xi_0,\zeta,\overline{\zeta}\}$ of $\mathbb{R}^3$ such that

$$\mathbf{L}\xi_0=0,\quad \mathbf{L}\zeta=i\omega\zeta,\quad \mathbf{L}\overline{\zeta}=-i\omega\overline{\zeta}. \tag{2.27}$$

Then by taking into account the action of $\mathbf{S}$ on these vectors we can choose this basis such that *we have one of the following two cases:*

(i) $0^+(i\omega)$ *when* $\mathbf{S}\xi_0=\xi_0$, $\mathbf{S}\zeta=\overline{\zeta}$, $\mathbf{S}\overline{\zeta}=\zeta$;
(ii) $0^-(i\omega)$ *when* $\mathbf{S}\xi_0=-\xi_0$, $\mathbf{S}\zeta=\overline{\zeta}$, $\mathbf{S}\overline{\zeta}=\zeta$.

Notice that in this case we can always choose $\zeta\in\mathbb{C}^2$ such that $\mathbf{S}\zeta=\overline{\zeta}$. Indeed, since $\mathbf{L}\zeta=i\omega\zeta$, we have that $\mathbf{L}\mathbf{S}\zeta=-i\omega\mathbf{S}\zeta$. Consequently, there is $k\in\mathbb{C}$ such that $\mathbf{S}\zeta=k\overline{\zeta}$. Since $\mathbf{S}^2=\mathbb{I}$ this implies that $|k|=1$, i.e., there exists $\alpha\in\mathbb{R}$ such that $\mathbf{S}\zeta=e^{i\alpha}\overline{\zeta}$. Replacing ζ by $\zeta'=e^{-i\alpha/2}\zeta$, we find that $\mathbf{S}\zeta'=\overline{\zeta'}$.

The normal forms in the two cases can be found as for the previous bifurcations. *Setting*

$$u=A\xi_0+B\zeta+\overline{B\zeta},$$

with A real-valued and B complex-valued we obtain the following normal forms:

(i) In the case $0^+(i\omega)$,

$$\frac{dA}{dt} = \rho_A$$
$$\frac{dB}{dt} = i\omega B + iBQ(A, |B|^2, \mu) + \rho_B, \tag{2.28}$$

where Q is a real-valued polynomial, and

$$\rho_A(A, \overline{B}, B, \mu) = -\rho_A(A, B, \overline{B}, \mu), \quad \rho_B(A, \overline{B}, B, \mu) = -\overline{\rho_B}(A, B, \overline{B}, \mu).$$

(ii) In the case $0^-(i\omega)$,

$$\frac{dA}{dt} = P(A^2, |B|^2, \mu) + \rho_A$$
$$\frac{dB}{dt} = i\omega B + iBQ_1(A^2, |B|^2, \mu) + ABQ_2(A^2, |B|^2, \mu) + \rho_B, \tag{2.29}$$

where P, Q_1, Q_2 are real-valued polynomials, and

$$\rho_A(-A, \overline{B}, B, \mu) = \rho_A(A, B, \overline{B}, \mu), \quad \rho_B(-A, \overline{B}, B, \mu) = -\overline{\rho_B}(A, B, \overline{B}, \mu).$$

Solutions of the Normal Form

The study of the dynamics in the case $0^+(i\omega)$ is rather straightforward. First, for the truncated normal form, obtained by removing the remainders ρ_A and ρ_B, we find the line of symmetric equilibria $(A_0, 0)$, $A_0 \in \mathbb{R}$. Notice that the linearization of the vector field at these equilibria has a simple 0 eigenvalue and a pair of purely imaginary eigenvalues $\pm i\omega_{A_0,\mu}$ for A_0 and μ small, just as $\mathbf{L}$. This is due to the reversibility symmetry. Furthermore, each plane $A = A_0$ is invariant under the dynamics of the normal form, and in this plane the orbits are circles. Indeed, in polar coordinates $B = re^{i\theta}$ we find the system

$$\frac{dr}{dt} = 0, \quad \frac{d\theta}{dt} = \omega + Q(A, r^2, \mu),$$

which in addition shows that the period in t of the periodic orbits depends upon μ, A, and the radius r. When adding the remainders ρ_A and ρ_B, the implicit function theorem, allows us to show that these dynamics persist for the full system.

In the case $0^-(i\omega)$, for the truncated normal form we can use polar coordinates $B = re^{i\theta}$, again, and then the dynamics reduce to that of the two-dimensional system for A and r. The number and nature of equilibria and bounded orbits depends upon the coefficients of the leading order terms in P and Q_2. When adding the remainders ρ_A and ρ_B, the implicit function theorem allows us to show the persistence of the equilibria and periodic orbits. The question of persistence of homoclinic and heteroclinic orbits in this case is much more difficult, since the invariant set under the action of $\mathbf{S}$ is only one-dimensional. In the case of analytic vector fields, one could use the optimal normal form in Theorem 5.8 in Chapter 2. Nevertheless, despite the exponentially small estimate of the remainder, the question of persistence is open in this case.

4.3 Dimension 4

In this section, we consider reversible systems in $\mathbb{R}^4$ of the form

$$\frac{du}{dt} = \mathbf{F}(u,\mu), \tag{3.1}$$

for which we assume that the vector field satisfies the following hypothesis.

Hypothesis 3.1 *Assume that the vector field* $\mathbf{F}$ *is of class* $\mathscr{C}^k$, $k \geq 3$, *in a neighborhood* $\mathscr{V}$ *of* $(0,0) \in \mathbb{R}^4 \times \mathbb{R}$, *satisfying*

$$\mathbf{F}(0,0) = 0. \tag{3.2}$$

We further assume that there is a symmetry $\mathbf{S}$ *with*

$$\mathbf{S}^2 = \mathbb{I},$$

which anticommutes *with* $\mathbf{F}$,

$$\mathbf{F}(\mathbf{S}u,\mu) = -\mathbf{S}\mathbf{F}(u,\mu) \textit{ for all } (u,\mu) \in \mathscr{V}. \tag{3.3}$$

As in the previous sections, we set

$$\mathbf{L} = D_u\mathbf{F}(0,0), \tag{3.4}$$

and we have that

$$\mathbf{SL} = -\mathbf{LS}.$$

Consequently, the spectrum of $\mathbf{L}$ is symmetric with respect to both the real and imaginary axis. If $\mathbf{L}$ has eigenvalues that do not lie on the imaginary axis, then there are either four or two such eigenvalues. In the first case, the situation is hyperbolic, and the dynamics are the same as that of the linear equation $du/dt = \mathbf{L}u$, whereas in the second case one can use the center manifold theorem and reduce the problem to a two-dimensional one, which enters in the setting of Section 4.1. We therefore consider only the situation in which the four eigenvalues all lie on the imaginary axis where we distinguish the following cases:

(i) $\sigma(\mathbf{L}) = \{0, \pm i\omega\}$ for some nonzero real number ω, with 0 an algebraically double eigenvalue;
(ii) $\sigma(\mathbf{L}) = \{\pm i\omega\}$ for some nonzero real number ω, with $\pm i\omega$ algebraically double eigenvalues;
(iii) $\sigma(\mathbf{L}) = \{\pm i\omega_1, \pm i\omega_2\}$ for some nonzero real numbers ω_1 and ω_2;
(iv) $\sigma(\mathbf{L}) = \{0\}$, with 0 an algebraically quadruple eigenvalue.

In case (i), the generic situation occurs when 0 is geometrically simple, the $0^2(i\omega)$ bifurcation, for which we distinguish between $0^{2+}(i\omega)$, when for ζ_0 in the kernel of $\mathbf{L}$ we have $\mathbf{S}\zeta_0 = \zeta_0$, and $0^{2-}(i\omega)$, when $\mathbf{S}\zeta_0 = -\zeta_0$. Similarly, in the case (ii) the generic situation occurs when $\pm i\omega$ are geometrically simple, which is the $(i\omega)^2$ bifurcation. This bifurcation has been extensively studied, and is often referred to as 1 : 1-resonance or Hamiltonian–Hopf bifurcation. We discuss in more detail the cases $0^{2+}(i\omega)$ and $(i\omega)^2$, which are most frequently met in physical examples, and give only some elements of the analysis in the case $0^{2-}(i\omega)$, and in case (iii), $(i\omega_1)(i\omega_2)$ bifurcation.

In case (iv), the type of bifurcations further depends upon the geometric multiplicity of the zero eigenvalue, 0^4, 00^3, 0^20^2, 000^2, 0000, and the action of $\mathbf{S}$ on the kernel of $\mathbf{L}$. The complete study of these bifurcations, which are of codimension higher than 1, is still open. We present here only some elements of analysis for the 0^{4+} bifurcation, and for an 0^20^2 bifurcation with an additional $SO(2)$ symmetry, which are of importance in some physical examples.

4.3.1 Reversible $0^{2+}(i\omega)$ *Bifurcation*

We consider in this section the $0^{2+}(i\omega)$ bifurcation. This bifurcation has been extensively studied in the literature, and we refer for instance to [59, 61, 93], and the references therein, for further details.

Hypothesis 3.2 *Assume that* $\sigma(\mathbf{L}) = \{0, \pm i\omega\}$, *with* 0 *an algebraically double and geometrically simple eigenvalue. Further assume that the eigenvector* ξ_0 *associated with the eigenvalue* 0 *satisfies*

$$\mathbf{S}\xi_0 = \xi_0,$$

in which $\mathbf{S}$ *is the symmetry anticommuting with* $\mathbf{L}$ *in Hypothesis 3.1.*

Normal Form

As in the previous cases, we start by constructing a suitable basis of $\mathbb{R}^4$.

Lemma 3.3 ($0^{2+}(i\omega)$ **basis of** $\mathbb{R}^4$) *Assume that Hypothesis 3.2 holds. Then there exists a basis* $\{\xi_0, \xi_1, \operatorname{Re}\zeta, \operatorname{Im}\zeta\}$ *of* $\mathbb{R}^4$ *consisting of generalized eigenvectors* $\xi_0, \xi_1 \in \mathbb{R}^4$ *and* $\zeta \in \mathbb{C}^4$ *of* $\mathbf{L}$, *such that*

$$\mathbf{L}\xi_0 = 0, \quad \mathbf{L}\xi_1 = \xi_0, \quad \mathbf{L}\zeta = i\omega\zeta,$$
$$\mathbf{S}\xi_0 = \xi_0, \quad \mathbf{S}\xi_1 = -\xi_1, \quad \mathbf{S}\zeta = \overline{\zeta}.$$

Proof The vectors ξ_0 and ξ_1 are constructed as in the proof of Lemma 1.4, and ζ is the eigenvector associated to $i\omega$ such that $\mathbf{S}\zeta = \overline{\zeta}$, as explained in Section 4.2.4. □

Notation 3.4 *In the basis above, we represent a vector in $u \in \mathbb{R}^4$ by $(A,B,C,\overline{C})$,*

$$u = A\xi_0 + B\xi_1 + C\zeta + \overline{C\zeta},$$

with $A,B \in \mathbb{R}$, and $C \in \mathbb{C}$. We identify $\mathbb{R}^4$ with the space $\mathbb{R}^2 \times \widetilde{\mathbb{R}^2}$ in which

$$\widetilde{\mathbb{R}^2} = \{(C,\overline{C})\ ;\ C \in \mathbb{C}\}.$$

Lemma 3.5 ($0^{2+}(i\omega)$ **normal form)** *Assume that Hypotheses 3.1 and 3.2 hold, and consider the basis $\{\xi_0,\xi_1,\zeta,\overline{\zeta}\}$ of $\mathbb{R}^2 \times \widetilde{\mathbb{R}^2}$ in Lemma 3.3. Then for any positive integer p, $2 \leq p \leq k$, there exist neighborhoods $\mathscr{V}_1$ and $\mathscr{V}_2$ of 0 in $\mathbb{R}^2 \times \widetilde{\mathbb{R}^2}$ and $\mathbb{R}$, respectively, a map $\Phi : \mathscr{V}_1 \times \mathscr{V}_2 \to \mathbb{R}^4$ with the following properties:*

(i) Φ is of class $\mathscr{C}^k$, satisfying

$$\Phi(0,0,0,0,0) = 0, \quad \partial_{(A,B,C,\overline{C})}\Phi(0,0,0,0,0) = 0, \tag{3.5}$$

and

$$\Phi(A,-B,\overline{C},C,\mu) = \mathbf{S}\Phi(A,B,C,\overline{C},\mu).$$

(ii) For $(A,B,C,\overline{C}) \in \mathscr{V}_1$, the change of variables

$$u = A\xi_0 + B\xi_1 + C\zeta + \overline{C\zeta} + \Phi(A,B,C,\overline{C},\mu) \tag{3.6}$$

transforms equation (3.1) into the normal form

$$\begin{aligned}
\frac{dA}{dt} &= B \\
\frac{dB}{dt} &= P(A,|C|^2,\mu) + \rho_B(A,B,C,\overline{C},\mu) \\
\frac{dC}{dt} &= i\omega C + iCQ(A,|C|^2,\mu) + \rho_C(A,B,C,\overline{C},\mu),
\end{aligned} \tag{3.7}$$

where P and Q are real-valued polynomials of degree p and $p-1$ in $(A,B,C,\overline{C})$, respectively. The remainders ρ_B and ρ_C are of class $\mathscr{C}^k$ and satisfy

$$\begin{aligned}
\rho_B(A,-B,\overline{C},C,\mu) &= \rho_B(A,B,C,\overline{C},\mu), \\
\rho_C(A,-B,\overline{C},C,\mu) &= -\overline{\rho_C}(A,B,C,\overline{C},\mu),
\end{aligned}$$

with the estimate

$$|\rho_B(A,B,C,\overline{C},\mu)| + |\rho_C(A,B,C,\overline{C},\mu)| = o((|A|+|B|+|C|)^p). \tag{3.8}$$

Proof According to Theorem 2.2 and Lemma 1.18 in Chapter 3, there is a polynomial $\Phi_1(\cdot,\mu)$ of degree p in $(A,B,C,\overline{C})$, with coefficients depending upon μ, which satisfies (3.5), and such that the change of variables

$$u = A\xi_0 + B\xi_1 + C\zeta + \overline{C\zeta} + \Phi_1(A,B,C,\overline{C},\mu)$$

transforms (3.1) into the system

$$\begin{aligned}
\frac{dA}{dt} &= B + AP_0(A,|C|^2,\mu) + \rho_1(A,B,C,\overline{C},\mu)\\
\frac{dB}{dt} &= BP_0(A,|C|^2,\mu) + P_1(A,|C|^2,\mu) + \rho_2(A,B,C,\overline{C},\mu)\\
\frac{dC}{dt} &= i\omega C + CP_2(A,|C|^2,\mu) + \rho_3(A,B,C,\overline{C},\mu),
\end{aligned}$$

where P_0 and P_2 are polynomials of degree $p-1$ in $(A,C,\overline{C})$, P_1 is a polynomial of degree p in $(A,C,\overline{C})$, and ρ_j, $j=1,2,3$, are functions of class $\mathscr{C}^k$ satisfying (3.8).

Now, using the reversibility symmetry $\mathbf{S}$, by Theorem 3.4 we have that

$$\boldsymbol{\Phi}_1(A,-B,\overline{C},C,\mu) = \mathbf{S}\boldsymbol{\Phi}_1(A,B,C,\overline{C},\mu),$$

and that the first component of the vector field satisfies

$$AP_0(A,|C|^2,\mu) + \rho_1(A,-B,\overline{C},C,\mu) = -AP_0(A,|C|^2,\mu) - \rho_1(A,B,C,\overline{C},\mu).$$

Consequently, $P_0 = 0$. For the second and the third components of the vector field we then find

$$\begin{aligned}
P_1(A,|C|^2,\mu) + \rho_2(A,-B,\overline{C},C,\mu) &= P_1(A,|C|^2,\mu) + \rho_2(A,B,C,\overline{C},\mu)\\
\overline{C}P_2(A,|C|^2,\mu) + \rho_3(A,-B,\overline{C},C,\mu) &= -\overline{CP_2}(A,|C|^2,\mu) - \overline{\rho_3}(A,B,C,\overline{C},\mu),
\end{aligned}$$

which imply that the polynomial P_2 is purely imaginary.

Finally, we make the change of variables

$$B' = B + \rho_1(A,B,C,\overline{C},\mu),$$

which preserves the reversibility symmetry, and gives the desired normal form. Notice that with this change of variables the polynomial $\boldsymbol{\Phi}_1$ is transformed into the function $\boldsymbol{\Phi}$ of class $\mathscr{C}^k$. □

Solutions of the Normal Form System

We analyze now the dynamics of the normal form in the generic situation in which the following assumption on the expansion of the vector field holds.

Hypothesis 3.6 *Assume that the expansions of P and Q in (3.7),*

$$\begin{aligned}
P(A,|C|^2,\mu) &= a\mu + bA^2 + c|C|^2 + O(|\mu|^2 + (|\mu| + |A| + |C|)^3),\\
Q(A,|C|^2,\mu) &= \alpha\mu + \beta A + \gamma|C|^2 + O((|\mu| + |A| + |C|^2)^2),
\end{aligned}$$

are such that a, b, and c do not vanish.

Truncated System

We start with the truncated system

$$\begin{aligned}\frac{dA}{dt} &= B \\ \frac{dB}{dt} &= P(A,|C|^2,\mu) \\ \frac{dC}{dt} &= i\omega C + iCQ(A,|C|^2,\mu).\end{aligned} \tag{3.9}$$

This system is integrable, with two first integrals

$$K = |C|^2, \quad H = B^2 - 2\int_0^A P(s,K,\mu)ds.$$

Further notice that the system is invariant under rotations in the C-plane. This shows that the projections on the C-plane of the orbits are all circles, whereas the projections on the (A,B)-plane are described by

$$B^2 = f_{H,K}(A,\mu), \quad f_{H,K}(A,\mu) = 2\int_0^A P(s,K,\mu)ds + H. \tag{3.10}$$

The leading order terms in the expansion of $f_{H,K}$ are

$$f_{H,K}(A,\mu) = \frac{2}{3}bA^3 + 2\widetilde{\mu}A + H, \quad \widetilde{\mu} = a\mu + cK,$$

which implies that the projections on the (A,B)-plane of the orbits are as in the case of the 0^{2+} bifurcation. In particular, there are no bounded orbits when $\widetilde{\mu}b > 0$, and the orbits are as shown in Figure 3.1 when $\widetilde{\mu}b < 0$.

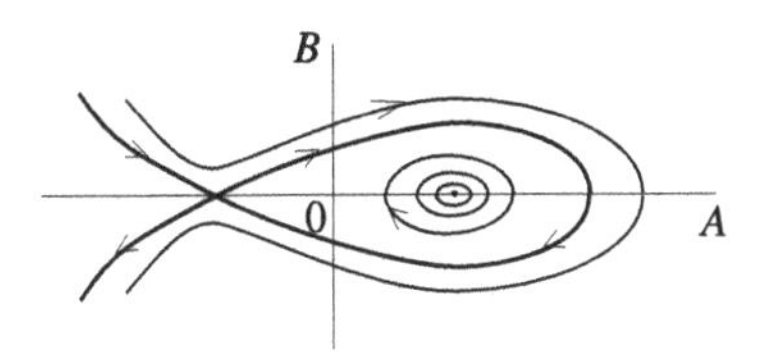

Fig. 3.1 Dynamics of the $0^{2+}(i\omega)$ bifurcation. Plot of the projections in the (A,B)-plane of the orbits of the truncated system (3.9) obtained by varying H in the case $\widetilde{\mu} > 0$ and $b < 0$. For $\widetilde{\mu} < 0$ and $b > 0$ the phase portrait is qualitatively the same, and for $\widetilde{\mu}b > 0$ there are no bounded orbits.

For the truncated system, we then find bounded orbits in the case $\widetilde{\mu}b < 0$, only.

(i) The two equilibria, the center $(A_c,0)$ and the saddle $(A_h,0)$ in the (A,B)-plane, give two equilibria, a center $(A_c,0,0)$ and a saddle-center $(A_h,0,0)$ of the truncated system, which are found for $K = 0$. For any $K > 0$, the equilibria

$(A_c, 0)$ and $(A_h, 0)$ give periodic orbits $(A_c, 0, C_{c,K}(t))$ and $(A_h, 0, C_{h,K}(t))$, respectively, with $|C_{c,K}(t)| = |C_{h,K}(t)| = \sqrt{K}$. Below, we refer to these periodic orbits as *periodic orbits of the first kind.*

(ii) The homoclinic orbit in the (A,B)-plane gives a homoclinic orbit to the equilibrium $(A_h, 0, 0)$ for $K = 0$, and for any $K > 0$ it gives a one-parameter family of homoclinic orbits connecting the periodic orbit $(A_h, 0, C_{h,K}(t))$ to itself. More precisely, we have a "circle" of homoclinic orbits parameterized by some $\phi \in \mathbb{R}/2\pi\mathbb{Z}$. We point out that when we restrict terms in the expansions of P and Q, to the quadratic order:

$$P(A, |C|^2, \mu) = a\mu + bA^2 + c|C|^2, \quad Q(A, |C|^2, \mu) = \alpha\mu + \beta A + \gamma|C|^2,$$

these homoclinic orbits can be computed explicitly. For instance, in the case $\widetilde{\mu} < 0$ and $b > 0$, one finds the solutions

$$A(t) = A_h\left(1 - \frac{3}{\cosh^2 \delta t}\right), \quad B(t) = A'(t), \quad C(t) = \sqrt{K}e^{i(\Omega t + \kappa \tanh \delta t + \phi)},$$

where

$$\delta = \sqrt{\frac{bA_h}{2}}, \quad \Omega = \omega + \alpha\mu + \gamma K + \beta A_h, \quad \kappa = -3\beta\sqrt{\frac{2A_h}{b}},$$

and $\phi \in \mathbb{R}/2\pi\mathbb{Z}$ is arbitrary. In particular, this gives an explicit description of the asymptotics at infinity, where we find a phase shift $\kappa\pi = O(|-\widetilde{\mu}|^{1/4})$ between $t = -\infty$ and $t = +\infty$.

(iii) The periodic orbits in Figure 3.1 correspond to a family of invariant 2-tori in the four-dimensional space. Depending on the ratio between the frequency in the C-plane and the frequency in the (A,B)-plane, on these tori we find *quasi-periodic or periodic orbits* of the truncated system. Below, we refer to these periodic orbits as *periodic orbits of the second kind*, or *elliptic*.

Remark 3.7 *Recall that $\widetilde{\mu} = a\mu + cK$ and $K \geq 0$. The condition $\widetilde{\mu}b < 0$ for the existence of bounded orbits is then $ab\mu + bcK < 0$, so that small bounded solutions exist for any small μ if $bc < 0$, and they exist only when $ab\mu < 0$ if $bc > 0$.*

Remark 3.8 *In applications there is often a trivial equilibrium that exists for all values of μ. Then in the expansion of the polynomial P the coefficient a vanishes, so that*

$$P(A, |C|^2, \mu) = a'\mu A + bA^2 + c|C|^2 + O(\mu^2|A| + |\mu|A^2 + |A|^3 + |C|^4). \quad (3.11)$$

It turns out that this situation is similar to the case $ab\mu < 0$. Indeed, assuming that $ab\mu < 0$, one can make in (3.9) the change of variables

$$A = A_0 + A',$$

where A_0 is a solution of $P(A_0, 0, \mu) = 0$,

$$A_0 = \operatorname{sign}(b)\sqrt{\frac{-a\mu}{b}} + h.o.t..$$

This leads to a similar system for (A', B, C), *with* $P(A, |C|^2, \mu)$ *replaced by*

$$P'(A', |C|^2, \mu) = 2\sqrt{-ab\mu}A' + bA'^2 + c|C|^2 + O(|\mu||A'| + |\mu|^{1/2}|A'|^2 + |A'|^3 + |C|^4),$$

which is of the form (3.11).

Full System

We discuss now the persistence of the bounded orbits found above for the full normal form (3.7). This question is a delicate problem here, which has been extensively studied.

First, the persistence of the equilibria and the periodic orbits as given in case (i) by the two equilibria $(A_c, 0)$ and $(A_h, 0)$ found in the (A, B)-plane, can be shown by an implicit function argument, with an adapted Lyapunov–Schmidt method [59, 93]. The persistence of the periodic orbits given in case (iii) by the periodic orbits found in the (A, B)-plane, and lying on the invariant 2-tori of the system (3.9), is more complicated but can also be proved in the same way [59]. The persistence of quasiperiodic solutions in case (iii), lying on the other invariant 2-tori of (3.9), leads to a small divisor problem. This question can be analyzed using a parameterization with the two first integrals by restricting the study to a region where these solutions exist for the system (3.9). In [59] it is roughly proved that for a fixed value of the bifurcation parameter μ, quasiperiodic solutions of (3.7) exist for (K, H) lying in a region that is locally the product of a line with a Cantor set.

The question of persistence of the orbits homoclinic to periodic orbits in case (ii) has received partial answers [6, 59, 91, 115, 116] and is more generally studied in [93]. In the truncated system each of these periodic orbits has a two-dimensional unstable manifold, which intersects transversally the two-dimensional plane $B = \operatorname{Im} C = 0$, which is invariant under the reversibility symmetry $\mathbf{S}$, in two points of coordinates $(A_0, 0, \pm C_0, \pm C_0)$, with $C_0^2 = K > 0$. Using reversibility, taking these points as initial conditions at $t = 0$ gives two orbits that are homoclinic to the periodic orbit and are invariant under the reversibility symmetry. For the full system (3.7), a transversality argument allows us to show the persistence of these homoclinic orbits, provided the size $\sqrt{K}$ of the periodic orbit is not too small.

Notice that these are just two orbits in the circle of homoclinics to these periodic orbits, and are precisely those which are reversible. The persistence of some homoclinic orbit that is not reversible is an open problem.

In the limit $K = 0$, the periodic orbits shrink to the saddle-center equilibrium $(A_h, 0, 0)$. This equilibrium has a one-dimensional unstable manifold, which misses in general the two-dimensional plane of symmetry. This is the challenging problem of asymptotics beyond all orders, since these homoclinic orbits exist for the normal form at any order (except that the normal form cannot be written up to infinite order, even for analytical vector fields).

Remark 3.9 (Analytic vector fields) *(i) We point out that for analytic vector fields, the homoclinics to periodic orbits persist for periodic orbits of size larger than an exponentially small quantity with respect to $|\mu|$ (e.g., see [93] and the references therein, and [61] for a shorter proof using the result in Theorem 5.8 of Chapter 3).*

(ii) For analytic vector fields, one could in principle compute an infinite expansion in powers of $|\mu|^{1/2}$ of a solution homoclinic to the saddle-center $(A_h, 0, 0)$. However, a delicate analysis of singularities in the complex time-plane shows that in general this expansion does not converge [92, 93].

(iii) As already mentioned, the center manifold reduction does not preserve analyticity, so the results mentioned above for analytic vector fields cannot be directly transfered to higher order systems. However, in this situation one can use the results in Chapter 3, Corollary 5.12 and Remark 5.13, together with Theorem 5.8 of that Chapter. We refer the reader to [61] for further details about this situation.

We summarize in the next theorem the results briefly described above for C^k-vector fields. We refer to [93] for the case of analytic vector fields and the persistence of the homoclinics to periodic orbits, and to [59] for more precise statements on the persistence of the periodic orbits of the second kind and of quasiperiodic solutions.

Theorem 3.10 **($0^{2+}(i\omega)$ bifurcation)** *Assume that Hypotheses 3.1, 3.2, and 3.6 hold. Then for differential equation (3.1) a reversible $0^{2+}(i\omega)$ bifurcation occurs at $\mu = 0$. More precisely, the following properties hold in a neighborhood of 0 in $\mathbb{R}^4$ for sufficiently small μ:*

(i) For $ab\mu < 0$ and $bc < 0$, there are two equilibria, a center and a saddle-center, together with two one-parameter families of periodic orbits of the first kind, parameterized by their size r, which tend to the two equilibria as $r \to 0$. For any periodic orbit in the family which tends to the saddle-equilibrium, with size r not too small, $r > r_*(\mu)$, there is a pair of reversible homoclinic orbits connecting this periodic orbit to itself.

(ii) For $ab\mu < 0$ and $bc > 0$, there are two equilibria, a center and a saddle-center, together with two families of periodic orbits of the first kind, parameterized by their size r, for $r < r^*(\mu) = O(|\mu|^{1/2})$, and which tend to the two equilibria as $r \to 0$. For any periodic orbit in the family which tends to the saddle-equilibrium, with size r not too small, $r > r_*(\mu)$, there is a pair of reversible homoclinic orbits connecting this periodic orbit to itself.

(iii) For $ab\mu > 0$ and $bc < 0$, there are two families of periodic orbits of the first kind, parameterized by their size r, for $r > r^*(\mu) = O(|\mu|^{1/2})$. To any periodic orbit in one of these families, there is a pair of reversible homoclinic orbits connecting the periodic orbit to itself.

(iv) For $ab\mu > 0$ and $bc > 0$, there are no bounded solutions.

Furthermore, in the cases (i), (ii), and (iii), there are periodic orbits of the second kind and quasiperiodic orbits.

Computation of $0^{2+}(i\omega)$ Bifurcations in Infinite Dimensions

We show below how to compute the principal coefficients in the normal form (3.7) when starting from an infinite-dimensional system of the form

$$\frac{du}{dt} = \mathbf{L}u + \mathbf{R}(u,\mu). \tag{3.12}$$

We assume that the parameter μ is real, and that the system (3.12) possesses a reversibility symmetry $\mathbf{S}$ and satisfies the hypotheses of Theorems 3.3 and 3.15 in Chapter 2. We further assume that the spectrum of the linear operator $\mathbf{L}$ is such that $\sigma_0 = \{0, \pm i\omega\}$, where 0 is an algebraically double and geometrically simple eigenvalue, with a symmetric eigenvector ξ_0 such that $\mathbf{S}\xi_0 = \xi_0$, and $\pm i\omega$ are simple eigenvalues. Then the four-dimensional reduced system satisfies the hypotheses of Lemma 3.5, so that its normal form is given by (3.7).

We proceed as in Section 3.4 and in the previous examples. In equality (4.3) in Chapter 3, we take $v_0 = A\xi_0 + B\xi_1 + C\zeta + \overline{C\zeta}$ and then write

$$u = A\xi_0 + B\xi_1 + C\zeta + \overline{C\zeta} + \widetilde{\Psi}(A,B,C,\overline{C},\mu), \tag{3.13}$$

where $\widetilde{\Psi}$ takes values in $\mathscr{Z}$. With the notations from Section 3.2.3, we consider the Taylor expansion (1.15) of $\mathbf{R}$, and the expansion of $\widetilde{\Psi}$,

$$\widetilde{\Psi}(A,B,C,\overline{C},\mu) = \sum_{1\le r+s+q+l+m\le p} A^r B^s C^q \overline{C}^l \mu^m \Psi_{rsqlm},$$

where

$$\Psi_{rsql0} = 0 \text{ for } r+s+q+l = 1.$$

Using the reversibility symmetry we find that

$$\mathbf{S}\Psi_{rsqlm} = (-1)^s \Psi_{rslqm}, \quad \Psi_{rsqlm} = \overline{\Psi}_{rslqm}.$$

Identity (4.4) in Chapter 3 is, in this case,

$$\begin{aligned} &B\partial_A\widetilde{\Psi} + i\omega C\partial_C\widetilde{\Psi} - i\omega\overline{C}\partial_{\overline{C}}\widetilde{\Psi} + (\xi_1 + \partial_B\widetilde{\Psi})P(A,|C|^2,\mu) \\ &+ (iC(\zeta + \partial_C\widetilde{\Psi}) - i\overline{C}(\overline{\zeta} + \partial_{\overline{C}}\widetilde{\Psi}))Q(A,|C|^2,\mu) \\ &\qquad = \mathbf{L}\widetilde{\Psi} + \mathbf{R}(A\xi_0 + B\xi_1 + C\zeta + \overline{C\zeta} + \widetilde{\Psi},\mu). \end{aligned}$$

Using the expansions of $\mathbf{R}$, $\widetilde{\Psi}$, P, and Q, we find at orders $O(\mu)$, $O(A^2)$, $O(C\overline{C})$, $O(AC)$, and $O(\mu C)$, the equalities

$$\begin{aligned}
a\xi_1 &= \mathbf{L}\Psi_{00001} + \mathbf{R}_{0,1},\\
b\xi_1 &= \mathbf{L}\Psi_{20000} + \mathbf{R}_{2,0}(\xi_0,\xi_0),\\
c\xi_1 &= \mathbf{L}\Psi_{00110} + 2\mathbf{R}_{2,0}(\zeta,\overline{\zeta}),\\
i\beta\zeta &= (\mathbf{L}-i\omega)\Psi_{10100} + 2\mathbf{R}_{2,0}(\xi_0,\zeta),\\
i\alpha\zeta &= (\mathbf{L}-i\omega)\Psi_{00101} + \mathbf{R}_{1,1}\zeta + 2\mathbf{R}_{2,0}(\zeta,\Psi_{00001}).
\end{aligned}$$

The different coefficients are now found from the solvability conditions for these five equations, just as in Section 3.4.4 of Chapter 3,

$$\begin{aligned}
&a = \langle \mathbf{R}_{0,1},\xi_1^*\rangle, \quad b = \langle \mathbf{R}_{2,0}(\xi_0,\xi_0),\xi_1^*\rangle, \quad c = \langle 2\mathbf{R}_{2,0}(\zeta,\overline{\zeta}),\xi_1^*\rangle,\\
&i\beta = \langle 2\mathbf{R}_{2,0}(\xi_0,\zeta),\zeta^*\rangle, \quad i\alpha = \langle \mathbf{R}_{1,1}\zeta + 2\mathbf{R}_{2,0}(\zeta,\Psi_{00001}),\zeta^*\rangle.
\end{aligned}$$

Here the vectors ξ_1^* and ζ^* are orthogonal to the ranges of $\mathbf{L}$ and $i\omega - \mathbf{L}$, respectively, and are found as in the Sections 3.4.2 and 3.4.4.

Exercise 3.11 *Using the reversibility symmetry, show from the formulas above that the coefficients* a, b, c, α, *and* β *are real.*

Exercise 3.12 *Consider system (3.12) under the assumptions above. Further assume that* $u = 0$ *is a solution for all* μ, *i.e.,* $\mathbf{R}(0,\mu) = 0$.

(i) *Show that*

$$\Psi(0,0,0,0,\mu) = 0, \quad P(0,0,\mu) = 0, \quad \rho_{B,C}(0,0,0,0,\mu) = 0.$$

(ii) *Consider the expansion of* P,

$$P(A,|C|^2,\mu) = a'\mu A + bA^2 + c|C|^2 + h.o.t..$$

Show that

$$a' = \langle \mathbf{R}_{1,1}\xi_0,\xi^*\rangle$$

and that the function $f_{H,K}$ *in (3.10) is*

$$f_{H,K}(A,\mu) = \frac{2}{3}bA^3 + a'\mu A^2 + 2cKA + H + h.o.t..$$

Compare with (2.16) and determine the phase portraits of the truncated normal form.

(iii) *Assume that* $b = 0$ *and take*

$$P(A,|C|^2,\mu) = a'\mu A + b'A^3 + c|C|^2 + h.o.t..$$

Show that

$$b' = \langle 2\mathbf{R}_{2,0}(\xi_0,\Psi_{20000}) + \mathbf{R}_{3,0}(\xi_0\xi_0,\xi_0),\xi_1^*\rangle,$$

where

$$\mathbf{L}\Psi_{20000} + \mathbf{R}_{2,0}(\xi_0,\xi_0) = 0.$$

Determine the leading order terms in the expansion of $f_{H,K}$ *and compare with (2.18). Determine the phase portraits of the truncated normal form.*

(iv) *Discuss the phase portraits of the full four-dimensional reduced system in cases (ii) and (iii).*

Example: Two-dimensional NLS-type Equation

Consider the following nonlinear Schrödinger equation (NLS) in the plane

$$i\partial_t U + \Delta U + V(x)U + Uf(|U|^2) = 0, \tag{3.14}$$

in which U is complex-valued, $U(t,x,y) \in \mathbb{C}$, $V(x)$ a given one-dimensional potential, i.e., independent of y, and f a smooth map. This kind of equation arises as a model in many different contexts, as for instance, wave formation in Bose–Einstein condensates, or waves in photorefractive media.

We are interested in solutions of the form

$$U(t,x,y) = e^{i\omega t} v(x,y),$$

with *real-valued* profiles v satisfying the steady equation

$$\Delta v - \omega v + V(x)v + vf(v^2) = 0. \tag{3.15}$$

Hypothesis 3.13 *(i) Assume that $f : \mathbb{R} \to \mathbb{R}$ is a smooth function with $f(0) = 0$ and*

$$f(w) = d_0 w + \mathrm{O}(w^2) \quad as \quad w \to 0,$$

with $d_0 \neq 0$.

(ii) Assume that the potential $V : \mathbb{R} \to \mathbb{R}$ is a smooth function, such that the one-dimensional operator

$$\mathbf{L}_V = \partial_{xx} + V(x)$$

acting in $L^2(\mathbb{R})$ has the spectrum

$$\sigma(\mathbf{L}_V) = \sigma_c(\mathbf{L}_V) \cup \{\gamma_n, \dots, \gamma_1\}, \quad \sigma_c(\mathbf{L}_V) \subset \{\lambda \in \mathbb{C}\,;\, \mathrm{Re}\,\lambda \leq \gamma_*\},$$

for some $n \geq 1$, where $\gamma_* < \gamma_n < \cdots < \gamma_1$, and γ_j, $j = 1, \dots, n$, are simple eigenvalues with associated eigenfunctions g_j normalized in the norm of $L^2(\mathbb{R})$.

Spatial Dynamics

We start by writing equation (3.15) as a first order system,

$$\begin{aligned} v_y &= w \\ w_y &= (\omega - \mathbf{L}_V)v - vf(v^2), \end{aligned} \tag{3.16}$$

where the time-like variable is the spatial variable y, in which the potential is homogeneous. This system is of the form (2.1) in Chapter 2, with $u = (v,w)$ and a linear part $\mathbf{L}_\omega$ depending upon ω. We assume that $\mathbf{L}_\omega$ acts in the Hilbert space $\mathscr{X} = H^1(\mathbb{R}) \times L^2(\mathbb{R})$, in which it is a closed operator with dense domain $\mathscr{Z} = H^2(\mathbb{R}) \times H^1(\mathbb{R})$.

Equation (3.15) possesses two discrete symmetries: a reflection in y, $v(\cdot,y) \mapsto v(\cdot,-y)$, and a reflection in v, $v \mapsto -v$, as a remnant of the phase rotation invariance of the original NLS equation. As a consequence, the first order system (3.16), possesses a reversibility symmetry $\mathbf{S}$ and is equivariant under a symmetry $\mathbf{T}$ acting through

$$\mathbf{S}(v,w) = (v,-w), \quad \mathbf{T}(v,w) = (-v,-w).$$

Reversible Bifurcations

We take ω as bifurcation parameter and look for the bifurcation points. For this, we compute the spectrum of $\mathbf{L}_\omega$, from Hypothesis 3.13(ii) which describes the spectral properties of $\mathbf{L}_V$. A direct calculation allows us to conclude that there is an increasing sequence of bifurcation points at $\omega = \gamma_1, \gamma_2, \dots, \gamma_n$. At each bifurcation point $\omega = \gamma_k$ the spectrum of $\mathbf{L}_\omega$ consists of

one eigenvalue in zero, geometrically simple and algebraically double;
$k-1$ pairs of simple complex conjugated eigenvalues

$$\pm \mathrm{i}\sqrt{\gamma_j - \gamma_k}, \quad j = 1, \dots, k-1;$$

and

the rest of the spectrum lies at a distance $\sqrt{\gamma_k - \gamma_{k+1}} > 0$ from the imaginary axis.

Furthermore, in each case the eigenvector $(g_k, 0)$ in the kernel of $\mathbf{L}_{\gamma_k}$ is invariant under the action of $\mathbf{S}$. In particular, this shows that we are in the presence of an 0^{2+} bifurcation at $\omega = \gamma_1$, and an $0^{2+}(i\omega)$ bifurcation at $\omega = \gamma_2$.

First Bifurcation at $\omega = \gamma_1$

We set $\omega = \gamma_1 + \mu$, and rewrite the first order system in form (1.13), with y replacing the time t and $\mathbf{L} = \mathbf{L}_{\gamma_1}$. This system satisfies the hypotheses of Theorems 3.3, 3.13, and 3.15 in Chapter 2, so that it possesses a two-dimensional center manifold. Furthermore, the two-dimensional space $\mathscr{E}_0$ is spanned by the vectors

$$\zeta_0 = \begin{pmatrix} g_1 \\ 0 \end{pmatrix}, \quad \zeta_1 = \begin{pmatrix} 0 \\ g_1 \end{pmatrix},$$

satisfying

$$\mathbf{L}\zeta_0 = 0, \quad \mathbf{L}\zeta_1 = \zeta_0, \quad \mathbf{S}\zeta_0 = \zeta_0, \quad \mathbf{S}\zeta_1 = -\zeta_1, \quad \mathbf{T}\zeta_0 = -\zeta_0, \quad \mathbf{T}\zeta_1 = -\zeta_1.$$

Then the reduced system satisfies Hypotheses 1.1 and 1.3, and, in addition, it is equivariant under the action of $\mathbf{T}$. Applying the result in Theorem 1.5, we conclude that the normal form of the reduced system is given by (1.9), in which the polynomial Q is odd in A, due to the symmetry $\mathbf{T}$ (see also Remark 1.9).

The computation of the principal coefficients in the normal form is now much easier than in the general case. We find that

$$Q(A,\mu) = \mu A + dA^3 + O(\mu^2|A| + |A|^5), \quad d = -d_0 \int_{\mathbb{R}} g_1^4(x)dx.$$

Consequently, the reduced system possesses bounded solutions in the cases $\mu < 0, d_0 > 0$, $\mu < 0, d_0 < 0$, and $\mu > 0, d_0 > 0$, when the phase portraits are as in Figure 1.3, with $c = 1$ and d as above. Summarizing we have:

(i) periodic orbits in the three cases, which correspond to solutions of (3.15) that are periodic in y and localized in x;
(ii) a pair of heteroclinic orbits in the case $\mu < 0$, $d_0 < 0$, which correspond to solutions of (3.15) that are asymptotically constant in y and localized in x;
(iii) a pair of homoclinic orbits in the case $\mu > 0$, $d_0 > 0$, which correspond to solutions of (3.15) that are fully localized.

Second Bifurcation at $\omega = \gamma_2$

Now we set $\omega = \gamma_2 + \mu$, and rewrite the first order system in form (3.12), with y replacing time t and $\mathbf{L} = \mathbf{L}_{\gamma_2}$. Applying the results in Theorems 3.3, 3.13, and 3.15 in Chapter 2, we obtain in this case a four-dimensional center manifold. The two-dimensional space $\mathscr{E}_0$ is spanned by the vectors

$$\xi_0 = \begin{pmatrix} g_2 \\ 0 \end{pmatrix}, \quad \xi_1 = \begin{pmatrix} 0 \\ g_2 \end{pmatrix}, \quad \zeta = \begin{pmatrix} g_1 \\ i\sqrt{\gamma_1 - \gamma_2} g_1 \end{pmatrix}, \quad \overline{\zeta} = \begin{pmatrix} g_1 \\ -i\sqrt{\gamma_1 - \gamma_2} g_1 \end{pmatrix},$$

satisfying

$$\mathbf{L}\xi_0 = 0, \quad \mathbf{L}\xi_1 = \xi_0, \quad \mathbf{S}\xi_0 = \xi_0, \quad \mathbf{S}\xi_1 = -\xi_1, \quad \mathbf{T}\xi_0 = -\xi_0, \quad \mathbf{T}\xi_1 = -\xi_1,$$

and

$$\mathbf{L}\zeta = i\sqrt{\gamma_1 - \gamma_2}\zeta, \quad \mathbf{S}\zeta = \overline{\zeta}, \quad \mathbf{T}\zeta = -\zeta.$$

Then the reduced system satisfies Hypotheses 3.1 and 3.2, and, in addition, it is equivariant under the action of $\mathbf{T}$. Applying the result in Theorem 3.5, we conclude that the normal form of the reduced system is given by (3.7), in which the polynomials P and Q are odd and even, respectively, in A, due to the symmetry $\mathbf{T}$ (see also Remark 1.9).

The computation of the principal coefficients in the normal form is again easier than in the general case. We find the truncated system

$$\frac{dA}{dt} = B, \quad \frac{dB}{dt} = \mu A + dA^3, \quad \frac{dC}{dt} = i\sqrt{\gamma_1 - \gamma_2}C, \tag{3.17}$$

in which

$$d = -d_0 \int_{\mathbb{R}} g_2^4(x)dx.$$

We can now proceed as in the general case and conclude that this system possesses bounded solutions in the cases $\mu < 0, d_0 > 0$, $\mu < 0, d_0 < 0$, and $\mu > 0, d_0 > 0$, and more precisely:

(i) periodic orbits, of the first and second kind, and quasiperiodic orbits, in the three cases, which correspond to solutions of (3.15) that are periodic and quasiperiodic, respectively, in y and localized in x;
(ii) a pair of heteroclinic orbits to the periodic orbits of the first kind, which are not too small, in the case $\mu < 0$, $d_0 < 0$, which correspond to solutions of (3.15) that have an asymptotically constant profile with relative small oscillations at infinity in y and are localized in x;
(iii) a pair of homoclinic orbits to the periodic orbits of the first kind, which are not too small, in the case $\mu > 0$, $d_0 > 0$, which correspond to solutions of (3.15) which have a localized profile with relative small oscillations at infinity in y and are localized in x.

We refer to [40] for further details, and extensions to the case of complex-valued solutions, nearly one-dimensional potentials, and systems of NLS-type equations.

4.3.2 Reversible $0^{2-}(i\omega)$ Bifurcation (Elements)

We consider in this section the $0^{2-}(i\omega)$ bifurcation, when 0 is an algebraically double and geometrically simple eigenvalue of $\mathbf{L}$, with associated eigenvector ξ_0 satisfying

$$\mathbf{S}\xi_0 = -\xi_0,$$

and $\pm i\omega$ are simple eigenvalues. The complete study of the dynamics of this bifurcation is an open problem. We shall only briefly outline here the normal form and some elementary facts about its dynamics. This case is roughly treated in [56], and with an additional reversibility symmetry in [124].

$0^{2-}(i\omega)$ *Normal Form*

First, by arguing as in the case of the $0^{2+}(i\omega)$ bifurcation, we can prove here that *there is a basis* $\{\xi_0, \xi_1, \zeta, \overline{\zeta}\}$ *of* $\mathbb{R}^2 \times \widetilde{\mathbb{R}^2}$ *such that*

$$\mathbf{L}\xi_0 = 0, \quad \mathbf{L}\xi_1 = \xi_0, \quad \mathbf{L}\zeta = i\omega\zeta,$$
$$\mathbf{S}\xi_0 = -\xi_0, \quad \mathbf{S}\xi_1 = \xi_1, \quad \mathbf{S}\zeta = \overline{\zeta}.$$

Using this basis, in the same way as in the proof of Lemma 3.5 we find in this case that *for any* $p < k$ *there exists a map* $\Phi : \mathscr{V}_1 \times \mathscr{V}_2 \to \mathbb{R}^4$ *of class* C^k, *defined in a neighborhood* $\mathscr{V}_1 \times \mathscr{V}_2$ *of* 0 *in* $\mathbb{R}^2 \times \widetilde{\mathbb{R}^2} \times \mathbb{R}$, *with*

$$\Phi(0,0,0,0,0) = 0, \quad \partial_{(A,B,C,\overline{C})}\Phi(0,0,0,0,0) = 0,$$

and

$$\Phi(-A,B,\overline{C},C,\mu) = \mathbf{S}\Phi(A,B,C,\overline{C},\mu),$$

such that the change of variables

$$u = A\xi_0 + B\xi_1 + C\zeta + \overline{C\zeta} + \Phi(A,B,C,\overline{C},\mu), \tag{3.18}$$

transforms the differential equation into the normal form

$$\begin{aligned}
\frac{dA}{dt} &= B \\
\frac{dB}{dt} &= ABP(A^2,|C|^2,\mu) + AQ(A^2,|C|^2,\mu) + \rho_B(A,B,C,\overline{C},\mu) \\
\frac{dC}{dt} &= i\omega C + iCR_1(A^2,|C|^2,\mu) + ACR_2(A^2,|C|^2,\mu) + \rho_C(A,B,C,\overline{C},\mu).
\end{aligned} \tag{3.19}$$

Here P, Q, R_1, and R_2 are real-valued polynomials of degrees $p-2$, $p-1$, $p-1$, and $p-2$, respectively, in $(A,B,C,\overline{C})$, and the remainders ρ_B and ρ_C are functions of class $\mathscr{C}^k$ satisfying

$$\begin{aligned}
\rho_B(-A,B,\overline{C},C,\mu) &= -\rho_B(A,B,C,\overline{C},\mu), \\
\rho_C(-A,B,\overline{C},C,\mu) &= -\overline{\rho_C}(A,B,C,\overline{C},\mu),
\end{aligned}$$

and

$$|\rho_B(A,B,C,\overline{C},\mu)| + |\rho_C(A,B,C,\overline{C},\mu)| = o((|A| + B| + |C|)^p).$$

Solutions of the Normal Form System

The complete description of the dynamics of the full normal form in this case is still open. We only mention below some preliminary properties.

Consider the truncated system

$$\begin{aligned}
\frac{dA}{dt} &= B \\
\frac{dB}{dt} &= a\mu A + cAB + bA^3 + dA|C|^2 \\
\frac{dC}{dt} &= i\omega C + fAC,
\end{aligned} \tag{3.20}$$

obtained by keeping only the leading order terms in the normal form (3.19). We assume that the coefficients a, b, and f in this system do not vanish.

First notice that the plane $C = 0$ is invariant, and that the dynamics in this plane are those of the 0^{2-} bifurcation discussed in Section 4.1.2. Next, for $C \neq 0$, we set

$$A^2 = v, \quad B = u(v), \quad C = \sqrt{w(v)}e^{i\theta},$$

which leads to the system

$$\begin{aligned} 2u\frac{du}{dv} &= a\mu + cu + bv + dw \\ u\frac{dw}{dv} &= fw \\ \frac{d\theta}{dt} &= \omega. \end{aligned}$$

Since $f \neq 0$, we may set

$$s = \ln w,$$

which leads to the system

$$\begin{aligned} \frac{du}{ds} &= \frac{1}{2f}(a\mu + de^s + cu + bv) \\ \frac{dv}{ds} &= \frac{u}{f}. \end{aligned}$$

This system is linear, and assuming that

$$2f^2 - fc - b \neq 0, \tag{3.21}$$

we obtain the general solution

$$\begin{aligned} u(s) &= f(d'e^s + \alpha\lambda^+ e^{\lambda^+ s} + \beta\lambda^- e^{\lambda^- s}) \\ v(s) &= -\frac{a\mu}{b} + d'e^s + \alpha e^{\lambda^+ s} + \beta e^{\lambda^- s} \\ w(s) &= e^s, \end{aligned} \tag{3.22}$$

in which α and β are arbitrary constants. Substituting s from the last equality into the first two equalities give

$$\begin{aligned} u &= f(d'w + \alpha\lambda^+ w^{\lambda^+} + \beta\lambda^- w^{\lambda^-}) \\ v &= -\frac{a\mu}{b} + d'w + \alpha w^{\lambda^+} + \beta w^{\lambda^-}, \end{aligned}$$

where

$$d' = d(2f^2 - fc - b)^{-1},$$

and $\lambda^{\pm}$ are the solutions of

$$2f^2\lambda^2 - fc\lambda - b = 0.$$

We only consider values of α, β such that u and v are real, $v(s) \geq 0$, and the parametric curve $(u(s), v(s), w(s))$ is bounded (recall that $v = A^2$, $u = A'$, $w = |C|^2$). Notice that the values $u = 0$, $v = -a\mu/b$, $w = 0$ correspond to a pair of equilibria $(A^{\pm}, 0)$ when $ab\mu < 0$, where $A^{\pm}$ are the two equilibria given Section 4.1.2.

In all cases, there is a set of (α, β) such that there exist s_0 and s_1 such that

$$v(s_0) = v(s_1) = 0, \quad v(s) > 0 \text{ for all } s \in (s_0, s_1).$$

It then follows that there exists $s_2 \in (s_0, s_1)$ such that $u(s_2) = 0$, and a straightforward symmetry argument allows us to show the existence of a closed orbit in the $(A, B, |C|)$-space. This gives a family of invariant 2-tori for the truncated system (3.20). For the full system (3.19), one would expect that only KAM [1] tori persist.

For $ab\mu < 0$, when we have the two equilibria $(A^{\pm}, 0)$, we can further distinguish the following cases, just as in Section 4.1.2:

(i) $a\mu < 0, b > 0$. Then $\lambda^- < 0$, $\lambda^+ > 0$, and $\lambda^+ \neq 1$. Take $\beta = 0$ in (3.22). Then one finds a two-parameter family of heteroclinic orbits connecting the equilibria $(A^+, 0)$ and $(A^-, 0)$, parameterized by α and the phase θ. These orbits intersect the symmetry plane $\{A = 0, \operatorname{Im} C = 0\}$ transversally, so that one expects the persistence of a one-parameter family of pairs of reversible heteroclinic orbits for the full system (3.19).

(ii) $a\mu > 0$, $b < 0$, $c^2 + 8b > 0$. Then λ^- and λ^+ are both real with the same sign, and $\lambda^{\pm} \neq 1$. For $fc < 0$ the stable and unstable manifolds of $(A^{\pm}, 0)$ are two-dimensional. Take $\alpha = \beta = 0$ in (3.22). Then one finds a one-parameter family of heteroclinic orbits, parameterized by the phase θ. Again, these orbits intersect transversally the symmetry plane $\{A = 0, \operatorname{Im} C = 0\}$, so that one expects the persistence of a pair of reversible heteroclinic orbits for the system (3.19). For $fc > 0$, the equilibria $(A^{\pm}, 0)$ are nodes, one stable, and the other one unstable. The orbits connecting $(A^+, 0)$ and $(A^-, 0)$ are expected to persist for system (3.19) by the same argument as before. In addition, there are now orbits connecting $(A^+, 0)$, or $(A^-, 0)$, to the origin. These are also expected to persist since the one-dimensional stable (resp., unstable) manifold of the origin, which stays close to that of the truncated system, ends necessarily at $(A^-, 0)$ (resp., $(A^+, 0)$).

(iii) $a\mu > 0$, $b < 0$, $c^2 + 8b < 0$. Then λ^- and λ^+ are complex conjugate. The discussion is the same as above for $fc < 0$, whereas for $fc > 0$, the nodes $(A^{\pm}, 0)$ are replaced by foci. The main difference concerns now the homoclinic orbit to the origin (see Figure 1.5(iv)), which is not expected to persist for the full system (3.19), due to the fact that the plane $C = 0$ is no longer invariant, and to the fact that the stable and unstable manifolds of the origin are only one-dimensional. One would expect here again to have homoclinic orbits to periodic orbits, (up to an exponentially small size, if the vector field is analytic), just as in the case $0^{2+}(i\omega)$.

---

[1] The Kolmogorov–Arnold–Moser (KAM) theorem allows one to prove the existence of invariant tori, in particular in Hamiltonian systems, for which a suitable formulation is performed. It is particularly used in celestial mechanics, where small divisor problems occur. It is related to the "strong implicit function theorem", which is out of the scope of this book.

4.3.3 Reversible $(i\omega)^2$ Bifurcation (1-1 resonance)

We consider in this section the $(i\omega)^2$ bifurcation. This bifurcation is also referred to as "1:1 resonance," or "reversible-Hopf bifurcation," or "Hamiltonian–Hopf bifurcation" in the case of a Hamiltonian system. We refer for instance to [68], and to [28] for the Hamiltonian case, for detailed proofs.

Hypothesis 3.14 *Assume that $\sigma(\mathbf{L}) = \{\pm i\omega\}$, with $\pm i\omega$ algebraically double and geometrically simple eigenvalues.*

Normal Form

As in the previous cases, we start by constructing a suitable basis of $\mathbb{R}^4$.

Lemma 3.15 (**$(i\omega)^2$ basis of $\mathbb{R}^4$**) *Assume that Hypotheses 3.1 and 3.2 hold. Then there exists a basis $\{\mathrm{Re}\,\zeta_0, \mathrm{Im}\,\zeta_0, \mathrm{Re}\,\zeta_1, \mathrm{Im}\,\zeta_1\}$ of $\mathbb{R}^4$, with $\zeta_0, \zeta_1 \in \mathbb{C}^4$ generalized eigenvectors of $\mathbf{L}$, such that*

$$(\mathbf{L} - i\omega)\zeta_0 = 0, \quad (\mathbf{L} - i\omega)\zeta_1 = \zeta_0, \quad (\mathbf{L} + i\omega)\overline{\zeta_0} = 0, \quad (\mathbf{L} + i\omega)\overline{\zeta_1} = \overline{\zeta_0},$$
$$\mathbf{S}\zeta_0 = \overline{\zeta_0}, \quad \mathbf{S}\zeta_1 = -\overline{\zeta_1}.$$

Proof Consider an eigenvector ζ_0' associated to the eigenvalue $i\omega$. Since $\mathbf{L}$ anticommutes with $\mathbf{S}$ we have

$$(\mathbf{L} + i\omega)\mathbf{S}\zeta_0' = 0,$$

so there exists $k \in \mathbb{C}$ such that

$$\mathbf{S}\zeta_0' = k\overline{\zeta_0'}.$$

Furthermore, since $\mathbf{S}^2 = \mathbb{I}$, we have $|k| = 1$. Set $k = e^{i\alpha}$ and

$$\zeta_0 = e^{-i\alpha/2}\zeta_0'.$$

Then

$$(\mathbf{L} - i\omega)\zeta_0 = 0, \quad \mathbf{S}\zeta_0 = \overline{\zeta_0}.$$

Next, take a generalized eigenvector ζ_1' such that

$$(\mathbf{L} - i\omega)\zeta_1' = \zeta_0.$$

Then

$$(\mathbf{L} + i\omega)\mathbf{S}\zeta_1' = -\mathbf{S}\zeta_0 = -\overline{\zeta_0},$$

and there exists $\beta_0 \in \mathbb{C}$ such that

$$\mathbf{S}\zeta_1' = -\overline{\zeta_1'} + \beta\overline{\zeta_0}.$$

Applying $\mathbf{S}$ to this equality, taking into account that $\mathbf{S}^2 = \mathbb{I}$, we obtain

$$\zeta_1' = -\mathbf{S}\overline{\zeta_1'} + \beta\zeta_0 = \zeta_1' + (\beta - \overline{\beta})\zeta_0.$$

Consequently, $\beta \in \mathbb{R}$, and we choose

$$\zeta_1 = \zeta_1' - \frac{\beta}{2}\zeta_0.$$

Then

$$(\mathbf{L} - i\omega)\zeta_1 = \zeta_0$$

and

$$\mathbf{S}\zeta_1 = \mathbf{S}\zeta_1' - \frac{\beta}{2}\overline{\zeta_0} = -\overline{\zeta_1'} + \frac{\beta}{2}\overline{\zeta_0} = -\overline{\zeta_1},$$

which proves the lemma. □

Notation 3.16 *In the basis above, we represent a vector in $u \in \mathbb{R}^4$ by $(A,B,\overline{A},\overline{B})$,*

$$u = A\zeta_0 + B\zeta_1 + \overline{A\zeta_0} + \overline{B\zeta_1},$$

with $A,B \in \mathbb{C}$. We identify $\mathbb{R}^4$ with the space $\widetilde{\mathbb{R}^4}$ in which

$$\widetilde{\mathbb{R}^4} = \{(A,B,\overline{A},\overline{B})\ ;\ A,B \in \mathbb{C}\}.$$

Lemma 3.17 ($(i\omega)^2$ normal form) *Assume that Hypotheses 3.1 and 3.14 hold, and consider the basis $\{\zeta_0, \zeta_1, \overline{\zeta_0}, \overline{\zeta_1}\}$ of $\widetilde{\mathbb{R}^4}$ in Lemma 3.15. Then for any positive integer p, $2 \leq p \leq k$, there exist neighborhoods $\mathscr{V}_1$ and $\mathscr{V}_2$ of 0 in $\widetilde{\mathbb{R}^4}$ and $\mathbb{R}$, respectively, and for any $\mu \in \mathscr{V}_2$ there is a polynomial $\boldsymbol{\Phi}(\cdot,\mu) : \widetilde{\mathbb{R}^4} \to \widetilde{\mathbb{R}^4}$ of degree p with the following properties:*

(i) The coefficients of the monomials of degree q in $\boldsymbol{\Phi}(\cdot,\mu)$ are functions of μ of class $\mathscr{C}^{k-q}$,

$$\boldsymbol{\Phi}(0,0,0,0,0) = 0, \quad \partial_{(A,B,\overline{A},\overline{B})}\boldsymbol{\Phi}(0,0,0,0,0) = 0, \tag{3.23}$$

and

$$\boldsymbol{\Phi}(\overline{A}, -\overline{B}, A, -B, \mu) = \mathbf{S}\boldsymbol{\Phi}(A,B,\overline{A},\overline{B},\mu).$$

(ii) For $(A,B,\overline{A},\overline{B}) \in \mathscr{V}_1$, the change of variables

$$u = A\zeta_0 + B\zeta_1 + \overline{A\zeta_0} + \overline{B\zeta_1} + \boldsymbol{\Phi}(A,B,\overline{A},\overline{B},\mu), \tag{3.24}$$

transforms the equation (3.1) into the normal form

$$\frac{dA}{dt} = i\omega A + B + iAP\Big(|A|^2, \frac{i}{2}(A\overline{B} - \overline{A}B), \mu\Big) + \rho_A(A, B, \overline{A}, \overline{B}, \mu)$$
$$\frac{dB}{dt} = i\omega B + iBP\Big(|A|^2, \frac{i}{2}(A\overline{B} - \overline{A}B), \mu\Big) + AQ\Big(|A|^2, \frac{i}{2}(A\overline{B} - \overline{A}B), \mu\Big)$$
$$+\rho_B(A, B, \overline{A}, \overline{B}, \mu), \tag{3.25}$$

where P and Q are real-valued polynomials of degree $p-1$ in $(A, B, \overline{A}, \overline{B})$. The remainders ρ_A and ρ_B are of class $\mathscr{C}^k$, and satisfy

$$\rho_A(\overline{A}, -\overline{B}, A, -B, \mu) = -\overline{\rho_A}(A, B, \overline{A}, \overline{B}, \mu),$$
$$\rho_B(\overline{A}, -\overline{B}, A, -B, \mu) = \overline{\rho_B}(A, B, \overline{A}, \overline{B}, \mu),$$

with the estimate

$$|\rho_A(A, B, \overline{A}, \overline{B}, \mu)| + |\rho_B(A, B, \overline{A}, \overline{B}, \mu)| = o((|A| + |B|)^p).$$

Proof According to Theorem 2.2 and Lemma 1.17 of Chapter 3, there is a polynomial $\Phi(\cdot, \mu)$ of degree p with coefficients of the monomials of degree q of class $\mathscr{C}^{k-q}$ in μ, which satisfies (3.23), and such that the change of variables (3.24) transforms (3.1) into

$$\frac{dA}{dt} = i\omega A + B + iAP\Big(|A|^2, \frac{i}{2}(A\overline{B} - \overline{A}B), \mu\Big) + \rho_A(A, B, \overline{A}, \overline{B}, \mu)$$
$$\frac{dB}{dt} = i\omega B + iBP\Big(|A|^2, \frac{i}{2}(A\overline{B} - \overline{A}B), \mu\Big) + AQ\Big(|A|^2, \frac{i}{2}(A\overline{B} - \overline{A}B), \mu\Big)$$
$$+\rho_B(A, B, \overline{A}, \overline{B}, \mu),$$

where P and Q are polynomials of degree $p-1$ and the remainders satisfy the estimates in the part (ii) of the lemma.

Now, we use the reversibility symmetry. According to Theorem 3.4 we have that

$$\Phi(\overline{A}, -\overline{B}, A, -B, \mu) = \mathbf{S}\Phi(A, B, \overline{A}, \overline{B}, \mu),$$

and for the first component in the normal form we find

$$i\overline{A}P\Big(|A|^2, \frac{i}{2}(A\overline{B} - \overline{A}B), \mu\Big) + \rho_A(\overline{A}, -\overline{B}, A, -B, \mu)$$
$$= \overline{iAP}\Big(|A|^2, \frac{i}{2}(A\overline{B} - \overline{A}B), \mu\Big) - \overline{\rho_A}(A, B, \overline{A}, \overline{B}, \mu).$$

This implies that P is real-valued and that the symmetry property for ρ_A in the part (ii) of the lemma holds. Next, for the second component we find

$$\overline{A}Q\Big(|A|^2, \frac{i}{2}(A\overline{B} - \overline{A}B), \mu\Big) + \rho_B(\overline{A}, -\overline{B}, A, -B, \mu)$$
$$= \overline{AQ}\Big(|A|^2, \frac{i}{2}(A\overline{B} - \overline{A}B), \mu\Big) + \overline{\rho_B}(A, B, \overline{A}, \overline{B}, \mu),$$

which implies that Q is real-valued, and the symmetry property for ρ_B. □

Solutions of the Normal Form System

We analyze now the dynamics of the normal form in the generic situation in which the following assumption on the expansion of the vector field holds.

Hypothesis 3.18 *Assume that the expansions of P and Q in (3.25),*

$$P\Big(|A|^2,\frac{i}{2}(A\overline{B}-\overline{A}B),\mu\Big) = \alpha\mu+\beta|A|^2+\frac{i\gamma}{2}(A\overline{B}-\overline{A}B)+O((|\mu|+(|A|+|B|)^2)^2),$$
$$Q\Big(|A|^2,\frac{i}{2}(A\overline{B}-\overline{A}B),\mu\Big) = a\mu+b|A|^2+\frac{ic}{2}(A\overline{B}-\overline{A}B)+O((|\mu|+(|A|+|B|)^2)^2).$$

are such that a and b do not vanish.

A first observation is that the system (3.25) always has an equilibrium at the origin.

Truncated System

We start with the truncated system

$$\begin{aligned}\frac{dA}{dt} &= i\omega A+B+iAP\Big(|A|^2,\frac{i}{2}(A\overline{B}-\overline{A}B),\mu\Big)\\ \frac{dB}{dt} &= i\omega B+iBP\Big(|A|^2,\frac{i}{2}(A\overline{B}-\overline{A}B),\mu\Big)+AQ\Big(|A|^2,\frac{i}{2}(A\overline{B}-\overline{A}B),\mu\Big).\end{aligned} \tag{3.26}$$

This system is integrable with first integrals

$$K=\frac{i}{2}(A\overline{B}-\overline{A}B),\quad H=|B|^2-\int_0^{|A|^2}Q(s,K,\mu)\,ds.$$

Also notice that in addition to the reversibility symmetry $\mathbf{S}$ acting through $\mathbf{S}(A,B)=(\overline{A},-\overline{B})$, this truncated system is equivariant under the $SO(2)$ group action,

$$\mathbf{R}_\phi(A,B)=(Ae^{i\phi},Be^{i\phi}),\quad \phi\in\mathbb{R}/2\pi\mathbb{Z}. \tag{3.27}$$

In addition, we have that

$$\mathbf{R}_\phi\mathbf{S}=\mathbf{S}\mathbf{R}_{-\phi},$$

so that the truncated normal actually possesses an $O(2)$ symmetry.

For the analysis of this system it is convenient to work in polar coordinates

$$A=r_0e^{i(\omega t+\theta_0)},\quad B=r_1e^{i(\omega t+\theta_1)},$$

in which the two first integrals are given by

$$K = r_0 r_1 \sin(\theta_1 - \theta_0), \quad H = r_1^2 - \int_0^{r_0^2} Q(s, K, \mu) ds, \tag{3.28}$$

and system (3.9) becomes

$$\begin{aligned}
\frac{dr_0}{dt} &= r_1 \cos(\theta_1 - \theta_0) \\
\frac{dr_1}{dt} &= r_0 \cos(\theta_1 - \theta_0) Q(r_0^2, K, \mu) \\
\frac{d\theta_0}{dt} &= \frac{r_1}{r_0} \sin(\theta_1 - \theta_0) + P(r_0^2, K, \mu) \\
\frac{d(\theta_1 - \theta_0)}{dt} &= -\frac{\sin(\theta_1 - \theta_0)}{r_0 r_1} \left(r_1^2 + r_0^2 Q(r_0^2, K, \mu) \right).
\end{aligned} \tag{3.29}$$

Remark 3.19 *(i) The projection of the phase portrait in the* (r_0, r_1)*-plane is similar to that in Figure 1.3, when allowing for negative values for* r_0 *and* r_1 *(which requires a careful redefinition of the phases* θ_0 *and* θ_1*). However, because of the dependence of* θ_0 *and* θ_1 *on* t*, only parts of these curves correspond to the present case. For the analysis of this situation we shall use different variables below.*

(ii) In the case when the system is Hamiltonian, the normal form has an additional relationship between polynomials P *and* Q *(e.g., see [67]). We point out that in [28] suitable symplectic polar coordinates (*r_0 *being one of them) lead to different, less singular, phase portraits.*

Next, we set

$$u_0 = r_0^2, \quad u_1 = r_1^2,$$

for which we find

$$\frac{du_1}{dt} = Q(u_0, K, \mu) \frac{du_0}{dt},$$

and

$$\left(\frac{du_0}{dt} \right)^2 = 4(u_0 u_1 - K^2), \quad u_1 = G(u_0, K, \mu) + H,$$

with

$$G(u_0, K, \mu) = \int_0^{u_0} Q(s, K, \mu) ds.$$

Equivalently, we have

$$\left(\frac{du_0}{dt} \right)^2 = 4 f_{H,K}(u_0, \mu), \quad f_{H,K}(u_0, \mu) = u_0 \left(G(u_0, K, \mu) + H \right) - K^2. \tag{3.30}$$

Furthermore,

$$\frac{d(\theta_1 - \theta_0)}{dt} = -\frac{K}{u_0 u_1} \frac{\partial}{\partial u_0} \left(f_{H,K}(u_0, \mu) \right) = -K \frac{\partial}{\partial u_0} \left(\ln(u_0 u_1) \right),$$

which gives

$$\frac{d(\theta_1-\theta_0)}{du_0} = -\text{sign}\left(\frac{du_0}{dt}\right)\frac{\partial}{\partial u_0}\left(\tan^{-1}\left(\frac{1}{K}f_{H,K}^{1/2}(u_0,\mu)\right)\right).$$

Integration with respect to u_0 gives

$$\theta_1-\theta_0 = -\text{sign}\left(\frac{du_0}{dt}\right)\tan^{-1}\left(\frac{1}{K}f_{H,K}^{1/2}(u_0,\mu)\right)+\theta_*,$$

with an arbitrary integration constant θ_*. Finally, θ_0 can be expressed in terms of u_0 by integrating the third equation in the system (3.29). Summarizing, we can solve (3.29) by first determining u_0 from the equality (3.30), and then obtaining successively u_1, $\theta_1-\theta_0$, and θ_0.

Restricting the system to $(u_0,u_1,\theta_1-\theta_0)$ equilibria are given by

$$f_{H,K}(u_0,\mu)=0, \quad \frac{\partial}{\partial u_0}f_{H,K}(u_0,\mu)=0. \tag{3.31}$$

Notice that this corresponds to either

$$\theta_1-\theta_0=\pm\frac{\pi}{2}, \quad r_1^2+r_0^2Q(r_0^2,K,\mu)=0, \quad r_0r_1\neq 0,$$

or to

$$r_1=0, \quad Q(r_0^2,K,\mu)=0.$$

Equations (3.31) give a curve in the (H,K)-plane. At leading orders we find

$$f_{H,K}(u_0,\mu)=\frac{b}{2}u_0^3+\widetilde{\mu}u_0^2+Hu_0-K^2, \quad \widetilde{\mu}=a\mu+cK,$$

which gives the parametric equations,

$$H=-\frac{3b}{2}s^2-2\widetilde{\mu}s, \quad K^2=-s^2(\widetilde{\mu}+bs).$$

We plot this curve in Figure 3.2, together with the shape of $f_{H,K}(\cdot,\mu)$ for different values of H and K, in cases (i) $\widetilde{\mu}<0$, $b<0$, (ii) $\widetilde{\mu}<0$, $b>0$, and (iii) $\widetilde{\mu}>0$, $b<0$. In the case $\widetilde{\mu}>0$, $b>0$, it is not difficult to check that the system has no bounded solutions, except for the equilibrium at the origin. In the cases (ii) and (iii), we have the particular points $(0,K_E)$ and $(H_c,0)$ given at leading orders by

$$(0,K_E)=\left(0,\sqrt{-\frac{4a^3\widetilde{\mu}^3}{27b^2}}\right), \quad (H_c,0)=\left(\frac{a^2\widetilde{\mu}^2}{2b},0\right).$$

The projections in the (u_0,u_0')-plane of the bounded orbits obtained by varying H for a fixed K in the three cases are shown in Figure 3.3. In order to obtain these curves we only need to consider the parts of the graphs of $f_{H,K}(\cdot,\mu)$ where $u_0\geq 0$ and $f_{H,K}(u_0,\mu)\geq 0$. Notice that for a fixed $K\neq 0$, the curves do not intersect each other as H varies, whereas for $H=0$ all these curves intersect at one point, which is

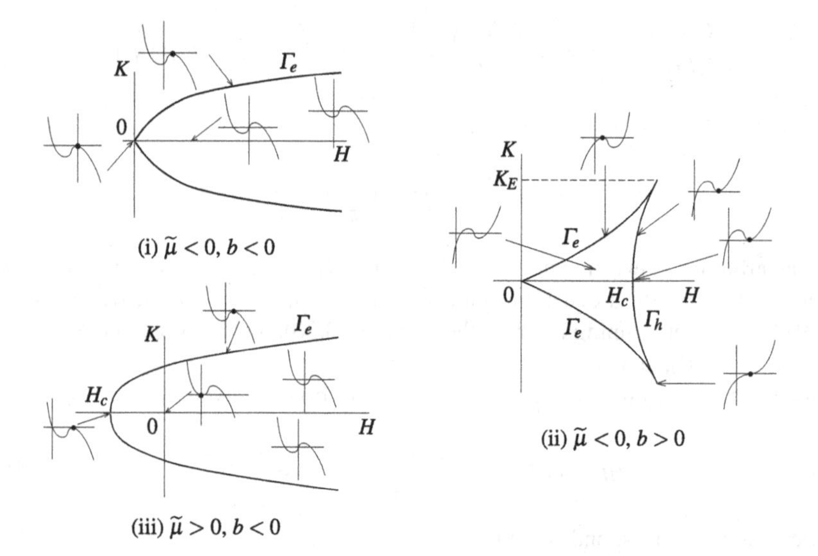

Fig. 3.2 Graphs of $f_{H,K}(\cdot,\mu)$ depending on (H,K) in the cases (i) $\widetilde{\mu} < 0, b < 0$, (ii) $\widetilde{\mu} < 0, b > 0$, (iii) $\widetilde{\mu} > 0, b < 0$, when the system has nonequilibrium bounded orbits. In case (ii), bounded orbits exist for values of (H,K) in the closed bounded set surrounded by the curves Γ_e and Γ_h, whereas in cases (i) and (iii) bounded orbits exist for values of (H,K) in the closed set bounded by the curves Γ_e and containing the positive H-axis.

the origin. In particular, this leads to the unusual behavior of the orbits close to the origin in the case $H = 0$ (see Figure 3.3).

(i) For $\widetilde{\mu} < 0$, $b < 0$, in the (u_0, u_0')-plane we have for any $K \neq 0$ one nontrivial equilibrium, which is surrounded by a one-parameter family of periodic orbits, and for $K = 0$ a one-parameter family of periodic orbits, which shrink to the origin as $H \to 0$. The equilibrium at the origin corresponds to the equilibrium at the origin for the truncated normal form (3.26), whereas each nontrivial equilibrium in the (u_0, u_0')-plane corresponds to an equilibrium of the $(r_0, r_1, \theta_1 - \theta_0)$-system, and to a periodic orbit for the truncated normal form. The periodic orbits in the (u_0, u_0')-plane correspond to periodic orbits of the $(r_0, r_1, \theta_1 - \theta_0)$-system, and to invariant tori with quasiperiodic or periodic solutions for the truncated normal form (3.26).

(ii) For $\widetilde{\mu} < 0$, $b > 0$, in addition to equilibria and periodic orbits which are similar to those found in case (i), in the (u_0, u_0')-plane we now find a homoclinic orbit for each K with $|K| < K_E$. For the truncated normal form (3.26), each of these homoclinic orbits corresponds to a circle of homoclinics to periodic orbits, which, it turns out, have a phase shift between the limits $t \to \pm\infty$, just as in the $0^{2+}(i\omega)$ bifurcation. For the system truncated at cubic order, we can compute explicitly the circle of homoclinic orbits found for $K = 0$,

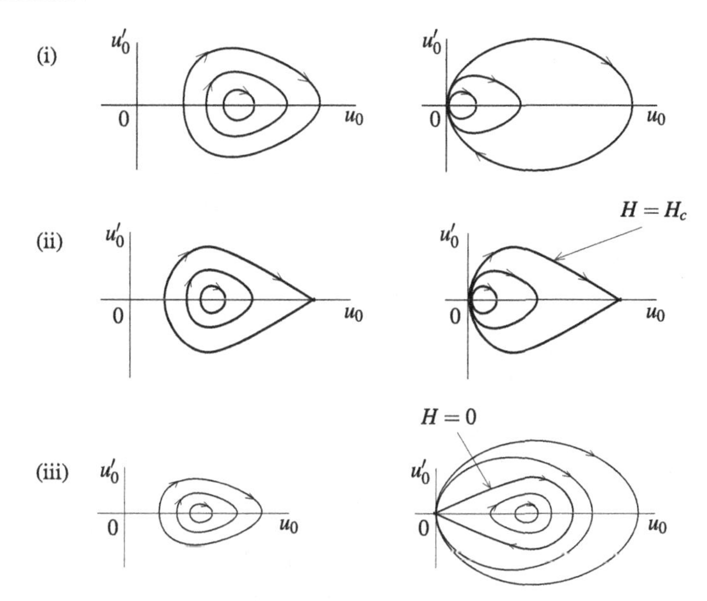

Fig. 3.3 Dynamics of the $(i\omega)^2$ bifurcation. Plot of the projections in the (u_0, u_0')-plane of the bounded orbits of the truncated system (3.26) obtained by varying H for a fixed $K \neq 0$ (left) and $K = 0$ (right), in the cases (i) $\widetilde{\mu} < 0$, $b < 0$, (ii) $\widetilde{\mu} < 0$, $b > 0$, (iii) $\widetilde{\mu} > 0$, $b < 0$, when the system has nonequilibrium bounded orbits.

$$r_0 = \sqrt{\frac{-a\mu}{b}} \tanh\left(\sqrt{\frac{-a\mu}{2}}|t|\right), \quad r_1 = |r_0'|,$$

$$\theta_1 - \theta_0 \in \{0, \pi\}, \quad \theta_0 = \left(\alpha - \frac{a\beta}{b}\right)\mu t - \frac{\beta\sqrt{-2a\mu}}{b} \tanh\left(\sqrt{\frac{-a\mu}{2}}t\right) + \theta_*.$$

In this expression, we have explicitly the phase shift $2\beta\sqrt{-2a\mu}/b$ between $t = +\infty$ and $t = -\infty$ on the asymptotic periodic orbit which is given by

$$r_0 = \sqrt{\frac{-a\mu}{b}}, \quad r_1 = 0, \quad \theta_0 = \left(\alpha\mu - \frac{a\beta\mu}{b}\right)t.$$

(iii) For $\widetilde{\mu} > 0$, $b < 0$, in the (u_0, u_0')-plane, we have again equilibria and periodic orbits similar to those found in case (i). In addition, there is in this case a homoclinic orbit to the origin, found for $H = K = 0$. For the truncated normal form (3.26), this homoclinic orbit corresponds to a circle of homoclinics to the origin. For the system truncated at cubic order, these homoclinic orbits are given by

$$r_0 = \sqrt{\frac{-2a\mu}{b}}\,(\cosh(\sqrt{a\mu}t)))^{-1}, \quad r_1 = |r_0'|,$$

$$\theta_1 - \theta_0 \in \{0, \pi\}, \quad \theta_0 = \alpha\mu t - \frac{2\beta\sqrt{a\mu}}{b}\tanh(\sqrt{a\mu}t) + \theta_*.$$

Finally, notice that the small bounded solutions are such that

$$K = O(|\widetilde{\mu}|^{3/2}), \quad H = O(|\widetilde{\mu}|^2),$$

hence $\widetilde{\mu} = O(\mu)$. This further implies that the leading order terms describe correctly the shape of $f_{H,K}(\cdot,\mu)$ in the neighborhood of 0, so that all these bounded solutions can be found for the truncated system (3.26).

Remark 3.20 *In literature, the case $b > 0$ is also called the "defocusing" case when related to the nonlinear Schrödinger equation, or "supercritical" case when it arises in viscous hydrodynamical applications. The solutions corresponding to the homoclinic orbits found in this case are sometimes called "dark solitary waves," because the amplitude of the corresponding localized solution is smaller in the central region than in the limits as $t \to \pm\infty$. The case $b < 0$ is also referred to as the "focusing" or "subcritical" case, for the same reasons as above, while the solutions corresponding to the homoclinics to 0 in this case are called "bright" solitary waves.*

Full System

The origin is an equilibrium of the normal form (3.25), and using the implicit function theorem it is easy to check that it corresponds to a symmetric equilibrium in the four-dimensional system (3.1).

The persistence of the periodic orbits corresponding to the nontrivial equilibria found in the (u_0, u_0')-plane in the three cases (i), (ii), and (iii) can be proved with the help of the implicit function theorem. The persistence of invariant tori corresponding to periodic orbits in the (u_0, u_0')-plane in the three cases (i), (ii), and (iii) is a question of persistence of quasiperiodic solutions and leads to a small divisor problem, just as in the case of the $0^2(i\omega)$ bifurcation. However, it can treated in a subset of the two-dimensional parameter space defined by the two first integrals. It is roughly proved in [64], in the context of a hydrodynamical application, that for a fixed value of μ quasiperiodic solutions of (3.25) typically exist for (H, K) lying in a region that is locally the product of a line with a Cantor set.

Next, the question of persistence of the orbits homoclinic to periodic solutions in the case (ii) turns out to be simpler than in the case of the $0^{2+}(i\omega)$ bifurcation, since here the family of asymptotic periodic orbits does not shrink to a point. Indeed, for the reduced system (3.26), the two-dimensional unstable manifold of any asymptotic periodic orbit, which is rotationally symmetric, intersects transversally the two-dimensional plane that is invariant under the reversibility symmetry in two points. Then a perturbation argument, controlling the size of the perturbation, which is linked to the size of the periodic orbit, allows us to show that these two intersection points persist for system (3.25), and a symmetry argument allows to conclude the existence of a pair of reversible homoclinic orbits to every periodic orbit (e.g., see [68]).

Finally, the persistence of the orbits homoclinic to 0 in case (iii) can be solved in the same way, since the intersection of the rotationally invariant two-dimensional unstable manifold of the origin for the reduced system (3.26) intersects transversally in two distinct points the two-dimensional plane of symmetry [68]. However, we point out that while the reduced system (3.26) possesses a circle of homoclinics, we just indicated above the persistence of two of them, those which are reversible. The persistence of nonsymmetric homoclinics is still an open problem, related to asymptotics beyond all orders.

Summarizing, we have the following result (see [68] and the references therein for further details).

Theorem 3.21 (**$(i\omega)^2$ bifurcation**) *Assume that Hypotheses 3.1, 3.14, and 3.18 hold. Then for differential equation (3.1) a reversible $(i\omega)^2$ bifurcation occurs at $\mu = 0$. More precisely, the following properties hold in a neighborhood of 0 in $\mathbb{R}^4$ for sufficiently small μ:*

(i) For $b < 0$, or $b > 0$ and $a\mu < 0$, there is one symmetric equilibrium, a one-parameter family of periodic orbits, and a two-parameter family of quasiperiodic orbits located on KAM tori.
(ii) For $a\mu < 0$ and $b > 0$, there is a one-parameter family of pairs of reversible homoclinic orbits to periodic orbits.
(iii) For $a\mu > 0$ and $b < 0$, there is a pair of reversible homoclinic orbits to the symmetric equilibrium.
(iv) For $a\mu > 0$ and $b > 0$, there is one symmetric equilibrium and no other bounded solutions.

Remark 3.22 *We show in Appendix D.2 how to compute the principal coefficients in the normal form (3.25) when starting from an infinite-dimensional system of the form (3.12).*

Example: Steady Solutions of the Swift–Hohenberg Equation

Consider the Swift–Hohenberg equation, discussed in Section 2.4.3 of Chapter 2. We are now interested in bounded steady solutions of (4.13), which arise as bifurcations from the trivial solution $u = 0$, i.e., small bounded solutions of the fourth order differential equation

$$\left(1+\frac{d^2}{dx^2}\right)^2 u - \mu u + u^3 = 0, \quad x \in \mathbb{R}. \tag{3.32}$$

Formulation as a First Order System

We start by writing equation (3.32) as a first order system

$$\frac{dv}{dx} = \mathbf{L}v + \mathbf{R}(v,\mu) \tag{3.33}$$

of the form (3.1), but now with the spatial variable x being the time t. We set

$$v_1 = u, \quad v_2 = \frac{du}{dx}, \quad v_3 = \frac{d^2u}{dx^2}, \quad v_4 = \frac{d^3u}{dx^3},$$

and then (3.32) is of the form (3.33) with

$$v = \begin{pmatrix} v_1 \\ v_2 \\ v_3 \\ v_4 \end{pmatrix}, \quad \mathbf{L} = \begin{pmatrix} 0 & 1 & 0 & 0 \\ 0 & 0 & 1 & 0 \\ 0 & 0 & 0 & 1 \\ -1 & 0 & -2 & 0 \end{pmatrix},$$

and

$$\mathbf{R}(v,\mu) = \mu\mathbf{R}_{1,1}v + \mathbf{R}_{3,0}(v,v,v), \quad \mathbf{R}_{1,1}v = \begin{pmatrix} 0 \\ 0 \\ 0 \\ v_1 \end{pmatrix}, \quad \mathbf{R}_{3,0}(u,v,w) = \begin{pmatrix} 0 \\ 0 \\ 0 \\ -u_1v_1w_1 \end{pmatrix}.$$

The system (3.33) is reversible, with reversibility symmetry $\mathbf{S}$ defined by

$$\mathbf{S}(v_1,v_2,v_3,v_4) = (v_1,-v_2,v_3,-v_4),$$

and it is easy to check that the vector field in this system satisfies Hypothesis 3.1.

Bifurcations

In order to determine the type of bifurcations which may arise in this system we look at the eigenvalues λ of the 4×4-matrix $\mathbf{L}+\mu\mathbf{R}_{1,1}$. A direct calculation gives that λ satisfies

$$(\lambda^2+1)^2 - \mu = 0,$$

so that the four eigenvalues of this matrix are given by

$$\lambda = \pm i\left(1 \pm \frac{1}{2}\sqrt{\mu} - \frac{1}{8}\mu + O(|\mu|^{3/2})\right). \tag{3.34}$$

In particular, we have that $\pm i$ are double eigenvalues when $\mu = 0$. Furthermore, the vectors

$$\zeta_0 = (1,i,-1,-i), \quad \zeta_1 = (0,1,2i,-3)$$

satisfy

$$\mathbf{L}\zeta_0 = i\zeta_0, \quad \mathbf{L}\zeta_1 = i\zeta_1 + \zeta_0, \quad \mathbf{S}\zeta_0 = \overline{\zeta_0}, \quad \mathbf{S}\zeta_1 = -\overline{\zeta_1}.$$

This implies that $\mathbf{L}$ satisfies Hypothesis 3.14 and that the vectors $\zeta_0, \zeta_1 \in \mathbb{C}^4$ provide a basis of $\mathbb{R}^4$ as in Lemma 3.15.

Normal Form

We are here in the presence of a $(i\omega)^2$ bifurcation, with $\omega = 1$, and according to Lemma 3.17 the system has the normal form (3.25). The computation of the principal coefficients in the normal form is in this particular case simpler than in the general case discussed in Appendix D.2, since the vector field is cubic.

First observe that the eigenvalues of the linearization at the origin of the normal form (3.25) are given by

$$i\left(1 \pm \sqrt{-a\mu} + \alpha\mu + O(|\mu|^{3/2})\right),$$

where α and a are the coefficients in the expansions of P and Q in Hypothesis 3.18. These eigenvalues are the same as those of $\mathbf{L} + \mu\mathbf{R}_{1,1}$ in (3.34), which implies that

$$\alpha = -\frac{1}{8}, \quad a = -\frac{1}{4}.$$

Actually, this argument can be used in general, which is often quicker than using formulas (D.45) and (D.46) (see Exercise 3.5 in Chapter 2).

With the notations from Appendix D.2, we have to solve now the following system

$$\begin{aligned}
b\zeta_1 + i\beta\zeta_0 &= (\mathbf{L} - i)\Psi_{20100} + 3\mathbf{R}_{3,0}(\zeta_0, \zeta_0, \overline{\zeta_0}) \\
\frac{ic}{2}\zeta_1 - \frac{\gamma}{2}\zeta_0 + \Psi_{20100} &= (\mathbf{L} - i)\Psi_{20010} \\
\left(i\beta - \frac{ic}{2}\right)\zeta_1 + \frac{\gamma}{2}\zeta_0 + 2\Psi_{20100} &= (\mathbf{L} - i)\Psi_{11100} \\
\frac{\gamma}{2}\zeta_1 + \Psi_{11100} &= (\mathbf{L} - i)\Psi_{02100} \\
-\frac{\gamma}{2}\zeta_1 + 2\Psi_{20010} + \Psi_{11100} &= (\mathbf{L} - i)\Psi_{11010} \\
\Psi_{11010} + \Psi_{02100} &= (\mathbf{L} - i)\Psi_{02010}.
\end{aligned}$$

The solvability conditions for these equations determine the coefficients b, c, β, and γ.

Following the procedure for solving these equations given in Appendix D.2, and also in the previous examples, we compute the vector ζ_1^* in the kernel of $(\mathbf{L} - i)^*$,

$$\zeta_1^* = -\frac{1}{4}(-i, 1, -i, 1),$$

which satisfies

$$\langle \zeta_1, \zeta_1^* \rangle = 1, \quad \langle \zeta_0, \zeta_1^* \rangle = 0, \quad \langle \overline{\zeta_1}, \zeta_1^* \rangle = 0, \quad \langle \overline{\zeta_0}, \zeta_1^* \rangle = 0.$$

Notice that $\mathbf{S}^* = \mathbf{S}$ and $\mathbf{S}\zeta_1^* = -\overline{\zeta_1^*}$. Then we obtain successively,

$$b = \langle 3\mathbf{R}_{3,0}(\zeta_0, \zeta_0, \overline{\zeta_0}), \zeta_1^* \rangle = \frac{3}{4}, \quad \widetilde{\Psi}_{20100} = \left(0, 0, \frac{3}{4}, \frac{9i}{4}\right),$$

$$i\beta + \frac{ic}{2} = \langle -\widetilde{\Psi}_{20100}, \zeta_1^* \rangle = \frac{3i}{4}, \quad 3i\beta - \frac{ic}{2} = \langle -2\widetilde{\Psi}_{20100}, \zeta_1^* \rangle = \frac{3i}{2},$$

and

$$\widetilde{\Psi}_{11100} = 2\widetilde{\Psi}_{20010} = \left(0, 0, \frac{3i}{2}, -3\right),$$

so that

$$\beta = \frac{9}{16}, \quad c = \frac{3}{8},$$

and

$$3\gamma = \langle 2\widetilde{\Psi}_{20010} - \widetilde{\Psi}_{11100}, \zeta_1^* \rangle = 0.$$

Consequently, the normal form truncated at cubic order reads

$$\begin{aligned}
\frac{dA}{dx} &= iA + B + iA\left(-\frac{1}{8}\mu + \frac{9}{16}|A|^2\right) \\
\frac{dB}{dx} &= iB + iB\left(-\frac{1}{8}\mu + \frac{9}{16}|A|^2\right) + A\left(-\frac{1}{4}\mu + \frac{3}{4}|A|^2 + \frac{3i}{16}(A\overline{B} - \overline{A}B)\right).
\end{aligned}$$

Solutions of the Normal Form

The symmetric equilibrium in Theorem 3.21 is here the origin. Since $a < 0$ and $b > 0$, from Theorem 3.21 we can conclude that there are no nontrivial small, bounded solutions when $\mu < 0$, and that for $\mu > 0$ we have here case (ii) in Figures 3.2 and 3.3. Notice that the steady 2π-periodic solutions found for $\mu > 0$ in the steady bifurcation discussed in Section 2.4.3 are contained in the set of steady periodic solutions found here.

4.3.4 Reversible $(i\omega_1)(i\omega_2)$ Bifurcation (Elements)

In this section, we consider the $(i\omega_1)(i\omega_2)$ bifurcation, when $\mathbf{L}$ has four simple purely imaginary eigenvalues $\pm i\omega_1$, $\pm i\omega_2$, such that $\omega_1/\omega_2 = r/s \in \mathbb{Q}$, with r and s positive integers, $r < s$, and the fraction is irreducible. Without the reversibility symmetry, this bifurcation was discussed in Section 3.4.5 of Chapter 3. In contrast to the reversible bifurcations discussed in the previous sections, we assume here that the parameter μ is two-dimensional, $\mu \in \mathbb{R}^2$.

$(i\omega_1)(i\omega_2)$ Normal Form

First, *we choose the eigenvectors ζ_1 and ζ_2, associated with the simple eigenvalues $i\omega_1$ and $i\omega_2$, respectively, of* $\mathbf{L}$ *such that*

$$\mathbf{L}\zeta_1 = i\omega_1\zeta_1, \quad \mathbf{S}\zeta_1 = \overline{\zeta_1}, \quad \mathbf{L}\zeta_2 = i\omega_2\zeta_2, \quad \mathbf{S}\zeta_2 = \overline{\zeta_2}.$$

Starting from the normal form found in Section 3.4.5 and using the reversibility symmetry, it is not difficult to show in this case that *for any positive integer p, $2 \leq p \leq k$, there exist neighborhoods $\widetilde{\mathscr{V}_1}$ and $\widetilde{\mathscr{V}_2}$ of 0 in $\mathbb{R}^4$ and $\mathbb{R}^2$, respectively, and for any $\mu \in \widetilde{\mathscr{V}_2}$ there is a polynomial $\Phi(\cdot,\mu) : \mathbb{R}^4 \to \mathbb{R}^4$ of degree p with*

$$\Phi(0,0,0,0,0) = 0, \quad \partial_{(A,B,\overline{A},\overline{B})}\Phi(0,0,0,0,0) = 0,$$

and

$$\Phi(\overline{A},\overline{B},A,B,\mu) = \mathbf{S}\Phi(A,B,\overline{A},\overline{B},\mu),$$

such that the change of variables

$$u = A\zeta_0 + B\zeta_1 + \overline{A}\overline{\zeta_0} + \overline{B}\overline{\zeta_1} + \Phi(A,B,\overline{A},\overline{B},\mu) \tag{3.35}$$

transforms the differential equation into the normal form

$$\begin{aligned}
\frac{dA}{dt} &= i\omega_1 A + iAP_1(|A|^2,|B|^2,A^s\overline{B}^r,\mu) + i\overline{A}^{s-1}B^rP_2(|A|^2,|B|^2,\overline{A}^sB^r,\mu)\\
&\quad +\rho_A(A,B,\overline{A},\overline{B},\mu)\\
\frac{dB}{dt} &= i\omega_2 B + iBQ_1(|A|^2,|B|^2,\overline{A}^sB^r,\mu) + iA^s\overline{B}^{r-1}Q_2(|A|^2,|B|^2,A^s\overline{B}^r,\mu)\\
&\quad +\rho_B(A,B,\overline{A},\overline{B},\mu).
\end{aligned} \tag{3.36}$$

Here P_j, Q_j, $j = 1,2$, are polynomials with real coefficients of degrees $p-1$ in $(A,B,\overline{A},\overline{B})$ for $j = 1$ and of degree $p-r-s+1$ for $j = 2$. The remainders ρ_A and ρ_B are functions of class $\mathscr{C}^k$ which satisfy

$$\begin{aligned}
\rho_A(\overline{A},\overline{B},A,B,\mu) &= -\overline{\rho_A}(A,B,\overline{A},\overline{B},\mu),\\
\rho_B(\overline{A},\overline{B},A,B,\mu) &= -\overline{\rho_B}(A,B,\overline{A},\overline{B},\mu),
\end{aligned}$$

and

$$|\rho_A(A,B,\overline{A},\overline{B},\mu)| + |\rho_B(A,B,\overline{A},\overline{B},\mu)| = o((|A|+|B|)^p).$$

Solutions of the Normal Form System

Recall that $\omega_1/\omega_2 = r/s \in \mathbb{Q}$, with r and s positive integers, such that $r < s$ and the fraction is irreducible, and that the cases $(r,s) = (1,2)$ and $(r,s) = (1,3)$ are called strongly resonant, whereas the cases of $r+s \geq 5$ are called weakly resonant (see Section 3.4.5).

In the *weakly resonant cases*, when $r+s \geq 5$, the normal form truncated at order $r+s-2$ is

$$\frac{dA}{dt} = i\omega_1 A + iAP(|A|^2,|B|^2,\mu)$$
$$\frac{dB}{dt} = i\omega_2 B + iBQ(|A|^2,|B|^2,\mu), \tag{3.37}$$

where P and Q are real-valued polynomials of degree $r+s-3 \geq 2$ in $(A,B,\overline{A},\overline{B})$, at most. In polar coordinates

$$A = r_1 e^{i\theta_1}, \quad B = r_2 e^{i\theta_2},$$

we find the system

$$\frac{dr_1}{dt} = 0, \quad \frac{dr_2}{dt} = 0, \quad \frac{d\theta_1}{dt} = \omega_1 + P(r_1^2,r_2^2,\mu), \quad \frac{d\theta_1}{dt} = \omega_2 + Q(r_1^2,r_2^2,\mu).$$

Consequently, the dynamics of the truncated system (3.37) lie on invariant 2-tori $r_1 = c_1$, $r_2 = c_2$, with real constants c_1 and c_2, and the solutions are either periodic or quasiperiodic, depending on whether the ratio

$$\frac{\omega_1 + P(r_1^2,r_2^2,\mu)}{\omega_2 + Q(r_1^2,r_2^2,\mu)}$$

is rational or irrational. For the full system (3.36) only KAM tori are expected to persist, on which the flow is quasiperiodic. This persistence problem is again a small divisor problem and is extensively studied (see for instance [111]). The dynamics between these KAM tori are expected to be very intricate and should mimic the corresponding problem for Hamiltonian systems.

We consider now the *strongly resonant case* $(r,s) = (1,2)$, also referred to as $1:2$ resonance. The case $(r,s) = (1,3)$, also called $1:3$ resonance, has common features, but its complete study is still quite open.

Assuming that $(r,s) = (1,2)$, the normal form truncated at order 3 is

$$\frac{dA}{dt} = iA\left(\omega_1 + \mu_1 + a_1|A|^2 + a_2|B|^2\right) + ia_3\overline{A}B$$
$$\frac{dB}{dt} = iB\left(\omega_2 + \mu_2 + b_1|A|^2 + b_2|B|^2\right) + ib_3A^2, \tag{3.38}$$

where μ_j, $j = 1, 2$, are small parameters, and a_j, b_j, $j = 1, 3$, are real coefficients. A key property of this truncated system is that it is integrable. Indeed, setting

$$A = r_1 e^{i\theta_1}, \quad B = r_2 e^{i\theta_2}, \quad \Theta = \theta_2 - 2\theta_1, \quad \nu = \mu_2 - 2\mu_1$$

leads to the system

$$\begin{aligned}
\frac{dr_1}{dt} &= -a_3 r_1 r_2 \sin\Theta \\
\frac{dr_2}{dt} &= b_3 r_1^2 \sin\Theta \\
\frac{d\Theta}{dt} &= \nu + \frac{1}{r_2}\cos\Theta (b_3 r_1^2 - 2a_3 r_2^2),
\end{aligned} \tag{3.39}$$

together with an equation for $d\theta_1/dt$, which can be solved afterwards. It is now straightforward to check that (3.39) has the two first integrals

$$K = b_3 r_1^2 + a_3 r_2^2, \quad H = r_1^2 r_2 \cos\Theta - \frac{\nu}{2a_3} r_1^2,$$

where we have assumed that $a_3 \neq 0$. Setting $u_0 = r_1^2$, one finds in this case an equation similar to (3.30),

$$\left(\frac{d}{dt}u_0\right)^2 = 4 f_{H,K}(u_0, \mu).$$

The solutions can be now analyzed as in the case of the $(i\omega)^2$ bifurcation studied in Section 4.3.3. Depending upon the sign of $a_3 b_3$, one finds in this case different periodic orbits and homoclinic orbits to periodic orbits. We refer to [5] for further details and the proofs of the persistence results.

4.3.5 *Reversible* 0^{4+} *Bifurcation (Elements)*

We briefly discuss in this section the 0^{4+} bifurcation, when 0 is an algebraically quadruple and geometrically simple eigenvalue of $\mathbf{L}$, with associated eigenvector ζ_0 satisfying

$$\mathbf{S}\zeta_0 = \zeta_0.$$

As in the previous section, we take here the parameter $\mu \in \mathbb{R}^m$.

0^{4+} *Normal Form*

First, by arguing as in the case of the 0^{3+} bifurcation, we can prove here that *there is a basis* $\{\zeta_0, \zeta_1, \zeta_2, \zeta_3\}$ *of* $\mathbb{R}^4$ *such that*

$$\begin{aligned}&\mathbf{L}\zeta_0 = 0, \quad \mathbf{L}\zeta_1 = \zeta_0, \quad \mathbf{L}\zeta_2 = \zeta_1, \quad \mathbf{L}\zeta_3 = \zeta_2,\\ &\mathbf{S}\zeta_0 = 0, \quad \mathbf{S}\zeta_1 = -\zeta_0, \quad \mathbf{S}\zeta_2 = \zeta_1, \quad \mathbf{S}\zeta_3 = -\zeta_2.\end{aligned} \tag{3.40}$$

The normal form in this case is proved in [54]. Using the basis above it is shown that *for any $p < k$ there exists a polynomial $\Phi(\cdot,\mu) : \mathbb{R}^4 \to \mathbb{R}^4$ of degree p, with*

$$\Phi(0,0,0,0,0) = 0, \quad \partial_{(A,B,C,D)}\Phi(0,0,0,0,0) = 0,$$

and

$$\Phi(A,-B,C,-D,\mu) = \mathbf{S}\Phi(A,B,C,D,\mu),$$

such that the change of variables

$$u = A\zeta_0 + B\zeta_1 + C\zeta_2 + D\zeta_3 + \Phi(A,B,C,\mu), \tag{3.41}$$

transforms the differential equation into the normal form

$$\begin{aligned}\frac{d}{dt}\begin{pmatrix}A\\B\\C\\D\end{pmatrix} &= \mathbf{L}\begin{pmatrix}A\\B\\C\\D\end{pmatrix} + P_1(A,p_2,p_4,\mu)\begin{pmatrix}0\\0\\0\\1\end{pmatrix} + P_2(A,p_2,p_4,\mu)\begin{pmatrix}0\\A\\B\\C\end{pmatrix}\\ &+P_3(p_2,p_4,\mu)\begin{pmatrix}0\\p_2\\q_2\\r_2\end{pmatrix} + P_4(A,p_2,p_4,\mu)\begin{pmatrix}p_3\\q_3\\r_3\\s_3\end{pmatrix}\\ &+P_5(A,p_2,p_4,\mu)\begin{pmatrix}0\\0\\p_3\\q_3\end{pmatrix} + \rho(A,B,C,D,\mu).\end{aligned}$$

Here

$$\begin{aligned}p_2 &= B^2 - 2AC, \quad q_2 = -3AD + BC, \quad r_2 = -3BD + 2C^2,\\ p_3 &= B^3 - 3ABC + 3A^2D, \quad q_3 = 3ABD - 2AC^2 + B^2C,\\ r_3 &= -3ACD + 3B^2D - BC^2, \quad s_3 = 3BCD - \tfrac{4}{3}C^3 - 3AD^2,\\ p_4 &= 3B^2C^2 - 6B^3D - 8AC^3 + 18ABCD - 9A^2D^2,\end{aligned}$$

P_j, $j = 1,\ldots,5$, are polynomials of degree $p - j + 1$ in (A,B,C,D), with

$$P_1(0,0,0,0) = 0, \quad \partial_A P_1(0,0,0,0) = 0, \quad P_2(0,0,0,0) = 0,$$

and the remainder ρ is of class $\mathscr{C}^k$ satisfying

$$\rho(A,-B,C,-D,\mu) = -\mathbf{S}\rho(A,B,C,D,\mu),$$

and the estimate

$$\|\rho(A,B,C,D,\mu)\| = o((|A|+|B|+|C|+|D|)^p).$$

Remark 3.23 *Notice that we have* $p_3^2 = p_2^3 - p_1^2 p_4$.

Solutions of the Normal Form

The study of this bifurcation is widely open. As for the other bifurcations, we can start from the truncated normal form, at order 2 here,

$$\begin{aligned}
\frac{dA}{dt} &= B \\
\frac{dB}{dt} &= C + \mu_2 A + cA^2 \\
\frac{dC}{dt} &= D + \mu_2 B + cAB \\
\frac{dD}{dt} &= \mu_0 + \mu_1 A + \mu_2 C + aA^2 + b(B^2 - 2AC) + cAC,
\end{aligned} \tag{3.42}$$

where a, b, and c are real coefficients, and μ_0, μ_1, μ_2 are small parameters coming from the terms in the expansion of the vector field that are linear in $\mu \in \mathbb{R}^m$. In particular, assuming that $a \neq 0$, it is not difficult to check that there is a saddle-node bifurcation for

$$\mu_0 \sim \frac{1}{4a}(\mu_1 - \mu_2^2)^2,$$

in which a pair of equilibria invariant under $\mathbf{S}$ bifurcates from 0.

We restrict ourselves below to the particular case in which *the origin is an equilibrium for all values of the parameter* μ, so that $\mu_0 = 0$. This is the case which arises more often in applications. Linearizing at the origin, we then find the matrix

$$\mathbf{L}_\mu = \begin{pmatrix} 0 & 1 & 0 & 0 \\ \mu_2 & 0 & 1 & 0 \\ 0 & \mu_2 & 0 & 1 \\ \mu_1 & 0 & \mu_2 & 0 \end{pmatrix},$$

with eigenvalues λ satisfying

$$\lambda^4 - 3\mu_2\lambda^2 + \mu_2^2 - \mu_1 = 0.$$

Clearly, the four eigenvalues are symmetric with respect to both axis in the complex plane, and their location depends upon the parameters μ_1 and μ_2.

We plot in Figure 3.4 the location of the four eigenvalues of $\mathbf{L}_\mu$, depending on the parameters μ_1 and μ_2. The four curves Γ_j, $j = 1,\dots,4$, are such that

(i) $\mu_1 = \mu_2^2$, $\mu_1 > 0$, for Γ_1, along which 0 is an algebraically double and geometrically simple eigenvalue, and $\mathbf{L}_\mu$ has a pair of simple real eigenvalues;

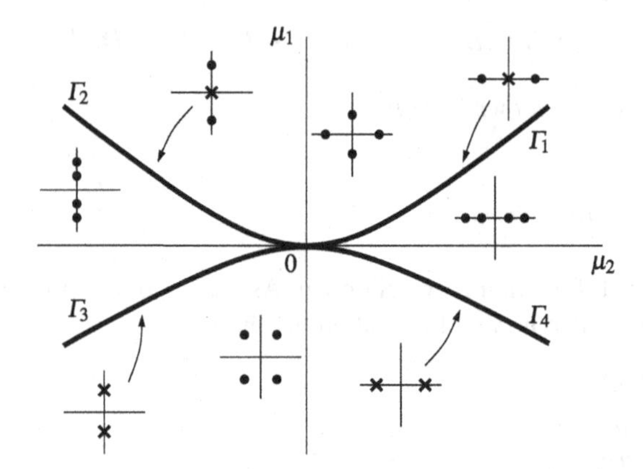

Fig. 3.4 Location of the four eigenvalues of $\mathbf{L}_\mu$, depending on the parameters (μ_1, μ_2).

(ii) $\mu_1 = \mu_2^2$, $\mu_1 < 0$, for Γ_2, along which 0 is an algebraically double and geometrically simple eigenvalue, and $\mathbf{L}_\mu$ has a pair of simple purely imaginary eigenvalues;
(iii) $\mu_1 = -5\mu_2^2/4$, $\mu_1 < 0$, for Γ_3, along which $\mathbf{L}_\mu$ has a pair of double purely imaginary eigenvalues;
(iv) $\mu_1 = -5\mu_2^2/4$, $\mu_1 > 0$, for Γ_4, along which $\mathbf{L}_\mu$ has a pair of double real eigenvalues.

It appears that each point on the curves Γ_1, Γ_2, and Γ_3 is a bifurcation point, with

(i) a reversible 0^{2+} bifurcation occuring when crossing Γ_1 in the parameter plane;
(ii) a reversible $0^{2+}(i\omega)$ bifurcation occuring when crossing Γ_2;
(iii) a reversible $(i\omega)^2$ bifurcation occuring when crossing Γ_3.

The computations of the bifurcations near Γ_1 and Γ_3 are done in [54]. It is a remarkable fact that for Γ_3 the coefficient of the cubic term of the normal form, which determines whether focusing or defocusing case occurs, always gives the focusing case here, provided that $a \neq 0$. In particular, this implies the existence of homoclinic orbits to 0. The results found for each of these bifurcations are valid in *little horn-like regions of the parameter plane*, due to the fact that when (μ_1, μ_2) tends to 0 along Γ_j, the nonzero eigenvalues all tend to 0, which gives a singularity in the coefficients of the normal forms. Though not a bifurcation curve, the dynamics in a neighborhood of Γ_4 turn out to be quite rich. In particular, a new type of bounded orbits can be found here, which are multipulse homoclinic orbits. We refer to the review papers [13, 49], and the references therein, for different results of existence of such orbits.

Exercise 3.24 *Consider the fourth order ODE*

$$u^{(4)} - 3\mu_2 u'' + (\mu_2^2 - \mu_1)u - au^2 = 0,$$

where $\mu_1, \mu_2 \in \mathbb{R}$ are two small parameters and a is a given real number. (See the example in Section 2.4.1 for the particular case $3\mu_2 = 1$ and $\mu_1 - \mu_2^2 = \mu$.)

(i) Write the equation as a first order reversible system of the form (3.42) with $b = c = 0$.

(ii) Consider the curve Γ_1 *in Figure 3.4 (*0^{2+} *bifurcation). Show that for* (μ_1, μ_2) *close to* Γ_1 *the system possesses a two-dimensional center manifold, and show that the reduced system has the normal form*

$$\frac{dA}{dt} = B$$
$$\frac{dB}{dt} = -\frac{\varepsilon}{3\mu_2}A - \frac{a}{3\mu_2}A^2 + h.o.t.,$$

where $\varepsilon = \mu_1 - \mu_2^2$, $\mu_2 > 0$, *and* A, B *are real-valued functions.*

(iii) Consider the curve Γ_2 *in Figure 3.4 (*$0^{2+}(i\omega)$ *bifurcation). Show that for* (μ_1, μ_2) *close to* Γ_2 *the system can be put in the normal form*

$$\frac{dA}{dt} = B$$
$$\frac{dB}{dt} = -\frac{1}{3\mu_2}\left(\varepsilon A + aA^2 + 2a|C|^2\right) + h.o.t.$$
$$\frac{dC}{dt} = \frac{i}{(-3\mu_2)^{1/2}}C\left(1 + \frac{\varepsilon}{18\mu_2} + \frac{aA}{9\mu_2^2}\right) + h.o.t.,$$

where $\varepsilon = \mu_1 - \mu_2^2$, $\mu_2 < 0$, A, B *are real-valued functions, and* C *is complex-valued.*

(iv) Consider the curve Γ_3 *in Figure 3.4 (*$(i\omega)^{2+}$ *bifurcation). Show that for* (μ_1, μ_2) *close to* Γ_3 *the system can be put in the normal form*

$$\frac{dA}{dt} = i\left(-\frac{3}{2}\mu_2\right)^{1/2} A + B + \frac{i\varepsilon}{(-6\mu_2)^{3/2}}A + h.o.t.$$
$$\frac{dB}{dt} = i\left(-\frac{3}{2}\mu_2\right)^{1/2} B + \frac{i\varepsilon}{(-6\mu_2)^{3/2}}B + \frac{\varepsilon}{-6\mu_2}A + \frac{76a^2}{243\mu_2^3}A|A|^2 + h.o.t.,$$

where $\varepsilon = -(5/4)\mu_2^2 - \mu_1$, $\mu_2 < 0$, *and* A, B *are complex-valued functions.*

Hint: *This exercise is solved in [56]. Notice that some of the coefficients in the normal form become unbounded as* $(\mu_1, \mu_2) \to (0,0)$. *This is due to the quadruple zero eigenvalue of the linearization for* $(\mu_1, \mu_2) = (0,0)$, *and shows that these normal forms are not valid in a neighborhood of the origin.*

Remark 3.25 *The equation*

$$\varepsilon^2 u^{(4)} + u'' - cu + 3u^2 = 0$$

arises as a traveling wave equation from a fifth order Korteweg–de Vries (KdV) type equation when we seek solutions which decay to 0 *at infinity [105]. Here* ε *is a small parameter and* c *the constant speed of propagation of the traveling waves. After a suitable scaling, this equation enters into the frame above, with small parameters close to the curve* Γ_2 *in Figure 3.4, where an* $0^{2+}(i\omega)$ *bifurcation occurs. From the results in Theorem 3.10, one can then conclude that for small* $\varepsilon \neq 0$ *there are no homoclinic orbits to* 0 *in this case, in contrast to the case* $\varepsilon = 0$ *in which such orbits do exist.*

4.3.6 Reversible $0^2 0^2$ Bifurcation with $SO(2)$ Symmetry

We end this chapter with a brief discussion of reversible $0^2 0^2$ bifurcations in the presence of a $SO(2)$ symmetry. We assume that the matrix $\mathbf{L}$ has an algebraically quadruple and geometrically double zero eigenvalue with a two 2×2-Jordan block. Besides the reversibility symmetry $\mathbf{S}$, we now further assume that the vector field $\mathbf{F}(\cdot, \mu)$ in (3.1) is $SO(2)$-equivariant, that is, there exists a one-parameter continuous family of linear maps $\mathbf{R}_\varphi$ which act nontrivially on $\mathbb{R}^4$, for $\varphi \in \mathbb{R}/2\pi\mathbb{Z}$, with the following properties:

(i) $\mathbf{R}_0 = \mathbb{I}$ and $\mathbf{R}_\varphi \circ \mathbf{R}_\psi = \mathbf{R}_{\varphi+\psi}$ for all $\varphi,\ \psi \in \mathbb{R}/2\pi\mathbb{Z}$;
(ii) $\mathbf{F}(\mathbf{R}_\varphi u, \mu) = \mathbf{R}_\varphi \mathbf{F}(u, \mu)$ for all $\varphi \in \mathbb{R}/2\pi\mathbb{Z}$.

In addition, we assume that

$$\mathbf{S}\mathbf{R}_\varphi = \mathbf{R}_\varphi \mathbf{S} \text{ for all } \varphi \in \mathbb{R}/2\pi\mathbb{Z}, \quad \text{or} \quad \mathbf{S}\mathbf{R}_\varphi = \mathbf{R}_{-\varphi}\mathbf{S} \text{ for all } \varphi \in \mathbb{R}/2\pi\mathbb{Z}. \quad (3.43)$$

Normal Form

First, we construct *a basis* $\{\zeta_0, \zeta_1, \overline{\zeta_0}, \overline{\zeta_1}\}$ *of* $\mathbb{R}^4$*, identified here with* $\widetilde{\mathbb{R}^4}$*, such that*

$$\mathbf{L}\zeta_0 = 0, \quad \mathbf{L}\zeta_1 = \zeta_0,$$
$$\mathbf{R}_\phi \zeta_0 = e^{im\phi}\zeta_0, \quad \mathbf{R}_\phi \zeta_1 = e^{im\phi}\zeta_1,$$

for some nonzero integer m. Then, depending upon the action of $\mathbf{S}$ *on* ζ_0*, we distinguish the following cases:*

(i) $\mathbf{S}\zeta_0 = \zeta_0$, $\mathbf{S}\zeta_1 = -\zeta_1$;
(ii) $\mathbf{S}\zeta_0 = -\zeta_0$, $\mathbf{S}\zeta_1 = \zeta_1$;
(iii) $\mathbf{S}\zeta_0 = \overline{\zeta_0}$, $\mathbf{S}\zeta_1 = -\overline{\zeta_1}$.

We have the cases (i) and (ii) when $\mathbf{R}_\phi\mathbf{S} = \mathbf{S}\mathbf{R}_\phi$ and the case (iii) when $\mathbf{R}_\phi\mathbf{S} = \mathbf{S}\mathbf{R}_{-\phi}$.

The construction of this basis is similar to that of the basis of $\mathbb{R}^2$ for the steady bifurcation with $O(2)$ symmetry in Section 1.2.4 of Chapter 1. By arguing as in the proof of (2.40)–(2.41) in Chapter 1, we find here the vector ζ_0 satisfying

$$\mathbf{L}\zeta_0 = 0, \quad \mathbf{R}_\phi \zeta_0 = e^{im\phi}\zeta_0.$$

Then we can choose ζ_1 such that

$$\mathbf{L}\zeta_1 = \zeta_0, \quad \mathbf{R}_\phi \zeta_1 = e^{im\phi}\zeta_1.$$

Since $\mathbf{S}\zeta_0$ belongs to the kernel of $\mathbf{L}$ we have

$$\mathbf{S}\zeta_0 = a\zeta_0 + b\overline{\zeta_0}$$

for some complex numbers a and b, and the cases (i)–(iii) are obtained from the equalities (3.43).

The normal form is obtained in the basis above, starting from the result in Lemma 1.19 in Chapter 3, using the Theorems 2.2, 3.2, and 3.4 in Chapter 3, with the symmetries $\mathbf{R}_\phi$ and $\mathbf{S}$. We obtain that *for any $p < k$ there exists a polynomial $\Phi(\cdot,\mu) : \widetilde{\mathbb{R}^4} \to \widetilde{\mathbb{R}^4}$ of degree p, with*

$$\Phi(0,0,0,0,0) = 0, \quad \partial_{(A,B,\overline{A},\overline{B})}\Phi(0,0,0,0,0) = 0,$$

and

$$\Phi(e^{im\phi}A, e^{im\phi}B, e^{-im\phi}\overline{A}, e^{-im\phi}\overline{B}, \mu) = \mathbf{R}_\phi \Phi(A,B,\overline{A},\overline{B},\mu) \text{ for all } \phi \in R/2\pi\mathbb{Z},$$

such that the change of variables

$$u = A\zeta_0 + B\zeta_1 + \overline{A\zeta_0} + \overline{B\zeta_1} + \Phi(A,B,\overline{A},\overline{B},\mu) \tag{3.44}$$

transforms equation (3.1) into the normal form

$$\begin{aligned} \frac{dA}{dt} &= B + AP(|A|^2, i(A\overline{B} - \overline{A}B), \mu) + \rho_A(A,B,\overline{A},\overline{B},\mu) \\ \frac{dB}{dt} &= BP(|A|^2, i(A\overline{B} - \overline{A}B), \mu) + AQ(|A|^2, i(A\overline{B} - \overline{A}B), \mu) \\ &\quad + \rho_B(A,B,\overline{A},\overline{B},\mu). \end{aligned} \tag{3.45}$$

Here P and Q are polynomials of degree $p-1$ in $(A,\overline{A},B,\overline{B},\mu)$ and the remainders ρ_A and ρ_B satisfy

$$\begin{aligned} \rho_A(e^{im\phi}A, e^{im\phi}B, e^{-im\phi}\overline{A}, e^{-im\phi}\overline{B}, \mu) &= e^{im\phi}\rho_A(A,B,\overline{A},\overline{B},\mu), \\ \rho_B(e^{im\phi}A, e^{im\phi}B, e^{-im\phi}\overline{A}, e^{-im\phi}\overline{B}, \mu) &= e^{im\phi}\rho_B(A,B,\overline{A},\overline{B},\mu) \end{aligned}$$

for all $\phi \in \mathbb{R}/2\pi\mathbb{Z}$, and the estimate

$$|\rho_A(A,B,\overline{A},\overline{B},\mu)| + |\rho_B(A,B,\overline{A},\overline{B},\mu)| = o\left((|A| + |B|)^p\right).$$

Moreover the following properties hold in the three cases:

(i) $\Phi(A,-B,\overline{A},-\overline{B},\mu) = \mathbf{S}\Phi(A,B,\overline{A},\overline{B},\mu)$;
P and Q are odd and even, respectively, in their second argument;
$\rho_A(A,-B,\overline{A},-\overline{B},\mu) = -\rho_A(A,B,\overline{A},\overline{B},\mu)$, *and*
$\rho_B(A,-B,\overline{A},-\overline{B},\mu) = \rho_B(A,B,\overline{A},\overline{B},\mu)$;

(ii) $\Phi(-A,B,-\overline{A},\overline{B},\mu) = \mathbf{S}\Phi(A,B,\overline{A},\overline{B},\mu)$;
P and Q are odd and even, respectively, in their second argument;
$\rho_A(-A,B,-\overline{A},\overline{B},\mu) = \rho_A(A,B,\overline{A},\overline{B},\mu)$, *and*
$\rho_B(-A,B,-\overline{A},\overline{B},\mu) = -\rho_B(A,B,\overline{A},\overline{B},\mu)$;

(iii) $\Phi(\overline{A},-\overline{B},A,-B,\mu) = \mathbf{S}\Phi(A,B,\overline{A},\overline{B},\mu)$;
P is imaginary-valued and Q is real-valued;
$\rho_A(\overline{A},-\overline{B},A,-B,\mu) = \overline{-\rho_A(A,B,\overline{A},\overline{B},\mu)}$, *and*
$\rho_B(\overline{A},-\overline{B},A,-B,\mu) = \overline{\rho_B(A,B,\overline{A},\overline{B},\mu)}$.

Solutions of the Normal Form

The cases (i) and (ii) have the same type of normal form, but the complete study of the dynamics in these cases is open. Some partial answers can be found in [66] where this situation arises in an infinite-dimensional problem.

In the case (iii), the normal form is a particular case of the normal form (3.25) for the $(i\omega)^2$ bifurcation when $\omega = 0$ and the remainders ρ_A and ρ_B have the additional rotational invariance given above. Consequently, the analysis done for the $(i\omega)^2$ bifurcation stays also valid here. In addition, the rotational invariance of the full vector field allows us to show in this case the persistence of a "circle" of homoclinics, instead of only a pair of reversible homoclinics.

Exercise 3.26 *Consider a system in* $\mathbb{R}^4$ *of the form (3.1) satisfying Hypothesis 3.1. Assume that the matrix* $\mathbf{L} = D_u\mathbf{F}(0,0)$ *has two geometrically double eigenvalues* $\pm i\omega$ *with eigenvectors* $\zeta_0, \zeta_1, \overline{\zeta_0}, \overline{\zeta_1} \in \mathbb{C}^4$ *such that*

$$\mathbf{L}\zeta_0 = i\omega\zeta_0, \quad \mathbf{L}\overline{\zeta_1} = i\omega\overline{\zeta_1}, \quad \mathbf{S}\zeta_0 = \zeta_1 \neq \overline{\zeta_0}.$$

Further assume that the vector field is $SO(2)$*-equivariant with respect to a group representation* $(\mathbf{R}_\phi)_{\phi\in\mathbb{R}/2\pi\mathbb{Z}}$*, which commutes with the reversibility symmetry* $\mathbf{S}$ *and acts on the vectors* ζ_0, ζ_1 *through*

$$\mathbf{R}_\phi\zeta_0 = e^{im\phi}\zeta_0, \quad \mathbf{R}_\phi\zeta_1 = e^{im\phi}\zeta_1.$$

(i) Show that the normal form of this system is

$$\begin{aligned}\frac{dA}{dt} &= i\omega A + AP(|A|^2,|B|^2,\mu) + \rho_A(A,B,\overline{A},\overline{B},\mu)\\ \frac{dB}{dt} &= -i\omega A - BP(|B|^2,|A|^2,\mu) + \rho_B(A,B,\overline{A},\overline{B},\mu),\end{aligned}$$

where P *is a complex polynomial in its arguments such that* $P(0,0,0) = 0$.

(ii) Write the truncated system, obtained by removing the remainders ρ_A *and* ρ_B*, in polar coordinates, and study the phase portraits for the radial components. (Proceed as in Section 3.4.3.)*

(iii) Determine the bounded solutions of the system.

Hint: *See [65].*

Exercise 3.27 *Consider the system of Korteweg–de Vries (KdV) equations*

$$\begin{aligned}u_t + \Delta_1 u_x + \mu_1 uu_x + \lambda_1 u_{xxx} + \kappa_1 v_x &= 0\\ v_t + \Delta_2 v_x + \mu_2 vv_x + \lambda_2 v_{xxx} + \kappa_2 u_x &= 0,\end{aligned}$$

where $\Delta_1 - \Delta_2$ *is a detuning parameter,* κ_1, κ_2 *are coupling parameters, and* λ_1, λ_2 *are linear dispersive coefficients.*

(i) Consider traveling wave solutions of the form

$$u(x,t) = u_1(x-ct), \quad v(x,t) = v_1(x-ct),$$

where c *is the constant speed of propagation, and assume that these solutions tend to* 0 *as* $|x| \to \infty$*. Set* $\xi = x - ct$*,* $u_2 = u_1'$*,* $v_2 = v_1'$*, and show that* u_1*,* u_2*,* v_1*,* v_2 *satisfy the following system of ODEs:*

$$
\begin{aligned}
u_1' &= u_2 \\
u_2' &= \frac{1}{\lambda_1}\left((c-\Delta_1)u_1 - \frac{\mu_1}{2}u_1^2 - \kappa_1 v_1\right) \\
v_1' &= v_2 \\
v_2' &= \frac{1}{\lambda_2}\left((c-\Delta_2)v_1 - \frac{\mu_2}{2}v_1^2 - \kappa_2 u_1\right).
\end{aligned}
\tag{3.46}
$$

(*ii*) *Show that the system (3.46) is reversible.*

(*iii*) *Show that the purely imaginary eigenvalues ik of the matrix obtained by linearizing the vector field at* 0 *satisfy the "dispersion relation"*

$$\lambda_1\lambda_2 k^4 + Dk^2 + H = 0,$$

in which

$$H = (c-\delta_1)^2 - \delta_2^2 - \kappa_1\kappa_2, \quad D = (\lambda_1+\lambda_2)(c-\delta_1) - (\lambda_1-\lambda_2)\delta_2, \quad \delta_{1,2} = \frac{1}{2}(\Delta_1 \pm \Delta_2).$$

(*iv*) *Assuming that* $\lambda_1\lambda_2\kappa_1\kappa_2 < 0$, *fix* δ_1 *and consider the curves* $H = 0$, $D = 0$, *and* $P = D^2 - 4\lambda_1\lambda_2 H = 0$ *in the* (δ_2, c)*-plane. Show that these curves are bifurcation curves for the system (3.46), and determine the nature of the corresponding reversible bifurcations* (0^{2+}, $0^{2+}(i\omega)$, *or* $(i\omega)^2$).

Hint: *See [31].*

Chapter 5
Applications

In this chapter we present several applications of the methods given in this book: the center manifold theorem in Chapter 2, the normal form theory in Chapter 3, and the results on reversible bifurcations in Chapter 4. We discuss hydrodynamic instabilities arising in the Navier–Stokes equations in Section 5.1, and we consider in Section 5.2 the question of existence of traveling waves for three different situations: the water-wave problem; reaction-diffusion systems in two dimensions; and one-dimensional lattices.

5.1 Hydrodynamic Instabilities

5.1.1 Hydrodynamic Problem

Consider a viscous incompressible fluid filling a domain Ω in $\mathbb{R}^2$ or $\mathbb{R}^3$. We present in this section the hydrodynamic problem corresponding to the following three types of domains:

(i) a smooth bounded domain $\Omega \subset \mathbb{R}^2$ or $\Omega \subset \mathbb{R}^3$;
(ii) an infinite cylindrical domain $\Omega = \Sigma \times \mathbb{R}$, where the section Σ is a smooth bounded domain in $\mathbb{R}^2$;
(iii) a domain situated between two planes $\Omega = \mathbb{R}^2 \times I$, where $I = (\alpha, \beta)$ is a bounded interval in $\mathbb{R}$.

The velocity V of fluid particles and the pressure p are functions of $(x,t) \in \Omega \times \mathbb{R}^+$ and satisfy the Navier–Stokes equations

$$\begin{aligned} \frac{\partial V}{\partial t} + (V \cdot \nabla)V + \frac{1}{\rho}\nabla p &= \nu \Delta V + f(x), \\ \nabla \cdot V &= 0. \end{aligned} \tag{1.1}$$

M. Haragus, G. Iooss, *Local Bifurcations, Center Manifolds, and Normal Forms in Infinite-Dimensional Dynamical Systems*, Universitext,
DOI 10.1007/978-0-85729-112-7_5, © EDP Sciences 2011

In this system $V(x,t)$ has two or three components, when $\Omega \subset \mathbb{R}^2$ or $\Omega \subset \mathbb{R}^3$, respectively, the volumic mass ρ is constant, ∇, $\nabla\cdot$, and Δ denote the gradient, divergence and Laplace operators, respectively, ν is the kinematic viscosity, and f represents an external massic force, independent of t. The first equation represents the momentum balance, while the second is the incompressibility condition.

Boundary Conditions

System (1.1) is completed by boundary conditions. In the three cases, we assume that we have fixed geometric boundaries.

The simplest situation occurs in case (i) of a smooth bounded domain Ω, when the boundary conditions are

$$V|_{\partial\Omega} = a, \qquad \int_{\partial\Omega} a\cdot n dS = 0, \tag{1.2}$$

where a is a given vector field, independent of t and having zero total flux, in order to be compatible with the incompressibility condition, and n is the exterior unit normal to $\partial\Omega$.

In case (ii) of a cylindrical domain $\Omega = \Sigma\times\mathbb{R}$, the boundary conditions are

$$V|_{\partial\Sigma\times\mathbb{R}} = a, \qquad \int_{\partial\Sigma} a\cdot n ds = 0, \tag{1.3}$$

to which one can add, for instance, the following periodicity conditions along the cylinder:

$$V(x,t) = V(x+he_z,t), \quad \nabla p(x,t) = \nabla p(x+he_z,t) \text{ for all } x=(X,z)\in\Sigma\times\mathbb{R}, \tag{1.4}$$

where h is the period in the direction $z\in\mathbb{R}$ along the cylinder, and $e_z = (0,1)\in\Sigma\times\mathbb{R}$. Notice that we require only ∇p to be periodic and not p, which would also be a possibility, but less realistic. These conditions are completed by the assumption

$$\int_\Sigma V\cdot n dS = D, \tag{1.5}$$

where D is a given constant, showing that V has a given flux through the section Σ of the cylinder. It is not difficult to check that this flux is independent of $z\in\mathbb{R}$. This implies that p is allowed to increase linearly in z over a period.

Finally, in case (iii) of a domain $\Omega = \mathbb{R}^2\times(\alpha,\beta)$ situated between two planes, the boundary conditions are

$$V|_{\mathbb{R}^2\times\{\alpha\}} = V|_{\mathbb{R}^2\times\{\beta\}} = a,$$

which imply that the total mass flux through the periodicity domain is zero, together with a biperiodicity condition,

$$V(x,t) = V(x+n_1e_1+n_2e_2,t), \quad \nabla p(x,t) = \nabla p(x+n_1e_1+n_2e_2,t) \tag{1.6}$$

for all $x = (X,z) \in \mathbb{R}^2 \times (\alpha,\beta)$, where $(n_1,n_2) \in \mathbb{Z}^2$, and the lattice of periods is generated by two noncolinear vectors e_1 and e_2 in $\mathbb{R}^2$. To these conditions we add two conditions on the flux of the velocity in the directions of two vectors k_1 and k_2 in the X-plane,

$$\int_{\Sigma_1} V\cdot k_2\,dS = D_1, \quad \int_{\Sigma_2} V\cdot k_1\,dS = D_2. \tag{1.7}$$

The vectors k_1 and k_2 are such that

$$\langle e_j, k_l\rangle = 2\pi\delta_{jl}, \tag{1.8}$$

and Σ_1 (resp., Σ_2) is the face orthogonal to k_2 (resp., to k_1) of the parallelepiped built with vectors e_1, e_2 and the interval (α,β) orthogonally to the X-plane, which constitutes the domain of periodicity.

Remark 1.1 (Free boundaries) *Sometimes the boundary, or part of the boundary, of the domain Ω is "free," which means that the fluid is in contact with another fluid, the common boundary being unknown. An example of a free-boundary problem is discussed in more detail in Section 5.2.1, where we consider the water-wave problem. Here, we only mention the simplified situation in which one assumes that the part of the boundary $\partial\Omega_1$, say, where the fluid is in contact with another fluid, is fixed. (This is acceptable for instance if the external fluid is mercury and the internal one is water.) Then, on this part of the boundary one has the following conditions:*

$$V\cdot n|_{\partial\Omega_1} = 0, \tag{1.9}$$

showing that no fluid crosses the boundary, and

$$(\nabla V + \nabla^t V)\cdot n|_{\partial\Omega_1} \times n = 0, \tag{1.10}$$

showing that the tangent stresses cancel.

Basic Solution

We assume that a smooth stationary solution $(V^{(0)}(x), p^{(0)}(x))$ is known for system (1.1), together with the corresponding boundary conditions. We set

$$V = V^{(0)} + U, \quad p = p^{(0)} + \rho q,$$

which leads to the system

$$\begin{aligned} \frac{\partial U}{\partial t} &= \nu\Delta U - (V^{(0)}\cdot\nabla)U - (U\cdot\nabla)V^{(0)} - (U\cdot\nabla)U - \nabla q \\ \nabla\cdot U &= 0. \end{aligned} \tag{1.11}$$

In case (i) the boundary condition (1.2) becomes

$$U|_{\partial\Omega} = 0, \tag{1.12}$$

whereas in case (ii), the boundary conditions (1.3), (1.4), and (1.5) become, respectively,

$$U|_{\partial\Sigma\times\mathbb{R}} = 0, \tag{1.13}$$

$$U(x,t) = U(x+he_z,t), \quad \nabla q(x,t) = \nabla q(x+he_z,t) \text{ for all } x = (X,z) \in \Sigma\times\mathbb{R},$$

and

$$\int_\Sigma U\cdot n dS = 0. \tag{1.14}$$

The boundary conditions in case (iii) are similar.

Analytical Set-up

We introduce now the basic Hilbert spaces in which system (1.11), together with the corresponding boundary conditions, is analyzed.

In case (i), we restrict to the case $\Omega\subset\mathbb{R}^3$, and define the Hilbert space

$$\mathscr{X} = \left\{U\in\left(L^2(\Omega)\right)^3 ; \nabla\cdot U = 0,\ U\cdot n|_{\partial\Omega} = 0\right\},$$

equipped with the scalar product of $\left(L^2(\Omega)\right)^3$. Notice that here the trace $U\cdot n|_{\partial\Omega}$ is well-defined in $H^{-1/2}(\partial\Omega)$ (e.g., see [118]). Next, we consider the subspace

$$\mathscr{Z} = \left\{U\in\left(H^2(\Omega)\cap H_0^1(\Omega)\right)^3 ; \nabla\cdot U = 0\right\}\subset\mathscr{X};$$

i.e., the functions in this subspace satisfy the boundary condition (1.12).

A key property of the Hilbert space $\mathscr{X}$ is that the kernel of the orthogonal projection Π_0 in $\left(L^2(\Omega)\right)^3$ on the subspace $\mathscr{X}$ can be identified with the space $\{\nabla\phi ; \phi\in H^1(\Omega)\}$ (e.g., see [129, 85, 118]). Then, using the projection Π_0, the pressure term ∇q in (1.11) can be eliminated, and we obtain a system of the form

$$\frac{dU}{dt} = \mathbf{L}U + \mathbf{R}(U) \tag{1.15}$$

posed in $\mathscr{X}$ for $U(\cdot,t)\in\mathscr{Z}$, where

$$\mathbf{L}U = \Pi_0\left(\nu\Delta U - (V^{(0)}\cdot\nabla)U - (U\cdot\nabla)V^{(0)}\right), \quad \mathbf{R}(U) = -\Pi_0\left((U\cdot\nabla)U\right). \tag{1.16}$$

The linear operator $\mathbf{L}$, acting in $\mathscr{X}$, may be regarded as a lower order perturbation of the self-adjoint operator $\Pi_0(\nu\Delta U)$. It is a closed operator in $\mathscr{X}$, with dense domain $\mathscr{Z}$ and a compact resolvent. The spectrum of $\mathbf{L}$ consists of isolated eigenvalues with finite multiplicities, situated in a sector of the complex plane centered on the real axis, and oriented on the negative side of this axis [129]. Its resolvent satisfies the estimate (2.9) in Chapter 2 (see [129, 85]), and in fact also the estimate (2.10) in Chapter 2, with $\alpha = 3/4$ (see [52, 9]), but this latter estimate is useless if Theorem 2.20 in Chapter 2 is applied. Actually, one can prove in this case that $\mathbf{L}$ is the generator of an analytic semigroup $e^{\mathbf{L}t}$ for $t > 0$ (see [76]).

The nonlinear term $\mathbf{R}(U)$ satisfies $\mathbf{R}(U) \in \mathscr{X} \cap \left(H^1(\Omega)\right)^3$ for $U \in \mathscr{Z}$, by the Sobolev embedding theorem, and the map $\mathbf{R} : \mathscr{Z} \to \mathscr{X}$ is quadratic and continuous.

Remark 1.2 *In the case of a free boundary, when a part of the boundary is subjected to conditions (1.9)–(1.10), we can use the same space $\mathscr{X}$, and replace $\left(H_0^1(\Omega)\right)^3$ in the definition of $\mathscr{Z}$ by the space*

$$\left\{U \in \left(H^1(\Omega)\right)^3 \; ; \; U|_{\partial\Omega_2} = 0, \; U \cdot n|_{\partial\Omega_1} = 0, \; (\nabla U + \nabla^t U) \cdot n\}|_{\partial\Omega_1} \times n = 0\right\},$$

where $\partial\Omega_1 \cup \partial\Omega_2 = \partial\Omega$.

In case (ii) of a cylindrical domain $\Omega = \Sigma \times \mathbb{R}$, we define the Hilbert space

$$\mathscr{X} = \left\{U \in \left(L^2(\Sigma \times (\mathbb{R}/h\mathbb{Z}))\right)^3 \; ; \; \nabla \cdot U = 0, \; U \cdot n|_{\partial\Sigma\times\mathbb{R}} = 0, \; \int_\Sigma U \cdot n dS = 0\right\}.$$

We point out that here the orthogonal complement of $\mathscr{X}$ in $\left(L^2(\Sigma \times (\mathbb{R}/h\mathbb{Z}))\right)^3$ is the space $\{\nabla\phi \; ; \; \phi \in H^1(\Sigma \times (\mathbb{R}/h\mathbb{Z})) + z\mathbb{R}\}$, i.e., $\nabla\phi$ is a periodic function, while ϕ is not periodic [15]. The space $\mathscr{Z}$ is defined as a subspace of $\left(H^2(\Sigma \times (\mathbb{R}/h\mathbb{Z}))\right)^3 \cap \mathscr{X}$, according to the boundary conditions. Using again the orthogonal projection Π_0 on $\mathscr{X}$, the Navier–Stokes system can be written in form (1.15) with $\mathbf{L}$ and $\mathbf{R}$ defined as in (1.16).

Similarly, in case (iii) for a domain $\Omega = \mathbb{R}^2 \times I$, we can define the spaces $\mathscr{X}$ and $\mathscr{Z}$ in an appropriate manner taking into account the boundary conditions, and use the orthogonal projection Π_0 to write the system in the form (1.15). In both cases, the properties of $\mathbf{L}$ and $\mathbf{R}$ mentioned above are still valid.

Summarizing, in the three cases we have a system of the form (1.15), for which Hypotheses 2.1 and 3.20 required by the center manifold theorem in Chapter 2 are verified, and in order to check Hypothesis 2.4 in Chapter 2, it is enough to locate the eigenvalues that have the largest real parts. In general, this is obtained by a careful study of their location for each specific physical situation. We point out that the parameter dependency comes from the viscosity ν and the boundary data, which influence the basic solution $(V^{(0)}, p^{(0)})$.

We present in the next two sections two classical examples where the theoretical tools developed in the previous chapters apply particularly well. A description of classical experiments and physical results connected to both examples may be found in the books [75, 83].

5.1.2 Couette–Taylor Problem

We briefly present in this section some results on the Couette–Taylor problem, which have been obtained with the help of the methods described in this book. We refer to the book [15] for details, and to [117] for the huge bibliography on this problem.

Hydrodynamic Problem

Consider two coaxial cylinders of radii R_1 (the inner cylinder), and R_2 (the outer cylinder), the gap between them being filled by an incompressible viscous fluid. Both cylinders rotate with constant rotation rates Ω_1 and Ω_2, respectively (see Figure 1.1(i)). For fixing ideas, we assume that $\Omega_1 > 0$. When the length of the cylinders is large with respect to the gap $R_2 - R_1$, it is physically reasonable, for a first study, to replace the rather complicated physically relevant boundary conditions at the ends of the cylinders by periodicity conditions, as this is also suggested by experimental observations. The mathematical problem consists then in solving the Navier–Stokes system (1.1) in the cylindrical domain

$$\Omega = \Sigma \times \mathbb{R}, \quad \Sigma = \{(x,y) \in \mathbb{R}^2 \, ; \, R_1^2 < x^2 + y^2 < R_2^2\},$$

with $f = 0$ and the boundary conditions (1.3), (1.4), (1.14). In these boundary conditions a is now the velocity $R_1\Omega_1$ or $R_2\Omega_2$ tangent to the inner or outer cylinder, respectively, and orthogonal to the axis of rotation, and the flux of the velocity through any section is $D = 0$.

Couette Flow

This problem possesses a basic steady solution $(V^{(0)}, p^{(0)})$, *the Couette flow*, given in cylindrical coordinates (r, θ, z) by

$$V^{(0)} = (0, v_0(r), 0), \quad p^{(0)} = \rho \int \frac{v_0^2}{r} \, dr$$

with

$$v_0(r) = \frac{\Omega_2 R_2^2 - \Omega_1 R_1^2}{R_2^2 - R_1^2} r + \frac{(\Omega_1 - \Omega_2) R_1^2 R_2^2}{R_2^2 - R_1^2} \frac{1}{r}.$$

Notice that this solution is independent of z, the coordinate along the cylinder, and θ, the angle around the axis, and that its streamlines are circles centered on the rotation axis.

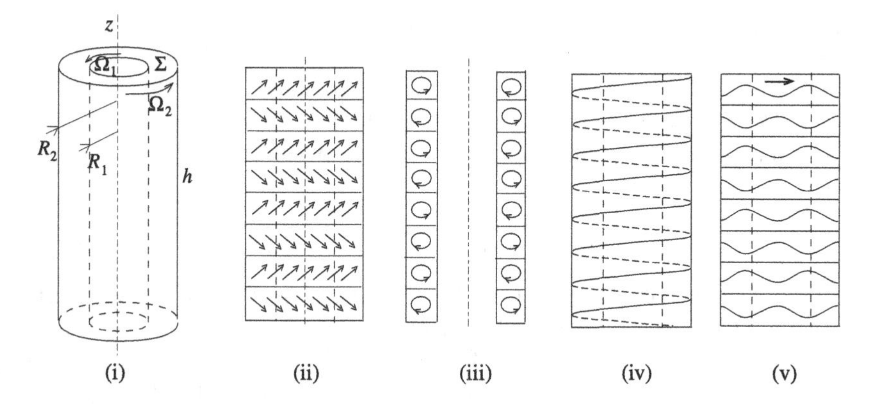

Fig. 1.1 (i) Domain of periodicity for the Couette–Taylor problem. (ii) Side view of the Taylor vortex flow. (iii) Meridian view of the Taylor cells. (iv) Helicoidal waves (traveling in both z and θ directions). (v) Ribbons (standing in z direction, traveling in θ direction).

Symmetries

A fundamental feature of this system consists in its symmetries. When $f = 0$, the Navier–Stokes system (1.1) possesses the Galilean invariance, which is typical to any physical system ruled by Newtonian laws. The result is the symmetries of the system are restricted to the symmetries of the boundary conditions. For the Couette–Taylor problem, the invariance under translations along the z-axis allied with the periodicity conditions, and the invariance under reflections through any plane orthogonal to this axis induce an $O(2)$ symmetry (the same as in the example in Section 2.4.3 of Chapter 2). Notice that gravity plays no role here, since it may be included in the gradient of the pressure. In addition, the system is invariant under rotations around the z-axis that induce a $SO(2)$-symmetry.

In cylindrical coordinates (r, θ, z), we have the following linear representations of these symmetries:

$$\begin{aligned}(\tau_a V)(r, \theta, z) &= V(r, \theta, z + a), \quad a \in \mathbb{R}/h\mathbb{Z},\\ (\mathbf{S}V)(r, \theta, z) &= (V_r(r, \theta, -z), V_\theta(r, \theta, -z), -V_z((r, \theta, -z)),\\ (\mathbf{R}_\phi V)(r, \theta, z) &= V(r, \theta + \phi, z), \quad \phi \in \mathbb{R}/2\pi\mathbb{Z},\end{aligned}$$

which satisfy

$$\tau_a \mathbf{S} = \mathbf{S}\tau_{-a}, \quad \tau_h = \mathbb{I}, \quad \tau_a \tau_b = \tau_{a+b}.$$

Consequently, $(\tau_a, \mathbf{S})$ is an $O(2)$ grouprepresentation, and $\mathbf{R}_\phi$ represents a $SO(2)$ action, which commutes with the $O(2)$ action. We point out that the basic Couette flow $(V^{(0)}, p^{(0)})$ is left invariant by all these symmetries, which are then inherited by the system (1.11).

Instabilities

As usual in any physical problem, we need to choose the scales. Here the length scale is $(R_2 - R_1)$ and the velocity scale is $R_1 \Omega_1$. Three dimensionless parameters appear in the equations of the problem, which we can choose as

$$\Omega_r = \frac{\Omega_2}{\Omega_1}, \quad \eta = \frac{R_1}{R_2}, \quad \mathscr{R} = \frac{R_1 \Omega_1 (R_2 - R_1)}{\nu},$$

where $\mathscr{R}$ is a Reynolds number. Consider the system (1.11) satisfied by perturbations of the basic Couette flow, and more precisely its formulation (1.15) as a first order system. Fixing the parameters Ω_r and η, we take $\mathscr{R}$ as bifurcation parameter, and denote the linear operator $\mathbf{L}$ in (1.15) by $\mathbf{L}_{\mathscr{R}}$. It turns out that the spectrum of $\mathbf{L}_{\mathscr{R}}$ is strictly contained in the left half-complex plane, i.e., the Couette flow is stable, for low values of $\mathscr{R}$, i.e. for small rotation rate of the inner cylinder, or high viscosity. Instabilities are obtained by increasing $\mathscr{R}$ (for instance by increasing the rotation rate of the inner cylinder). This may be interpreted by the fact that for Ω_1 large enough, the excess of centrifugal forces acting on particles close to the inner cylinder, with respect to those near the outer cylinder, becomes dominant if we diminish the viscosity ν. The nature of these instabilities now depends upon the values of Ω_r.

The Case $\Omega_r > 0$ or $\Omega_r < 0$ Close to 0

In this case it has been shown numerically that as $\mathscr{R}$ increases, there is a critical value $\mathscr{R}_c$ for which an eigenvalue of $\mathbf{L}_{\mathscr{R}}$ crosses the imaginary axis, passing through 0 from the left to the right, and all other eigenvalues remain in the left half-complex plane. We are here in the presence of a steady $O(2)$ bifurcation in which 0 is a double eigenvalue with complex conjugated eigenvectors

$$\zeta = e^{ik_c z} \widehat{U}(r), \quad \overline{\zeta} = \mathbf{S}\zeta,$$

where the wave number k_c is such that there is an integer n with

$$k_c h = 2n\pi,$$

and

$$\tau_a \zeta = e^{ik_c a} \zeta \text{ for all } a \in \mathbb{R}.$$

Applying the center manifold Theorems 3.23 and 3.13 in Chapter 2, one finds a two-dimensional center manifold, and the reduced vector field commutes with the restrictions of τ_a and $\mathbf{S}$ on the two-dimensional subspace $\mathscr{E}_0$ spanned by ζ and $\overline{\zeta}$. We point out that $\mathbf{R}_\phi$ acts trivially on $\mathscr{E}_0$, which means that all solutions on the center manifold are invariant under $\mathbf{R}_\phi$. Consequently, for the reduced system we are in the situation described in Section 1.2.4 of Chapter 1 (see also the first part of example in Section 2.4.3, Chapter 2). The reduced dynamics are ruled by the following ordinary differential equation:

$$\frac{dA}{dt} = Ag(|A|^2, \mu), \quad g(|A|^2, \mu) = a\mu + b|A|^2 + h.o.t., \quad a, b \in \mathbb{R}, \tag{1.17}$$

in which A is complex-valued, $\mu = \mathscr{R} - \mathscr{R}_c$, and a, b are real numbers depending upon Ω_r. Equation (1.17) is called the *Landau equation* in the physics literature, as it was first formally derived by Landau [86].

According to the results in [15] the coefficients a and b are such that $a > 0$, $b < 0$ when $\Omega_r > 0$, and b changes sign for a certain small value of $\Omega_r < 0$. We can now apply Theorem 2.18 in Chapter 1 and conclude that for $b < 0$ (resp., for $b > 0$) we have a *supercritical (resp., subcritical) pitchfork bifurcation to a circle of steady stable (resp., unstable) solutions.* In the infinite-dimensional phase space of the full system (1.15), this circle of solutions corresponds to solutions that are shifted along the z direction, i.e., obtained by the action of τ_a. In addition, the action of $\tau_{2\pi/k_c}$ is trivial, which means that the period in z of the bifurcating solutions is $2\pi/k_c = h/n$, and the solutions are invariant under the action of $\mathbf{R}_\phi$. Two of the shifted solutions are also invariant under $\mathbf{S}$, which means that the corresponding flow does not cross the planes $z = k\pi/k_c$, $k \in \mathbb{Z}$, thus forming axisymmetric toroidal cells. This constitutes the *Taylor vortex flow* (see Figure 1.1(ii)–(iii)).

The Case $\Omega_r < 0$, not too close to 0

In this case, numerical results show that the Couette flow first becomes unstable at a critical value $\mathscr{R}_c$ of $\mathscr{R}$, when a pair of complex conjugate eigenvalues of $\mathbf{L}_{\mathscr{R}}$ crosses the imaginary axis, from the left to the right, as $\mathscr{R}$ is increased, and the rest of the spectrum stays in the left half-complex plane. These two eigenvalues are both double, as this case is generic for $O(2)$ equivariant systems, with two eigenvectors of the form

$$\zeta_0 = e^{i(k_c z + m\theta)}\widehat{U}(r), \quad \zeta_1 = e^{i(-k_c z + m\theta)}\mathbf{S}\widehat{U}(r),$$

where $m \neq 0$, and the critical wave number k_c is determined as in the previous case.

Applying the center manifold Theorems 3.23 and 3.13 of Chapter 2, we find a four-dimensional center manifold, and the reduced vector field commutes with the actions of the induced symmetries τ_a, $\mathbf{S}$, and $\mathbf{R}_\phi$, found from

$$\tau_a\zeta_0 = e^{ik_c a}\zeta_0, \quad \tau_a\zeta_1 = e^{-ik_c a}\zeta_1, \quad \mathbf{S}\zeta_0 = \zeta_1, \quad \mathbf{S}\zeta_1 = \zeta_0,$$
$$\mathbf{R}_\phi\zeta_0 = e^{im\phi}\zeta_0, \quad \mathbf{R}_\phi\zeta_1 = e^{im\phi}\zeta_1.$$

We are here in the presence of a Hopf bifurcation with $O(2)$ symmetry, as discussed in Section 3.4.3, but with an additional $SO(2)$ symmetry represented by $\mathbf{R}_\phi$. With the notations from Section 3.4.3, it turns out that the dynamics are ruled by a system in $\mathbb{C}^2$ of the form

$$\frac{dA}{dt} = AP(|A|^2, |B|^2, \mu)$$
$$\frac{dB}{dt} = BP(|B|^2, |A|^2, \mu),$$

where $\mu = \mathscr{R} - \mathscr{R}_c$, and

$$P(|A|^2, |B|^2, \mu) = i\omega + a\mu + b|A|^2 + c|B|^2 + h.o.t.$$

is a smooth function of its arguments, and with no "remainder ρ."

The coefficients a, b, and c are complex, and their explicit values can be found in [15]. The bifurcating solutions corresponding to $A = 0$ or to $B = 0$ travel along and around the z-axis with constant velocities. These are *helicoidal waves*, also called *spirals*, and they are axially periodic just as the Taylor vortex flow (see Figure 1.1(iv)). The bifurcating solutions obtained for $|A| = |B|$ are *standing waves* located in fixed horizontal periodic cells, as they are for the Taylor vortex flow, but with a non-axisymmetric internal structure rotating around the axis with a constant velocity. These solutions are also called *ribbons* (see Figure 1.1(v)). We point out that both types of waves may be observed, depending upon the other parameters (see [15] for the predicted parameter values, and [117] for the corresponding experimental observations).

Further Bifurcations

The next step consists in considering the circle of solutions corresponding to the Taylor vortex flow and to study the resulting bifurcation, which is a symmetry-breaking bifurcation. Here, one may proceed as indicated in Section 2.3.3 of Chapter 2, for systems possessing a continuous symmetry and a one-parameter family of equilibria. Theorem 3.19 in Chapter 2 applies, provided we know the "critical" eigenvalues, in addition to the eigenvalue 0, of the operator obtained by linearizing at one point of the "circle" of Taylor vortex solutions where the solution is invariant under symmetry **S**. It is shown in [15] that when $\mathscr{R}$ passes a new critical value $\mathscr{R}_2$, depending on the parameters Ω_r and η, a Hopf bifurcation occurs. To one purely imaginary eigenvalue corresponds a non-axisymmetric eigenvector, which is either symmetric or antisymmetric, with the same or the double axial periodicity as the Taylor flow, and leading to *twisted vortices*, *wavy vortices*, *wavy inflow boundaries*, or *wavy outflow boundaries*. All these flows are rotating waves around the z-axis, due to the Hopf bifurcation with the $SO(2)$ symmetry broken by the eigenvectors (see also Section 3.3.1), but with various cell structures, the two first having the same axial periodicity as the Taylor vortex flow, the last two having a double period. One can proceed in the same way when starting with spirals or ribbons instead of the Taylor vortex flow [15].

Finally, we point out that these tools can also be used to study imperfect situations such as when cylinders are slightly eccentric, which breaks the $SO(2)$ symmetry, or in the presence of a little flux of fluid downwards, e.g., due to a leak in the apparatus, which breaks the reflection symmetry **S**, or in the presence of a small bump on one cylinder, which breaks the translation invariance (see also the example in Section 2.4.3 of Chapter 2).

5.1.3 Bénard–Rayleigh Convection Problem

Hydrodynamic Problem

Consider a viscous fluid filling the region between two horizontal planes. Each planar boundary may be a rigid plane, or a "free" boundary in the sense explained in Remark 1.1. In addition, we assume that the lower and upper planes are at temperatures T_0 and T_1, respectively, with $T_0 > T_1$ (see Figure 1.2(i)). The difference of temperature between the two planes modifies the fluid density, tending to place the lighter fluid below the heavier one. The gravity then induces, through the Archimedian force, an instability of the "conduction regime" where the fluid is at rest, while the temperature depends linearly on the vertical coordinate z. This instability is prevented up to a certain level by viscosity, so that there is a critical value of the temperature difference, below which nothing happens and above which a "convective regime" appears.

The Navier–Stokes system (1.1) is not sufficient to describe this situation. An additional equation for energy conservation is needed, where the internal energy is proportional to temperature. In the Boussinesq approximation, the dependency of the density ρ in function of the temperature T,

$$\rho = \rho_0 \left(1 - \alpha(T - T_0)\right),$$

where α is the volume expansion coefficient, is taken into account in the momentum equation, only in the external volumic gravity force $-\rho g e_z$, introducing the coupling between (V,p) and T. We refer to [75, Vol. II] for a very complete discussion and bibliography on various geometries and boundary conditions in this problem.

Several different scalings are used in literature. We adopt here the one in [83], which consists in choosing the length, time, velocity, and temperature scales respectively as d, d^2/κ, κ/d, $\nu\kappa/\alpha g d^3$, where d is the distance between the planes, κ is the thermal diffusivity, and ν, α, and g are as above. This leads to the system

$$\begin{aligned} \frac{\partial V}{\partial t} + V \cdot \nabla V + \nabla p &= \mathscr{P}(\theta e_z + \Delta V) \\ \nabla \cdot V &= 0 \\ \frac{\partial \theta}{\partial t} + V \cdot \nabla \theta &= \Delta \theta + \mathscr{R}(V \cdot e_z), \end{aligned} \tag{1.18}$$

replacing the Navier–Stokes system (1.11). Here θ is the deviation of the temperature from the conduction profile, which satisfies the boundary conditions, and $V = (V_1, V_2, V_z)$, p, and θ are functions of (x,t), $x = (X,z)$, with $X = (x_1,x_2) \in \mathbb{R}^2$

the horizontal coordinates and $z \in (0,1)$ the vertical coordinate, e_z being the unitary ascendent vector. There are two dimensionless numbers in this problem: the Prandtl number $\mathscr{P}$ and the Rayleigh number $\mathscr{R}$ defined respectively as

$$\mathscr{P} = \frac{\nu}{\kappa}, \quad \mathscr{R} = \frac{\alpha g d^3 (T_0 - T_1)}{\nu \kappa}.$$

System (1.18) is completed by the boundary conditions

$$V_z = \theta = 0, \quad z = 0, 1,$$

together with either a "rigid surface" condition

$$V_1 = V_2 = 0, \tag{1.19}$$

or a "free surface" condition

$$\frac{\partial V_1}{\partial z} = \frac{\partial V_2}{\partial z} = 0 \tag{1.20}$$

on the planes $z = 0$ or $z = 1$. Notice that here the kinematic viscosity is independent of the temperature T. If this is not the case, some qualitative results change. Also, adding a solute with a certain concentration, satisfying an equation and boundary conditions of the same form as θ, gives richer results [75, Vol. II].

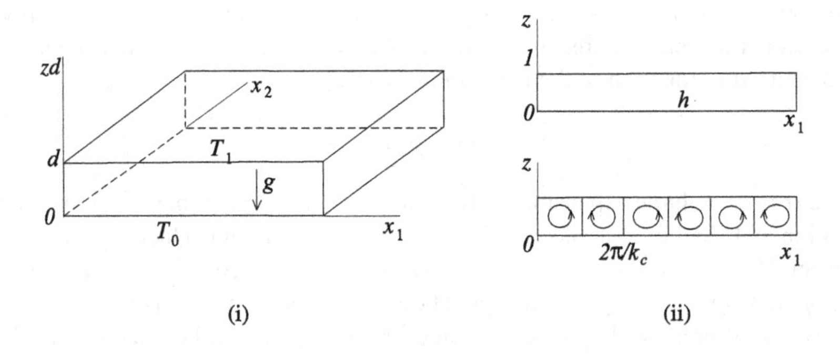

Fig. 1.2 (i) Bénard–Rayleigh problem. (ii) Domain of periodicity for bidimensional convection (above) and convection rolls (below).

Bidimensional Convection

We restrict ourselves first to the case of bidimensional flows, i.e., we assume that $V_2 = 0$, and $V = (V_1, V_z)$, p, and θ are only functions of x_1, z, and t.

Formulation as a First Order System

We set $U = (V,\theta)$, and then the system is of the form (1.15) in the space $\mathscr{X}$ of h-periodic functions in x_1, defined by

$$\mathscr{X} = \left\{ U \in \left(L^2((\mathbb{R}/h\mathbb{Z}) \times (0,1))\right)^3 ; \nabla \cdot V = 0,\ V_z|_{z=0,1} = 0,\ \int_0^1 V_1 dz = 0 \right\}.$$

In the case of rigid boundary conditions (1.19) on both planes $z = 0$ and $z = 1$, the domain of $\mathbf{L}$ is defined by

$$\mathscr{Z}_{(r,r)} = \left\{ U \in \left(H^2((\mathbb{R}/h\mathbb{Z}) \times (0,1))\right)^3 ; \nabla \cdot V = 0, \right.$$
$$\left. V|_{z=0,1} = \theta|_{z=0,1} = 0,\ \int_0^1 V_1 dz = 0 \right\},$$

and similarly we define $\mathscr{Z}_{(r,f)}$, $\mathscr{Z}_{(f,r)}$, and $\mathscr{Z}_{(f,f)}$ by replacing the rigid boundary condition $V_1 = 0$ by the free boundary condition $\partial V_1/\partial z = 0$ on $z = 1$, $z = 0$, and $z = 0, 1$, respectively (see Figure 1.2(ii)). Here we have

$$\mathbf{L}U = (\Pi_0 \mathscr{P}(\Delta V + \theta e_z), \Delta\theta + \mathscr{R}V_z), \quad \mathbf{R}(U) = (-\Pi_0(V \cdot \nabla V), -V \cdot \nabla\theta) \tag{1.21}$$

with $\mathbf{R} : \mathscr{Z} \to \mathscr{Y} = \mathscr{X} \cap \left(H^1((\mathbb{R}/h\mathbb{Z}) \times (0,1))\right)^3$ quadratic and continuous. Here $\mathscr{Z}$ represents one of the spaces $\mathscr{Z}_{(r,r)}$, $\mathscr{Z}_{(r,f)}$, $\mathscr{Z}_{(f,r)}$, and $\mathscr{Z}_{(f,f)}$ above, depending upon the choice of boundary conditions. Notice that the pressure p is not necessarily periodic in x_1, and that the orthogonal projection Π_0 in $\left(L^2((\mathbb{R}/h\mathbb{Z}) \times (0,1))\right)^3$ on the subspace $\mathscr{X}$ eliminates the periodic gradient ∇p, as in Section 5.1.1.

A specific property of $\mathbf{L}$ in this case is that there is a special scalar product in the Hilbert space $\mathscr{X}$, with corresponding norm equivalent to the usual one, such that $\mathbf{L}$ is *self-adjoint*. This scalar product is defined by

$$\langle U^{(1)}, U^{(2)} \rangle = \langle V^{(1)}, V^{(2)} \rangle|_{(L^2((\mathbb{R}/h\mathbb{Z})\times(0,1)))^2} + \frac{\mathscr{P}}{\mathscr{R}} \langle \theta^{(1)}, \theta^{(2)} \rangle|_{L^2((\mathbb{R}/h\mathbb{Z})\times(0,1))}.$$

As a consequence, the spectrum of $\mathbf{L}$ is now located on the real axis. Notice that $\mathbf{L}$ is a relatively compact perturbation of the uncoupled self-adjoint negative operator

$$\mathbf{L}'U = (\Pi_0 \mathscr{P} \Delta V, \Delta\theta),$$

and that it has a compact resolvent, since its domain is compactly embedded in $\mathscr{X}$ (see [76]). The spectrum of $\mathbf{L}$ consists then of isolated semisimple real eigenvalues of finite multiplicities, accumulating at $-\infty$, only. Furthermore, the resolvent estimate (2.9) in Chapter 2 is straightforward, and the estimate (2.10) in Chapter 2 also holds with $\alpha = 3/4$ (see [52]). As for the case considered in Section 5.1.1, the hypotheses required by the center manifold theorem in Chapter 2 are all satisfied.

Symmetries

This problem is invariant under translations parallel to the x_1-axis and under the reflection $x_1 \mapsto -x_1$. Then the system (1.15) possesses an $O(2)$ symmetry group represented by τ_a and $\mathbf{S}$ defined through

$$\begin{aligned}(\tau_a U)(x_1,z) &= U(x_1+a,z), \quad a \in \mathbb{R}/h\mathbb{Z}\\ (\mathbf{S}U)(x_1,z) &= (-V_1(-x_1,z),V_z(-x_1,z),\theta(-x_1,z)),\end{aligned} \tag{1.22}$$

where $\tau_h = \mathbb{I}$, because of the periodicity assumption. In addition, in the cases of "rigid-rigid" and "free-free" boundary conditions, i.e., with $\mathscr{Z}_{(r,r)}$ and $\mathscr{Z}_{(f,f)}$, respectively, there is the additional symmetry with respect to the half-plane $z = 1/2$,

$$(\mathbf{S}_z U)(x_1,z) = (V_1(x_1,1-z),-V_z(x_1,1-z),-\theta(x_1,1-z)). \tag{1.23}$$

Bifurcations

We fix the Prandtl number $\mathscr{P}$ and take the Reynolds number $\mathscr{R}$ as bifurcation parameter. As before, we denote by $\mathbf{L}_{\mathscr{R}}$ the linear operator $\mathbf{L}$ in (1.15). Then upon increasing $\mathscr{R}$ from 0, there is a critical value $\mathscr{R}_c$ for which the largest real eigenvalue of $\mathbf{L}_{\mathscr{R}}$ crosses the imaginary axis from the left to the right [113, 119] (see also [75, Vol. II]). The eigenvalue 0 of $\mathbf{L}_{\mathscr{R}_c}$ is double, as it is generic for $O(2)$ equivariant operators, and the corresponding eigenvectors are of the form

$$\zeta = e^{ik_c x_1}\widehat{U}(z), \quad \overline{\zeta} = \mathbf{S}\zeta,$$

where k_c is a positive critical wavenumber. In the case of "free-free" boundary conditions, the eigenvectors are explicit and k_c is easily obtained. In other cases, the existence of such a positive k_c may be proved analytically [123] (or following the method in [126]); see also [75, Vol. II], but its uniqueness is, so far, only a numerical evidence. Notice that the action of τ_a on the eigenvector ζ is

$$\tau_a\zeta = e^{ik_c a}\zeta,$$

so that we are in the presence of a steady bifurcation with $O(2)$ symmetry.

Applying the center manifold Theorems 3.23 and 3.13 in Chapter 2, we find a two-dimensional center manifold and a reduced system, which commutes with the restrictions of τ_a and $\mathbf{S}$ on the two-dimensional subspace $\mathscr{E}_0$ spanned by ζ and $\overline{\zeta}$. The reduced equation is a Landau equation (1.17), and we find precisely the situation described in Section 1.2.4 of Chapter 1 (see also the first part of the example in Section 2.4.3, Chapter 2). Here $\mu = \mathscr{R} - \mathscr{R}_c$, and $a > 0$, $b < 0$ ([119, 127]; see also [75, Vol. II]). Notice that in the cases of "rigid-rigid" or "free-free" boundary conditions, the reduced system also commutes with the restriction on $\mathscr{E}_0$ of the symmetry $\mathbf{S}_z$. However, the action of this symmetry is $\pm\mathbb{I}$ on $\mathscr{E}_0$, which does not influence the Landau equation, already odd in $(A,\overline{A})$. Applying Theorem 2.18 in

Chapter 1, we find a pitchfork bifurcation of a "circle" of stable steady solutions, obtained by translating with τ_a a symmetric, periodic solution. All these solutions have the period $2\pi/k_c$ and, as in the previous section, appear in cells of size π/k_c, the velocity being tangent to the boundaries of the rectangular cells. These solutions are the *convection rolls* (see Figure 1.2(iii)).

Tridimensional Convection

Consider now the three-dimensional case, in which V_2 is not identically 0, and V, p, and θ are functions of X, z, and t, $X = (x_1, x_2)$. Here, we assume the biperiodicity condition (1.6), where the lattice of periods Γ is generated by two independent horizontal vectors $\{e_1, e_2\}$, and the dual lattice of wave vectors is generated by the two vectors $\{k_1, k_2\}$ defined by (1.8). It turns out that in this case the critical wavenumber found in the bidimensional case, is now the radius of a critical circle in the Fourier plane. It was shown in [78] that the only possible forms of periodic patterns are rolls, hexagons, regular triangles, and rectangles (see also [30]). Since experimental evidence mostly show convection in rolls and convection in hexagonal cells, we choose a lattice compatible with both patterns, as initiated in [109].

Formulation as a First Order System

We choose

$$e_1 = h\left(\frac{\sqrt{3}}{2}, \frac{1}{2}\right), \quad e_2 = h(0,1), \quad k_1 = k_c(1,0), \quad k_2 = k_c\left(-\frac{1}{2}, \frac{\sqrt{3}}{2}\right),$$

where h is determined by the critical wavelength k_c,

$$hk_c = \frac{4\pi}{\sqrt{3}}.$$

It is not difficult to check that this lattice is invariant under rotations of angle $\pi/3$ (see Figure 1.3(i)).

According to the flux conditions (1.7), we choose the Hilbert spaces

$$\mathscr{X} = \Big\{ U \in \left(L^2((\mathbb{R}^2/\Gamma) \times (0,1))\right)^4 ; \nabla \cdot V = 0, \; V_z|_{z=0,1} = 0,$$
$$\int_{\Sigma_1} V \cdot k_2 dS = \int_{\Sigma_2} V \cdot k_1 dS = 0 \Big\},$$

and in the case of "rigid-rigid" boundary conditions

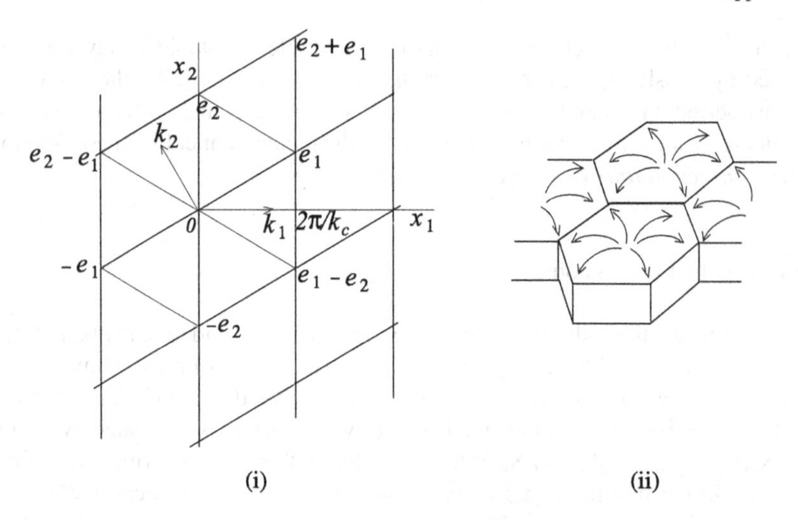

Fig. 1.3 (i) Lattice Γ in the X-plane, for 3-D convection. (ii) Flow in a hexagonal cell.

$$\mathscr{X}_{(r,r)} = \left\{ U \in \left(H^2((\mathbb{R}^2/\Gamma) \times (0,1))\right)^4 ; \nabla \cdot V = 0, \, V|_{z=0,1} = \theta|_{z=0,1} = 0, \right.$$

$$\left. \int_{\Sigma_1} V \cdot k_2 dS = \int_{\Sigma_2} V \cdot k_1 dS = 0 \right\},$$

and similarly $\mathscr{X}_{(r,f)}$, $\mathscr{X}_{(f,r)}$, and $\mathscr{X}_{(f,f)}$, by replacing the rigid boundary conditions $V_1 = V_2 = 0$ by the free boundary conditions $\partial V_1/\partial z = \partial V_2/\partial z = 0$ on $z = 1$, $z = 0$, and $z = 0, 1$, respectively. We set $U = (V, \theta)$, just as in the two-dimensional case, and then the system is of the form (1.15), with $\mathbf{L}$ and $\mathbf{R}$ defined as in (1.21). The linear operator $\mathbf{L}$ and the quadratic map $\mathbf{R}$ have the same properties as in the two-dimensional case.

Symmetries

This problem is invariant under horizontal translations, represented by the operators τ_a when replacing $x_1 + a$ by $X + a$ for any $a \in \mathbb{R}^2/\Gamma$, and invariant under the mirror symmetry $\mathbf{S}$ defined as in (1.22). In addition, it is invariant under the rotation

$$(\mathbf{R}_{2\pi/3} U)(X, z) = \left(R_{2\pi/3}(V(R_{-2\pi/3}X, z)), \theta(R_{-2\pi/3}X, z)\right), \tag{1.24}$$

where $R_{2\pi/3}$ is the horizontal rotation, in the X-plane, of angle $2\pi/3$. The group generated by $\mathbf{S}$ and $\mathbf{R}_{2\pi/3}$ is denoted by D_6, consisting of rotations on a circle of angle $\pi/3$ together with the symmetries through a diameter. In the cases of "rigid-rigid" and "free-free" boundary conditions, we still have the symmetry $\mathbf{S}_z$, defined by (1.23) with x_1 replaced by X.

Bifurcations

We fix the Prandtl number $\mathscr{P}$ and take the Reynolds number $\mathscr{R}$ as bifurcation parameter. As before, we denote by $\mathbf{L}_{\mathscr{R}}$ the linear operator $\mathbf{L}$ in (1.15). Upon increasing $\mathscr{R}$, there is a critical value $\mathscr{R}_c$ for which the largest real eigenvalue of $\mathbf{L}_{\mathscr{R}}$ crosses the imaginary axis from the left to the right, which is now of multiplicity six. The associated eigenvectors are now of the form

$$\zeta_j = e^{ik_j \cdot X}\widehat{U}_j(z), \quad j = 1, \dots, 6,$$

and satisfy

$$\zeta_2 = \mathbf{R}_{2\pi/3}\zeta_1, \quad \zeta_3 = \mathbf{R}_{-2\pi/3}\zeta_1, \quad \zeta_{j+3} = \mathbf{S}\zeta_j = \overline{\zeta_j}, \quad j = 1,2,3,$$

where

$$k_3 = -(k_1 + k_2), \quad k_{j+3} = -k_j, \quad j = 1,2,3.$$

Furthermore

$$\tau_a\zeta_j = e^{ik_j \cdot a}\zeta_j, \quad e^{ik_3 \cdot a} = e^{-i(k_1+k_2)\cdot a},$$

and the action of the symmetry $\mathbf{S}_z$ is either the identity $\mathbb{I}$ or $-\mathbb{I}$, when it is relevant.

Applying the center manifold Theorems 3.23 and 3.13 in Chapter 2, we find a six-dimensional center manifold. For $U_0 \in \mathscr{E}_0$, the eigenspace associated to the eigenvalue 0 of $\mathbf{L}_{\mathscr{R}_c}$, we set

$$U_0 = A\zeta_1 + B\zeta_2 + C\zeta_3 + \overline{A\zeta_1} + \overline{B\zeta_2} + \overline{C\zeta_3}, \tag{1.25}$$

and then we have the induced symmetries

$$\tau_a(A,B,C) = (Ae^{ik_1 \cdot a}, Be^{ik_2 \cdot a}, Ce^{ik_3 \cdot a}) \text{ for all } a \in \mathbb{R}^2/\Gamma,$$
$$\mathbf{S}(A,B,C) = (\overline{A}, \overline{B}, \overline{C}), \quad \mathbf{R}_{2\pi/3}(A,B,C) = (C,A,B),$$

and when $\mathbf{S}_z$ is relevant,

$$\mathbf{S}_z(A,B,C) = \pm(A,B,C).$$

The general form of vector fields commuting with these symmetries is given in [30, Chap. XIII]. When the symmetry $\mathbf{S}_z$ is irrelevant, or when it is the identity on $\mathscr{E}_0$, it is sufficient to consider the six-dimensional system truncated at order 3, of the form

$$\begin{aligned}
\frac{dA}{dt} &= a\mu A + c\overline{BC} + bA|A|^2 + dA(|B|^2 + |C|^2)\\
\frac{dB}{dt} &= a\mu B + c\overline{CA} + bB|B|^2 + dB(|C|^2 + |A|^2)\\
\frac{dC}{dt} &= a\mu C + c\overline{AB} + bC|C|^2 + dC(|A|^2 + |B|^2).
\end{aligned} \tag{1.26}$$

Here $\mu = \mathscr{R} - \mathscr{R}_c$, $a > 0$, and the other coefficients are all real. The coefficient b is the same as in the two-dimensional case, hence we have $b < 0$. In general the presence of quadratic terms changes drastically the stability of the steady solutions of (1.26) (see [30, Chap. XIII]). However in the present case, *a specific property of the Navier–Stokes equation implies that* $c = 0$. This comes from the fact that for any U in the domain of $\mathbf{L}$, we have

$$\langle \mathbf{R}(U), U \rangle = 0,$$

where $\langle \cdot, \cdot \rangle$ is the usual scalar product in $(L^2)^4$, and this scalar product arises in the computation of c, with $U = U_0$ given by (1.25).

When $B = C = 0$ we recover the Landau equation (1.17) for A, which gives the circle of steady solutions

$$a\mu + b|A|^2 = 0, \quad B = C = 0,$$

corresponding to the steady convection rolls found in the two-dimensional case. In addition, we have here the solutions obtained through the actions of $\mathbf{R}_{2\pi/3}$ and $\mathbf{S}$, which correspond to convection rolls obtained by $\pi/3$-rotations of the two-dimensional rolls above, so altogether we have *three "circles" of rolls*. In contrast to the two-dimensional case, in which these rolls are stable, here they may also be unstable. Indeed, since we have a "circle" of bifurcating solutions, one eigenvalue of the linearized operator is 0, and the other eigenvalues are now $2b|A|^2$, the same as in the two-dimensional case, and a quadruple eigenvalue $(d-b)|A|^2$. Consequently, the condition for stability of these rolls is

$$d < b < 0.$$

Another class of steady solutions of the system (1.26), with $c = 0$, is

$$A = re^{i\theta_1}, \quad B = re^{i\theta_2}, \quad C = re^{i\theta_3},$$

where $r > 0$ satisfies

$$a\mu + (b + 2d)r^2 = 0,$$

and the phases θ_j are arbitrary. For $\theta_j = 0$, this solution is invariant under the actions of $\mathbf{R}_{2\pi/3}$ and $\mathbf{S}$, and corresponds to *hexagonal convection cells* [30, Chap. XIII] (see Figure 1.3(ii)). It should be noticed by the same argument as in the two-dimensional convection, by using the periodicity and the symmetry $\mathbf{S}$, that the velocity field is tangent to the planes $x_1 = 2\pi n/k_c$ for any $n \in \mathbb{Z}$. Hence, by the D_6 rotational invariance, the velocity field is tangent to all the vertical planes deduced from this family, by rotations of angles $\pi/3$ and $2\pi/3$. This means that the fluid particles are confined in vertical triangular prisms, and a basic hexagonal prism for the pattern is formed with six of these triangular prisms. The linearized operator at these hexagonal convection cells has a triple eigenvalue 0, a simple eigenvalue $2(b+2d)r^2$, and a double eigenvalue $2(b-d)r^2$. This latter eigenvalue implies that the hexagonal convection cells and the convection rolls cannot be both stable. In the case of "rigid-rigid" boundary conditions it is shown in [128] that $b+2d<0$. Actually, the result in [128] shows that hexagonal cells are stable under perturbations with hexagonal symmetry, in which case only the simple eigenvalue $2(b+2d)r^2$ is present. We also point out that if $c \neq 0$ in system (1.26), then the phases of the steady solutions above lose one degree of freedom, and the bifurcation is two-sided. In particular, the hexagonal cells are then unstable [109, 30], but this might only apply to a different physical situation, since here $c=0$.

In the absence of the symmetry $\mathbf{S}_z$ we need to include the fourth order terms in (1.26), in order to avoid the occurrence of a three-parameter family of hexagonal cells: Only two arbitrary phases are relevant because of the action of τ_a, and this leads to a degenerescence shown by the triple 0 eigenvalue. Adding fourth order terms (see [30] for their structure) allows us to fix $\theta_1+\theta_2+\theta_3 \in \{0,\pi\}$, and to obtain another, in general nonzero, simple eigenvalue decreasing by one the multiplicity of the 0 eigenvalue, for the linearized operator.

It appears that the symmetry $\mathbf{S}_z$ acts as $-\mathbb{I}$ on $\mathscr{E}_0$ in the case of the "free-free" boundary conditions, because of a factor $\sin(\pi z)$ in the components V_z and θ, and of a factor $\cos(\pi z)$ in the components V_1 and V_2 of $\widehat{U}_j(z)$, in the formula of the eigenvector ζ_j. It is a priori not automatic, but it is shown numerically that it is also the case for "rigid-rigid" boundary conditions, since for $\mathscr{R}=\mathscr{R}_c$ the components V_z and θ in $\widehat{U}_j(z)$ are invariant under the symmetry $z \mapsto 1-z$ (see [14]). With such a symmetry, the vector field in (1.26) is odd in $(A,B,C,\overline{A},\overline{B},\overline{C})$, so that there are no terms of even orders. Consequently, one has to consider the fifth order terms in order to solve the degenerescence and find all steady solutions. For further details we refer to [30, Chap. XIII], where the problem is treated using the Lyapunov–Schmidt method, but the results can be adapted to the present approach. It is shown that there are four types of steady solutions: *rolls, hexagons, regular triangles,* and *patchwork quilts*, which all may be stable, depending on the coefficients, but not simultaneously. This confirms the prediction in [78], though only the first two types of solutions are usually observed.

Tridimensional Convection in an Elongated Cylindrical Domain

Finally, we briefly discuss the case of a long horizontal cylindrical container, with rectangular section in the (x_2,z)-plane, and small sides compared to the length of the cylinder along the x_1-axis. Physically, to satisfy the a priori periodicity in x_1 which we impose to the solutions, it might be convenient to take a thin ring-shaped container (a torus) having a radius large with respect to the sides of the rectangular meridian section. This problem also possesses an $O(2)$ symmetry, and it turns out to be similar to the case of two-dimensional convection [75, Vol. II]. The same approach as above can be used, showing the existence of a "circle" of *stable convection rolls*, bifurcating for $\mathscr{R}>\mathscr{R}_c$, which are periodic in x_1, the cells being parallel to the x_2-axis.

Second Bifurcation

We are interested here in the next bifurcation, when $\mathscr{R}$ crosses a second critical value $\mathscr{R}_2$, at which the stable convection rolls for $\mathscr{R} > \mathscr{R}_c$ become unstable.

The "circle" of convection rolls is given by $\tau_a U_*$, $a \in \mathbb{R}$, where U_* is a symmetric solution, $\mathbf{S}U_* = U_*$. Notice that there are two such symmetric solutions on the "circle," and that all these solutions are of class $\mathscr{C}^\infty$. The generator of the group $(\tau_a)_{a\in\mathbb{R}}$ is the derivative $\partial_{x_1} \in \mathscr{L}(\mathscr{X}, \mathscr{Y})$, and then $\partial_{x_1} U_*$, the Goldstone mode, satisfies

$$\partial_{x_1} U_* \in \mathscr{Z}, \quad (\mathbf{L} + D_U \mathbf{R}(U_*))(\partial_{x_1} U_*) = 0, \quad \mathbf{S}(\partial_{x_1} U_*) = -\partial_{x_1} U_*.$$

In particular, this shows that the operator $\mathbf{L} + D_U \mathbf{R}(U_*)$ has an eigenvalue 0 with eigenvector $\partial_{x_1} U_*$. It turns out, that experimental evidence suggests that this eigenvalue is actually algebraically double and geometrically simple when $\mathscr{R} = \mathscr{R}_2$. Indeed, for $\mathscr{R}$ close to $\mathscr{R}_2$ there are bifurcating solutions which are slow traveling waves, and, as we shall see below, correspond to the situation in which there is a generalized antisymmetric eigenvector ξ_0, such that

$$(\mathbf{L} + D_u \mathbf{R}(U_*))\xi_0 = \partial_{x_1} U_*, \quad \mathbf{S}\xi_0 = -\xi_0,$$

(see also [15, p. 102], for an analogue for the Couette–Taylor problem).

Following the method of construction of center manifolds near a line of equilibria in Section 2.3.3 of Chapter 2, we consider the new coordinates (α, v) defined through

$$U = \tau_\alpha (U_* + v), \quad \langle v, \partial_{x_1} U_* \rangle = 0,$$

where $\langle \cdot, \cdot \rangle$ is the scalar product in $(L^2)^4$. Then the linear operator $\mathbf{L}'$ defined in (3.18) in Chapter 2, acting on v, which commutes with $\mathbf{S}$ due to the choice of U_*, has a simple eigenvalue crossing the imaginary axis through 0, when $\mathscr{R}$ crosses $\mathscr{R}_2$. Applying the center manifold Theorems 3.19 and 3.13 in Chapter 2, we conclude that a *pitchfork bifurcation* occurs in the equation for v when $\mathscr{R} = \mathscr{R}_2$ (see also the general study of the ten possible solutions generically bifurcating from a one-dimensional periodic pattern in [19]). Since $\alpha(t)$ has a small constant derivative, the bifurcating solutions are *traveling waves with speeds close to* 0, which arise in pairs exchanged by the symmetry $\mathbf{S}$, i.e., traveling in opposite directions. This type of flow is indeed observed in experiments [8].

5.2 Existence of Traveling Waves

Traveling waves are particular solutions of partial differential equations in domains with at least one unbounded coordinate, which describe propagation phenomena at constant speeds. We present three different examples of traveling-wave problems for which the existence question is treated with the help of the methods described in this book. The analysis relies upon a formulation of the existence problem as a first order system of the form (2.1) in Chapter 2, in which the timelike variable t is an unbounded coordinate of the domain on which the PDE is posed. This idea goes back to K. Kirchgässner [80], and this type of approach is often referred to as "spatial dynamics."

5.2.1 Gravity-Capillary Water-Waves

The Hydrodynamic Problem

The classical gravity-capillary water-wave problem concerns the three-dimensional irrotational flow of a perfect fluid of constant density subject to the forces of gravity and surface tension. The fluid motion is governed by the Euler equations in a domain bounded below by a rigid horizontal bottom and above by a free surface.

Denote by (x,y,z) the usual Cartesian coordinates. We assume that the fluid occupies the domain

$$D_\eta = \{(x,y,z)\ ;\ x,z \in \mathbb{R},\ y \in (0, h+\eta(x,z,t))\},$$

where $\eta > -h$ is a function of the unbounded horizontal spatial coordinates x, z and of time t, and h represents the depth of the fluid in its undisturbed state. Consider the Eulerian velocity potential ϕ. The mathematical problem consists in solving Laplace's equation

$$\phi_{xx} + \phi_{yy} + \phi_{zz} = 0 \qquad \text{in } D_\eta, \tag{2.1}$$

with kinematic boundary conditions

$$\phi_y = 0, \qquad \text{on } y = 0 \tag{2.2}$$

$$\phi_y = \eta_t + \eta_x\phi_x + \eta_z\phi_z, \qquad \text{on } y = h+\eta \tag{2.3}$$

showing that the water cannot permeate the rigid bottom at $y = 0$ or the free surface at $y = h+\eta(x,z,t)$, and the dynamic boundary condition

$$\begin{aligned} \phi_t = {} & -\frac{1}{2}(\phi_x^2 + \phi_y^2 + \phi_z^2) - g\eta \\ & + \frac{T}{\rho}\left(\frac{\eta_x}{\sqrt{1+\eta_x^2+\eta_z^2}}\right)_x + \frac{T}{\rho}\left(\frac{\eta_z}{\sqrt{1+\eta_x^2+\eta_z^2}}\right)_z + B \quad \text{on } Y = h+\eta, \end{aligned} \tag{2.4}$$

at the free surface. Here g is the acceleration due to gravity, T is the coefficient of surface tension, ρ the density of the fluid, and B is a constant called the Bernoulli constant.

Traveling Waves

Traveling waves are particular solutions (ϕ, η) of the problem (2.1)–(2.4) of the form $\eta = \eta(x+ct, z)$, $\phi = \phi(x+ct, y, z)$, i.e., waves that are uniformly translating in the horizontal direction x with speed $-c$. We point out that there is no loss of generality in choosing x as the direction of propagation since the system (2.1)–(2.4) is invariant under rotations in the (x,z)-plane. Substituting this form of η, ϕ into (2.1)–(2.4), and introducing the dimensionless variables

$$(x', y', z') = \frac{1}{h}(x, y, z), \qquad \eta'(x', z') = \frac{1}{h}\eta(x, z), \qquad \phi'(x', y', z') = \frac{1}{ch}\phi(x, y, z),$$

we find the equations

$$\phi_{xx} + \phi_{yy} + \phi_{zz} = 0, \qquad 0 < y < 1 + \eta \tag{2.5}$$

$$\phi_y = 0, \qquad y = 0 \tag{2.6}$$

$$\phi_y = \eta_x + \eta_x \phi_x + \eta_z \phi_z, \qquad y = 1 + \eta \tag{2.7}$$

$$\begin{aligned} &\phi_x + \frac{1}{2}(\phi_x^2 + \phi_y^2 + \phi_z^2) + \lambda \eta \\ &\quad - b\left(\frac{\eta_x}{\sqrt{1+\eta_x^2+\eta_z^2}}\right)_x - b\left(\frac{\eta_z}{\sqrt{1+\eta_x^2+\eta_z^2}}\right)_z = 0, \qquad y = 1 + \eta, \end{aligned} \tag{2.8}$$

in which the primes have been dropped, x is a shorthand for the variable $x + ct$, and the Bernoulli constant has been set to zero. The dimensionless numbers

$$\lambda = \frac{gh}{c^2}, \qquad b = \frac{T}{\rho h c^2}$$

are respectively the inverse square of the Froude number and the Weber number.

Symmetries

The time-dependent water-wave equations (2.1)–(2.4) possess several symmetries. Of importance in the present approach to traveling waves are the following continuous symmetries: the invariance under translations in ϕ,

$$\phi \mapsto \phi + A, \quad A \in \mathbb{R}; \tag{2.9}$$

the horizontal Galilean invariance

$$x \mapsto x - Ct, \quad \phi \mapsto \phi + Cx, \quad \eta \mapsto \eta - \frac{1}{2g}C^2, \quad C \in \mathbb{R}; \tag{2.10}$$

and the invariance under rotations in the (x,z)-plane. As a remnant of this rotation invariance, the problem (2.5)–(2.8) possesses the two discrete symmetries

$$\begin{aligned} x \mapsto -x, \quad z \mapsto z, \quad \eta \mapsto \eta, \quad \phi \mapsto -\phi; \\ x \mapsto x, \quad z \mapsto -z, \quad \eta \mapsto \eta, \quad \phi \mapsto \phi. \end{aligned} \tag{2.11}$$

We point out that the equations (2.5)–(2.8) also possess a Hamiltonian structure, which has been used in many different studies of traveling water waves (see the review paper [33] and the references therein).

Two-Dimensional Waves

The question of existence of two-dimensional traveling water waves, i.e., solutions of (2.5)–(2.8) that are independent of z, has a long history, and has been studied by many different authors in many different works. We present here an approach to this question, which relies upon the methods described in this book. For further details we refer to the review paper [23] and the references therein. We point out that this approach also works for pure gravity waves when $b = 0$. We refer for instance to [81] for an analysis of this case.

Spatial Dynamics

We restrict ourselves to two-dimensional waves, i.e., solutions of the system (2.5)–(2.8) that are independent of z. A very convenient way of formulating the system (2.5)–(2.8) as a first order system of the form (2.1) in Chapter 2, in which x is the timelike variable, is with the help of a variables and coordinates transformation due to Levi–Civita [90] (see also [81] for a different formulation).

Consider the complex velocity potential defined through

$$w(x+iy) = \varkappa + i\zeta, \quad \varkappa = x + \phi(x,y), \quad \zeta = \psi(x,y),$$

where here $x+\phi$ is the velocity potential in the moving frame, in contrast to ϕ used above which is the velocity potential in the original reference frame, and ψ is the stream function. Notice that in the moving frame, the two-dimensional velocity vector (u,v) satisfies the following equations, due to incompressibility and irrotationality of the flow,

$$u = 1 + \phi_x = \psi_y, \quad v = \phi_y = -\psi_x,$$

showing that $x+\phi$ and ψ satisfy the Cauchy–Riemann equations.

We define the new variables (α,β) by

$$w'(x+iy) = u - iv = e^{-i(\alpha+i\beta)},$$

where $\alpha = \arg(v/u)$ is the slope of the streamline and $\beta = (1/2)\ln(u^2+v^2)$, and introduce the change of coordinates defined by

$$dx + idy = e^{i(\alpha+i\beta)}(d\varkappa + id\zeta).$$

Then the bottom of the domain $y = 0$ corresponds to $\zeta = 0$ and the free surface $y = 1 + \eta(x)$ corresponds to $\zeta = 1$, because the Bernoulli constant B has been set to 0. Furthermore, $(\alpha, \beta) = 0$ corresponds to the rest state $(\phi, \eta) = 0$ in (2.5)–(2.8). With these new variables we regard $\alpha + i\beta$ as an analytic function of $\varkappa + i\zeta$.

A key property of this choice of variables is that we still have the Cauchy–Riemann equations for (α, β):

$$\alpha_\varkappa = \beta_\zeta, \quad \alpha_\zeta = -\beta_\varkappa,$$

but now for $(\varkappa, \zeta)$ in a fixed strip $(\varkappa, \zeta) \in \mathbb{R} \times (0,1)$. Then the Cauchy–Riemann equations above replace the equation (2.5), and the boundary conditions (2.6)–(2.8) become

$$\begin{aligned}
&\alpha = 0, \quad \zeta = 0,\\
&\widetilde{\eta}_\varkappa = e^{-\beta}\sin\alpha, \quad \zeta = 1,\\
&\frac{1}{2}(e^{2\beta}-1) + \lambda\widetilde{\eta} - be^{\beta}\alpha_\varkappa = 0, \quad \zeta = 1,
\end{aligned}$$

where $\widetilde{\eta}(\varkappa) = \eta(x)$. Notice that we recover the shape of the free surface from the expression

$$\widetilde{\eta}(\varkappa) = \int_0^1 (e^{-\beta}\cos\alpha - 1)\,d\zeta.$$

We now set

$$U(\varkappa,\zeta) = (\alpha_0(\varkappa), \alpha(\varkappa,\zeta), \beta(\varkappa,\zeta)), \quad \alpha_0(\varkappa) = \alpha(\varkappa, 1),$$

and then the system (2.5)–(2.8) is transformed into a system of the form

$$\frac{dU}{d\varkappa} = \mathbf{F}(U, \lambda, b), \tag{2.12}$$

with

$$\mathbf{F}(U,\lambda,b) = \begin{pmatrix} \frac{1}{b}\sinh\beta_0 + \frac{\lambda}{b}e^{-\beta_0}\int_0^1(e^{-\beta}\cos\alpha - 1)d\zeta \\ \frac{\partial\beta}{\partial\zeta} \\ -\frac{\partial\alpha}{\partial\zeta} \end{pmatrix}, \quad \beta_0(\varkappa) = \beta(\varkappa, 1).$$

We also consider the spaces

$$\mathscr{X} = \mathbb{R} \times \left(L^2(0,1)\right)^2, \quad \mathscr{Z} = \{U \in \mathbb{R} \times \left(H^1(0,1)\right)^2 \; ; \; \alpha(0) = 0, \; \alpha_0 = \alpha(1)\},$$

so that $\mathbf{F}(\cdot,\lambda,b) : \mathscr{Z} \to \mathscr{X}$ is a smooth map.

Notice that the system (2.12) is reversible, with the reversibility symmetry $\mathbf{S}$ defined by

$$\mathbf{S}(\alpha_0,\alpha,\beta) = (-\alpha_0,-\alpha,\beta).$$

We also point out that we can write

$$\mathbf{F}(U,\lambda,b) = \mathbf{L}_{\lambda,b}U + \mathbf{R}(U,\lambda,b), \quad \mathbf{L}_{\lambda,b} = D_U\mathbf{F}(0,\lambda,b),$$

with $\mathbf{L}_{\lambda,b}$ a closed linear operator in $\mathscr{X}$ with domain $\mathscr{Z}$, and the nonlinearity $\mathbf{R}(\cdot,\lambda,b)$ having the last two components identically 0, so that $\mathbf{R}(\cdot,\lambda,b)$ is a smooth map from $\mathscr{Z}$ into $\mathbb{R} \times \{0\} \subset \mathscr{X}$.

Bifurcation Analysis

For the system (2.12) the bifurcations are determined by the purely imaginary spectrum of the linear operator

$$\mathbf{L}_{\lambda,b} = D_U\mathbf{F}(0,\lambda,b), \quad \mathbf{L}_{\lambda,b}U = \begin{pmatrix} \frac{1}{b}\beta_0 - \frac{\lambda}{b}\int_0^1 \beta d\zeta \\ \frac{\partial\beta}{\partial\zeta} \\ -\frac{\partial\alpha}{\partial\zeta} \end{pmatrix}.$$

Since $\mathscr{Z}$ is compactly embedded in $\mathscr{X}$, the operator $\mathbf{L}_{\lambda,b}$ has a compact resolvent, so that its spectrum consists of isolated eigenvalues with finite algebraic multiplicities, only accumulating at infinity. It is straightforward to check that a nonzero purely imaginary number $i\kappa \neq 0$ is an eigenvalue if and only if it satisfies the dispersion relation

$$(\lambda + b\kappa^2)\tanh(\kappa) - \kappa = 0$$

and that 0 is an eigenvalue only when $\lambda = 1$. The resulting bifurcation diagram, and the location of the four eigenvalues closest to the imaginary axis is shown in Figure 2.1. We conclude that there are three bifurcation curves: the half-line $\{\lambda = 1,\ b > 1/3\}$, the segment $\{\lambda = 1,\ 0 < b < 1/3\}$, and the curve C_0, along which we find a 0^{2+}, $0^{2+}(i\omega)$, and $(i\omega)^2$ bifurcation, respectively. The point $(\lambda,b) = (1,1/3)$ is a codimension two bifurcation point, where a 0^{4+} bifurcation occurs.

The center manifold Theorems 3.3 and 3.15 in Chapter 2 together with the results on reversible bifurcations in Chapter 4: Theorem 1.8 (see also Remark 1.9(ii)), Theorem 3.10 (see also Exercise 3.12), and Theorem 3.21(iii), can be applied to study each of these bifurcations. Besides periodic and quasi-periodic orbits, for the reduced systems one finds homoclinic orbits in the case of the 0^{2+} and $(i\omega)^2$ bifurcations, and homoclinic orbits to periodic orbits in the case of the $0^{2+}(i\omega)$ bifurcation. The solutions corresponding to all these homoclinics are *solitary waves*, and are of particular interest in the water-wave problem (see Figure 2.2).

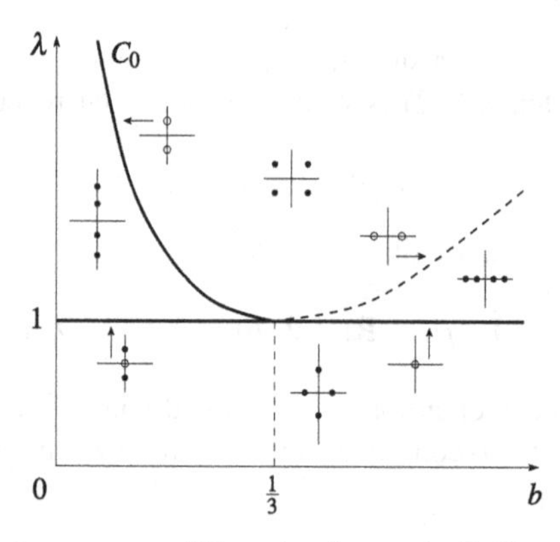

Fig. 2.1 Two-dimensional water waves: bifurcation diagram in (b,λ)-parameter plane, and behavior of the critical eigenvalues of $\mathbf{L}_{\lambda,b}$. The solid lines represent the bifurcation curves, whereas the solid and hollow dots represent simple and double eigenvalues, respectively.

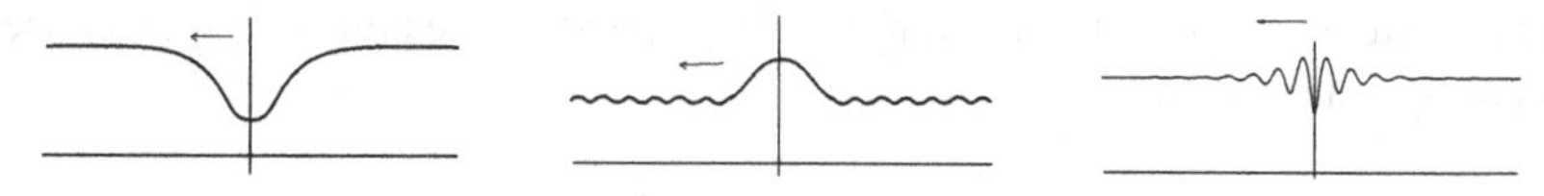

Fig. 2.2 Two-dimensional solitary waves found in the 0^{2+}, $0^{2+}(i\omega)$, and $(i\omega)^2$ bifurcations (from left to the right). The arrows indicate the direction of propagation.

Three-Dimensional Waves

In contrast to the case of two-dimensional traveling waves, there are relatively few mathematical results on the existence of three-dimensional traveling waves, which are all quite recent. Most of these results rely upon a formulation of the equations (2.5)–(2.8) as a first order system of the form (2.1) in Chapter 2, and a center manifold reduction. A major difficulty, which seems to be specific to the three-dimensional problem, is that the formulation of equations (2.5)–(2.8) as a first order system is not explicit, due to the nonlinear boundary conditions at the free surface. This difficulty has been first overcome in [35]. In addition, in the three-dimensional problem the domain has infinitely many unbounded directions, any horizontal direction being unbounded, whereas there is only one unbounded direction in the two-dimensional problem. Then any of these unbounded directions can be taken as a timelike variable in the formulation of the equations as a first order system.

Another particularity of the three-dimensional problem is that, to be able to apply the center manifold reduction, one may address only particular wave-profiles in a direction transverse to the direction which is taken as timelike variable. A natural choice is to address only waves that are periodic in such a direction, but one could also impose some boundary conditions, as Dirichlet or Neumann. Without this restriction in formulation (2.1) in Chapter 2, as a first order system the linear operator $\mathbf{L}$ has a purely continuous spectrum, with no gap around the imaginary axis, so that the center manifold theorem does not apply. Furthermore, even with this restriction, in the particular case of gravity waves, i.e. in the absence of surface tension when $b = 0$, it turns out that $\mathbf{L}$ has infinitely many imaginary eigenvalues, so that the center manifold theorem does not apply either. It is therefore essential when using this approach to the three-dimensional problem to assume that $b \neq 0$.

Here, we restrict ourselves to the simpler, particular case in which the timelike variable in the formulation of the equations as a first order system of the form (2.1) in Chapter 2 is the direction x of propagation, and when the waves are periodic in the perpendicular direction z. This is the case considered in [35]. While the formulation in [35] relies upon the Hamiltonian structure of the equations, we present below a different formulation, which does not use this Hamiltonian structure (see also [41] for a similar formulation).

Spatial Dynamics

Choosing the direction of propagation x as timelike variable, and restricting to waves which are periodic of period $L = 2\pi/\ell$ in z, our purpose is to write equations (2.5)–(2.8) as a first order system of the form

$$\frac{dU}{dx} = \mathbf{F}(U, \lambda, b, \ell). \tag{2.13}$$

Notice that in contrast to the two-dimensional problem we have now the additional parameter ℓ, which is the wavenumber in z.

We introduce the new variables

$$u = \phi_x, \quad \xi = \frac{\eta_x}{\left(1+\eta_x^2+\eta_z^2\right)^{1/2}},$$

and rewrite equations (2.5)–(2.8) as

$$\begin{aligned}
&\phi_x = u, \qquad 0 < y < 1+\eta \\
&u_x = -\phi_{yy} - \phi_{zz}, \qquad 0 < y < 1+\eta \\
&\eta_x = \xi \left(\frac{1+\eta_z^2}{1-\xi^2}\right)^{1/2} \\
&\xi_x = \frac{1}{b}u + \frac{\lambda}{b}\eta + \frac{1}{2b}\left(u^2+\phi_y^2+\phi_z^2\right) - \left(\eta_z\left(\frac{1-\xi^2}{1+\eta_z^2}\right)^{1/2}\right)_z, \qquad y = 1+\eta,
\end{aligned}$$

to which we add the boundary conditions

$$\begin{aligned}
&\phi_y = 0, \qquad y = 0, \\
&\phi_y = \xi(1+u)\left(\frac{1+\eta_z^2}{1-\xi^2}\right)^{1/2} + \eta_z\phi_z, \qquad y = 1+\eta.
\end{aligned}$$

We now flatten the free boundary at $y = 1+\eta$ and normalize the period in z to 2π, by taking the new coordinates

$$y' = \frac{y}{1+\eta(x,z)}, \qquad z' = \ell z, \quad \ell = \frac{2\pi}{L},$$

so that $y' \in (0,1)$ and the waves are 2π-periodic in z'. In the coordinate system (x,y',z'), after suppressing the primes, we find the equations

$$\begin{aligned}
\phi_x &= u + \frac{y\phi_y\xi}{1+\eta}\left(\frac{1+\ell^2\eta_z^2}{1-\xi^2}\right)^{1/2}\\
u_x &= -\ell^2\left(\phi_z - \frac{y\eta_z}{1+\eta}\phi_y\right)_z + \frac{\ell^2 y\eta_z}{1+\eta}\left(\phi_z - \frac{y\eta_z}{1+\eta}\phi_y\right)_y - \frac{\phi_{yy}}{(1+\eta)^2}\\
&\quad + \frac{yu_y\xi}{1+\eta}\left(\frac{1+\ell^2\eta_z^2}{1-\xi^2}\right)^{1/2}\\
\eta_x &= \xi\left(\frac{1+\ell^2\eta_z^2}{1-\xi^2}\right)^{1/2}\\
\xi_x &= \frac{1}{b}(u|_{y=1} + \lambda\eta) + \frac{1}{2b}\left[u^2 + \ell^2\left(\phi_z - \frac{y\eta_z}{1+\eta}\phi_y\right)^2 + \frac{\phi_y^2}{(1+\eta)^2}\right]_{y=1}\\
&\quad - \ell^2\left(\eta_z\left(\frac{1-\xi^2}{1+\ell^2\eta_z^2}\right)^{1/2}\right)_z
\end{aligned}$$

and the boundary conditions

$$\begin{aligned}
\phi_y|_{y=0} &= 0,\\
\phi_y|_{y=1} &= \left(\frac{1+\eta}{1+\ell^2\eta_z^2}\right)\left(\xi(1+u|_{y=1})\left(\frac{1+\ell^2\eta_z^2}{1-\xi^2}\right)^{1/2} + \ell^2\eta_z\phi_z|_{y=1}\right).
\end{aligned}$$

Next, consider the following spaces of functions which are 2π-periodic in z,

$$\begin{aligned}
H^s_{\rm per}(S) &= \{f \in H^s_{\rm loc}(\mathbb{R})\,;\, f(\cdot + 2\pi) = f(\cdot)\},\\
H^s_{\rm per}(\Sigma) &= \{f \in H^s_{\rm loc}((0,1)\times\mathbb{R})\,;\, f(\cdot,\cdot + 2\pi) = f(\cdot,\cdot)\},
\end{aligned}$$

where $H^s_{\rm loc}$ is the classical Sobolev space, $S = (0,2\pi)$, $\Sigma = (0,1)\times(0,2\pi)$, and we assume that $0 < s < 1/2$. We set

$$\mathscr{H}_s = H^{s+1}_{\rm per}(\Sigma) \times H^s_{\rm per}(\Sigma) \times H^{s+1}_{\rm per}(S) \times H^s_{\rm per}(S),$$

and $V = (\phi, u, \eta, \xi)$. Then the system above is of the form

$$\frac{dV}{dx} = \widetilde{\mathbf{F}}(V, \lambda, b, \ell), \tag{2.14}$$

with $\widetilde{\mathbf{F}}(\cdot,\lambda,b,\ell) : D(\widetilde{\mathbf{F}}) \subset \mathscr{H}_{s+1} \to \mathscr{H}_s$ a smooth map with domain a codimension-two manifold in a neighborhood of 0,

$$D(\widetilde{\mathbf{F}}) = \Big\{ (\phi,u,\eta,\xi) \in \mathscr{H}_{s+1}\ ;\ |\xi| < 1,\ \eta > -1,\ \phi_y|_{y=0} = 0,$$
$$\phi_y|_{y=1} = \left(\tfrac{1+\eta}{1+\ell^2\eta_z^2}\right)\left(\xi(1+u|_{y=1})\left(\tfrac{1+\ell^2\eta_z^2}{1-\xi^2}\right)^{1/2} + \ell^2\eta_z\phi_z|_{y=1}\right)\Big\},$$

defined by the boundary conditions. Notice that $H^s_{\mathrm{per}}(S)$ and $H^t_{\mathrm{per}}(\Sigma)$ are Banach algebras for $s > 1/2$ and $t > 1$, respectively, and that $H^{s+1}_{\mathrm{per}}(S) \cdot H^s_{\mathrm{per}}(S) \subset H^s_{\mathrm{per}}(S)$ and $H^{s+1}_{\mathrm{per}}(\Sigma) \cdot H^s_{\mathrm{per}}(\Sigma) \subset H^s_{\mathrm{per}}(\Sigma)$ if $s > 0$.

One of the difficulties here is due to nonlinear boundary condition in the definition of the domain of $\widetilde{\mathbf{F}}$. In order to transform this boundary condition into a linear one, we follow the arguments in [35].

Consider the smooth map $H : \mathscr{H}_{s+1} \to H^{s+1}_{\mathrm{per}}(\Sigma)$ defined by

$$H(\phi,u,\eta,\xi) = y\left(\frac{1+\eta}{1+\ell^2\eta_z^2}\right)\left(\xi(1+u)\left(\frac{1+\ell^2\eta_z^2}{1-\xi^2}\right)^{1/2} + \ell^2\eta_z\phi_z\right),$$

so that the two boundary conditions become

$$\phi_y = H(\phi,u,\eta,\xi), \qquad y = 0,1.$$

We construct a suitable change of variables in the neighborhood of the origin with the help of the smooth map $G : \mathscr{H}_{s+1} \to \mathscr{H}_{s+1}$ defined by

$$G(\phi,u,\eta,\xi) = (\phi - \varphi_y, u, \eta, \xi),$$

where $\varphi \in H^{s+3}_{\mathrm{per}}(\Sigma)$ is the unique solution of the linear boundary value problem

$$\Delta\varphi = H(\phi,u,\eta,\xi), \qquad (y,z) \in \Sigma,$$
$$\varphi = 0, \qquad y = 0,1.$$

Setting

$$\psi = \phi - \varphi_y,$$

the boundary conditions become

$$\psi_y = 0, \quad y = 0,1. \tag{2.15}$$

By arguing as in [35], it can be shown that G has a bounded inverse at the origin, $(dG(0))^{-1} = \mathbb{I}$, and that the operator $dG(U) : \mathscr{H}_{s+1} \to \mathscr{H}_{s+1}$ extends to an isomorphism $\widetilde{dG}(U) : \mathscr{H}_s \to \mathscr{H}_s$ for every $U \in \mathscr{H}_{s+1}$ sufficiently small.

Now we set $V = G^{-1}(U)$ and the system (2.14) becomes

$$\frac{dU}{dx} = dG(G^{-1}(U))\left(\widetilde{\mathbf{F}}(G^{-1}(U),\lambda,b,\ell)\right),$$

where $U = G(\phi,u,\eta,\xi) = (\psi,u,\eta,\xi)$, as defined above, satisfies the linear boundary conditions (2.15). This system is of the form (2.13), with $\mathbf{F}$ a smooth map, and $\mathbf{F}(\cdot,\lambda,b,\ell) : D(\mathbf{F}) \subset \mathscr{H}_{s+1} \to \mathscr{H}_s$ defined on the linear subspace

$$D(\mathbf{F}) = \{(\psi,u,\eta,\xi) \in \mathscr{H}_{s+1} \; ; \; \psi_y|_{y=0,1} = 0\} \subset \mathscr{H}_{s+1}.$$

Notice that since $0 < s < 1/2$ the domain $D(\mathbf{F})$ is dense in $\mathscr{H}_s$, which is useful to avoid technical complications in the reduction procedure, and that $\mathscr{H}_{s+1}$ is compactly embedded in $\mathscr{H}_s$.

The two discrete symmetries (2.11) of equations (2.5)–(2.8) induce a reversibility symmetry $\mathbf{S}$ and an equivariance symmetry $\mathbf{T}$ of the system (2.13), which are defined through

$$\mathbf{S}(\psi,u,\eta,\xi) = (-\psi,u,\eta,-\xi),$$
$$\mathbf{T}(\psi(y,z),u(y,z),\eta(z),\xi(z)) = (\psi(y,-z),u(y,-z),\eta(-z),\xi(-z)).$$

Bifurcation Analysis

For the system (2.13) the bifurcations are determined by the purely imaginary spectrum of the linear map

$$\mathbf{L}_{\lambda,b,\ell} = D_U\mathbf{F}(0,\lambda,b,\ell),$$

which is a closed linear operator in $\mathscr{X} = \mathscr{H}_s$, with dense domain $\mathscr{Z} = D(\mathbf{F})$. Since $D(\mathbf{F})$ is compactly embedded in $\mathscr{H}_s$, the operator $\mathbf{L}_{\lambda,b,\ell}$ has a compact resolvent, so that its spectrum consists of isolated eigenvalues, only accumulating at infinity. Using Fourier series in z, a direct calculation shows that a purely imaginary number $i\kappa$ is an eigenvalue of $\mathbf{L}_{\lambda,b,\ell}$, with corresponding eigenvectors in the nth Fourier mode, if and only if

$$\kappa^2 = (\lambda + b\gamma_n^2)\gamma_n \tanh\gamma_n, \quad \gamma_n^2 = \kappa^2 + n^2\ell^2, \quad n \in \mathbb{Z}. \tag{2.16}$$

Fixing the wavenumber ℓ in the direction z, and taking λ and b as bifurcation parameters one finds an infinite sequence of bifurcation curves C_n, $n = 0,1,2,\ldots$, in the (b,λ)-parameter plane, which accumulate at the half-axis $b = 0$, $\lambda > 0$, as $n \to \infty$ (see Figure 2.3).

The center manifold theorem can be applied for values of (b,λ) close to each of these bifurcation curves, the resulting center manifolds being of dimension $8n+4$ for (b,λ) close to C_n when $\lambda > 1$, and of dimension $8n+2$ for (b,λ) close to C_n when $\lambda < 1$. We point out that the center manifold has actually two additional dimensions, due to a double zero eigenvalue of $\mathbf{L}_{\lambda,b,\ell}$ which exists for all values of (b,λ). This eigenvalue results from the continuous symmetries (2.9)–(2.10) of the equations (2.1)–(2.4). An appropriate use of these invariances allows us to eliminate these two additional dimensions in the reduction procedure [34, 40]. Furthermore,

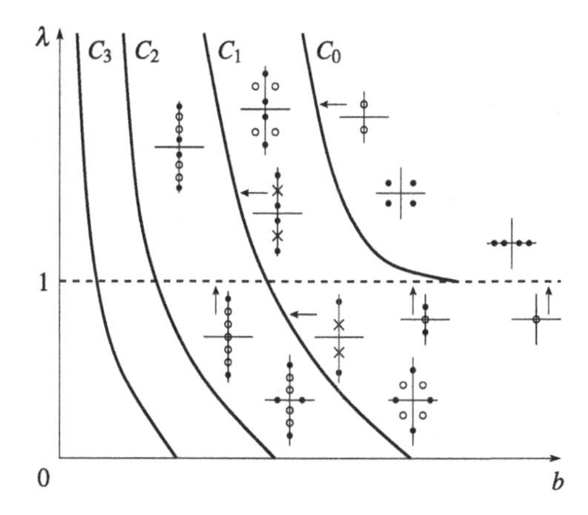

Fig. 2.3 Three-dimensional water waves: first four bifurcation curves in (b,λ)-parameter plane for a fixed ℓ, and behavior of the critical eigenvalues of $\mathbf{L}_{\lambda,b,\ell}$. The solid dots, hollow dots, and crosses represent simple, double, and quadruple eigenvalues, respectively.

using the equivariance symmetry $\mathbf{T}$, one can restrict to the study of symmetric solutions satisfying $\mathbf{T}U = U$, and then the reduced systems of ODEs are of lower dimension, $4n+4$ for (b,λ) close to C_n when $\lambda > 1$, and $4n+2$ for (b,λ) close to C_n when $\lambda < 1$.

The first bifurcation is found when crossing the curve C_0 from right to the left, when the reduced center manifold is four-dimensional, and we have a reversible $(i\omega)^2$ bifurcation. But this is precisely the $(i\omega)^2$ bifurcation found in the two-dimensional problem, because the purely imaginary eigenvalues correspond to the Fourier mode $n = 0$ in z, and therefore the associated eigenvectors are constant in z. The bifurcating solutions are all two-dimensional.

The simplest bifurcation of truly three-dimensional waves occurs when crossing the curve C_1 for $\lambda < 1$. Restricting to symmetric solutions, $\mathbf{T}U = U$, we have a reversible $(i\omega_1)^2(i\omega_2)$ bifurcation. This bifurcation has been analyzed in [35], where for the reduced system the existence of periodic orbits and of orbits homoclinic to periodic orbits has been shown. These orbits correspond to *water waves that are even in both x and z, are periodic in z, and have the profile of a periodic wave or a generalized solitary wave in the direction of propagation x* (see Figure 2.4). The other bifurcations are more complicated and their analysis is open.

Further Waves

As already mentioned, one may choose any horizontal direction as timelike variable in the formulation of the equations as a first order system, and then take any transverse direction as direction in which the waves are periodic. This freedom introduces two additional parameters in the bifurcation problem, making the analysis

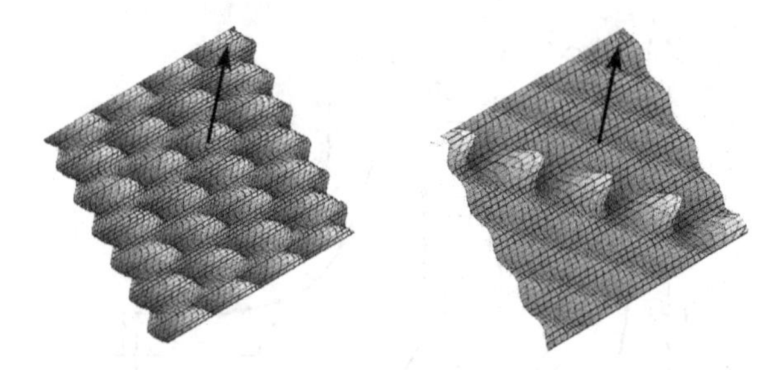

Fig. 2.4 Three-dimensional water-waves having the profile of a periodic wave (left), and of a generalized solitary wave (right) in the direction of propagation; they are periodic in the perpendicular direction. The arrows indicate the direction of propagation.

much more involved, but allowing us to find additional classes of traveling waves [34] (see also [36]).

This approach to three-dimensional waves excludes the fully localized waves, and also the case of gravity waves, when the surface tension vanishes. We refer to [37] and [69] for two recent existence results of fully localized waves in the case of large surface tension, and of doubly periodic waves in the case of zero surface tension, respectively.

5.2.2 Almost-Planar Waves in Reaction-Diffusion Systems

We consider the reaction-diffusion system

$$U_t = D\Delta U + f(U), \quad (x,y) \in \mathbb{R}^2,\, t \in \mathbb{R}, \tag{2.17}$$

where $U(x,y,t) \in \mathbb{R}^N$ is a vector of N chemical species, $D = \mathrm{diag}\,(D_1,\ldots,D_N) > 0$ is a positive, diagonal diffusion matrix, Δ represents the laplacian in $\mathbb{R}^2$, and the reaction kinetics f are assumed to be smooth. We are interested in two-dimensional traveling waves propagating with speed c in the direction y, i.e., bounded solutions of the form $U(x,y,t) = u(x,y-ct)$. Notice that there is no loss of generality in taking y as direction of propagation, because of the rotational invariance of the system (2.17). In a moving coordinate system $(x',y') = (x,y-ct)$, the traveling wave u satisfies the stationary equation

$$D\Delta u + c\partial_y u + f(u) = 0, \tag{2.18}$$

in which the primes have been dropped.

Almost-Planar Waves

We focus on the existence of *almost-planar traveling waves*, which are solutions of (2.18) that are close to a given one-dimensional wave $q_*(y)$ *(planar wave)*. We restrict here to the particular case when the planar wave $q_*(y)$ is either a front or a pulse (see Figure 2.5), and rely upon the approach to almost-planar waves in [42].

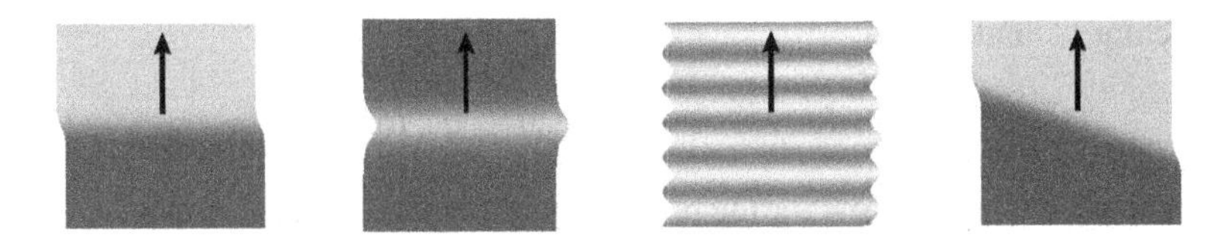

Fig. 2.5 Planar waves: front, pulse, periodic wave, and a rotated front (from left to the right). The arrows indicate the direction of propagation y.

This approach can be adapted to the case of periodic waves in a rather straightforward way.

Assuming that $q_*(y)$ is a one-dimensional traveling wave with speed c_* satisfying (2.18), a solution $u(x,y)$ of (2.18) is an almost planar traveling wave δ-close to q_*, if u is of the form

$$u(x,y) = q_*(y+\xi(x)) + \widetilde{u}(x,y), \tag{2.19}$$

with $\xi \in C^2(\mathbb{R})$, and

$$\sup_{x\in\mathbb{R}} |\xi'(x)| < \delta, \qquad \sup_{(x,y)\in\mathbb{R}^2} |\widetilde{u}(x,y)| < \delta, \qquad |c-c_*| < \delta. \tag{2.20}$$

The wave u is a trivial almost planar wave if u is a *rotated planar wave* $u = q_*((\cos\vartheta)x + (\sin\vartheta)y)$, for some $\vartheta \in \mathbb{R}$ (see Figure 2.5). In particular, in this case we have that $\xi'' = 0$. Depending upon the shape of the derivative ξ', and its limits $\xi'(x) \to \eta_\pm \in \mathbb{R}$, as $x \to \pm\infty$, we distinguish several classes of almost planar waves: *periodically modulated waves* when ξ' is periodic, *interior corners* when $\eta_+ < \eta_-$, *exterior corners* when $\eta_+ > \eta_-$, *steps* when $\eta_+ = \eta_- \neq 0$, *holes* when $\eta_+ = \eta_- = 0$ (see Figure 2.6)

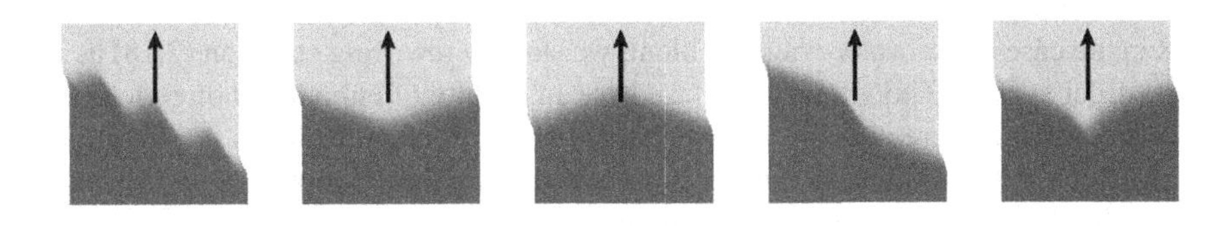

Fig. 2.6 Almost-planar waves close to a planar front: periodically modulated wave, interior corner, exterior corner, step, and hole (from left to the right). The arrows indicate the direction of propagation y.

Hypotheses

We assume the existence of a traveling one-dimensional front, or pulse. More precisely, we assume that there exists a speed $c_* > 0$ and asymptotic states $q_\pm$ such that there exists an x-independent traveling-wave solution $q_*(y)$ of (2.18),

$$Dq_*'' + c_* q_*' + f(q_*) = 0, \tag{2.21}$$

connecting q_- and q_+, i.e.,

$$q_*(y) \to q_+ \text{ for } y \to +\infty, \quad q_*(y) \to q_- \text{ for } y \to -\infty. \tag{2.22}$$

We emphasize that we allow for the possibility of pulses, when $q_+ = q_-$.

The key assumption is concerned with the *stability* of the above traveling-wave solution. We discuss here the simplest situation, in which the traveling wave is asymptotically stable with respect to both one- and two-dimensional perturbations.

Consider the linearized operator $\mathscr{M}_* : H^2(\mathbb{R}^2, \mathbb{R}^N) \subset L^2(\mathbb{R}^2, \mathbb{R}^N) \to L^2(\mathbb{R}^2, \mathbb{R}^N)$, defined through

$$\mathscr{M}_* u = D\Delta u + c_* \partial_y u + f'(q_*(\cdot))u \tag{2.23}$$

and obtained by linearizing equation (2.18) at the one-dimensional wave q_* and its Fourier conjugates $\mathscr{M}_k : H^2(\mathbb{R}, \mathbb{R}^N) \subset L^2(\mathbb{R}, \mathbb{R}^N) \to L^2(\mathbb{R}, \mathbb{R}^N)$,

$$\mathscr{M}_k u = D(\partial_{yy} - k^2)u + c_* \partial_y u + f'(q_*(\cdot))u. \tag{2.24}$$

We assume that

(i) zero is an isolated eigenvalue of $\mathscr{M}_0$ and the rest of the spectrum is strictly contained in the left half-complex plane;
(ii) the eigenvalue zero is simple with associated eigenvector the derivative q_*', i.e., the traveling wave q_* is asymptotically stable in one-space dimension;
(iii) the spectra of $\mathscr{M}_k$, for $k \neq 0$ are strictly contained in the left half-complex plane and that the unique eigenvalue $\lambda_{\mathrm{d}}(k)$, $k \sim 0$ with $\lambda_{\mathrm{d}}(0) = \lambda_{\mathrm{d}}'(0) = 0$ satisfies $\lambda_{\mathrm{d}}''(0) < 0$ i.e., the traveling wave q_* is asymptotically stable in two space dimensions.

Bifurcation Analysis

As in the case of the water-wave problem, we start by rewriting equation (2.18) as a first order system. Though we also have here an infinity of unbounded directions in the (x,y)-plane, there is in this case a distinguished direction, which is the direction x in which the one-dimensional wave q_* is homogeneous. Choosing x as timelike variable we write the equation (2.18) in the form

$$\frac{dU}{dx} = \mathbf{L}_c U + \mathbf{F}(U), \tag{2.25}$$

where $U = (u,v)$,

$$\mathbf{L}_c = \begin{pmatrix} 0 & \mathbb{I} \\ -\partial_{yy} - D^{-1}c\partial_y & 0 \end{pmatrix}, \quad \mathbf{F}(U) = \begin{pmatrix} 0 \\ -D^{-1}f(u) \end{pmatrix}. \tag{2.26}$$

The system (2.25) possesses a reversibility symmetry $\mathbf{S}$, defined through $\mathbf{S}(u,v) = (u,-v)$, and a continuous translation symmetry, induced by the y-shift $\xi : U(\cdot) \mapsto U(\cdot + \xi)$.

An important consequence of the choice of x as timelike variable is that the one-dimensional wave q_* is now an equilibrium of (2.25), and due to the continuous translation symmetry above we have a family of equilibria

$$Q_*^\xi = \begin{pmatrix} q_*^\xi(\cdot) \\ 0 \end{pmatrix} = \begin{pmatrix} q_*(\cdot + \xi) \\ 0 \end{pmatrix}. \tag{2.27}$$

The linearization of (2.25) about Q_*^0 is given by the operator

$$\mathbf{L}^0 = \mathbf{L}_{c_*} + D_U\mathbf{F}(Q_*^0),$$

which is a closed linear operator in $\mathscr{X} = (H^1 \times L^2)(\mathbb{R},\mathbb{R}^N)$ with dense domain $\mathscr{Z} = (H^2 \times H^1)(\mathbb{R},\mathbb{R}^N)$. Using the stability properties (i)–(iii) of q_*, one can show in this case that 0 is an algebraically double and geometrically simple eigenvalue of $\mathbf{L}^0$, and that the rest of the spectrum lies away from the imaginary axis in a region $\{\lambda \in \mathbb{C}\,;\, |\mathrm{Re}\,\lambda| > \gamma\}$, for some $\gamma > 0$. Consequently, $\mathbf{L}^0$ satisfies Hypothesis 2.4 required by the center manifold theorem in Chapter 2. The space $\mathscr{E}_0$ is two-dimensional and spanned by the vectors

$$\zeta_0 = \begin{pmatrix} q_*'(\cdot) \\ 0 \end{pmatrix}, \quad \zeta_1 = \begin{pmatrix} 0 \\ q_*'(\cdot) \end{pmatrix},$$

with $\mathbf{L}^0\zeta_0 = 0$ and $\mathbf{L}^0\zeta_1 = \zeta_0$. Similarly, to the shifted equilibria Q_*^ξ we consider the shifted linear operator $\mathbf{L}^\xi$, the shifted vectors $\zeta_0^\xi = ((q_*')^\xi(\cdot),0)$, and $\zeta_1^\xi = (0,(q_*')^\xi(\cdot))$, and the shifted space $\mathscr{E}_0^\xi$.

Following the general strategy for construction of a local center manifold along a line of equilibria described in Section 2.3.3, and taking into account the above properties of the linearization $\mathbf{L}^\xi$ around Q_*^ξ, we decompose

$$U = Q_*^\xi + \eta\zeta_1^\xi + W^\xi \text{ with } \mathbf{P}^\xi W^\xi = 0. \tag{2.28}$$

Here ξ and η are real functions depending upon x, and $\mathbf{P}^\xi$ represents the spectral projection onto $\mathscr{E}_0^\xi$. Substituting (2.28) into (2.25), and then projecting successively on ζ_0^ξ, ζ_1^ξ, and the complement $(\mathbb{I} - \mathbf{P}^\xi)\mathscr{X}$ of $\mathscr{E}_0^\xi$ we find the equation for ξ,

$$\xi_x = \eta + O(|\eta|\,\|W\|_{\mathscr{Z}}), \tag{2.29}$$

which decouples, and the quasilinear system

$$\eta_x = \frac{2}{\lambda_{\rm d}''(0)}(c-c_*) - \frac{c_*}{\lambda_{\rm d}''(0)}\eta^2 + O(|c-c_*|\,\|W\|_{\mathscr{X}} + |\eta|\,\|W\|_{\mathscr{X}} + \|W\|_{\mathscr{X}}^2)$$
$$W_x = \mathbf{L}^0 W + O(|c-c_*| + |\eta|^2 + \|W\|_{\mathscr{X}}^2 + |\eta|\,\|W\|_{\mathscr{Y}}) \qquad (2.30)$$

posed on the Hilbert space $\mathbb{R}\times\mathscr{X}_h$, with $\mathscr{X}_h = (\mathrm{id}-\mathbf{P})\mathscr{X}$.

Applying the center manifold theorem for quasilinear systems, Theorem 2.20 in Chapter 2, to system (2.30), we find a one-dimensional center manifold on which the dynamics is governed by the scalar equation

$$\eta_x = \frac{2}{\lambda_{\rm d}''(0)}\left(c-c_*-\frac{c_*}{2}\eta^2\right) + O(|c-c_*|^2 + |\eta|^4). \qquad (2.31)$$

The vector field in (2.31) is even in η, due to reversibility, and one can show that a *saddle-node bifurcation* occurs at $c=c_*$. For $c>c_*$ there are two equilibria, which in the reaction-diffusion system (2.18) correspond to rotations of the one-dimensional wave q_*, and a heteroclinic orbit connecting these two equilibria, which corresponds to an *interior corner*.

Further Waves

We discussed above the simplest situation, in which the one-dimensional wave q_* is stable in both one and two dimensions. Another situation that may occur is the one-dimensional wave q_* still stable in one dimension, but unstable in two dimensions. For a parameter-dependent reaction-diffusion system

$$U_t = D\Delta U + f(U,\mu),$$

in which μ is a small real parameter, one can consider the *onset of two-dimensional instability* where there exists a smooth family of waves $q_*(y,\mu)$, with speeds $c_*(\mu)$ connecting the asymptotic states $q_\pm(\mu)$ for $\mu\sim 0$, which are stable in two dimensions for $\mu<0$, say, and unstable for $\mu>0$. Using the same approach, one finds in this case a three-dimensional center manifold on which the dynamics are governed by a steady Kuramoto–Sivashinsky equation. Known results on existence of heteroclinic and homoclinic orbits for this equation allow us to show existence of both *interior* and *exterior corners*, and of *steps*, for $\mu>0$, where the one-dimensional wave is unstable [42, Section 4].

The same approach can be used in several different contexts. For example, the traveling one-dimensional wave q_* may be replaced by a *modulated wave* $U(x,y,t) = q_*(y-c_*t,\omega t)$ of the reaction-diffusion system (2.17), connecting two homogeneous equilibria $q_\pm$ as $y\to\pm\infty$, and with q_* being 2π-periodic in its second argument. In this case one finds a two-dimensional center manifold on which the dynamics are governed by a coupled system of steady conservation laws. In addition to the almost-planar waves found close to a traveling wave, one can also show now the existence of *holes* [42, Section 5].

Next, the isotropic system (2.17) can be replaced by an *anisotropic model*

$$U_t = \nabla \cdot (a\nabla U) + f(U, \nabla U), \tag{2.32}$$

in which the diffusion matrices are supposed to be elliptic, $a = (a_{ij}^m)_{1\leq i,j\leq 2}^{1\leq m\leq N}$ such that

$$(\nabla \cdot (a\nabla u))^m = a_{ij}^m \partial_{ij} u^m, \quad \sum_{ij} a_{ij}^m y_i y_j \geq M \sum_i y_i^2,$$

for some positive constant $M > 0$ independent of $y = (y_1, y_2)$. We refer to [43] for an analysis of this situation, and to [44, 45] for further extensions.

5.2.3 Waves in Lattices

A large variety of physical problems, such as crystal dislocation, localized excitations in ionic crystals, and thermal denaturation of DNA, may be described by a one-dimensional lattice for which the dynamics satisfy the system

$$\frac{d^2 u_n}{dt^2} + W'(u_n) = V'(u_{n+1} - u_n) - V'(u_n - u_{n-1}), \quad n \in \mathbb{Z},, \tag{2.33}$$

where u_n is the displacement of the nth particle from an equilibrium position. In particular, this system describes a chain of particles nonlinearly coupled to their first neighbors. The interaction potential V and the on-site potential W are assumed to be analytic in a neighborhood of $u = 0$, with

$$V'(0) = W'(0) = 0, \quad V''(0) > 0, \quad W''(0) > 0 \text{ or } W = 0.$$

The system (2.33) is referred to as the Fermi–Pasta–Ulam (FPU) lattice if $W = 0$, and the Klein–Gordon (KG) lattice if V is harmonic, $V(x) = (\gamma/2)x^2$, for some $\gamma > 0$.

A wide class of wave solutions may be found, which satisfy

$$u_n(t) = u_{n-p}(t - p\tau), \quad p \geq 1, \quad \tau \in \mathbb{R}. \tag{2.34}$$

The case $p = 1$ corresponds to traveling waves with speed $1/\tau$. For $p > 1$ we obtain pulsating traveling waves, which are translated by p sites after a fixed propagation time $p\tau$, and oscillate as they propagate through the lattice. Solutions satisfying (2.34) and such that $u_n(t) \to 0$ as $n \to \pm\infty$ are known as *solitary waves* when $p = 1$ and *exact breathers* when $p > 1$. Such systems have been intensively studied in the literature. We refer to [57] for a review of mathematical results using center manifold theorems. Here, we briefly discuss the case of traveling waves for the FPU lattice, i.e., when $W = 0$ in (2.33) and $p = 1$ in (2.34).

Formulation as a First Order System

Consider the system (2.33) with $W = 0$, and assume that

$$V'(x) = x + N(x), \quad N(x) = \alpha x^2 + \beta x^3 + O(|x^4|),$$

in a neighborhood of 0. We set

$$u_n(t) = y(x), \quad x = n - \frac{t}{\tau}, \tag{2.35}$$

so that the system (2.33) leads to the following scalar *advance-delay differential equation*, for the scalar function $y(x)$,

$$\frac{1}{\tau^2}\frac{d^2y}{dx^2} = V'\left(y(x+1) - y(x)\right) - V'\left(y(x) - y(x-1)\right). \tag{2.36}$$

We are interested in small bounded solutions of this equation.

Notice that the equation (2.36) possesses the reversibility symmetry

$$x \mapsto -x, \quad y \mapsto -y, \tag{2.37}$$

and has the first integral

$$J = \frac{dy}{dx} - \tau^2 \int_0^1 V'(y(x+v) - y(x+v-1))\,dv. \tag{2.38}$$

We start by writing the equation in the form (2.1) in Chapter 2. Consider the Banach spaces

$$\mathscr{X} = \mathbb{R}^2 \times \mathscr{C}^0[-1,1], \quad \mathscr{Z} = \{(y,\xi,Y) \in \mathbb{R}^2 \times \mathscr{C}^1[-1,1]\,;\, Y(0) = y\},$$

equipped with usual norms. We set $\xi = dy/dx$, $Y(x,v) = y(x+v)$, and $U = (y,\xi,Y)$. Then the equation (2.36) is of the form

$$\frac{dU}{dx} = \mathbf{L}_\tau U + \tau^2 \mathbf{R}(U), \tag{2.39}$$

with

$$\mathbf{L}_\tau U = \begin{pmatrix} \xi \\ \tau^2(Y|_{v=1} - 2y + Y|_{v=-1}) \\ \frac{\partial Y}{\partial v} \end{pmatrix}, \; \mathbf{R}(U) = \begin{pmatrix} 0 \\ N(Y|_{v=1} - y) - N(y - Y|_{v=-1}) \\ 0 \end{pmatrix}.$$

Here $\mathbf{L}_\tau \in \mathscr{L}(\mathscr{Z}, \mathscr{X})$ and the map $\mathbf{R} : \mathscr{X} \to \mathscr{Z}$ is analytic.

The system (2.39) is reversible, with the reversibility symmetry defined by

$$\mathbf{S}(y,\xi,Y) = (-y,\xi,-Y \circ s), \quad (Y \circ s)(v) = Y(-v),$$

as a consequence of (2.37), and it has the first integral

$$J = \xi(x) - \tau^2 \int_0^1 V'(Y(x,v) - Y(x,v-1))\,dv, \tag{2.40}$$

due to (2.38). Finally, notice that equation (2.39) possesses a two-parameter family of solutions,

$$U_{a,b}(x) = (ax+b)\chi_0 + a\chi_1, \quad \chi_0 = (1,0,1), \quad \chi_1 = (0,1,v), \quad a,b \in \mathbb{R}, \tag{2.41}$$

corresponding to uniformly extended or compressed lattices, depending on the sign of a.

Bifurcation Analysis

As in most of the previous examples, the operator $\mathbf{L}_\tau$ has a compact resolvent, so that its spectrum consists of isolated eigenvalues with finite algebraic multiplicities, only accumulating at infinity. Purely imaginary eigenvalues $i\kappa$ are solutions of the dispersion relation

$$\kappa^2 + 2\tau^2(\cos\kappa - 1) = 0,$$

from which it is not difficult to check that there are only a finite number of purely imaginary eigenvalues. The remaining eigenvalues are located at a positive distance from the imaginary axis, in a region of the complex plane with exponential shape, centered on the imaginary axis [55]. Notice that 0 is always an eigenvalue of $\mathbf{L}_\tau$, at least double, due to the presence of the two-parameter family of solutions mentioned above.

The lowest value of the parameter τ at which a bifurcation occurs is $\tau = 1$, corresponding to the "sound speed" in the physics literature. For $\tau < 1$, the linear operator $\mathbf{L}_\tau$ has, besides the double eigenvalue 0, a pair of simple real eigenvalues, with opposite signs, the rest of the spectrum being away from the imaginary axis. These two nonzero eigenvalues collide in 0, when $\tau = 1$, so that 0 is a quadruple eigenvalue of $\mathbf{L}_1$. For $\tau > 1$, not too large, there is a pair of purely imaginary eigenvalues close to 0, in addition to the double eigenvalue 0, and the remaining eigenvalues are all bounded away from the imaginary axis.

A particularity of the present case is that the resolvent estimates in the Hypothesis 2.15 in Chapter 2 are not verified, but one can prove directly that Hypothesis 2.7 of the center manifold theorem in Chapter 2 holds. The method of proof relies upon Fourier analysis on a specific space of distributions [60, Appendix]. Applying center manifold Theorems 3.3 and 3.15 in Chapter 2, we obtain a four-dimensional center manifold. However, it is possible to reduce the dimension of this center manifold by two. First, using the equivariance of (2.39) under the group action $U \mapsto U + a\chi_0$, $a \in \mathbb{R}$, as indicated in Theorem 3.19 in Chapter 2, the dimension of the center manifold is reduced by one, and one finds a 0^{3+} reversible bifurcation at $\tau = 1$ (see [55]). However, using in addition to the first integral (2.40), we are then left with a two-dimensional manifold, and a 0^{2+} *reversible bifurcation* at $\tau = 1$ (see [57]). Applying the results in Theorem 1.8 of Chapter 4, in the case $\alpha \neq 0$ one finds, in particular, homoclinic orbits, which correspond to *solitary waves* connecting a stretched or a compressed state to itself. In the case $\alpha = 0$ and $\beta \neq 0$, we can in fact use the phase portraits in Figure 2.4 of Chapter 4, and conclude, in particular, the existence of heteroclinic orbits, which correspond to *fronts* connecting a compressed state to a stretched state [55].

Appendix

A Elements of Functional Analysis

We collect in this section a number of notions and results from the theory of linear operators in Banach spaces. We refer for instance to the book [76] for further details and proofs. In addition, in Section A.6 we give the definitions and some basic results on Sobolev spaces, for which we refer to the books [1, 2].

A.1 Bounded and Closed Operators

Consider the Banach spaces $\mathscr{X}$ and $\mathscr{Z}$ equipped with the norms $\|\cdot\|_{\mathscr{X}}$ and $\|\cdot\|_{\mathscr{Z}}$, respectively, and a linear map (or linear operator) $\mathbf{L} : \mathscr{Z} \to \mathscr{X}$. We denote by $\operatorname{im}\mathbf{L}$ the *range* of $\mathbf{L}$,

$$\operatorname{im}\mathbf{L} = \{\mathbf{L}u \in \mathscr{X}\ ;\ u \in \mathscr{Z}\} \subset \mathscr{X},$$

by $\ker\mathbf{L}$ its *kernel*,

$$\ker\mathbf{L} = \{u \in \mathscr{Z}\ ;\ \mathbf{L}u = 0\} \subset \mathscr{Z},$$

and by $\mathrm{G}(\mathbf{L})$ its *graph*,

$$\mathrm{G}(\mathbf{L}) = \{(u, \mathbf{L}u)\ ;\ u \in \mathscr{Z}\} \subset \mathscr{Z} \times \mathscr{X}.$$

Definition A.1 (Bounded operator) *A linear operator* $\mathbf{L} : \mathscr{Z} \to \mathscr{X}$ *is called* a bounded linear operator, *or simply* a bounded operator, *if* $\mathbf{L}$ *is continuous. The set of bounded linear operators is denoted by* $\mathscr{L}(\mathscr{Z}, \mathscr{X})$, *and by* $\mathscr{L}(\mathscr{X})$ *if* $\mathscr{Z} = \mathscr{X}$.

Properties A.2 [76, Chapter III, §2.2, §3.1]

(i) For a linear operator $\mathbf{L} : \mathscr{Z} \to \mathscr{X}$ the following properties are equivalent:

a. $\mathbf{L}$ is continuous, i.e., $\mathbf{L}$ is a bounded linear operator;
b. $\mathbf{L}$ is continuous in 0;
c. $\sup\{\|\mathbf{L}u\|_{\mathscr{X}}\ ;\ u \in \mathscr{Z},\ \|u\|_{\mathscr{Z}} = 1\} < \infty$.

M. Haragus, G. Iooss, *Local Bifurcations, Center Manifolds, and Normal Forms in Infinite-Dimensional Dynamical Systems*, Universitext,
DOI 10.1007/978-0-85729-112-7, © EDP Sciences 2011

(ii) For a bounded linear operator $\mathbf{L}$, the real number

$$\|\mathbf{L}\|_{\mathscr{L}(\mathscr{Z},\mathscr{X})} \stackrel{\text{def}}{=} \sup_{\|u\|_{\mathscr{Z}}=1} \|\mathbf{L}u\|_{\mathscr{X}} = \sup_{0<\|u\|_{\mathscr{Z}}\leq 1} \frac{\|\mathbf{L}u\|_{\mathscr{X}}}{\|u\|_{\mathscr{Z}}} = \sup_{\|u\|_{\mathscr{Z}}\neq 0} \frac{\|\mathbf{L}u\|_{\mathscr{X}}}{\|u\|_{\mathscr{Z}}}$$

is called *norm* of $\mathbf{L}$.

(iii) The set of bounded linear operators $\mathscr{L}(\mathscr{Z},\mathscr{X})$ is a Banach space when equipped with the norm $\|\cdot\|_{\mathscr{L}(\mathscr{Z},\mathscr{X})}$.

Definition A.3 (Closed operator) *A linear operator* $\mathbf{L}: D(\mathbf{L}) \subset \mathscr{X} \to \mathscr{X}$ *defined on a linear subspace* $D(\mathbf{L}) \subset \mathscr{X}$ *is called* a closed linear operator, *or simply* a closed operator, *if its graph* $G(\mathbf{L})$ *is a closed set in* $\mathscr{X} \times \mathscr{X}$. *The set of closed linear operators is denoted by* $\mathscr{C}(\mathscr{X})$.

Properties A.4 [76, Chapter III, §5.2, Theorem 5.20]

(i) A linear operator $\mathbf{L}: D(\mathbf{L}) \subset \mathscr{X} \to \mathscr{X}$ is closed if and only if for any sequence $(u_n)_{n\in\mathbb{N}} \subset D(\mathbf{L})$ such that $u_n \to u$ in $\mathscr{X}$ and $\mathbf{L}u_n \to v$ in $\mathscr{X}$, we have that $u \in D(\mathbf{L})$ and $\mathbf{L}u = v$.

(ii) The sum of a closed operator with a bounded operator is a closed operator. However, the sum of two closed operators is not always a closed operator.

(iii) A closed operator with domain $D(\mathbf{L}) = \mathscr{X}$ is bounded (closed graph theorem).

(iv) For a closed operator $\mathbf{L}: D(\mathbf{L}) \subset \mathscr{X} \to \mathscr{X}$ the domain $D(\mathbf{L})$ equipped with the norm

$$\|u\|_{\mathbf{L}} = \left(\|u\|_{\mathscr{X}}^2 + \|\mathbf{L}u\|_{\mathscr{X}}^2\right)^{1/2}$$

is a Banach space, and the injection $i: D(\mathbf{L}) \to \mathscr{X}$ is bounded. This norm is also called *the graph norm.*

(v) A closed operator $\mathbf{L}: D(\mathbf{L}) \subset \mathscr{X} \to \mathscr{X}$ belongs to $\mathscr{L}(D(\mathbf{L}),\mathscr{X})$ when $D(\mathbf{L})$ is equipped with the graph norm $\|\cdot\|_{\mathbf{L}}$.

A.2 *Resolvent and Spectrum*

Definition A.5 *Consider a linear operator* $\mathbf{L}: D(\mathbf{L}) \subset \mathscr{X} \to \mathscr{X}$.

(i) We call resolvent set *of* $\mathbf{L}$ *the set of complex numbers*

$$\rho(\mathbf{L}) = \{\lambda \in \mathbb{C}\,;\, (\lambda\mathbb{I} - \mathbf{L}) \textit{ invertible and } (\lambda\mathbb{I} - \mathbf{L})^{-1} \in \mathscr{L}(\mathscr{X})\}.$$

The operator $(\lambda\mathbb{I} - \mathbf{L})^{-1}$, *for* $\lambda \in \rho(\mathbf{L})$, *is called* the resolvent *of* $\mathbf{L}$.

(ii) We call the spectrum *of* $\mathbf{L}$ *the complement of the resolvent set,*

$$\sigma(\mathbf{L}) = \mathbb{C} \setminus \rho(\mathbf{L}).$$

(iii) A complex number $\lambda \in \mathbb{C}$ *is called* an eigenvalue *of* $\mathbf{L}$ *if* $\ker(\lambda\mathbb{I} - \mathbf{L}) \neq \{0\}$. *The kernel* $\ker(\lambda\mathbb{I} - \mathbf{L})$ *is called* the eigenspace *associated with the eigenvalue*

λ, and any element $u \in \ker(\lambda\mathbb{I} - \mathbf{L})\setminus\{0\}$ is called an eigenvector *associated with the eigenvalue λ.*

(iv) For an eigenvalue $\lambda \in \sigma(\mathbf{L})$, the dimension of $\ker(\lambda\mathbb{I} - \mathbf{L})$ *is called* the geometric multiplicity *of λ. An eigenvalue with geometric multiplicity one is called* geometrically simple.

(v) For an isolated eigenvalue $\lambda \in \sigma(\mathbf{L})$, the dimension of the largest subspace $\mathscr{X}_\lambda \subset D(\mathbf{L})$ which is invariant under the action of $\mathbf{L}$ and such that $\sigma(\mathbf{L}|_{\mathscr{X}_\lambda}) = \{\lambda\}$ is called the algebraic multiplicity *of λ. An eigenvalue with algebraic multiplicity one is called* algebraically simple *or* simple.

(vi) An eigenvalue is called semisimple *if its algebraic and geometric multiplicities are the same.*

Properties A.6 [76, Chapter III, §6.1, §6.3]

(i) The spectrum of a closed operator $\mathbf{L} \in \mathscr{C}(\mathscr{X})$ is a closed set.

(ii) The spectrum of a bounded operator $\mathbf{L} \in \mathscr{L}(\mathscr{X})$ is a closed, bounded, nonempty set.

(iii) For a closed operator $\mathbf{L} \in \mathscr{C}(\mathscr{X})$, if $\lambda \in \rho(\mathbf{L})$, then the resolvent $(\lambda\mathbb{I} - \mathbf{L})^{-1} : \mathscr{X} \to D(\mathbf{L})$ is a bounded operator, when $D(\mathbf{L})$ is equipped with the graph norm, i.e., $(\lambda\mathbb{I} - \mathbf{L})^{-1} \in \mathscr{L}(\mathscr{X}, D(\mathbf{L}))$.

(iv) The map $\lambda \mapsto (\lambda\mathbb{I} - \mathbf{L})^{-1}$ is holomorphic from $\rho(\mathbf{L})$ into $\mathscr{L}(\mathscr{X})$.

(v) For $\lambda \in \rho(\mathbf{L})$, the resolvent $(\lambda\mathbb{I} - \mathbf{L})^{-1} : \mathscr{X} \to D(\mathbf{L})$ commutes with $\mathbf{L}$, i.e.,

$$\mathbf{L}(\lambda\mathbb{I} - \mathbf{L})^{-1}u = (\lambda\mathbb{I} - \mathbf{L})^{-1}\mathbf{L}u \text{ for all } u \in D(\mathbf{L}).$$

(vi) For $\lambda, \mu \in \rho(\mathbf{L})$, the resolvents $(\lambda\mathbb{I} - \mathbf{L})^{-1}$ and $(\mu\mathbb{I} - \mathbf{L})^{-1}$ commute and

$$(\lambda\mathbb{I} - \mathbf{L})^{-1} - (\mu\mathbb{I} - \mathbf{L})^{-1} = (\mu - \lambda)(\lambda\mathbb{I} - \mathbf{L})^{-1}(\mu\mathbb{I} - \mathbf{L})^{-1}.$$

(vii) For an operator $\mathbf{L}$, we call *the extended spectrum* the set $\sigma_\infty(\mathbf{L}) \subset \mathbb{C} \cup \{\infty\}$ defined by

$$\sigma_\infty(\mathbf{L}) = \begin{cases} \sigma(\mathbf{L}) & \text{if } \sigma(\mathbf{L}) \text{ is bounded,} \\ \sigma(\mathbf{L}) \cup \{\infty\} & \text{if } \sigma(\mathbf{L}) \text{ is unbounded.} \end{cases}$$

Then for $\lambda, \mu \in \rho(\mathbf{L})$ the spectrum of the resolvent satisfies

$$\sigma\left((\lambda\mathbb{I} - \mathbf{L})^{-1}\right) = \left\{(\lambda - \mu)^{-1} \,;\, \mu \in \sigma_\infty(\mathbf{L})\right\}. \tag{A.1}$$

Theorem A.7 (Spectral decomposition [76, Chapter III, Theorem 6.17]) *Consider a closed operator $\mathbf{L} : D(\mathbf{L}) \subset \mathscr{X} \to \mathscr{X}$. Assume that $\sigma(\mathbf{L}) = F \cup G$, with $F \cap G = \varnothing$ and $F \subset \mathbb{C}$ a closed, bounded set, such that there exists a rectifiable, simple, closed curve Γ which encloses an open set containing F in its interior and G in its exterior. Then there exists a decomposition of $\mathscr{X} = \mathscr{X}_F \oplus \mathscr{X}_G$, with $\mathscr{X}_F$ and $\mathscr{X}_G$ invariant under the action of $\mathbf{L}$, such that the spectra of the restrictions $\mathbf{L}|_{\mathscr{X}_F}$ and $\mathbf{L}|_{\mathscr{X}_G}$ coincide with F and G, respectively. Furthermore, $\mathbf{L}|_{\mathscr{X}_F}$ is a bounded operator, $\mathbf{L}|_{\mathscr{X}_F} \in \mathscr{L}(\mathscr{X}_F)$, and the unique spectral projection $\mathbf{P}_F : \mathscr{X} \to \mathscr{X}_F$ which* *commutes with* $\mathbf{L}$ *is given by the Dunford integral formula*

$$\mathbf{P}_F = \frac{1}{2\pi i}\int_\Gamma (\lambda\mathbb{I}-\mathbf{L})^{-1}d\lambda.$$

Remark A.8 *(i) The result in the above theorem still holds when the curve* Γ *is replaced by a finite number of rectifiable, simple, closed curves.*

(ii) In the particular case when $F=\{\lambda\}$ *is reduced to one point,* λ *is an isolated point of the spectrum. If the dimension of* $\mathscr{X}_F$ *is finite, then* λ *is an eigenvalue of* $\mathbf{L}$*, and the dimension of* $\mathscr{X}_F$ *is the algebraic multiplicity of* λ *[76, Chapter III §6.5].*

A.3 Compact Operators and Operators with Compact Resolvent

Definition A.9 (Compact operator) *A linear operator* $\mathbf{L}:\mathscr{Z}\to\mathscr{X}$ *is called* a compact operator *if for any bounded sequence* $(u_n)_{n\in\mathbb{N}}\subset\mathscr{Z}$*, the sequence* $(\mathbf{L}u_n)_{n\in\mathbb{N}}\subset\mathscr{X}$ *contains a convergent subsequence.*

Properties A.10 [76, Chapter III §4.2]

(i) A compact operator is bounded.

(ii) The sum of two compact operators is a compact operator. For a bounded operator $\mathbf{L}:\mathscr{Y}\to\mathscr{Z}$ and a compact operator $\mathbf{K}:\mathscr{Z}\to\mathscr{X}$, the composed operator $\mathbf{K}\circ\mathbf{L}:\mathscr{Y}\to\mathscr{X}$ is compact. A similar property holds for $\mathbf{L}\circ\mathbf{K}$ (adapt the spaces).

Theorem A.11 (Spectrum [76, Chapter III, Theorem 6.26]) *Consider a compact operator* $\mathbf{L}:\mathscr{X}\to\mathscr{X}$*. Then the following properties hold:*

(i) $0\in\sigma(\mathbf{L})$ *if* $\mathscr{X}$ *is infinite-dimensional;*

(ii) any $\lambda\in\sigma(\mathbf{L})$ *with* $\lambda\neq 0$ *is an isolated eigenvalue with finite algebraic multiplicity;*

(iii) $\sigma(\mathbf{L})$ *is a countable set with at most one accumulation point in* 0.

Definition A.12 (Operator with compact resolvent) *A linear operator* $\mathbf{L}:D(\mathbf{L})\subset\mathscr{X}\to\mathscr{X}$ *is called* an operator with compact resolvent *if* $\rho(\mathbf{L})\neq\varnothing$ *and for some* $\lambda\in\rho(\mathbf{L})$ *the operator* $(\lambda\mathbb{I}-\mathbf{L})^{-1}:\mathscr{X}\to\mathscr{X}$ *is compact.*

Properties A.13 [76, Chapter III §6.8]

(i) If $(\lambda\mathbb{I}-\mathbf{L})^{-1}:\mathscr{X}\to\mathscr{X}$ is a compact operator for some $\lambda\in\rho(\mathbf{L})$, then it is a compact operator for any $\lambda\in\rho(\mathbf{L})$.

(ii) The spectrum of an operator with compact resolvent is a countable set consisting of isolated eigenvalues with finite algebraic multiplicities, with no accumulation point in $\mathbb{C}$.

A.4 Adjoint Operator

For a Banach space $\mathscr{X}$, denote by $\mathscr{X}^*$ the dual space, i.e., the space of all continuous linear forms on $\mathscr{X}$, and by $\langle \cdot, \cdot \rangle$ the duality product defined by

$$\langle u, u^* \rangle = \overline{u^*}(u) \text{ for all } u \in \mathscr{X},\ u^* \in \mathscr{X}^*.$$

Recall that $\mathscr{X} = \mathscr{X}^*$ when $\mathscr{X}$ is a Hilbert space.

Definition A.14 (Adjoint operators) *(i) Two linear operators* $\mathbf{L} : D(\mathbf{L}) \subset \mathscr{X} \to \mathscr{X}$ *and* $\mathbf{M} : D(\mathbf{M}) \subset \mathscr{X}^* \to \mathscr{X}^*$ *are called* adjoint *to each other if*

$$\langle \mathbf{L}u, v \rangle = \langle u, \mathbf{M}v \rangle \textit{ for all } u \in D(\mathbf{L}),\ v \in D(\mathbf{M}).$$

(ii) If there exists a unique maximal operator $\mathbf{L}^*$ *which is adjoint to* $\mathbf{L}$*, then* $\mathbf{L}^*$ *is called* the adjoint of $\mathbf{L}$.

Properties A.15 ([76, Chapter III, §5.5, Theorem 6.22])

(i) If $D(\mathbf{L})$ is dense in $\mathscr{X}$, for an operator $\mathbf{L} : D(\mathbf{L}) \subset \mathscr{X} \to \mathscr{X}$ there is a unique adjoint operator, but this property is not true in general. The adjoint operator $\mathbf{L}^* : D(\mathbf{L}^*) \subset \mathscr{X}^* \to \mathscr{X}^*$ is constructed in the following way. The domain $D(\mathbf{L}^*)$ consists of all $v \in \mathscr{X}^*$ such that $u \mapsto \langle \mathbf{L}u, v \rangle$ is a continuous linear form on $\mathscr{X}$. Then there exists $w \in \mathscr{X}^*$ such that

$$\langle \mathbf{L}u, v \rangle = \langle u, w \rangle \text{ for all } u \in D(\mathbf{L}),$$

and w is unique because $D(\mathbf{L})$ is dense in $\mathscr{X}$. We define $\mathbf{L}^* v = w$.

(ii) For an operator $\mathbf{L} : D(\mathbf{L}) \subset \mathscr{X} \to \mathscr{X}$ with $D(\mathbf{L})$ dense in $\mathscr{X}$, the adjoint operator $\mathbf{L}^* : D(\mathbf{L}^*) \subset \mathscr{X}^* \to \mathscr{X}^*$ is closed. In addition, if $\mathbf{L}$ is closed and the Banach space $\mathscr{X}$ is reflexive, then $\mathbf{L}^*$ is densely defined, i.e., $D(\mathbf{L}^*)$ is dense in $\mathscr{X}^*$.

(iii) For an operator $\mathbf{L} : D(\mathbf{L}) \subset \mathscr{X} \to \mathscr{X}$ with $D(\mathbf{L})$ dense in $\mathscr{X}$, we have

$$\ker \mathbf{L}^* = (\operatorname{im} \mathbf{L})^{\perp} \overset{\text{def}}{=} \{ v \in \mathscr{X}^* \,;\, \langle u, v \rangle = 0 \text{ for all } u \in \operatorname{im} \mathbf{L} \}.$$

(iv) For an operator $\mathbf{L} : D(\mathbf{L}) \subset \mathscr{X} \to \mathscr{X}$ the resolvent set and the spectrum of the adjoint operator $\mathbf{L}^* : D(\mathbf{L}^*) \subset \mathscr{X}^* \to \mathscr{X}^*$ satisfy

$$\rho(\mathbf{L}^*) = \overline{\rho(\mathbf{L})}, \quad \sigma(\mathbf{L}^*) = \overline{\sigma(\mathbf{L})}.$$

Furthermore,

$$(\overline{\lambda}\mathbb{I} - \mathbf{L}^*)^{-1} = \left((\lambda \mathbb{I} - \mathbf{L})^{-1} \right)^* \text{ for all } \lambda \in \rho(\mathbf{L}).$$

Definition A.16 (Self-adjoint operator) *A linear operator* $\mathbf{L} : D(\mathbf{L}) \subset \mathscr{X} \to \mathscr{X}$ *in a Hilbert space* $\mathscr{X}$*, with domain* $D(\mathbf{L})$ *dense in* $\mathscr{X}$*, is called self-adjoint if its adjoint* $\mathbf{L}^* : D(\mathbf{L}^*) \subset \mathscr{X} \to \mathscr{X}$ *satisfies* $D(\mathbf{L}) = D(\mathbf{L}^*)$ *and* $\mathbf{L}u = \mathbf{L}^*u$ *for all* $u \in D(\mathbf{L})$.

Properties A.17 [76, Chapter V, §3.4, §3.5]

(i) The spectrum of a self-adjoint operator is real.
(ii) The algebraic and geometric multiplicities of an isolated eigenvalue $\lambda \in \sigma(\mathbf{L})$ of a self-adjoint operator are the same, i.e., the eigenvalue is semisimple.

A.5 Fredholm Operators

Definition A.18 *A bounded operator* $\mathbf{L} \in \mathscr{L}(\mathscr{Y}, \mathscr{X})$ *is called* a Fredholm operator *if its kernel* $\ker \mathbf{L}$ *is finite-dimensional and its range* $\operatorname{im} \mathbf{L}$ *is closed and has finite codimension. The integer*

$$\operatorname{ind}(\mathbf{L}) \stackrel{def}{=} \dim(\ker \mathbf{L}) - \operatorname{codim}(\operatorname{im} \mathbf{L}),$$

is called the Fredholm index.

Remark A.19 *The above definition is easily extended to closed operators* $\mathbf{L} : D(\mathbf{L}) \subset \mathscr{X} \to \mathscr{X}$*, since* $\mathbf{L} \in \mathscr{L}(D(\mathbf{L}), \mathscr{X})$.

Properties A.20 [76, Chapter IV, §5.1, Theorem 5.22, Theorem 5.28, §5.2]

(i) The set of Fredholm operators is open in $\mathscr{L}(\mathscr{Y}, \mathscr{X})$ and the map $\mathbf{L} \mapsto \operatorname{ind}(\mathbf{L})$ is continuous.
(ii) For a closed operator $\mathbf{L} : D(\mathbf{L}) \subset \mathscr{X} \to \mathscr{X}$, if $\lambda \in \sigma(\mathbf{L})$ is an eigenvalue with finite algebraic multiplicity which is isolated in the spectrum of $\mathbf{L}$, then $\lambda \mathbb{I} - \mathbf{L}$ is a Fredholm operator with index 0.
(iii) If $\mathbf{L} : D(\mathbf{L}) \subset \mathscr{X} \to \mathscr{X}$ is a densely defined closed operator, then $\mathbf{L}$ is a Fredholm operator if and only if $\mathbf{L}^*$ is a Fredholm operator. Furthermore,

$$\operatorname{ind}(\mathbf{L}^*) = -\operatorname{ind}(\mathbf{L}).$$

A.6 Basic Sobolev Spaces

We recall in this section some basic properties of the Sobolev spaces $L^2(\Omega)$ and $H^m(\Omega)$, $m \in \mathbb{N}^*$ which are used in this book. We refer to [1, Chapters II, III, V, and VI] for more general statements, and for the case of the spaces $L^p(\Omega)$ and $W^{m,p}(\Omega)$.

The Space $L^2(\Omega)$

Consider a domain $\Omega \subset \mathbb{R}^n$, and the space of complex-valued square-integrable functions on Ω

$$\mathscr{L}^2(\Omega) = \left\{ f : \Omega \to \mathbb{C} \, ; \, f \text{ measurable and } \int_\Omega |f(x)|^2 dx < \infty \right\}.$$

This set is a linear space with respect to the natural operations (sum and multiplication by a complex number). For $f \in \mathscr{L}^2(\Omega)$, we set

$$\|f\| = \left(\int_\Omega |f(x)|^2 dx \right)^{1/2}. \tag{A.2}$$

Then $\|\cdot\|$ is a seminorm on $\mathscr{L}^2(\Omega)$, but not a norm, since if $\|f\| = 0$, then $f = 0$ almost everywhere, only. However, $\|\cdot\|$ can be transformed into a norm by replacing $\mathscr{L}^2(\Omega)$ by the quotient with respect to the kernel of $\|\cdot\|$, i.e., by

$$L^2(\Omega) = \mathscr{L}^2(\Omega) / \ker(\|\cdot\|).$$

Clearly, the kernel of $\|\cdot\|$ consists of functions that are equal to 0 almost everywhere, so that $L^2(\Omega)$ consists of classes of functions that are equal almost everywhere. Then $\|\cdot\|$ is a norm on $L^2(\Omega)$, or, in other words *$L^2(\Omega)$ is a normed space with norm $\|\cdot\|$ defined by (A.2)*. Furthermore, this norm corresponds to the scalar product defined by

$$\langle f, g \rangle = \int_\Omega f(x)\overline{g(x)} dx \text{ for all } f, \, g \in L^2(\Omega). \tag{A.3}$$

A key property of the space $L^2(\Omega)$ is that it is complete; more precisely we have the following result.

Properties A.21 The space $L^2(\Omega)$ equipped with the scalar product $\langle \cdot, \cdot \rangle$ defined by (A.3) is a Hilbert space.

The Spaces $H^m(\Omega)$ and $H_0^m(\Omega)$

Definition A.22 *Consider $m \in \mathbb{N}^*$.*

(i) We define the space

$$H^m(\Omega) = \left\{ u \in L^2(\Omega) \, ; \, D^\alpha u \in L^2(\Omega) \text{ for all } \alpha \in \mathbb{N}^n, \, |\alpha| \leq m \right\},$$

in which $D^\alpha u$ is the distributional partial derivative of u,

$$D^\alpha u = \frac{\partial^{|\alpha|} u}{\partial x_1^{\alpha_1} \ldots \partial x_n^{\alpha_n}},$$

for a multi-index $\alpha = (\alpha_1, \ldots, \alpha_n) \in \mathbb{N}^n$ and $|\alpha| = \alpha_1 + \cdots + \alpha_n$.

(ii) On $H^m(\Omega)$ we define the scalar product

$$\langle f, g \rangle_m = \sum_{|\alpha| \leq m} \langle D^\alpha f, D^\alpha g \rangle,$$

where $\langle\cdot,\cdot\rangle$ represents the scalar product in $L^2(\Omega)$, and the corresponding norm

$$\|u\|_m = \langle u,u\rangle_m^{1/2}.$$

(iii) We define $H_0^m(\Omega)$ as the closure of $C_0^\infty(\Omega)$ in $H^m(\Omega)$, where $C_0^\infty(\Omega)$ is the space of functions of class C^∞ which have compact support in Ω.

Properties A.23 [1, Theorem 3.2]

(i) $H^m(\Omega)$ equipped with the scalar product $\langle\cdot,\cdot\rangle_m$ is a Hilbert space.
(ii) $H^{m+j}(\Omega)$ is a dense subspace of $H^m(\Omega)$ and the imbedding $H^{m+j}(\Omega) \hookrightarrow H^m(\Omega)$ is continuous, for any $j \in \mathbb{N}^*$.

Properties A.24 (Sobolev imbedding theorem [1, Theorem 5.4]) Assume that either $\Omega = \mathbb{R}^n$ or Ω is a bounded domain in $\mathbb{R}^n$ having a locally Lipschitz boundary, i.e., for each point x on the boundary $\partial\Omega$ of Ω there exists a neighborhood U_x such that $\partial\Omega \cap U_x$ is the graph of a Lipschitz continuous function.

(i) For any $j \in \mathbb{N}^*$ such that $j > n/2$, we have that $H^{m+j}(\Omega) \subset C^m(\overline{\Omega})$, and the imbedding is continuous.
(ii) If Ω is an arbitrary domain in $\mathbb{R}^n$, the result (i) holds with $H_0^{m+j}(\Omega)$ instead of $H^{m+j}(\Omega)$.

Properties A.25 (Rellich–Kondrachov theorem [2, Theorem 3.8], [1, Theorem 6.2]) Assume that Ω is a bounded domain in $\mathbb{R}^n$ having a locally Lipschitz boundary.

(i) For any $j \in \mathbb{N}^*$, the imbedding $H^{m+j}(\Omega) \subset H^m(\Omega)$ is compact.
(ii) For any $j \in \mathbb{N}^*$ such that $j > n/2$, the imbedding $H^{m+j}(\Omega) \subset C^m(\overline{\Omega})$ is compact.
(iii) If Ω is an arbitrary domain in $\mathbb{R}^n$, the results (i) and (ii) hold with $H_0^{m+j}(\Omega)$ instead of $H^{m+j}(\Omega)$.

Spaces of Periodic Functions

An important particular case is that of spaces of periodic functions on the real line. Consider the space $L^2_{\text{loc}}(\mathbb{R})$ of measurable functions $f:\mathbb{R}\to\mathbb{C}$ satisfying $f \in L^2(a,b)$ for any bounded interval $(a,b) \subset \mathbb{R}$. We define the space of square-integrable ℓ-periodic functions by

$$L^2_{\text{per}}(0,\ell) = \{f \in L^2_{\text{loc}}(\mathbb{R})\,;\, f(\cdot+\ell) = f(\cdot)\},$$

and for $m \in \mathbb{N}^*$, the Sobolev spaces H^m consisting of ℓ-periodic functions by

$$H^m_{\text{per}}(0,\ell) = \{f \in L^2_{\text{per}}(\mathbb{R})\,;\, f^{(k)} \in L^2_{\text{per}}(\mathbb{R}) \text{ for all } k \le m\}.$$

Notice that $L^2_{\text{per}}(0,\ell)$ can be identified with $L^2(0,\ell)$, but this is not true for the spaces H^m, $m \ge 1$. The results above, in particular the imbedding theorems, hold for these spaces, as well.

B Center Manifolds

The references in this section are to theorems, hypotheses, formulas, and remarks in Chapter 2.

B.1 Proof of Theorem 2.9 (Center Manifolds)

Consider system (2.1), and assume that Hypotheses 2.1, 2.4, and 2.7 hold. For any $u \in \mathscr{X}$ we set

$$u = u_0 + u_h \in \mathscr{X}, \quad u_0 = \mathbf{P}_0 u \in \mathscr{E}_0, \quad u_h = \mathbf{P}_h u \in \mathscr{X}_h,$$

and rewrite the system (2.1) as

$$\begin{aligned} \frac{du_0}{dt} - \mathbf{L}_0 u_0 &= \mathbf{P}_0 \mathbf{R}(u) \\ \frac{du_h}{dt} - \mathbf{L}_h u_h &= \mathbf{P}_h \mathbf{R}(u). \end{aligned} \tag{B.1}$$

Modified System

We take a cut-off function $\chi : \mathscr{E}_0 \to \mathbb{R}$ of class $\mathscr{C}^\infty$ such that

$$\chi(u_0) = \begin{cases} 1 \text{ for } \|u_0\| \leq 1 \\ 0 \text{ for } \|u_0\| \geq 2 \end{cases}, \qquad \chi(u_0) \in [0,1] \text{ for all } u_0 \in \mathscr{E}_0.$$

Since $\mathscr{E}_0$ is finite-dimensional such a function always exists. We use this function to modify the nonlinear terms $\mathbf{R}(u)$ outside a neighborhood of the origin, in order to be able to control the norm of the u_0-component of the system (B.1) in the space of exponentially growing functions $\mathscr{C}_\eta(\mathbb{R}, \mathscr{E}_0)$.

We set

$$\mathbf{R}^\varepsilon(u) = \chi\left(\frac{u_0}{\varepsilon}\right) \mathbf{R}(u) \text{ for all } \varepsilon \in (0, \varepsilon_0),$$

where ε_0 is chosen such that

$$\left\{u = u_0 + u_h \,;\, \|u_0\|_{\mathscr{E}_0} \leq 2\varepsilon_0,\ \|u_h\|_{\mathscr{X}_h} \leq \varepsilon_0\right\} \subset \mathscr{V},$$

with $\mathscr{V}$ the neighborhood of the origin in Hypothesis 2.1. Then $\mathbf{R}^\varepsilon$ is well defined in the closed set

$$\mathscr{O}_\varepsilon = \mathscr{E}_0 \times B_\varepsilon(\mathscr{X}_h), \quad B_\varepsilon(\mathscr{X}_h) = \{u_h \in \mathscr{X}_h \,;\, \|u_h\| \leq \varepsilon\},$$

and satisfies

$$\mathbf{R}^\varepsilon(u) = \mathbf{R}(u) \text{ for all } u \in \mathscr{O}_\varepsilon,\ \|u_0\| \leq \varepsilon.$$

Consider the modified system

$$\begin{aligned}\frac{du_0}{dt} - \mathbf{L}_0 u_0 &= \mathbf{P}_0 \mathbf{R}^{\varepsilon}(u)\\ \frac{du_h}{dt} - \mathbf{L}_h u_h &= \mathbf{P}_h \mathbf{R}^{\varepsilon}(u).\end{aligned} \tag{B.2}$$

The nonlinear terms in this system now satisfy

$$\begin{aligned}\delta_0(\varepsilon) &\stackrel{\text{def}}{=} \sup_{u\in\mathscr{O}_\varepsilon} \left(\|\mathbf{P}_0\mathbf{R}^{\varepsilon}(u)\|_{\mathscr{E}_0}, \|\mathbf{P}_h\mathbf{R}^{\varepsilon}(u)\|_{\mathscr{Y}_h}\right) = O(\varepsilon^2)\\ \delta_1(\varepsilon) &\stackrel{\text{def}}{=} \sup_{u\in\mathscr{O}_\varepsilon} \left(\|D_u\mathbf{P}_0\mathbf{R}^{\varepsilon}(u)\|_{\mathscr{L}(\mathscr{Z},\mathscr{E}_0)}, \|D_u\mathbf{P}_h\mathbf{R}^{\varepsilon}(u)\|_{\mathscr{L}(\mathscr{Z},\mathscr{Y}_h)}\right) = O(\varepsilon).\end{aligned} \tag{B.3}$$

We prove below the existence of a "global" center manifold for this system which, due to the fact that $\mathbf{R}^{\varepsilon}$ and $\mathbf{R}$ coincide for $\|u_0\|_{\mathscr{E}_0} \leq \varepsilon$, will give the local center manifold for the system (2.1) in the theorem.

Integral Formulation

We replace system (B.2) by the integral formulation

$$\begin{aligned}u_0(t) &= \mathbf{S}_{0,\varepsilon}(u,t,u_0(0)) \stackrel{\text{def}}{=} e^{\mathbf{L}_0 t}u_0(0) + \int_0^t e^{\mathbf{L}_0(t-s)}\mathbf{P}_0\mathbf{R}^{\varepsilon}(u(s))ds\\ u_h &= \mathbf{S}_{h,\varepsilon}(u) \stackrel{\text{def}}{=} \mathbf{K}_h\mathbf{P}_h\mathbf{R}^{\varepsilon}(u).\end{aligned} \tag{B.4}$$

The first equation in this system is obtained by the variation of constant formula from the first equation in (B.1). Here $u_0(0) \in \mathscr{E}_0$ is arbitrary, and the exponential $e^{\mathbf{L}_0 t}$ exists since $\mathscr{E}_0$ is finite-dimensional. The second equation in (B.4) is obtained from Hypothesis 2.7, used with $f \in \mathscr{C}_0(\mathbb{R},\mathscr{Y}_h)$. It is now straightforward to check that this integral system is equivalent to (B.2) for

$$u = (u_0,u_h) \in \mathscr{N}_{\eta,\varepsilon} \stackrel{\text{def}}{=} \mathscr{C}_\eta(\mathbb{R},\mathscr{E}_0)\times\mathscr{C}_0(\mathbb{R},B_\varepsilon(\mathscr{Z}_h)),$$

with $0<\eta\leq\gamma$ and $\varepsilon\in(0,\varepsilon_0)$. Notice that $\mathscr{N}_{\eta,\varepsilon}$ is a closed subspace of $\mathscr{C}_\eta(\mathbb{R},\mathscr{Z})$, so that it is complete when equipped with the norm of $\mathscr{C}_\eta(\mathbb{R},\mathscr{Z})$.

Fixed Point Argument

Our aim now is to show that (B.4) has a unique solution $u=(u_0,u_h)\in\mathscr{N}_{\eta,\varepsilon}$ for any $u_0(0)\in\mathscr{E}_0$. For this we use a fixed point argument for the map

$$\mathbf{S}_\varepsilon(u,u_0(0)) \stackrel{\text{def}}{=} (\mathbf{S}_{0,\varepsilon}(u,\cdot,u_0(0)),\mathbf{S}_{h,\varepsilon}(u)), \quad \mathbf{S}_\varepsilon(\cdot,u_0(0)) : \mathscr{N}_{\eta,\varepsilon}\to\mathscr{N}_{\eta,\varepsilon}.$$

We show that $\mathbf{S}_\varepsilon(\cdot, u_0(0))$ is well defined and that it is a contraction with respect to the norm of $\mathscr{C}_\eta(\mathbb{R}, \mathscr{X})$ for $\eta \in (0, \gamma]$, with γ the constant in Hypothesis 2.7, and ε sufficiently small.

First, Hypothesis 2.4 implies that for any $\delta > 0$ there is a constant $c_\delta > 0$ such that

$$\|e^{\mathbf{L}_0 t}\|_{\mathscr{L}(\mathscr{E}_0)} \leq c_\delta e^{\delta|t|} \text{ for all } t \in \mathbb{R}. \tag{B.5}$$

Using this equality with $\delta = \eta$, we find

$$\sup_{t\in\mathbb{R}} \left(e^{-\eta|t|} \left\|e^{\mathbf{L}_0 t} u_0(0)\right\|_{\mathscr{E}_0} \right) \leq c_\eta \|u_0(0)\|_{\mathscr{E}_0},$$

which shows that the first term in $\mathbf{S}_{0,\varepsilon}(u, \cdot, u_0(0))$ belongs to $\mathscr{C}_\eta(\mathbb{R}, \mathscr{E}_0)$, for any $\eta > 0$. Next, for any $u \in \mathscr{N}_{\eta,\varepsilon}$, we have the estimates

$$\|\mathbf{P}_0 \mathbf{R}^\varepsilon(u(t))\|_{\mathscr{E}_0} \leq \delta_0(\varepsilon), \quad \|\mathbf{P}_h \mathbf{R}^\varepsilon(u(t))\|_{\mathscr{Y}_h} \leq \delta_0(\varepsilon),$$

which together with (B.5) for $\delta = \eta/2$, and Hypothesis 2.7 imply

$$\begin{aligned}\sup_{t\in\mathbb{R}} \left(e^{-\eta|t|} \left\| \int_0^t e^{\mathbf{L}_0(t-s)} \mathbf{P}_0 \mathbf{R}^\varepsilon(u(s)) ds \right\|_{\mathscr{E}_0} \right) &\leq c_\delta \delta_0(\varepsilon) \sup_{t\in\mathbb{R}} \left(e^{-\eta|t|} \int_0^t e^{\delta|t-s|} ds \right) \\ &\leq \frac{2c_{\eta/2}\delta_0(\varepsilon)}{\eta},\end{aligned}$$

and

$$\|\mathbf{K}_h \mathbf{P}_h \mathbf{R}^\varepsilon(u)\|_{\mathscr{C}_0(\mathbb{R}, \mathscr{Z}_h)} \leq C(0)\delta_0(\varepsilon).$$

This shows that $\mathbf{S}_\varepsilon(u, u_0(0)) \in \mathscr{N}_{\eta,\varepsilon}$, provided $C(0)\delta_0(\varepsilon) \leq \varepsilon$, which holds for ε sufficiently small since $\delta_0(\varepsilon) = O(\varepsilon^2)$.

Now we show that the map $\mathbf{S}_\varepsilon(\cdot, u_0(0))$ is a contraction with respect to the norm of $\mathscr{C}_\eta(\mathbb{R}, \mathscr{X})$ for $\eta \in (0, \gamma]$ and sufficiently small ε. From equality (B.3) we find that

$$\begin{aligned}\|\mathbf{R}^\varepsilon(u_1) - \mathbf{R}^\varepsilon(u_2)\|_{\mathscr{C}_\eta(\mathbb{R},\mathscr{Y})} &= \sup_{t\in\mathbb{R}} \left(e^{-\eta|t|} \|\mathbf{R}^\varepsilon(u_1(t)) - \mathbf{R}^\varepsilon(u_2(t))\|_{\mathscr{Y}} \right) \\ &\leq \delta_1(\varepsilon) \sup_{t\in\mathbb{R}} \left(e^{-\eta|t|} \|u_1(t) - u_2(t)\|_{\mathscr{X}} \right) \\ &\leq \delta_1(\varepsilon) \|u_1 - u_2\|_{\mathscr{C}_\eta(\mathbb{R},\mathscr{X})}\end{aligned}$$

for any $u_1, u_2 \in \mathscr{N}_{\eta,\varepsilon}$. Now, using (B.5) with $\delta = \eta/2$ we obtain

$$\begin{aligned}&\|\mathbf{S}_{0,\varepsilon}(u_1, \cdot, u_0(0)) - \mathbf{S}_{0,\varepsilon}(u_2, \cdot, u_0(0))\|_{\mathscr{C}_\eta(\mathbb{R},\mathscr{E}_0)} \\ &\qquad \leq c_\delta \delta_1(\varepsilon) \sup_{t\in\mathbb{R}} \left(e^{-\eta|t|} \left| \int_0^t e^{\eta|s| + \delta|t-s|} ds \right| \right) \|u_1 - u_2\|_{\mathscr{C}_\eta(\mathbb{R},\mathscr{X})} \\ &\qquad \leq \frac{2c_{\eta/2}\delta_1(\varepsilon)}{\eta} \|u_1 - u_2\|_{\mathscr{C}_\eta(\mathbb{R},\mathscr{X})},\end{aligned}$$

and using the estimate in Hypothesis 2.7 we find

$$\|\mathbf{S}_{h,\varepsilon}(u_1) - \mathbf{S}_{h,\varepsilon}(u_2)\|_{\mathscr{C}_\eta(\mathbb{R},\mathscr{Z}_h)} \le C(\eta)\delta_1(\varepsilon)\|u_1 - u_2\|_{\mathscr{C}_\eta(\mathbb{R},\mathscr{Z})}.$$

Since $\delta_1(\varepsilon) = O(\varepsilon)$ for any $\eta \in (0,\gamma]$, we can choose ε small enough such that

$$\|\mathbf{S}_\varepsilon(u_1,u_0(0)) - \mathbf{S}_\varepsilon(u_2,u_0(0))\|_{\mathscr{C}_\eta(\mathbb{R},\mathscr{Z})} \le \frac{1}{2}\|u_1 - u_2\|_{\mathscr{C}_\eta(\mathbb{R},\mathscr{Z})}.$$

Consequently, the map $\mathbf{S}_\varepsilon(\cdot,u_0(0))$ is a contraction in the complete metric space $\mathscr{N}_{\eta,\varepsilon}$.

Applying the fixed point theorem we now have the existence of a unique solution of (B.4),

$$u \stackrel{\text{def}}{=} \Phi(u_0(0)) \in \mathscr{N}_{\eta,\varepsilon}$$

for any $u_0(0) \in \mathscr{E}_0$, for any $\eta \in (0,\gamma]$, and ε sufficiently small. Clearly, this is also a solution of (B.2).

Properties of Φ

Recall that ε is chosen such that

$$C(0)\delta_0(\varepsilon) \le \varepsilon, \quad \frac{2c_{\eta/2}\delta_1(\varepsilon)}{\eta} \le \frac{1}{2}, \quad C(\eta)\delta_1(\varepsilon) \le \frac{1}{2}.$$

Then the continuity on $[0,\gamma]$ of the map $\eta \to C(\eta)$ in Hypothesis 2.7 implies that for any $\widetilde{\eta} \in (0,\gamma)$, we can choose $\varepsilon > 0$ such that these inequalities hold for all $\eta \in [\widetilde{\eta},\gamma]$. Consequently, for any $\widetilde{\eta} \in (0,\gamma)$, there exists $\varepsilon > 0$ such that the unique fixed point $\Phi(u_0(0))$ belongs to $\mathscr{N}_{\eta,\varepsilon}$ for any $\eta \in [\widetilde{\eta},\gamma]$. This property is used later when showing that the center manifold is of class $\mathscr{C}^k$.

Next, notice that the map $u_0(0) \mapsto \mathbf{S}_{0,\varepsilon}(u,\cdot,u_0(0))$ is Lipschitz from $\mathscr{E}_0$ into $\mathscr{C}_\eta(\mathbb{R},\mathscr{E}_0)$, so that the map $u_0(0) \mapsto \mathbf{S}_\varepsilon(u,u_0(0))$ is also Lipschitz. Consequently, Φ is a Lipschitz map. In addition, the uniqueness of the fixed point implies that

$$\Phi(0) = 0.$$

Construction of Ψ

We define now the map $\Psi : \mathscr{E}_0 \to \mathscr{Z}_h$ in the theorem, through

$$(u_0(0),\Psi(u_0(0)) \stackrel{\text{def}}{=} \Phi(u_0(0))(0) \text{ for all } u_0(0) \in \mathscr{E}_0,$$

i.e., by taking the component in $\mathscr{Z}_h$ of the fixed point $\Phi(u_0(0))$ at $t = 0$. Since Φ is a Lipschitz map, we have that Ψ is also a Lipschitz map, and since $\Phi(0) = 0$, we have

$$\Psi(0) = 0.$$

We prove now that Ψ has the properties (i) and (ii) in the theorem.

First, we show that the manifold

$$\mathcal{M}_{\eta,\varepsilon} = \{(u_0, \Psi(u_0)) \; ; \; u_0 \in \mathcal{E}_0\}$$

is a global invariant manifold for the flow defined by (B.2). We define the shift operator Γ_s through

$$(\Gamma_s u)(t) = u(t+s) \text{ for all } t, \; s \in \mathbb{R}.$$

Since system (B.2) is autonomous, it is equivariant under the action Γ_s for any $s \in \mathbb{R}$, so that if u is a solution of (B.2), then $\Gamma_s u$ is also a solution of (B.2). Moreover, $\Gamma_s u \in \mathcal{N}_{\eta,\varepsilon}$ when $u \in \mathcal{N}_{\eta,\varepsilon}$.

Consider a solution u of (B.2) with $u(0) = (u_0(0), \Psi(u_0(0)))$ for some $u_0(0) \in \mathcal{E}_0$. Then $u = \Phi(u_0(0)) \in \mathcal{N}_{\eta,\varepsilon}$, and since $\Gamma_s u \in \mathcal{N}_{\eta,\varepsilon}$ is also a solution, from the uniqueness of the fixed point we conclude that

$$\Gamma_s u = \Phi(u_0(s)) \text{ for all } s \in \mathbb{R}.$$

Consequently,

$$u(s) = (u_0(s), \Psi(u_0(s)) \text{ for all } s \in \mathbb{R},$$

which shows that $\mathcal{M}_{\eta,\varepsilon}$ is globally invariant under the flow defined by (B.2). Since the system (B.1) coincides with (B.2) in

$$\mathcal{O}_\varepsilon = B_\varepsilon(\mathcal{E}_0) \times B_\varepsilon(\mathcal{Z}_h),$$

this proves part (i) of the theorem with $\mathcal{M}_0 = \mathcal{M}_{\eta,\varepsilon}$ and $\mathcal{O} = \mathcal{O}_\varepsilon$. Indeed, assume that u is a solution of (B.1) such that $u(0) \in \mathcal{M}_0 \cap \mathcal{O}$ and $u(t) \in \mathcal{O}$ for all $t \in [0,T]$. Then u satisfies (B.2) for all $t \in [0,T]$, and since $u(0) \in \mathcal{M}_{\eta,\varepsilon}$ and $\mathcal{M}_{\eta,\varepsilon}$ is an invariant manifold, we have $u(t) \in \mathcal{M}_{\eta,\varepsilon} = \mathcal{M}_0$ for all $t \in [0,T]$.

Consider now a solution u of (B.1) which belongs to $\mathcal{O} = \mathcal{O}_\varepsilon$ for all $t \in \mathbb{R}$. Then $u \in \mathcal{N}_{\eta,\varepsilon}$ and it is also a solution of (B.2). Consequently, $u = \Phi(u_0(0))$, so that $u(0) \in \mathcal{M}_{\eta,\varepsilon} = \mathcal{M}_0$, which proves part (ii) of the theorem.

Regularity of Ψ

We have proved so far that Ψ is a Lipschitz map. Notice that for this proof we have only used the fact that $\mathbf{R}$ is of class $\mathscr{C}^1$. It remains to show that Ψ is of class $\mathscr{C}^k$ when $\mathbf{R}$ is of class $\mathscr{C}^k$. For this, it is enough to prove that Φ is of class $\mathscr{C}^k$.

The major difficulty in proving this property comes from the fact that the *Nemitsky operator*

$$\mathbf{R}^\varepsilon : \mathscr{C}_\eta(\mathbb{R}, \mathcal{X}) \to \mathscr{C}_\eta(\mathbb{R}, \mathcal{Y})$$

is not continuously differentiable, due to the growth of $u \in \mathscr{C}_\eta(\mathbb{R}, \mathscr{X})$ as $t \to \pm\infty$. The following properties of this operator are proved in [120, Lemma 3.7]:

(i) $\mathbf{R}^\varepsilon : \mathscr{C}_\eta(\mathbb{R}, \mathscr{X}) \to \mathscr{C}_\zeta(\mathbb{R}, \mathscr{Y})$ is continuous for any $\eta \geq 0$ and $\zeta > 0$;
(ii) $\mathbf{R}^\varepsilon : \mathscr{C}_\eta(\mathbb{R}, \mathscr{X}) \to \mathscr{C}_\zeta(\mathbb{R}, \mathscr{Y})$ is of class $\mathscr{C}^k$ for any $0 \leq \eta < \zeta/k$ and $\zeta > 0$.

We point out that the kth order derivative exists for $\eta = \zeta/k$, but this derivative is continuous only if $\eta < \zeta/k$.

Following [120], the integral system (B.4) is written as

$$u = \mathbf{S}u_0(0) + \mathbf{K}\mathbf{R}^\varepsilon(u), \tag{B.6}$$

with $\mathbf{S}$ and $\mathbf{K}$ linear maps defined by

$$(\mathbf{S}u_0(0))(t) = e^{\mathbf{L}_0 t}u_0(0),$$

and

$$(\mathbf{K}v)(t) = \int_0^t e^{\mathbf{L}_0(t-s)}\mathbf{P}_0(v(s))ds + (\mathbf{K}_h\mathbf{P}_h(v))(t).$$

We already showed that

$$\mathbf{S} \in \mathscr{L}(\mathscr{E}_0, \mathscr{C}_{\widetilde{\eta}}(\mathbb{R}, \mathscr{E}_0)), \quad \|\mathbf{S}u_0(0)\|_{\mathscr{C}_\eta(\mathbb{R},\mathscr{E}_0)} \leq c_{\eta/2}\|u_0(0)\|_{\mathscr{E}_0},$$

and that $\mathbf{K}\mathbf{R}^\varepsilon : \mathscr{N}_{\eta,\varepsilon} \to \mathscr{N}_{\eta,\varepsilon}$ is a contraction for any $\eta \in [\widetilde{\eta}, \gamma]$, when $\widetilde{\eta} \in (0, \gamma)$ and ε is sufficiently small.

The idea is to consider the fixed point $u = \Phi(u_0(0)) \in \mathscr{N}_{\eta,\varepsilon} \subset \mathscr{C}_\eta(\mathbb{R}, \mathscr{X})$ of (B.4) found for $\eta \in [\widetilde{\eta}, \gamma]$, with $\widetilde{\eta}$ taken such that $0 < \widetilde{\eta} < \gamma/k$, and to show that the map $\Phi : \mathscr{E}_0 \to \mathscr{C}_\eta(\mathbb{R}, \mathscr{X})$ is of class $\mathscr{C}^k$ for all $\eta \in (k\widetilde{\eta}, \gamma]$, with

$$D^p\Phi(u_0(0)) \in \mathscr{L}^p(\mathscr{E}_0, \mathscr{C}_{k\widetilde{\eta}}(\mathbb{R}, \mathscr{X})).$$

Here $\mathscr{L}^p(\mathscr{E}_0, \mathscr{C}_{k\widetilde{\eta}}(\mathbb{R}, \mathscr{X}))$ denotes the Banach space of p-linear continuous maps from $\mathscr{E}_0$ into $\mathscr{C}_{k\widetilde{\eta}}(\mathbb{R}, \mathscr{X})$. Several proofs of this result are available in the literature, all being quite long and technical. In particular, in [121] an abstract theorem for contractions on embedded Banach spaces is used, whereas in [120] a fiber contraction theorem due to Hirsch and Pugh [48] is used. While we refer to these works for further details, we only point out that the derivative $D\Phi(u_0(0))$ is the fixed point in $\mathscr{L}(\mathscr{E}_0, \mathscr{C}_{\widetilde{\eta}}(\mathbb{R}, \mathscr{X}))$ of the linear equation

$$D\Phi(u_0(0)) = \mathbf{S} + \mathbf{K}D_u\mathbf{R}^\varepsilon(\Phi(u_0(0))D\Phi(u_0(0)),$$

which may be differentiated up to order k. In particular, this implies that $D\mathbf{P}_h\Phi(0) = 0$ and $D\Psi(0) = 0$, and ends the proof of Theorem 2.9. □

B.2 Proof of Theorem 2.17 (Semilinear Case)

We prove here the first part of Theorem 2.17. The second part is an immediate consequence of this and of Theorem 2.9.

Estimates on the Resolvent of $\mathbf{L}_h$

First, the estimates on the resolvent (2.9) and (2.10) together with the fact that $\mathbf{L}_h$ has no spectrum on the imaginary axis, i.e., $i\omega\mathbb{I}-\mathbf{L}_h$ is invertible for any $\omega\in\mathbb{R}$, imply that there exists a positive constant c_1 such that for any $\omega\in\mathbb{R}$, the following estimates hold:

$$\|(i\omega\mathbb{I}-\mathbf{L}_h)^{-1}\|_{\mathscr{L}(\mathscr{X}_h)}\leq\frac{c_1}{1+|\omega|},\quad \|(i\omega\mathbb{I}-\mathbf{L}_h)^{-1}\|_{\mathscr{L}(\mathscr{Z}_h)}\leq\frac{c_1}{1+|\omega|},\tag{B.7}$$

$$\|(i\omega\mathbb{I}-\mathbf{L}_h)^{-1}\|_{\mathscr{L}(\mathscr{X}_h,\mathscr{Z}_h)}\leq c_1,\quad \|(i\omega\mathbb{I}-\mathbf{L}_h)^{-1}\|_{\mathscr{L}(\mathscr{Y}_h,\mathscr{Z}_h)}\leq\frac{c_1}{(1+|\omega|)^{1-\alpha}}.\tag{B.8}$$

Next, we claim that there exist $\delta>0$ and $M>0$ such that any $\lambda\in\mathbb{C}$ satisfying

$$\lambda=\mu+i\omega,\quad |\mu|\leq\delta(1+|\omega|)$$

belongs to the resolvent set of $\mathbf{L}_h$, and that the following estimates hold:

$$\|(\lambda\mathbb{I}-\mathbf{L}_h)^{-1}\|_{\mathscr{L}(\mathscr{X}_h)}\leq\frac{M}{1+|\lambda|},\quad \|(\lambda\mathbb{I}-\mathbf{L}_h)^{-1}\|_{\mathscr{L}(\mathscr{X}_h,\mathscr{Z}_h)}\leq M,\tag{B.9}$$

$$\|(\lambda\mathbb{I}-\mathbf{L}_h)^{-1}\|_{\mathscr{L}(\mathscr{Y}_h,\mathscr{Z}_h)}\leq\frac{M}{(1+|\lambda|)^{1-\alpha}}.\tag{B.10}$$

Indeed, we can write

$$\lambda\mathbb{I}-\mathbf{L}_h=\left(\mathbb{I}+\mu(i\omega\mathbb{I}-\mathbf{L}_h)^{-1}\right)(i\omega\mathbb{I}-\mathbf{L}_h)=(i\omega\mathbb{I}-\mathbf{L}_h)\left(\mathbb{I}+\mu(i\omega\mathbb{I}-\mathbf{L}_h)^{-1}\right),$$

and choosing $\delta=1/2c_1$ from equalities (B.7) we find

$$\|\mu(i\omega\mathbb{I}-\mathbf{L}_h)^{-1}\|_{\mathscr{L}(\mathscr{X}_h)}\leq\frac{1}{2},\quad \|\mu(i\omega\mathbb{I}-\mathbf{L}_h)^{-1}\|_{\mathscr{L}(\mathscr{Z}_h)}\leq\frac{1}{2}.$$

This shows that the operator $\mathbb{I}+\mu(i\omega\mathbb{I}-\mathbf{L}_h)^{-1}$ has a bounded inverse in both $\mathscr{L}(\mathscr{X}_h)$ and $\mathscr{L}(\mathscr{Z}_h)$, so that $\lambda\mathbb{I}-\mathbf{L}_h$ is invertible. Furthermore, inequalities (B.7) and (B.8) imply the inequalities above on the norms of $(\lambda\mathbb{I}-\mathbf{L}_h)^{-1}$.

We set

$$\beta\stackrel{\text{def}}{=}\min\{|\operatorname{Re}\lambda|\,;\,\lambda\in\sigma(\mathbf{L}_h)\}\geq\delta>0.\tag{B.11}$$

(Recall that $\beta>\gamma>0$, according to Hypothesis 2.4(i).)

Construction of $\mathbf{S}_\pm$

Consider the curves Γ_+ and Γ_- in $\mathbb{C}$ defined by

$$\Gamma_+ = \{-\delta|\omega| + i\omega\ ;\ \omega \in \mathbb{R}\}, \quad \Gamma_- = \{\delta|\omega| + i\omega\ ;\ \omega \in \mathbb{R}\},$$

and oriented such that ω increases along Γ_+ and decreases along Γ_-. The results above imply that these two curves lie in the resolvent set of $\mathbf{L}_h$, and that for any λ on one of these two curves the estimates (B.9) and (B.10) hold.

For $t > 0$ we define

$$\mathbf{S}_+(t) = \frac{1}{2i\pi}\int_{\Gamma_+} e^{\lambda t}(\lambda\mathbb{I} - \mathbf{L}_h)^{-1} d\lambda \in \mathscr{L}(\mathscr{X}_h, \mathscr{Z}_h),$$

for which the estimates (B.9) and the dominated convergence theorem allows us to show that it is well defined and that the map $t \mapsto \mathbf{S}_+(t)$ is differentiable with

$$\frac{d^n\mathbf{S}_+(t)}{dt^n} = (\mathbf{L}_h)^n\mathbf{S}_+(t) \text{ for all } n \geq 1. \tag{B.12}$$

Similarly, for $t < 0$ we set

$$\mathbf{S}_-(t) = \frac{1}{2i\pi}\int_{\Gamma_-} e^{\lambda t}(\lambda\mathbb{I} - \mathbf{L}_h)^{-1} d\lambda \in \mathscr{L}(\mathscr{X}_h, \mathscr{Z}_h),$$

for which we have that

$$\frac{d^n\mathbf{S}_-(t)}{dt^n} = (\mathbf{L}_h)^n\mathbf{S}_-(t) \text{ for all } n \geq 1. \tag{B.13}$$

Furthermore, the commutativity property

$$\mathbf{L}_h(\lambda\mathbb{I} - \mathbf{L}_h)^{-1} = (\lambda\mathbb{I} - \mathbf{L}_h)^{-1}\mathbf{L}_h$$

implies that

$$\mathbf{L}_h\mathbf{S}_+(t)u = \mathbf{S}_+(t)\mathbf{L}_h u \text{ for all } u \in \mathscr{Z}_h,\ t > 0,$$
$$\mathbf{L}_h\mathbf{S}_-(t)u = \mathbf{S}_-(t)\mathbf{L}_h u \text{ for all } u \in \mathscr{Z}_h,\ t < 0$$

and using the estimate (B.10) we show that for any fixed $\beta' < \beta$ and for $0 < \gamma' < \beta'$, there exists $M' > 0$ such that the estimates

$$\|\mathbf{S}_+(t)\|_{\mathscr{L}(\mathscr{Y}_h, \mathscr{Z}_h)} \leq M'(1 + t^{-\alpha})e^{-\gamma' t} \text{ for all } t > 0,$$
$$\|\mathbf{S}_-(t)\|_{\mathscr{L}(\mathscr{Y}_h, \mathscr{Z}_h)} \leq M'(1 + |t|^{-\alpha})e^{-\gamma'|t|} \text{ for all } t < 0, \tag{B.14}$$

hold. The following lemma is proved at the end of this section.

Lemma B.1 *The limits*

$$\mathbf{P}_- = \lim_{t\to 0^+} \mathbf{S}_+(t)|_{\mathscr{Y}_h} \in \mathscr{L}(\mathscr{Y}_h, \mathscr{X}_h), \quad \mathbf{P}_+ = \lim_{t\to 0^-} \mathbf{S}_-(t)|_{\mathscr{Y}_h} \in \mathscr{L}(\mathscr{Y}_h, \mathscr{X}_h), \tag{B.15}$$

exist, and

$$(\mathbf{P}_+ + \mathbf{P}_-)u = u \text{ for all } u \in \mathscr{Y}_h. \tag{B.16}$$

Checking Hypothesis 2.7

We now use the operators $\mathbf{S}_+(t)$ and $\mathbf{S}_-(t)$ above to solve the linear differential equation

$$\frac{du_h}{dt} = \mathbf{L}_h u_h + f(t). \tag{B.17}$$

We show that for any $f \in \mathscr{C}_\eta(\mathbb{R}, \mathscr{Y}_h)$ with $\eta \in [0, \gamma]$ this equation has a unique solution $u_h \in \mathscr{C}_\eta(\mathbb{R}, \mathscr{Z}_h)$ given by

$$u_h(t) = (\mathbf{K}_h f)(t) \stackrel{\text{def}}{=} \int_{-\infty}^{t} \mathbf{S}_+(t-s)f(s)ds - \int_{t}^{\infty} \mathbf{S}_-(t-s)f(s)ds, \tag{B.18}$$

with the properties in Hypothesis 2.7.

We assume that γ' in (B.14) is such that $\beta > \gamma' > \gamma$. Then using these two estimates and the dominated convergence theorem it is straightforward to check that $u_h = \mathbf{K}_h f \in \mathscr{C}_\eta(\mathbb{R}, \mathscr{Z}_h)$, and that the linear map $\mathbf{K}_h \in \mathscr{L}(\mathscr{C}_\eta(\mathbb{R}, \mathscr{Y}_h), \mathscr{C}_\eta(\mathbb{R}, \mathscr{Z}_h))$, with norm satisfying the inequality in Hypothesis 2.7. Moreover using (B.12), (B.13), and Lemma B.1, we obtain that in $\mathscr{X}_h$ the following holds

$$\begin{aligned}
\frac{du_h}{dt} &= \lim_{s\to t^-} \mathbf{S}_+(t-s)f(s) + \lim_{s\to t^+} \mathbf{S}_-(t-s)f(s) \\
&\quad + \int_{-\infty}^{t} \mathbf{L}_h \mathbf{S}_+(t-s)f(s)ds - \int_{t}^{\infty} \mathbf{L}_h \mathbf{S}_-(t-s)f(s)ds \\
&= \mathbf{L}_h u_h(t) + (\mathbf{P}_+ + \mathbf{P}_-)f(t) \\
&= \mathbf{L}_h u_h(t) + f(t).
\end{aligned}$$

Consequently, u_h is a solution of equation (B.17). It remains to prove the uniqueness of this solution.

Assume that $\widetilde{u}_h(t) \in \mathscr{C}_\eta(\mathbb{R}, \mathscr{Z}_h)$ is a solution of the homogeneous equation

$$\frac{d\widetilde{u}_h}{dt} = \mathbf{L}_h \widetilde{u}_h.$$

We show that $\widetilde{u}_h = 0$. Take any $t_0 \in \mathbb{R}$, and define

$$\widetilde{u}_+(t) \stackrel{\text{def}}{=} \mathbf{S}_+(t_0 - t)\widetilde{u}_h(t) \text{ for all } t < t_0,$$
$$\widetilde{u}_-(t) \stackrel{\text{def}}{=} \mathbf{S}_-(t_0 - t)\widetilde{u}_h(t) \text{ for all } t > t_0.$$

Then we have

$$\frac{d\widetilde{u}_+(t)}{dt} = -\mathbf{S}_+(t_0-t)\mathbf{L}_h\widetilde{u}_h(t) + \mathbf{S}_+(t_0-t)\frac{d\widetilde{u}_h}{dt}(t) = 0$$

in $\mathscr{X}_h$ for all $t < t_0$, hence

$$\widetilde{u}_+(t) = \lim_{s\to-\infty} \widetilde{u}_+(s) \text{ for all } t < t_0.$$

Using (B.21) and the continuous embedding from $\mathscr{Y}_h$ into $\mathscr{X}_h$, it follows that for $\eta < \gamma' < \beta$, there is a constant $C_{\gamma'}$ such that

$$\|\mathbf{S}_+(t)\|_{\mathscr{L}(\mathscr{Z}_h,\mathscr{X}_h)} \leq C_{\gamma'} e^{-\gamma' t} \text{ for all } t > 0.$$

Consequently,

$$\|\widetilde{u}_+(s)\|_{\mathscr{X}_h} \leq \|\mathbf{S}_+(t_0-s)\|_{\mathscr{L}(\mathscr{Z}_h,\mathscr{X}_h)}\|\widetilde{u}_h(s)\|_{\mathscr{Z}_h} \leq C_{\gamma'} e^{-\gamma'(t_0-s)} e^{\eta|s|}\|\widetilde{u}_h\|_{\mathscr{C}_\eta},$$

so that $\|\widetilde{u}_+(s)\|_{\mathscr{X}_h} \to 0$ as $s \to -\infty$. This implies that $\widetilde{u}_+(t) = 0$ for all $t < t_0$, and similarly we find that $\widetilde{u}_-(t) = 0$ for all $t > t_0$. From the definitions of $\mathbf{P}_+$ and $\mathbf{P}_-$, and from Lemma B.1, we conclude that $\widetilde{u}_h(t_0) = 0$. Since t_0 is arbitrary we have that $\widetilde{u}_h = 0$, which completes the proof of Theorem 2.17. □

Remark B.2 *In the particular case when $\sigma^+ = \varnothing$, we can define the bounded projection $\mathbf{P}_- = \mathbb{I} - \mathbf{P}_0 = \mathbf{P}_h$. The estimates (B.9) imply in this case that the linear operator $\mathbf{L}_- = \mathbf{L}_h$ is the infinitesimal generator of an analytic semigroup $(\mathbf{S}_+(t))_{t\geq 0}$ in $\mathscr{X}_h$, which satisfies*

$$\|\mathbf{S}_+(t)\|_{\mathscr{L}(\mathscr{Z}_h)} \leq ce^{-\gamma' t} \textit{ for all } t \geq 0.$$

This allows us to give a simpler proof of Theorem 2.17 in this case.

Proof (of Lemma B.1) For any $\eta > 0$ we define the paths Γ_+^η and Γ_-^η in $\mathbb{C}$ by

$$\Gamma_+^\eta = \{-\delta|\omega| + i\omega\,;\, \omega \in \mathbb{R},\, |\omega| \geq \delta^{-1}\eta\} \cup \{-\eta + i\omega\,;\, \omega \in \mathbb{R},\, |\omega| \leq \delta^{-1}\eta\},$$
$$\Gamma_-^\eta = \{\delta|\omega| + i\omega\,;\, \omega \in \mathbb{R},\, |\omega| \geq \delta^{-1}\eta\} \cup \{\eta + i\omega\,;\, \omega \in \mathbb{R},\, |\omega| \leq \delta^{-1}\eta\},$$

and orient them such that ω increases along Γ_+^η, and decreases along Γ_-^η (see Figure B.1).

For any $\eta \in (0,\beta)$ we can rewrite $\mathbf{S}_\pm(t)$ as

$$\mathbf{S}_+(t) = \frac{1}{2i\pi}\int_{\Gamma_+^\eta} e^{\lambda t}(\lambda\mathbb{I} - \mathbf{L}_h)^{-1} d\lambda \text{ for all } t > 0,$$
$$\mathbf{S}_-(t) = \frac{1}{2i\pi}\int_{\Gamma_-^\eta} e^{\lambda t}(\lambda\mathbb{I} - \mathbf{L}_h)^{-1} d\lambda \text{ for all } t < 0. \tag{B.19}$$

Using the identity

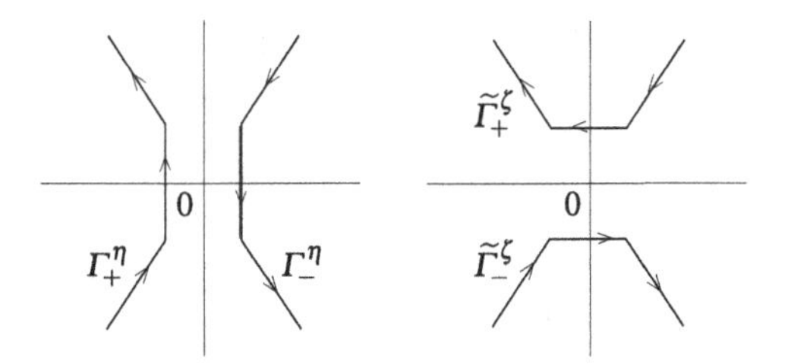

Fig. B.1 Plot in the complex plane of the paths $\Gamma_\pm^\eta$ (left) and $\widetilde{\Gamma}_\pm^\zeta$ (right).

$$\mathbf{L}_h(\lambda\mathbb{I}-\mathbf{L}_h)^{-1} = -\mathbb{I}_{\mathscr{X}_h} + \lambda(\lambda\mathbb{I}-\mathbf{L}_h)^{-1}, \tag{B.20}$$

which holds for any $\lambda \in \rho(\mathbf{L}_h)$, we obtain

$$\mathbf{S}_+(t) = \frac{1}{2i\pi}\left(\int_{\Gamma_+^\eta}\frac{e^{\lambda t}}{\lambda}d\lambda\right)\mathbb{I}_{\mathscr{X}_h} + \frac{1}{2i\pi}\int_{\Gamma_+^\eta}\frac{e^{\lambda t}}{\lambda}\mathbf{L}_h(\lambda\mathbb{I}-\mathbf{L}_h)^{-1}d\lambda.$$

The first integral in the right hand side of this equality is independent of η, and by taking the limit as $\eta\to\infty$ we conclude that this integral vanishes.

Next, using (B.9) and the fact that $\alpha\in[0,1)$ we find that

$$\|\mathbf{S}_+(t)\|_{\mathscr{L}(\mathscr{Y}_h,\mathscr{X}_h)} \le C_\eta e^{-\eta t} \text{ for all } t>0, \tag{B.21}$$

for any $\eta\in(0,\beta)$, and we conclude that

$$\mathbf{P}_- = \lim_{t\to 0^+}\mathbf{S}_+(t)|_{\mathscr{Y}_h} = \frac{1}{2i\pi}\int_{\Gamma_+^\eta}\frac{\mathbf{L}_h}{\lambda}(\lambda\mathbb{I}-\mathbf{L}_h)^{-1}d\lambda \in \mathscr{L}(\mathscr{Y}_h,\mathscr{X}_h)$$

is well defined. Similarly, we find

$$\mathbf{P}_+ = \lim_{t\to 0^-}\mathbf{S}_-(t)|_{\mathscr{Y}_h} = \frac{1}{2i\pi}\int_{\Gamma_-^\eta}\frac{\mathbf{L}_h}{\lambda}(\lambda\mathbb{I}-\mathbf{L}_h)^{-1}d\lambda \in \mathscr{L}(\mathscr{Y}_h,\mathscr{X}_h)$$

for any $\eta\in(0,\beta)$, which proves the first part of the lemma.

In order to prove (B.16), we define for $\zeta>0$ the paths $\widetilde{\Gamma}_\pm^\zeta$ by

$$\widetilde{\Gamma}_+^\zeta = \{\mu+i\delta^{-1}|\mu|\,;\,\mu\in\mathbb{R},\ |\mu|\ge\delta\zeta\}\cup\{\mu+i\zeta\,;\,\mu\in\mathbb{R},\ |\mu|\le\delta\zeta\},$$
$$\widetilde{\Gamma}_-^\zeta = \{\mu-i\delta^{-1}|\mu|\,;\,\mu\in\mathbb{R},\ |\mu|\ge\delta\zeta\}\cup\{\mu-i\zeta\,;\,\mu\in\mathbb{R},\ |\mu|\le\delta\zeta\},$$

oriented such that μ decreases along $\widetilde{\Gamma}_+^\zeta$, and increases along $\widetilde{\Gamma}_-^\zeta$. Now observe that the operators

$$\mathbf{B}_+ \stackrel{\text{def}}{=} \frac{1}{2i\pi} \int_{\widetilde{\Gamma}_+^\zeta} \frac{\mathbf{L}_h}{\lambda} (\lambda \mathbb{I} - \mathbf{L}_h)^{-1} d\lambda,$$

$$\mathbf{B}_- \stackrel{\text{def}}{=} \frac{1}{2i\pi} \int_{\widetilde{\Gamma}_-^\zeta} \frac{\mathbf{L}_h}{\lambda} (\lambda \mathbb{I} - \mathbf{L}_h)^{-1} d\lambda,$$

are independent of ζ, and that the dominated convergence theorem shows their limit in $\mathscr{L}(\mathscr{Y}_h, \mathscr{X}_h)$, as $\zeta \to \infty$, vanishes. Consequently, $\mathbf{B}_\pm = 0$.

Next, for $\eta = \delta\zeta$, we define the oriented clockwise rectangular path

$$\Gamma_\eta = \Gamma_+^\eta + \Gamma_-^\eta - \widetilde{\Gamma}_+^\zeta - \widetilde{\Gamma}_-^\zeta.$$

Then we have

$$\begin{aligned}
\mathbf{P}_+ + \mathbf{P}_- &= \mathbf{B}_+ + \mathbf{B}_- + \frac{1}{2i\pi} \int_{\Gamma_\eta} \frac{\mathbf{L}_h}{\lambda} (\lambda \mathbb{I} - \mathbf{L}_h)^{-1} d\lambda \\
&= \frac{1}{2i\pi} \int_{\Gamma_\eta} \frac{\mathbf{L}_h}{\lambda} (\lambda \mathbb{I} - \mathbf{L}_h)^{-1} d\lambda
\end{aligned}$$

in $\mathscr{L}(\mathscr{Y}_h, \mathscr{X}_h)$, and by (B.20),

$$\mathbf{P}_+ + \mathbf{P}_- = \frac{1}{2i\pi} \int_{\Gamma_\eta} (\lambda \mathbb{I} - \mathbf{L}_h)^{-1} d\lambda - \frac{1}{2i\pi} \left(\int_{\Gamma_\eta} \frac{d\lambda}{\lambda} \right) \mathbb{I}_{\mathscr{L}(\mathscr{Y}_h, \mathscr{X}_h)},$$

where $\mathbb{I}_{\mathscr{L}(\mathscr{Y}_h, \mathscr{X}_h)}$ denotes the continuous embedding from $\mathscr{Y}_h$ into $\mathscr{X}_h$. The first integral on the right hand side of this equality vanishes, since the interior of the rectangle Γ_η does not contain any singularity of $(\lambda \mathbb{I} - \mathbf{L}_h)^{-1}$, which then proves (B.16), and completes the proof of the lemma. □

B.3 Proof of Theorem 3.9 (Nonautonomous Vector Fields)

The proof of Theorem 3.9 in Chapter 2 follows exactly the proof of Theorem 2.9 given in Section B.1. The main difference is that we replace the arbitrary data $u_0(0)$ in the integral formulation (B.4) by $u_0(s) = v_0$, which modifies trivially (B.4). The fixed point is now denoted by $\Phi(v_0, s)$, so that the corresponding solution of the system (3.9), modified by the cut-off function, is

$$u(v_0, s, t) = u_0(v_0, s, t) + (\Phi(v_0, s))(t),$$

where $u_0(v_0, s, s) = v_0 \in \mathscr{E}_0$. We set

$$\Psi(u_0, t) \stackrel{\text{def}}{=} (\Phi(u_0, t))(t).$$

Then the uniqueness of the fixed point implies that

$$(\Phi(u_0(v_0,s,\tau),\tau))(t) = (\Phi(v_0,s))(t),$$

hence

$$\Psi(u_0(v_0,s,t),t) = (\Phi(v_0,s))(t).$$

This proves that

$$u_0(v_0,s,t) + \Psi(u_0(v_0,s,t),t) = u(v_0,s,t),$$

i.e., the set

$$\{(t, u_0 + \Psi(u_0,t))\ ;\ (u_0,t) \in \mathscr{E}_0 \times \mathbb{R}\}$$

is an integral manifold for (3.9) modified by the cut-off function. Restricting to the ball $B_\varepsilon(\mathscr{E}_0)$, this implies property (i) of the theorem. Property (ii) is obtained as for Theorem 2.9. □

We conclude this section with a brief proof of the particular cases in Corollary 3.11.

Property (i) results from a standard property of τ-periodic systems, which implies here that

$$u(v_0,s,t) = u(v_0,s+\tau,t+\tau).$$

This leads directly to

$$\Psi(u_0,t) = \Psi(u_0,t+\tau).$$

Part (ii) is obtained from the integral formulations (B.4) for both $u(v_0,s,t)$ and $u_\infty(v_0,s,t)$ by estimating, in a straightforward way, their difference as $t \to \infty$. □

B.4 Proof of Theorem 3.13 (Equivariant Systems)

The uniqueness of the global center manifold for the modified system (B.2) in Section B.1 implies that this manifold is invariant under $\mathbf{T}$, provided system (B.2) is equivariant under $\mathbf{T}$. Since equation (2.1) is equivariant under $\mathbf{T}$, the modified is also equivariant when the cut-off function χ satisfies

$$\chi(\mathbf{T}_0 u_0) = \chi(u_0) \text{ for all } u_0 \in \mathscr{E}_0. \tag{B.22}$$

Taking the Euclidean norm in $\mathscr{E}_0$, which is finite-dimensional, for any isometry $\mathbf{T}_0$ on $\mathscr{E}_0$ we can choose χ to be a smooth function of $\|u_0\|^2$, such that (B.22) holds. Consequently, the result in the theorem follows from the fact that $\mathbf{T}_0$ is an isometry on $\mathscr{E}_0$. □

B.5 Proof of Theorem 3.22 (Empty Unstable Spectrum)

With the notations from Section B.1, assume that $u(\cdot;u(0)) \in \mathscr{C}^0(\mathbb{R}^+,\mathscr{X})$ is a solution of (2.1), which belongs to $\mathscr{O}_{\varepsilon/4}$ for all $t \geq 0$. Consider $\overline{u}(t,u(0))$ defined through

$$\overline{u}(t;u(0)) = \begin{cases} u(t;u(0)) \text{ for } t \geq 0 \\ \widetilde{u}(t;u(0)) \text{ for } t \leq 0, \end{cases}$$

where $\widetilde{u}(\cdot;u(0))$ is the solution of the modified equation

$$\frac{d\widetilde{u}}{dt} = \mathbf{L}_0\widetilde{u} + \mathbf{P}_0\mathbf{R}^\varepsilon(\widetilde{u}), \quad \widetilde{u}(0) = u(0).$$

Notice that $\mathbf{P}_-\widetilde{u}(t) = \mathbf{P}_-u(0)$, where $\mathbf{P}_- = \mathbb{I} - \mathbf{P}_0 = \mathbf{P}_h$. Then we find that

$$\sup_{t\leq 0}\left(e^{\eta t}\|\mathbf{P}_0\overline{u}(t;u(0))\|_{\mathscr{E}_0}\right) < \infty,$$

and since

$$\|\mathbf{P}_0 u(t;u(0))\|_{\mathscr{E}_0} \leq \frac{\varepsilon}{4}, \quad \|\mathbf{P}_- u(t;u(0))\|_{\mathscr{X}} \leq \frac{\varepsilon}{4} \text{ for all } t \geq 0,$$

we have that $\overline{u}(\cdot;u(0) \in \mathscr{N}_{\eta,\varepsilon/4}$. Moreover, for all $t \in \mathbb{R}$,

$$\mathbf{P}_0\overline{u}(t;u(0)) = e^{\mathbf{L}_0 t}\mathbf{P}_0 u(0) + \int_0^t e^{\mathbf{L}_0(t-\tau)}\mathbf{P}_0\mathbf{R}^\varepsilon(\overline{u}(\tau;u(0)))d\tau. \tag{B.23}$$

Now assume that we have found a solution

$$z \in \mathscr{F}_\eta(\mathbb{R},\mathscr{E}_0) \times \left(\mathscr{C}_0(\mathbb{R},B_{\varepsilon/2}(\mathscr{X}_-)) \cap \mathscr{F}_\eta(\mathbb{R},\mathscr{X}_-)\right),$$

which means that $z(t) \to 0$ exponentially as $t \to +\infty$, of the equation

$$\begin{aligned} z(t) &= -\mathbf{P}_-\overline{u}(t;u(0)) + \mathbf{K}_h\mathbf{P}_-\mathbf{R}^\varepsilon(\overline{u}(\cdot;u(0)) + z)(t) \\ &\quad - \int_t^\infty e^{\mathbf{L}_0(t-\tau)}\mathbf{P}_0\left[\mathbf{R}^\varepsilon(\overline{u}(\tau;u(0)) + z(\tau)) - \mathbf{R}^\varepsilon(\overline{u}(\tau;u(0)))\right]d\tau. \end{aligned} \tag{B.24}$$

We show below that $t \mapsto \overline{u}(t;u(0)) + z(t)$ is solution of (B.2), which belongs to $\mathscr{N}_{\eta,\varepsilon}$. As a consequence, $u(0) + z(0) \in \mathscr{M}_0$, and by construction

$$u(t,u(0)+z(0)) = u(t;u(0)) + z(t) \text{ for all } t > 0.$$

This is precisely the assertion in Theorem 3.22, since we have found a solution lying on $\mathscr{M}_0$ which is exponentially asymptotic to the solution $u(\cdot;u(0))$ of the initial value problem.

To end the proof, we show that $t \mapsto \overline{u}(t;u(0)) + z(t)$ is solution of (B.2). First, from (B.24) we observe that

$$\mathbf{P}_0 z(0) = -\int_0^\infty e^{-\mathbf{L}_0\tau}\mathbf{P}_0\left[\mathbf{R}^\varepsilon(\overline{u}(\tau;u(0))+z(\tau)) - \mathbf{R}^\varepsilon(\overline{u}(\tau;u(0)))\right]d\tau,$$

and using (B.23) we obtain

$$\begin{aligned}\overline{u}(t;u(0))+z(t) &= e^{\mathbf{L}_0 t}P_0(u(0)+z(0)) + (\mathbf{K}_h\mathbf{P}_-\mathbf{R}^\varepsilon(\overline{u}(\cdot;u(0))+z))(t)\\ &\quad + \int_0^t e^{\mathbf{L}_0(t-\tau)}\mathbf{P}_0\mathbf{R}^\varepsilon(\overline{u}(\tau;u(0))+z(\tau))\,d\tau.\end{aligned}$$

This is equivalent to the fact that $\overline{u}(\cdot;u(0))+z \in \mathscr{N}_{\eta,\varepsilon}$ satisfies (B.2). It remains then to prove the existence of a solution

$$z \in \mathscr{F}_\eta(\mathbb{R},\mathscr{E}_0) \times \left(\mathscr{C}_0(\mathbb{R},B_{\varepsilon/2}(\mathscr{Z}_-)) \cap \mathscr{F}_\eta(\mathbb{R},\mathscr{Z}_-)\right)$$

of (B.24).

The argument is similar to the proof of the existence of the fixed point in the proof of the center manifold theorem in Section B.1. The main difference is that the space $\mathscr{C}_\eta$ is replaced by the space $\mathscr{F}_\eta$. Then for $z \in \mathscr{F}_\eta(\mathbb{R},\mathscr{E}_0) \times \left(\mathscr{C}_0(\mathbb{R},B_{\varepsilon/2}(\mathscr{Z}_-)) \cap \mathscr{F}_\eta(\mathbb{R},\mathscr{Z}_-)\right)$, examining all terms in (B.24), we find that:

(i) $\mathbf{R}^\varepsilon(\overline{u}(\tau;u(0))+z(\tau)) - \mathbf{R}^\varepsilon(\overline{u}(\tau;u(0))) \in \mathscr{F}_\eta(\mathbb{R},\mathscr{Y})$, and

$$t \mapsto -\int_t^\infty e^{\mathbf{L}_0(t-\tau)}\mathbf{P}_0\left[\mathbf{R}^\varepsilon(\overline{u}(\tau;u(0))+z(\tau)) - \mathbf{R}^\varepsilon(\overline{u}(\tau;u(0)))\right]d\tau \in \mathscr{F}_\eta(\mathbb{R},\mathscr{E}_0);$$

(ii) $\mathbf{K}_h\mathbf{P}_-\left(\mathbf{R}^\varepsilon(\overline{u}(\cdot;u(0))+z) - \mathbf{R}^\varepsilon(\overline{u}(\tau;u(0)))\right) \in \mathscr{F}_\eta(\mathbb{R},\mathscr{Z}_-) \cap \mathscr{C}_0(\mathbb{R},B_{\varepsilon/8}(\mathscr{Z}_-))$, by construction and from Hypothesis 3.20, for ε sufficiently small;

(iii) $v \overset{\text{def}}{=} \mathbf{K}_h\mathbf{P}_-(\mathbf{R}^\varepsilon(\overline{u}(\cdot;u(0)))) - \mathbf{P}_-\overline{u}(t;u(0)) \in \mathscr{F}_\eta(\mathbb{R},\mathscr{Z}_-) \cap \mathscr{C}_0(\mathbb{R},B_{3\varepsilon/8}(\mathscr{Z}_-))$.

The last property follows from the fact that $v \in \mathscr{C}_\eta(\mathbb{R},\mathscr{Z}_-)$, and v is by construction a solution of

$$\frac{dv}{dt} = \mathbf{L}_- v$$

for $t \geq 0$, which by Hypothesis 3.20 implies the exponential convergence to 0 as $t \to \infty$, that $\mathbf{K}_h\mathbf{P}_-(\mathbf{R}^\varepsilon(\overline{u}(\cdot;u(0)))) \in \mathscr{C}_0(\mathbb{R},B_{\varepsilon/8}(\mathscr{Z}_-))$, and that $\mathbf{P}_-\overline{u}(t;u(0)) = \mathbf{P}_- u(0) \in B_{\varepsilon/4}(\mathscr{Z}_-)$ for $t < 0$. This completes the proof of Theorem 3.22. □

C Normal Forms

The references in this section are to theorems, hypotheses, formulas, and remarks in Chapter 3.

C.1 Proof of Lemma 1.13 (0^3 Normal Form)

The proof below can be found in [25] (see also [20] for a different proof).

We set

$$\mathbf{N}(u) = (\Phi_1(A,B,C), \Phi_2(A,B,C), \Phi_3(A,B,C)), \quad u = (A,B,C),$$

where Φ_1, Φ_2, and Φ_3 are polynomials in (A,B,C). Then we have $\mathbf{L}^*u = (0,A,B)$ and the characterization (1.5) leads to

$$\begin{aligned} A\frac{\partial \Phi_1}{\partial B} + B\frac{\partial \Phi_1}{\partial C} &= 0 \\ A\frac{\partial \Phi_2}{\partial B} + B\frac{\partial \Phi_2}{\partial C} &= \Phi_1 \\ A\frac{\partial \Phi_3}{\partial B} + B\frac{\partial \Phi_3}{\partial C} &= \Phi_2. \end{aligned} \tag{C.1}$$

Since A and $B^2 - 2AC$ are first integrals of the linear vector field $\mathbf{L}^*$, we choose the new variables

$$\widetilde{A} = A, \quad \widetilde{B} = B^2 - 2AC, \quad \widetilde{C} = B,$$

where the change of variables is nonsingular as soon as $A \neq 0$, and define

$$\widetilde{\Phi}_j(\widetilde{A}, \widetilde{B}, \widetilde{C}) = \Phi_j(A,B,C), \quad j = 1,2,3.$$

Then the equations (C.1) become

$$\widetilde{A}\frac{\partial \widetilde{\Phi}_1}{\partial \widetilde{C}} = 0, \quad \widetilde{A}\frac{\partial \widetilde{\Phi}_2}{\partial \widetilde{C}} = \widetilde{\Phi}_1, \quad \widetilde{A}\frac{\partial \widetilde{\Phi}_3}{\partial \widetilde{C}} = \widetilde{\Phi}_2, \tag{C.2}$$

so that

$$\widetilde{\Phi}_1(\widetilde{A}, \widetilde{B}, \widetilde{C}) = \phi_1(\widetilde{A}, \widetilde{B}).$$

We claim that ϕ_1 is a polynomial in its arguments. Indeed, by construction we have that

$$\Phi_1(A,B,C) = \Phi_1\left(\widetilde{A}, \widetilde{C}, \frac{\widetilde{C}^2 - \widetilde{B}}{2\widetilde{A}}\right),$$

which shows that ϕ_1 is a polynomial in $\widetilde{B}$ with rational coefficients in $\widetilde{A}$. Consequently, we can write

$$\Phi_1(A,B,C) = \sum_{k\in\mathbb{N}} f_k(\widetilde{A})\widetilde{B}^k = \sum_{i\in\mathbb{Z},\, k\in\mathbb{N}} f_{ik} A^i \widetilde{B}^k.$$

Assume that $\iota = \min\{i \in \mathbb{Z}\,;\, f_{ik} \neq 0 \text{ for some } k \in \mathbb{N}\} < 0$. Multiplying by $A^{-\iota}$ and setting $A = 0$ yields

$$\sum_{k\in\mathbb{N}} f_{\iota k} B^{2k} = 0,$$

which implies that $f_{\iota k} = 0$. This contradicts the assumption $\iota < 0$, so that $\iota \geq 0$. Consequently, ϕ_1 is a polynomial of $(\widetilde{A},\widetilde{B})$, which proves the claim.

We write now

$$\phi_1(\widetilde{A},\widetilde{B}) = \widetilde{A}P_1(\widetilde{A},\widetilde{B}) + \psi_1(\widetilde{B}),$$

where P_1 and ψ_1 are polynomials, $\psi_1(\widetilde{B}) = \phi_1(0,\widetilde{B})$. Solving the second equation in (C.2) leads to

$$\widetilde{\Phi}_2(\widetilde{A},\widetilde{B},\widetilde{C}) = BP_1(\widetilde{A},\widetilde{B}) + \frac{B}{\widetilde{A}}\psi_1(\widetilde{B}) + \frac{\phi_2(\widetilde{A},\widetilde{B})}{\widetilde{A}},$$

where the same proof as above shows that ϕ_2 is a polynomial in its arguments. Multiplying this equality by $\widetilde{A}$ and setting $\widetilde{A} = 0$, we obtain

$$B\psi_1(B^2) + \phi_2(0,B^2) = 0 \text{ for all } B.$$

The two terms on the right hand side of this equality have different parity, so that $\psi_1(B^2) = \phi_2(0,B^2) = 0$, and then

$$\psi_1(\widetilde{B}) = 0, \quad \phi_2(0,\widetilde{B}) = 0 \text{ for all } \widetilde{B}.$$

Consequently, the polynomial $\phi_2(\widetilde{A},\widetilde{B})$ is divisible by $\widetilde{A}$ and we can write

$$\begin{aligned}
\widetilde{\Phi}_1(\widetilde{A},\widetilde{B},\widetilde{C}) &= AP_1(\widetilde{A},\widetilde{B}), \\
\widetilde{\Phi}_2(\widetilde{A},\widetilde{B},\widetilde{C}) &= BP_1(\widetilde{A},\widetilde{B}) + AP_2(\widetilde{A},\widetilde{B}) + \psi_2(\widetilde{B}),
\end{aligned}$$

where P_2 and ψ_2 are polynomials.

Finally, solving the last equation in (C.2) leads to

$$\widetilde{\Phi}_3(\widetilde{A},\widetilde{B},\widetilde{C}) = CP_1(\widetilde{A},\widetilde{B}) + BP_2(\widetilde{A},\widetilde{B}) + \frac{B}{A}\psi_2(\widetilde{B}) + \frac{\phi_3(\widetilde{A},\widetilde{B})}{A},$$

and the same arguments as the ones above for ψ_1 and ϕ_2 apply here for ψ_2 and ϕ_3. Then we conclude that $\psi_2(\widetilde{B}) = 0$, and $\phi_3(\widetilde{A},\widetilde{B})/A = P_3(\widetilde{A},\widetilde{B})$, which is a polynomial. This completes the proof of Lemma 1.13. □

C.2 *Proof of Lemma 1.17 ($(i\omega)^2$ Normal Form)*

We define

$$\mathbf{N}(u) = (\Phi_1(A,B,\overline{A},\overline{B}), \Phi_2(A,B,\overline{A},\overline{B}), \overline{\Phi_1(A,B,\overline{A},\overline{B})}, \overline{\Phi_1(A,B,\overline{A},\overline{B})}),$$

for $u = (A,B,\overline{A},\overline{B})$, and consider the differential operator

$$\mathscr{D}^* = -i\omega A\frac{\partial}{\partial A} + (A - i\omega B)\frac{\partial}{\partial B} + i\omega\overline{A}\frac{\partial}{\partial \overline{A}} + (\overline{A} + i\omega\overline{B})\frac{\partial}{\partial \overline{B}}.$$

Then using the characterization (1.5) in its complex form (see Remark 1.5), we find the system

$$\mathscr{D}^* \Phi_1 = -i\omega\Phi_1, \quad \mathscr{D}^* \Phi_2 = -i\omega\Phi_2 + \Phi_1.$$

First, notice that

$$\mathscr{D}^* A = -i\omega A, \quad \mathscr{D}^* B = A - i\omega B,$$

and that the equation $\mathscr{D}^* u = 0$ has the following three independent first integrals:

$$u_1 = A\overline{A}, \quad u_2 = i(A\overline{B} - \overline{A}B), \quad u_3 = i\omega\frac{B}{A} + \ln A.$$

Since $\mathscr{D}^*(\Phi_1/A) = 0$, we have that Φ_1/A is a first integral of $\mathscr{D}^* u = 0$, as well. Consequently,

$$\Phi_1(A, B, \overline{A}, \overline{B}) = A\phi(u_1, u_2, u_3) \tag{C.3}$$

for some function ϕ.

We claim that ϕ is a polynomial in u_1, u_2, and that it is independent of u_3. Indeed, we have

$$\begin{aligned}
\frac{\partial \phi}{\partial u_1} &= \frac{1}{A^2}\frac{\partial \Phi_1}{\partial \overline{A}} - \frac{B}{A^3}\frac{\partial \Phi_1}{\partial \overline{B}} \\
\frac{\partial \phi}{\partial u_2} &= \frac{-i}{A^2}\frac{\partial \Phi_1}{\partial \overline{B}} \\
\frac{\partial \phi}{\partial u_3} &= \frac{1}{i\omega}\frac{\partial \Phi_1}{\partial B} + \frac{\overline{A}}{i\omega A}\frac{\partial \Phi_1}{\partial \overline{B}}.
\end{aligned}$$

Assume that Φ_1 is of degree $n-1$. Then

$$\frac{\partial^n \phi}{\partial u_j^n} = 0, \quad j = 1, 2, 3,$$

so that

$$\frac{\partial^k \phi}{\partial u_1^\alpha \partial u_2^\beta \partial u_3^\gamma} = 0 \quad \text{for } \alpha + \beta + \gamma = k \geq 3n.$$

This shows that ϕ is a polynomial in u_1, u_2, u_3. Next, assume that ϕ depends upon u_3. Comparing the behavior of Φ_1 and $A\phi$ in the equality (C.3), as $A \to \infty$, we obtain a contradiction between the polynomial behavior in the left hand one side and the logarithmic behavior in the right hand side. Consequently, ϕ does not depend upon u_3 and we can write

$$\Phi_1(A, B, \overline{A}, \overline{B}) = AP(u_1, u_2),$$

where P is a polynomial in its arguments.

Finally, notice that $BP(u_1, u_2)$ is a particular solution of the equation

$$\mathscr{D}^* \Phi_2 = -i\omega\Phi_2 + \Phi_1.$$

Then proceeding as above for Φ_1, we obtain that

$$\Phi_2(A,B,\overline{A},\overline{B}) = BP(u_1,u_2) + AQ(u_1,u_2),$$

which ends the proof of Lemma 1.17. □

C.3 *Proof of Lemma 1.18* $(0^2(\mathrm{i}\omega)$ *Normal Form)*

We define

$$\mathbf{N}(u) = (\Phi_1(A,B,C,\overline{C}), \Phi_2(A,B,C,\overline{C}), \Phi_3(A,B,C,\overline{C}), \overline{\Phi_3}(A,B,C,\overline{C}))$$

for $u = (A,B,C,\overline{C})$, and consider the differential operator

$$\mathscr{D}^* = A\frac{\partial}{\partial B} - i\omega C\frac{\partial}{\partial C} + i\omega\overline{C}\frac{\partial}{\partial \overline{C}}.$$

Using characterization (1.5), we have to solve the system

$$\mathscr{D}^*\Phi_1 = 0, \quad \mathscr{D}^*\Phi_2 = \Phi_1, \quad \mathscr{D}^*\Phi_3 = -i\omega\Phi_3. \tag{C.4}$$

First, notice that

$$x = A, \quad y = |C|^2, \quad z = A\ln C + i\omega B$$

are three independent first integrals of the linear equation $\mathscr{D}^*u = 0$. Using the local diffeomorphism $(A,B,C,\overline{C}) \mapsto (x,y,z,B)$, with Jacobian determinant $-A$, it is easy to show that Φ_1 expressed in the new variables, $\widetilde{\Phi_1}(x,y,z,B)$, satisfies

$$A\frac{\partial\widetilde{\Phi_1}}{\partial B} = 0.$$

Consequently, there is a function ϕ, which is smooth, except at the origin, such that

$$\Phi_1(A,B,C,\overline{C}) = \phi(x,y,z). \tag{C.5}$$

We claim that ϕ is a polynomial in x, y, and that it is independent of z. Indeed, we have the equalities

$$\begin{aligned}
\frac{\partial\phi}{\partial x} &= \frac{\partial\Phi_1}{\partial A} + \frac{i\ln C}{\omega}\frac{\partial\Phi_1}{\partial B} \\
\frac{\partial\phi}{\partial y} &= \frac{1}{C}\frac{\partial\Phi_1}{\partial\overline{C}} \\
\frac{\partial\phi}{\partial z} &= \frac{1}{i\omega}\frac{\partial\Phi_1}{\partial B},
\end{aligned}$$

so that

$$\frac{\partial^n \phi}{\partial x^n} = \frac{\partial^n \Phi_1}{\partial A^n} + \left(\frac{i \ln C}{\omega}\right)^n \frac{\partial^n \Phi_1}{\partial B^n}$$
$$\frac{\partial^n \phi}{\partial y^n} = \frac{1}{C^n} \frac{\partial^n \Phi_1}{\partial \overline{C}^n}$$
$$\frac{\partial^n \phi}{\partial z^n} = \frac{1}{(i\omega)^n} \frac{\partial^n \Phi_1}{\partial B^n}.$$

The right hand sides of these equalities vanish for n sufficiently large, which implies that ϕ is a polynomial in its arguments. Next, assume that ϕ depends upon z. Comparing the behavior of Φ_1 and ϕ in the equality (C.5), as $A \to \infty$, we obtain a contradiction between the polynomial behavior in the left hand one side and the logarithmic behavior on the right hand side. Consequently, ϕ does not depend upon z and we can write

$$\Phi_1(A,B,C,\overline{C}) = \phi_0(A,|C|^2),$$

where ϕ_0 is a polynomial in its arguments.

Next, the second equation in (C.4) leads to

$$\Phi_2(A,B,C,\overline{C}) = \frac{B}{A}\phi_0(A,|C|^2) + \phi_1(A,|C|^2,z),$$

and using the fact that $A\Phi_2$ and ϕ_0 are polynomials, it follows that $A\phi_1 = \psi(A,|C|^2)$, with ψ a polynomial satisfying

$$B\phi_0(0,|C|^2) = \psi(0,|C|^2).$$

This implies that $\phi_0(0,|C|^2) = 0$, and that ϕ_1 is a polynomial in A, and $|C|^2$. Summarizing, there are two polynomials P_0 and P_1 in A and $|C|^2$, such that

$$\Phi_1(A,B,C,\overline{C}) = AP_0(A,|C|^2),$$
$$\Phi_2(A,B,C,\overline{C}) = BP_0(A,|C|^2) + P_1(A,|C|^2).$$

Finally, the equation for Φ_3 in (C.4) leads to

$$\mathscr{D}^*(\overline{C}\Phi_3) = 0.$$

Consequently,

$$\overline{C}\Phi_3(A,B,C,\overline{C}) = \phi_2(A,|C|^2),$$

where ϕ_2 is a polynomial such that $\phi_2(A,0) = 0$, and we conclude that there is a polynomial P_2 such that

$$\Phi_3(A,B,C,\overline{C}) = CP_2(A,|C|^2).$$

This completes the proof of Lemma 1.18. □

C.4 Proof of Lemma 1.19 (0^20^2 *Normal Form)*

We define

$$\mathbf{N}(u) = (\Phi_1(A,B,C,D), \Phi_2(A,B,C,D), \Phi_3(A,B,C,D), \Phi_4(A,B,C,D))$$

for $u = (A,B,C,D)$, and consider the differential operator

$$\mathscr{D}^* = A\frac{\partial}{\partial B} + C\frac{\partial}{\partial D}.$$

Using characterization (1.5), we obtain that

$$\mathscr{D}^*\Phi_1 = 0, \quad \mathscr{D}^*\Phi_2 = \Phi_1, \quad \mathscr{D}^*\Phi_3 = 0, \quad \mathscr{D}^*\Phi_4 = \Phi_3. \tag{C.6}$$

First, notice that

$$A, \quad C, \quad \widetilde{B} = BC - AD$$

are three independent first integrals of the equation $\mathscr{D}^*u = 0$. Then it is not difficult to show that there is a function ϕ_1 that is smooth, except at the origin, such that

$$\Phi_1(A,B,C,D) = \phi_1(A,C,\widetilde{B}).$$

Furthermore, we have the identities

$$\begin{aligned}
\frac{\partial^n \phi_1}{\partial A^n} &= \left(\frac{\partial}{\partial A} + \frac{D}{C}\frac{\partial}{\partial B}\right)^n \Phi_1 \\
\frac{\partial^n \phi_1}{\partial C^n} &= \left(\frac{\partial}{\partial C} + \frac{B}{A}\frac{\partial}{\partial D}\right)^n \Phi_1 \\
\frac{\partial^n \phi_1}{\partial \widetilde{B}^n} &= \left(\frac{1}{C}\frac{\partial}{\partial B}\right)^n \Phi_1,
\end{aligned}$$

and the right hand sides of these equalities vanish for n sufficiently large, since Φ_1 is a polynomial. Consequently, ϕ_1 is a polynomial in its arguments.

We decompose ϕ_1 as

$$\phi_1(A,C,\widetilde{B}) = AQ_1(A,C,\widetilde{B}) + CQ_2(A,C,\widetilde{B}) + Q_3(\widetilde{B}),$$

where Q_j are polynomials in their arguments. Notice that this decomposition is not unique. Now we set

$$\Phi_2(A,B,C,D) = BQ_1(A,C,\widetilde{B}) + DQ_2(A,C,\widetilde{B}) + \frac{B}{A}Q_3(\widetilde{B}) + \frac{1}{A}\widetilde{\Phi}_2(A,B,C,D) \tag{C.7}$$

so that the second equation in (C.6) leads to

$$\mathscr{D}^*\widetilde{\Phi}_2 = 0,$$

where $\widetilde{\Phi}_2$ is a polynomial in its arguments. The arguments used for ϕ_1 above, imply that

$$\widetilde{\Phi}_2(A,B,C,D) = \phi_2(A,C,\widetilde{B}),$$

with ϕ_2 polynomial in its arguments. Now multiplying (C.7) by A and taking $A = 0$ gives

$$BQ_3(BC) + \phi_2(0,C,BC)$$

for any $(B,C) \in \mathbb{R}^2$. This proves that

$$Q_3 = 0, \quad \phi_2(0,C,BC) = 0.$$

As a consequence, since B and BC are independent variables, the polynomial $\phi_2(A,C,\widetilde{B})$ is divisible by A, and there is a polynomial Q_4 such that

$$\Phi_2(A,B,C,D) = BQ_1(A,C,\widetilde{B}) + DQ_2(A,C,\widetilde{B}) + Q_4(A,C,\widetilde{B}).$$

Finally, notice that we can write

$$\begin{aligned} Q_4(A,C,\widetilde{B}) &= P_3(A,C) + \widetilde{B}Q_5(A,C,\widetilde{B}) \\ &= P_3(A,C) + BCQ_5(A,C,\widetilde{B}) - DAQ_5(A,C,\widetilde{B}), \end{aligned}$$

hence we obtain the final form

$$\begin{aligned} \Phi_1(A,B,C,D) &= AP_1(A,C,\widetilde{B}) + CP_2(A,C,\widetilde{B}) \\ \Phi_2(A,B,C,D) &= BP_1(A,C,\widetilde{B}) + DP_2(A,C,\widetilde{B}) + P_3(A,C), \end{aligned} \tag{C.8}$$

where

$$P_1 = Q_1 + CQ_5, \quad P_2 = Q_2 - AQ_5.$$

In the same way, from the last two equalities in (C.6) we obtain

$$\begin{aligned} \Phi_3(A,B,C,D) &= AP_4(A,C,\widetilde{B}) + CP_5(A,C,\widetilde{B}) \\ \Phi_4(A,B,C,D) &= BP_4(A,C,\widetilde{B}) + DP_5(A,C,\widetilde{B}) + P_6(A,C), \end{aligned} \tag{C.9}$$

which completes the proof of Lemma 1.19. □

C.5 Proof of Theorem 2.2 (Perturbed Normal Forms)

We have to determine two polynomials $\mathbf{\Phi}_\mu$ and $\mathbf{N}_\mu$ of degree p in $\mathbb{R}^n$ with coefficients depending upon μ, which satisfy the equality

$$\mathscr{F}(\mathbf{\Phi}_\mu, \mathbf{N}_\mu, \mu) = 0. \tag{C.10}$$

With the notations from Hypothesis 2.1, the map $\mathscr{F} : \mathscr{H}^2 \times \mathscr{V}_\mu \to \mathscr{H}$ defined by

$$(\mathscr{F}(\boldsymbol{\Phi},\mathbf{N},\mu))(v) \stackrel{\text{def}}{=} (\mathscr{A}_{\mathbf{L}}\boldsymbol{\Phi})(v) + \mathbf{N}(v) + \Pi_p(D\boldsymbol{\Phi}(v)\mathbf{N}(v) - \mathbf{R}(v+\boldsymbol{\Phi}(v),\mu)) \tag{C.11}$$

is of class $\mathscr{C}^k$, where we recall that $\mathscr{H}$ is the space of polynomials of degree p, and Π_p the linear map that associates to a map of class $\mathscr{C}^p$ the polynomial of degree p in its Taylor expansion. For notational simplicity, we suppress the indices μ and write $\boldsymbol{\Phi}$ and $\mathbf{N}$ instead of $\boldsymbol{\Phi}_\mu$ and $\mathbf{N}_\mu$, respectively.

For $\mu = 0$ we recover exactly the situation treated in Theorem 1.2. This means that we have a solution $(\boldsymbol{\Phi},\mathbf{N}) = (\boldsymbol{\Phi}^{(0)},\mathbf{N}^{(0)})$ of (C.10) for $\mu = 0$,

$$\mathscr{F}(\boldsymbol{\Phi}^{(0)},\mathbf{N}^{(0)},0) = 0,$$

which is unique when we restrict to

$$\boldsymbol{\Phi} \in (\ker \mathscr{A}_{\mathbf{L}})^{\perp}, \quad \mathbf{N} \in \ker(\mathscr{A}_{\mathbf{L}^*}). \tag{C.12}$$

In order to determine the polynomials $\boldsymbol{\Phi}$ and $\mathbf{N}$ for μ close to zero, we use the implicit function theorem to solve (C.10), together with (C.12), i.e., with

$$\mathscr{F} : (\ker \mathscr{A}_{\mathbf{L}})^{\perp} \times \ker(\mathscr{A}_{\mathbf{L}^*}) \times \mathscr{V}_\mu \to \mathscr{H}.$$

First, we compute the differential

$$\mathscr{D}_0 \stackrel{\text{def}}{=} D_{(\boldsymbol{\Phi},\mathbf{N})}\mathscr{F}(\boldsymbol{\Phi}^{(0)},\mathbf{N}^{(0)},0) : (\ker \mathscr{A}_{\mathbf{L}})^{\perp} \times \ker(\mathscr{A}_{\mathbf{L}^*}) \to \mathscr{H}$$

of $\mathscr{F}$ with respect to $(\boldsymbol{\Phi},\mathbf{N})$ at $(\boldsymbol{\Phi}^{(0)},\mathbf{N}^{(0)},0)$. A direct calculation gives

$$\begin{aligned}(\mathscr{D}_0(\Psi,\mathbf{M}))(v) = (\mathscr{A}_{\mathbf{L}}\Psi)(v) + \mathbf{M}(v) + \Pi_p\Big(&D\Psi(v)\mathbf{N}^{(0)}(v) \\ &+ D\boldsymbol{\Phi}^{(0)}(v)\mathbf{M}(v) - D\mathbf{R}(v+\boldsymbol{\Phi}^{(0)}(v),0)\Psi(v)\Big),\end{aligned} \tag{C.13}$$

and we prove now that this linear map is invertible.

Denote by π_q the linear map on $\mathscr{H}$, which associates to a polynomial $\mathbf{P}$ the polynomial $\mathbf{P}_q$ obtained by suppressing the monomials in $\mathbf{P}$ of degree different of q. With this notation, according to Theorem 1.2 we have that

$$(\boldsymbol{\Phi}_0^{(0)},\mathbf{N}_0^{(0)}) = (\boldsymbol{\Phi}_1^{(0)},\mathbf{N}_1^{(0)}) = 0,$$

since the polynomials $(\boldsymbol{\Phi}^{(0)},\mathbf{N}^{(0)})$ are at least quadratic. Identifying the homogeneous polynomials of degrees $0,\ldots,p$ on the right hand side of (C.13) we obtain, successively,

$$\begin{aligned}&(\mathscr{A}_{\mathbf{L}}\Psi_0)(v) + \mathbf{M}_0(v), \\ &(\mathscr{A}_{\mathbf{L}}\Psi_1)(v) + \mathbf{M}_1(v) + D\boldsymbol{\Phi}_2^{(0)}(v)\mathbf{M}_0 - D\mathbf{R}_2(v)\Psi_0,\end{aligned}$$

$$(\mathscr{A}_{\mathbf{L}}\Psi_2)(v)+\mathbf{M}_2(v)+D\Phi_3^{(0)}(v)\mathbf{M}_0+D\Phi_2^{(0)}(v)\mathbf{M}_1(v)-D\mathbf{R}_2(v)\Psi_1(v)$$
$$-D^2\mathbf{R}_2(\Psi_0,\Phi_2^{(0)}(v))-D\mathbf{R}_3(v)\Psi_0,$$
$$(\mathscr{A}_{\mathbf{L}}\Psi_p)(v)+\mathbf{M}_p(v)+\sum_{1\leq l\leq p} D\Phi_{l+1}^{(0)}(v)\mathbf{M}_{p-l}(v)+\sum_{1\leq l\leq p-1} D\Psi_{p-l}(v)\mathbf{N}_{l+1}^{(0)}(v)$$
$$-\sum_{2\leq q\leq p}\pi_p\left(D\mathbf{R}_q(v+\Phi^{(0)}(v))\Psi(v)\right),$$

where $\mathbf{R}_s(u)=D_u^s\mathbf{R}(0,0)(u^{(s)})/s!$. Now notice that for any degree q between 0 and p, the formulas above are of the form

$$\mathscr{A}_{\mathbf{L}}\Psi_q+\mathbf{M}_q+\mathbf{G}_q,$$

with $\mathbf{G}_q$ depending only upon $\mathbf{M}_j$ and Ψ_j with $j=0,\ldots,q-1$. We have seen in the proof of Theorem 1.2 that the equation

$$\mathscr{A}_{\mathbf{L}}\Psi+\mathbf{M}=\mathbf{Q}$$

has a unique solution

$$\mathbf{M}=\mathbf{P}_{\ker(\mathscr{A}_{\mathbf{L}^*})}\mathbf{Q}\in\ker(\mathscr{A}_{\mathbf{L}^*}),\quad \Psi\in(\ker\mathscr{A}_{\mathbf{L}})^{\perp},$$

for any homogeneous polynomial $\mathbf{Q}$ of degree q. This implies that the differential $\mathscr{D}_0$ is invertible, and by the implicit function theorem we conclude that (C.10) has a unique solution $(\Phi_\mu,\mathbf{N}_\mu)$ satisfying (C.12). Furthermore the map $\mu\mapsto(\Phi_\mu,\mathbf{N}_\mu)$ is of class $\mathscr{C}^k$, which implies that the coefficients of monomials of degree q are functions of μ of class $\mathscr{C}^{k-q}$. This completes the proof of Theorem 2.2. □

Remark C.1 *In the case where $\mathbf{R}(\cdot,\mu)$ is linear, equation (C.10) is affine in Φ, and we can look directly for a solution $(\Phi,\mathbf{N})\in(\mathscr{L}(\mathbb{R}^n))^2$, i.e., in the space of polynomials of degree 1.*

D Reversible Bifurcations

The references in this section are to theorems, hypotheses, formulas, and remarks in Chapter 4.

D.1 0^{3+} *Normal Form in Infinite Dimensions*

We show below how to compute the principal coefficients in the normal form (2.9) when starting from an infinite-dimensional system

$$\frac{du}{dt} = \mathbf{L}u + \mathbf{R}(u,\mu), \tag{D.1}$$

just like the system (4.1) in Chapter 3. We assume that the parameter μ is real, and that the system (D.1) possesses a reversibility symmetry $\mathbf{S}$ and satisfies the hypotheses of Theorems 3.3 and 3.15 in Chapter 2. We further assume that the spectrum of the linear operator $\mathbf{L}$ is such that $\sigma_0 = \{0\}$, where 0 is an algebraically triple and geometrically simple eigenvalue, with a symmetric eigenvector ζ_0 such that

$$\mathbf{S}\zeta_0 = \zeta_0.$$

Then the three-dimensional reduced system satisfies the hypotheses of Lemma 2.4, so that its normal form is given by (2.9).

For the computation of the coefficients in the expansion of P we proceed as in Section 3.4, and as for the reversible bifurcations in Section 4.1. We start from the equality (4.3) in Chapter 3, in which we take $v_0 = A\zeta_0 + B\zeta_1 + C\zeta_2$, and then write

$$u = A\zeta_0 + B\zeta_1 + C\zeta_2 + \widetilde{\Psi}(A,B,C,\mu), \tag{D.2}$$

where $\widetilde{\Psi}$ takes values in $\mathscr{Y}$. With the notations from Section 3.2.3, we consider the Taylor expansions of $\mathbf{R}$ in (1.15) and of $\widetilde{\Psi}$,

$$\widetilde{\Psi}(A,B,C,\mu) = \sum_{1\le r+s+q+n\le p} A^r B^s C^q \mu^n \Psi_{rsqn}, \quad \Psi_{rsq0} = 0 \text{ for } r+s+q=1.$$

Using the reversibility symmetry we find that

$$\mathbf{R}_{r,q}((\mathbf{S}u)^{(r)}) = -\mathbf{S}\mathbf{R}_{r,q}(u^{(r)}), \quad \mathbf{S}\Psi_{rsqn} = (-1)^s\Psi_{rsqn}.$$

Now we identify the different powers of (A,B,C,μ) in the identity (4.4) in Chapter 3, which is here

$$\begin{aligned}(\partial_A\widetilde{\Psi})B + (\partial_B\widetilde{\Psi})C + (\zeta_1 + \partial_B\widetilde{\Psi})AP(A,\widetilde{B},\mu) + (\zeta_2 + \partial_C\widetilde{\Psi})BP(A,\widetilde{B},\mu)\\ = \mathbf{L}\widetilde{\Psi} + \mathbf{R}(A\zeta_0 + B\zeta_1 + C\zeta_2 + \widetilde{\Psi},\mu).\end{aligned}$$

This leads to the following equalities, found successively at orders μ, $A\mu$, $B\mu$, $C\mu$:

$$0 = \mathbf{L}\Psi_{0001} + \mathbf{R}_{0,1}, \tag{D.3}$$
$$a\zeta_1 = \mathbf{L}\Psi_{1001} + \mathbf{R}_{1,1}\zeta_0 + 2\mathbf{R}_{2,0}(\zeta_0,\Psi_{0001}), \tag{D.4}$$
$$a\zeta_2 + \Psi_{1001} = \mathbf{L}\Psi_{0101} + \mathbf{R}_{1,1}\zeta_1 + 2\mathbf{R}_{2,0}(\zeta_1,\Psi_{0001}), \tag{D.5}$$
$$\Psi_{0101} = \mathbf{L}\Psi_{0011} + \mathbf{R}_{1,1}\zeta_2 + 2\mathbf{R}_{2,0}(\zeta_2,\Psi_{0001}); \tag{D.6}$$

at orders A^2, AB, AC, BC, C^2:

$$b\zeta_1 = \mathbf{L}\Psi_{2000} + \mathbf{R}_{2,0}(\zeta_0,\zeta_0), \tag{D.7}$$
$$b\zeta_2 + 2\Psi_{2000} = \mathbf{L}\Psi_{1100} + 2\mathbf{R}_{2,0}(\zeta_0,\zeta_1), \tag{D.8}$$

$$\Psi_{1100} = \mathbf{L}\Psi_{0200} + \mathbf{R}_{2,0}(\zeta_1, \zeta_1), \tag{D.9}$$
$$\Psi_{1100} = \mathbf{L}\Psi_{1010} + 2\mathbf{R}_{2,0}(\zeta_0, \zeta_2), \tag{D.10}$$
$$2\Psi_{0200} + \Psi_{1010} = \mathbf{L}\Psi_{0110} + 2\mathbf{R}_{2,0}(\zeta_1, \zeta_2), \tag{D.11}$$
$$\Psi_{0110} = \mathbf{L}\Psi_{0020} + \mathbf{R}_{2,0}(\zeta_2, \zeta_2); \tag{D.12}$$

at orders A^3, A^2B, AB^2, B^3, A^2C, ABC:

$$d\zeta_1 + b\Psi_{1100} = \mathbf{L}\Psi_{3000} + 2\mathbf{R}_{2,0}(\zeta_0, \Psi_{2000}) + \mathbf{R}_{3,0}(\zeta_0, \zeta_0, \zeta_0), \tag{D.13}$$
$$d\zeta_2 + 3\Psi_{3000} + 2b\Psi_{0200} + b\Psi_{1010} = \mathbf{L}\Psi_{2100} + 2\mathbf{R}_{2,0}(\zeta_0, \Psi_{1100}) + 2\mathbf{R}_{2,0}(\zeta_1, \Psi_{2000}) + 3\mathbf{R}_{3,0}(\zeta_0, \zeta_0\zeta_1), \tag{D.14}$$
$$c\zeta_1 + 2\Psi_{2100} + b\Psi_{0110} = \mathbf{L}\Psi_{1200} + 2\mathbf{R}_{2,0}(\zeta_0, \Psi_{0200}) + 2\mathbf{R}_{2,0}(\zeta_1, \Psi_{1100}) + 3\mathbf{R}_{3,0}(\zeta_0, \zeta_1, \zeta_1), \tag{D.15}$$
$$c\zeta_2 + 2\Psi_{1200} = \mathbf{L}\Psi_{0300} + 2\mathbf{R}_{2,0}(\zeta_1, \Psi_{0200}) + \mathbf{R}_{3,0}(\zeta_1, \zeta_1, \zeta_1), \tag{D.16}$$
$$-2c\zeta_1 + \Psi_{2100} + b\Psi_{0110} = \mathbf{L}\Psi_{2010} + 2\mathbf{R}_{2,0}(\zeta_0, \Psi_{1010}) + 2\mathbf{R}_{2,0}(\zeta_2, \Psi_{2000}) + 3\mathbf{R}_{3,0}(\zeta_0, \zeta_0, \zeta_2), \tag{D.17}$$
$$-2c\zeta_2 + 2\Psi_{2010} + 2\Psi_{1200} + 2b\Psi_{0020} = \mathbf{L}\Psi_{1110} + 2\mathbf{R}_{2,0}(\zeta_0, \Psi_{0110}) + 2\mathbf{R}_{2,0}(\zeta_1, \Psi_{1010}) + 2\mathbf{R}_{2,0}(\zeta_2, \Psi_{1100}) + 6\mathbf{R}_{3,0}(\zeta_0, \zeta_1, \zeta_2); \tag{D.18}$$

and at orders AC^2, B^2C, BC^2, C^3:

$$\Psi_{1110} = \mathbf{L}\Psi_{1020} + 2\mathbf{R}_{2,0}(\zeta_0, \Psi_{0020}) + 2\mathbf{R}_{2,0}(\zeta_2, \Psi_{1010}) + 3\mathbf{R}_{3,0}(\zeta_0, \zeta_2, \zeta_2), \tag{D.19}$$
$$\Psi_{1110} + 3\Psi_{0300} = \mathbf{L}\Psi_{0210} + 2\mathbf{R}_{2,0}(\zeta_1, \Psi_{0110}) + 2\mathbf{R}_{2,0}(\zeta_2, \Psi_{0200}) + 3\mathbf{R}_{3,0}(\zeta_1, \zeta_1, \zeta_2), \tag{D.20}$$
$$2\Psi_{0210} = \mathbf{L}\Psi_{0120} + 2\mathbf{R}_{2,0}(\zeta_1, \Psi_{0020}) + 2\mathbf{R}_{2,0}(\zeta_2, \Psi_{0110}) + 3\mathbf{R}_{3,0}(\zeta_1, \zeta_2, \zeta_2), \tag{D.21}$$
$$\Psi_{0120} = \mathbf{L}\Psi_{0030} + 2\mathbf{R}_{2,0}(\zeta_2, \Psi_{0020}) + \mathbf{R}_{3,0}(\zeta_2, \zeta_2, \zeta_2). \tag{D.22}$$

Notice that due to the symmetry properties of $\mathbf{R}_{p,q}$, ζ_0, ζ_1, ζ_2, and since $\mathbf{S}\Psi_{rsqn} = (-1)^s\Psi_{rsqn}$, we have that

$$\mathbf{S}\mathbf{R}_{0,1} = -\mathbf{R}_{0,1}, \quad \mathbf{S}\mathbf{R}_{1,1}\zeta_j = (-1)^{j+1}\mathbf{R}_{1,1}\zeta_j, \quad j = 0,1,2,$$

and

$$\mathbf{SR}_{2,0}(\zeta_j,\Psi_{rsqn}) = (-1)^{j+s+1}\mathbf{R}_{2,0}(\zeta_j,\Psi_{rsqn}), \quad j=0,1,2,$$
$$\mathbf{SR}_{3,0}(\zeta_j,\zeta_k,\zeta_l) = (-1)^{j+k+l+1}\mathbf{R}_{3,0}(\zeta_j,\zeta_k,\zeta_l), \quad j,k,l\in\{0,1,2\}.$$

The solvability conditions for these equations are now obtained by taking the duality product of each equation with the vector ζ_2^* orthogonal to the range of $\mathbf{L}$. Proceeding as for the other examples, we find that this vector is given by

$$\zeta_2^* = \mathbf{P}_0^*\zeta_{02}^* \in \mathscr{X}^*,$$

where $\mathbf{P}_0^*$ is the adjoint of the projection $\mathbf{P}_0$ onto the three-dimensional space $\mathscr{E}_0$, and ζ_{02}^* is the eigenvector associated with the eigenvalue 0 of the adjoint of $\mathbf{L}_0$, the restriction of $\mathbf{L}$ to $\mathscr{E}_0$, satisfying $\langle\zeta_2,\zeta_2^*\rangle = 1$. In addition, we have that

$$\langle\zeta_0,\zeta_2^*\rangle = 0, \quad \langle\zeta_1,\zeta^*\rangle = 0, \quad \langle\zeta_2,\zeta_2^*\rangle = 1,$$

and since $\mathbf{S}\zeta_2 = \zeta_2)$, that

$$\mathbf{S}^*\zeta_2^* - \zeta_2^*.$$

We can now solve the system (D.3)–(D.22). Since any antisymmetric vector of $\mathscr{X}$ lies in the range of $\mathbf{L}$, it is straightforward to check that there is no solvability condition for the equations (D.3), (D.4), (D.6), (D.7), (D.9), (D.10), (D.12), and (D.13), (D.15), (D.17), (D.19), (D.20), (D.22). The solvability conditions for the remaining equations allow us to determine the coefficients a, c, and d.

First, the equation (D.3) gives Ψ_{0001}, defined up to an arbitrary multiple of ζ_0. Then from (D.4) we obtain

$$\Psi_{1001} = \widetilde{\Psi}_{1001} + a\zeta_2,$$

where

$$\mathbf{L}\widetilde{\Psi}_{1001} + \mathbf{R}_{1,1}\zeta_0 + 2\mathbf{R}_{2,0}(\zeta_0,\Psi_{0001}) = 0.$$

For (D.5) we find the solvability condition

$$2a = \langle\mathbf{R}_{1,1}\zeta_1 + 2\mathbf{R}_{2,0}(\zeta_1,\Psi_{0001}) - \widetilde{\Psi}_{1001},\zeta_2^*\rangle,$$

which gives a, and from this equation we can determine Ψ_{0101} up to an arbitrary multiple of ζ_0. From (D.6) we can find Ψ_{0011} up to an arbitrary multiple of ζ_0, again.

Next, equation (D.7) gives

$$\Psi_{2000} = \widetilde{\Psi}_{2000} + b\zeta_2 + \psi_{2000}\zeta_0,$$

where

$$\mathbf{L}\widetilde{\Psi}_{2000} + \mathbf{R}_{2,0}(\zeta_0,\zeta_0) = 0.$$

For (D.8) we find the solvability condition

$$3b = \langle 2\mathbf{R}_{2,0}(\zeta_0,\zeta_1) - 2\widetilde{\Psi}_{2000},\zeta_2^*\rangle,$$

which gives b, and we can determine

$$\Psi_{1100} = \widetilde{\Psi}_{1100} + 2\psi_{2000}\zeta_1,$$

where

$$\mathbf{L}\widetilde{\Psi}_{1100} + 2\mathbf{R}_{2,0}(\zeta_0, \zeta_1) = 2\widetilde{\Psi}_{2000}.$$

Similarly, from (D.9) and (D.10) we obtain,

$$\Psi_{0200} = \widetilde{\Psi}_{0200} + 2\psi_{2000}\zeta_2, \quad \Psi_{1010} = \widetilde{\Psi}_{1010} + 2\psi_{2000}\zeta_2,$$

where

$$\mathbf{L}\widetilde{\Psi}_{0200} + \mathbf{R}_{2,0}(\zeta_1, \zeta_1) = \widetilde{\Psi}_{1100}, \quad \mathbf{L}\widetilde{\Psi}_{1010} + 2\mathbf{R}_{2,0}(\zeta_0, \zeta_2) = \widetilde{\Psi}_{1100}.$$

Now for (D.11) we find the solvability condition

$$6\psi_{2000} = \langle 2\mathbf{R}_{2,0}(\zeta_1, \zeta_2) - 2\widetilde{\Psi}_{0200} - \widetilde{\Psi}_{1010}, \zeta_2^* \rangle,$$

and from (D.11) and (D.12) we can determine Ψ_{0110} and Ψ_{0020}.

Finally, from (D.13) and (D.14) we find

$$\Psi_{3000} = \widetilde{\Psi}_{3000} + d\zeta_2 + \psi_{3000}\zeta_0, \quad \Psi_{2100} = \widetilde{\Psi}_{2100} + 3\psi_{3000}\zeta_1,$$

and the coefficient d,

$$\begin{aligned} 4d = &\langle 2\mathbf{R}_{2,0}(\zeta_0, \Psi_{1100}) + 2\mathbf{R}_{2,0}(\zeta_1, \Psi_{2000}) + 3\mathbf{R}_{3,0}(\zeta_0, \zeta_0\zeta_1), \zeta_2^* \rangle \\ &- \langle 3\widetilde{\Psi}_{3000} + 2b\Psi_{0200} + b\Psi_{1010}, \zeta_2^* \rangle. \end{aligned} \quad \text{(D.23)}$$

Now (D.15) and (D.16) give

$$\Psi_{1200} = \widetilde{\Psi}_{1200} + (c + 6\psi_{3000})\zeta_2, \quad \Psi_{0300} = \widetilde{\Psi}_{0300},$$

and

$$3c + 12\psi_{3000} = \langle 2\mathbf{R}_{2,0}(\zeta_1, \Psi_{0200}) + \mathbf{R}_{3,0}(\zeta_1, \zeta_1, \zeta_1) - 2\widetilde{\Psi}_{1200}, \zeta_2^* \rangle.$$

From (D.17) and (D.18) we obtain

$$\Psi_{2010} = \widetilde{\Psi}_{2010} + (3\psi_{3000} - 2c)\zeta_2 + \psi_{2010}\zeta_0, \quad \Psi_{1110} = \widetilde{\Psi}_{1110} + 2\psi_{2010}\zeta_1,$$

and

$$\begin{aligned} 12\psi_{3000} = &\langle 2\mathbf{R}_{2,0}(\zeta_0, \Psi_{0110}) + 2\mathbf{R}_{2,0}(\zeta_1, \Psi_{1010}) + 2\mathbf{R}_{2,0}(\zeta_2, \Psi_{1100}), \zeta_2^* \rangle \\ &+ \langle 6\mathbf{R}_{3,0}(\zeta_0, \zeta_1, \zeta_2), \zeta_2^* \rangle - \langle 2\widetilde{\Psi}_{2010} + 2\widetilde{\Psi}_{1200} + 2b\Psi_{0020}, \zeta_2^* \rangle. \end{aligned}$$

We conclude that

$$3c = \langle 2\mathbf{R}_{2,0}(\zeta_1, \Psi_{0200}) - 2\mathbf{R}_{2,0}(\zeta_0, \Psi_{0110}) - 2\mathbf{R}_{2,0}(\zeta_1, \Psi_{1010}), \zeta_2^* \rangle$$
$$-\langle 2\mathbf{R}_{2,0}(\zeta_2, \Psi_{1100}) - \mathbf{R}_{3,0}(\zeta_1, \zeta_1, \zeta_1) + 6\mathbf{R}_{3,0}(\zeta_0, \zeta_1, \zeta_2), \zeta_2^* \rangle$$
$$+\langle 2\widetilde{\Psi}_{2010} + 2b\Psi_{0020}, \zeta_2^* \rangle. \tag{D.24}$$

Furthermore, from the equations (D.19) and (D.20) we can find

$$\Psi_{1020} = \widetilde{\Psi}_{1020} + 2\psi_{2010}\zeta_2, \quad \Psi_{0210} = \widetilde{\Psi}_{0210} + 2\psi_{2010}\zeta_2,$$

the solvability condition for (D.21) gives

$$4\psi_{2010} = \langle 2\mathbf{R}_{2,0}(\zeta_1, \Psi_{0020}) + 2\mathbf{R}_{2,0}(\zeta_2, \Psi_{0110}) + 3\mathbf{R}_{3,0}(\zeta_1, \zeta_2, \zeta_2), \zeta_2^* \rangle,$$

and from (D.21) and (D.22) we can also determine Ψ_{0120} and Ψ_{0030}.

D.2 $(i\omega)^2$ *Normal Form in Infinite Dimensions*

We show below how to compute the principal coefficients in the normal form (3.25) when starting from an infinite-dimensional system of the form (3.12). We assume that the parameter μ is real and that the system (3.12) possesses a reversibility symmetry $\mathbf{S}$ and satisfies the hypotheses of Theorems 3.3 and 3.15 in Chapter 2. We further assume that the spectrum of the linear operator $\mathbf{L}$ is such that $\sigma_0 = \{\pm i\omega\}$, where $\pm i\omega$ are algebraically double and geometrically simple eigenvalues. Then the four-dimensional reduced system satisfies the hypotheses of Lemma 3.17, so that its normal form is given by (3.25).

We proceed as in Section 3.4, and in the previous cases. In equality (4.3) in Chapter 3, we take $v_0 = A\zeta_0 + B\zeta_1 + \overline{A\zeta_0} + \overline{B\zeta_1}$, and then write

$$u = A\zeta_0 + B\zeta_1 + \overline{A\zeta_0} + \overline{B\zeta_1} + \widetilde{\Psi}(A, B, \overline{A}, \overline{B}, \mu) \tag{D.25}$$

where $\widetilde{\Psi}$ takes values in $\mathscr{Z}$. With the notations from Section 3.2.3, we consider the Taylor expansion (1.15) of $\mathbf{R}$, and the expansion of $\widetilde{\Psi}$,

$$\widetilde{\Psi}(A, B, \overline{A}, \overline{B}, \mu) = \sum_{1 \leq r+s+q+l+m \leq p} A^r B^s \overline{A}^q \overline{B}^l \mu^m \Psi_{rsqlm},$$

where

$$\Psi_{rsql0} = 0, \quad \text{for } r+s+q+l = 1.$$

Using the reversibility symmetry we find that

$$\widetilde{\Psi}(\overline{A}, -\overline{B}, A, B, \mu) = \mathbf{S}\widetilde{\Psi}(A, B, \overline{A}, \overline{B}, \mu),$$

and

$$\mathbf{S}\Psi_{rsqlm} = (-1)^{s+l}\Psi_{qlrsm}, \quad \Psi_{rsqlm} = \overline{\Psi}_{qlrsm}.$$

Identity (4.4) in Chapter 3, is in this case

$$\begin{aligned}
&(i\omega A+B)\partial_A\Psi+i\omega B\partial_B\Psi+(-i\omega\overline{A}+\overline{B})\partial_{\overline{A}}\Psi-i\omega\overline{B}\partial_{\overline{B}}\Psi\\
&+\Big(iA(\zeta_0+\partial_A\Psi)-i\overline{A}(\overline{\zeta}_0+\partial_{\overline{A}}\Psi)\Big)P\\
&+(\zeta_1+\partial_B\Psi)(iBP+AQ)+(\overline{\zeta}_1+\partial_{\overline{B}}\Psi)(-i\overline{B}P+\overline{A}Q)\\
&\qquad\qquad=\mathbf{L}\Psi+\mathbf{R}(A\zeta_0+B\zeta_1+\overline{A}\overline{\zeta_0}+\overline{B}\overline{\zeta_1}+\Psi,\mu).
\end{aligned}$$

Using the expansions of $\mathbf{R}$, $\widetilde{\Psi}$, P, and Q, we find at orders μ, $A\mu$, and $B\mu$, the equalities

$$0=\mathbf{L}\Psi_{00001}+\mathbf{R}_{0,1}, \tag{D.26}$$
$$a\zeta_1+i\alpha\zeta_0=(\mathbf{L}-i\omega)\Psi_{10001}+\mathbf{R}_{1,1}\zeta_0+2\mathbf{R}_{2,0}(\zeta_0,\Psi_{00001}), \tag{D.27}$$
$$i\alpha\zeta_1+\Psi_{10001}=(\mathbf{L}-i\omega)\Psi_{01001}+\mathbf{R}_{1,1}\zeta_1+2\mathbf{R}_{2,0}(\zeta_1,\Psi_{00001}), \tag{D.28}$$

and at orders A^2, $A\overline{A}$, AB, $A\overline{B}$, B^2, $B\overline{B}$ we have:

$$0=(\mathbf{L}-2i\omega)\Psi_{20000}+\mathbf{R}_{2,0}(\zeta_0,\zeta_0), \tag{D.29}$$
$$0=\mathbf{L}\Psi_{10100}+2\mathbf{R}_{2,0}(\zeta_0,\overline{\zeta_0}), \tag{D.30}$$
$$2\Psi_{20000}=(\mathbf{L}-2i\omega)\Psi_{11000}+2\mathbf{R}_{2,0}(\zeta_0,\zeta_1), \tag{D.31}$$
$$\Psi_{10100}=\mathbf{L}\Psi_{10010}+2\mathbf{R}_{2,0}(\zeta_0,\overline{\zeta_1}), \tag{D.32}$$
$$\Psi_{11000}=(\mathbf{L}-2i\omega)\Psi_{02000}+\mathbf{R}_{2,0}(\zeta_1,\zeta_1), \tag{D.33}$$
$$\Psi_{10010}+\Psi_{01100}=\mathbf{L}\Psi_{01010}+2\mathbf{R}_{2,0}(\zeta_1,\overline{\zeta_1}). \tag{D.34}$$

At orders A^3, A^2B, AB^2, and B^3 we obtain:

$$0=(\mathbf{L}-3i\omega)\Psi_{30000}+2\mathbf{R}_{2,0}(\zeta_0,\Psi_{20000})+\mathbf{R}_{3,0}(\zeta_0,\zeta_0,\zeta_0), \tag{D.35}$$
$$\begin{aligned}3\Psi_{30000}=&(\mathbf{L}-3i\omega)\Psi_{21000}+2\mathbf{R}_{2,0}(\zeta_0,\Psi_{11000})+2\mathbf{R}_{2,0}(\zeta_1,\Psi_{20000})\\&+3\mathbf{R}_{3,0}(\zeta_0,\zeta_0,\zeta_1),\end{aligned} \tag{D.36}$$
$$\begin{aligned}2\Psi_{21000}=&(\mathbf{L}-3i\omega)\Psi_{12000}+2\mathbf{R}_{2,0}(\zeta_1,\Psi_{11000})+2\mathbf{R}_{2,0}(\zeta_0,\Psi_{02000})\\&+3\mathbf{R}_{3,0}(\zeta_0,\zeta_1,\zeta_1),\end{aligned} \tag{D.37}$$
$$\Psi_{12000}=(\mathbf{L}-3i\omega)\Psi_{03000}+2\mathbf{R}_{2,0}(\zeta_1,\Psi_{02000})+\mathbf{R}_{3,0}(\zeta_1,\zeta_1,\zeta_1), \tag{D.38}$$

and finally at orders $A^2\overline{A}$, $A^2\overline{B}$, $A\overline{A}B$, $\overline{A}B^2$, $AB\overline{B}$, and $B^2\overline{B}$ we find:

$$\begin{aligned}b\zeta_1+i\beta\zeta_0=&(\mathbf{L}-i\omega)\Psi_{20100}+2\mathbf{R}_{2,0}(\zeta_0,\Psi_{10100})\\&+2\mathbf{R}_{2,0}(\overline{\zeta_0},\Psi_{20000})+3\mathbf{R}_{3,0}(\zeta_0,\zeta_0,\overline{\zeta_0}),\end{aligned} \tag{D.39}$$
$$\begin{aligned}\frac{ic}{2}\zeta_1-\frac{\gamma}{2}\zeta_0+\Psi_{20100}=&(\mathbf{L}-i\omega)\Psi_{20010}+2\mathbf{R}_{2,0}(\zeta_0,\Psi_{10010})\\&+2\mathbf{R}_{2,0}(\overline{\zeta_1},\Psi_{20000})+3\mathbf{R}_{3,0}(\zeta_0,\zeta_0,\overline{\zeta_1}),\end{aligned} \tag{D.40}$$

$$\left(i\beta - \frac{ic}{2}\right)\zeta_1 + \frac{\gamma}{2}\zeta_0 + 2\Psi_{20100} = (\mathbf{L} - i\omega)\Psi_{11100} + 2\mathbf{R}_{2,0}(\zeta_0, \Psi_{01100})$$
$$+2\mathbf{R}_{2,0}(\overline{\zeta_0}, \Psi_{11000}) + 2\mathbf{R}_{2,0}(\zeta_1, \Psi_{10100})$$
$$+6\mathbf{R}_{3,0}(\zeta_0, \zeta_1, \overline{\zeta_0}), \tag{D.41}$$
$$\frac{\gamma}{2}\zeta_1 + \Psi_{11100} = (\mathbf{L} - i\omega)\Psi_{02100} + 2\mathbf{R}_{2,0}(\overline{\zeta_0}, \Psi_{02000})$$
$$+2\mathbf{R}_{2,0}(\zeta_1, \Psi_{01100}) + 3\mathbf{R}_{3,0}(\zeta_1, \zeta_1, \overline{\zeta_0}), \tag{D.42}$$
$$-\frac{\gamma}{2}\zeta_1 + 2\Psi_{20010} + \Psi_{11100} = (\mathbf{L} - i\omega)\Psi_{11010} + 2\mathbf{R}_{2,0}(\zeta_0, \Psi_{01010})$$
$$+2\mathbf{R}_{2,0}(\overline{\zeta_1}, \Psi_{11000}) + 2\mathbf{R}_{2,0}(\zeta_1, \Psi_{10010})$$
$$+6\mathbf{R}_{3,0}(\zeta_0, \zeta_1, \overline{\zeta_1}), \tag{D.43}$$
$$\Psi_{11010} + \Psi_{02100} = (\mathbf{L} - i\omega)\Psi_{02010} + 2\mathbf{R}_{2,0}(\zeta_1, \Psi_{01010})$$
$$+2\mathbf{R}_{2,0}(\overline{\zeta_1}, \Psi_{02000}) + 3\mathbf{R}_{3,0}(\zeta_1, \zeta_1, \overline{\zeta_1}). \tag{D.44}$$

Notice that due to the symmetry properties of $\mathbf{R}_{p,q}$, ζ_0, and ζ_1 we have

$$\mathbf{S}\mathbf{R}_{0,1} = -\mathbf{R}_{0,1}, \quad \mathbf{S}\mathbf{R}_{1,1}\zeta_0 = -\overline{\mathbf{R}_{1,1}\zeta_0}, \quad \mathbf{S}\mathbf{R}_{1,1}\zeta_1 = \overline{\mathbf{R}_{1,1}\zeta_1},$$

which, together with the symmetry properties for Ψ_{qlrsm}, imply that applying the symmetry $\mathbf{S}$ to both sides of (D.27), (D.39), (D.42), and (D.43), we obtain the opposite of the complex conjugate of these equalities, whereas by applying it to (D.28), (D.40), (D.41), and (D.44), we find the complex conjugate of these equalities.

First, notice that the invertibility of the operators $\mathbf{L}$, $(\mathbf{L} - 2i\omega)$, and $(\mathbf{L} - 3i\omega)$, lets us solve the equations (D.26), (D.29)–(D.34), and (D.35)–(D.38), and determine Ψ_{00001}, Ψ_{20000}, Ψ_{10100}, Ψ_{11000}, Ψ_{10010}, Ψ_{02000}, Ψ_{01010}, Ψ_{30000}, Ψ_{21000}, Ψ_{12000}, and Ψ_{03000}.

Next, consider the vector ζ_1^* orthogonal to the range of $\mathbf{L} - i\omega$, constructed as in the other cases, such that

$$\langle \zeta_0, \zeta_1^* \rangle = 0, \quad \langle \overline{\zeta_0}, \zeta_1^* \rangle = 0, \quad \langle \zeta_1, \zeta_1^* \rangle = 1, \quad \langle \overline{\zeta_1}, \zeta_1^* \rangle = 0,$$

and

$$\mathbf{S}^*\zeta_1^* = -\overline{\zeta_1^*}.$$

Then from the equations (D.27) and (D.28) we find

$$a = \langle \mathbf{R}_{1,1}\zeta_0 + 2\mathbf{R}_{2,0}(\zeta_0, \Psi_{00001}), \zeta_1^* \rangle \tag{D.45}$$

and

$$\Psi_{10001} = \widetilde{\Psi}_{10001} + i\alpha\zeta_1,$$
$$2i\alpha = \langle \mathbf{R}_{1,1}\zeta_1 + 2\mathbf{R}_{2,0}(\zeta_1, \Psi_{00001}) - \widetilde{\Psi}_{10001}, \zeta_1^* \rangle. \tag{D.46}$$

Taking into account the fact that $\langle \mathbf{S}u, \mathbf{S}^*v \rangle = \langle u, v \rangle$ for any $u \in \mathscr{X}$ and $v \in \mathscr{X}^*$, it is not difficult to check that a and α are real numbers.

Next, equation (D.39) gives

$$b = \langle 2\mathbf{R}_{2,0}(\zeta_0, \Psi_{10100}) + 2\mathbf{R}_{2,0}(\overline{\zeta_0}, \Psi_{20000}) + 3\mathbf{R}_{3,0}(\zeta_0, \zeta_0, \overline{\zeta_0}), \zeta_1^* \rangle, \qquad \text{(D.47)}$$

and

$$\Psi_{20100} = \widetilde{\Psi}_{20100} + i\beta\zeta_1 + \psi_{20100}\zeta_0,$$

with a constant $\psi_{20100} \in \mathbb{R}$, which will be determined later. The equations (D.40) and (D.41) now give

$$\begin{aligned} i\beta + \frac{ic}{2} &= \langle 2\mathbf{R}_{2,0}(\zeta_0, \Psi_{10010}) + 2\mathbf{R}_{2,0}(\overline{\zeta_1}, \Psi_{20000}), \zeta_1^* \rangle \\ &\quad + \langle 3\mathbf{R}_{3,0}(\zeta_0, \zeta_0, \overline{\zeta_1}) - \widetilde{\Psi}_{20100}, \zeta_1^* \rangle, \\ 3i\beta - \frac{ic}{2} &= \langle 2\mathbf{R}_{2,0}(\zeta_0, \Psi_{01100}) + 2\mathbf{R}_{2,0}(\overline{\zeta_0}, \Psi_{11000}) + 2\mathbf{R}_{2,0}(\zeta_1, \Psi_{10100}), \zeta_1^* \rangle \\ &\quad + \langle 6\mathbf{R}_{3,0}(\zeta_0, \zeta_1, \overline{\zeta_0}) - 2\widetilde{\Psi}_{20100}, \zeta_1^* \rangle, \end{aligned}$$

which determine the real coefficients β and c, and

$$\begin{aligned} \Psi_{20010} &= \widetilde{\Psi}_{20010} + \left(\psi_{20100} - \frac{\gamma}{2}\right)\zeta_1, \\ \Psi_{11100} &= \widetilde{\Psi}_{11100} + \left(2\psi_{20100} + \frac{\gamma}{2}\right)\zeta_1 + \psi_{11100}\zeta_0, \end{aligned}$$

with another constant ψ_{11100} determined later.

Finally, the equations (D.42)–(D.44) let us determine γ, ψ_{20100}, and ψ_{11100}. Indeed, (D.42) gives

$$\begin{aligned} \gamma + 2\psi_{20100} &= \langle 2\mathbf{R}_{2,0}(\overline{\zeta_0}, \Psi_{02000}) + 2\mathbf{R}_{2,0}(\zeta_1, \Psi_{01100}), \zeta_1^* \rangle \\ &\quad + \langle 3\mathbf{R}_{3,0}(\zeta_1, \zeta_1, \overline{\zeta_0}) - \widetilde{\Psi}_{11100}, \zeta_1^* \rangle, \end{aligned}$$

and

$$\Psi_{02100} = \widetilde{\Psi}_{02100} + \psi_{11100}\zeta_1,$$

whereas (D.43) leads to

$$\begin{aligned} 4\psi_{20100} - \gamma &= \langle 2\mathbf{R}_{2,0}(\zeta_0, \Psi_{01010}) + 2\mathbf{R}_{2,0}(\overline{\zeta_1}, \Psi_{11000}) + 2\mathbf{R}_{2,0}(\zeta_1, \Psi_{10010}), \zeta_1^* \rangle \\ &\quad + \langle 6\mathbf{R}_{3,0}(\zeta_0, \zeta_1, \overline{\zeta_1}) - 2\widetilde{\Psi}_{20010} - \widetilde{\Psi}_{11100}, \zeta_1^* \rangle, \end{aligned}$$

and

$$\Psi_{11010} = \widetilde{\Psi}_{11010} + \psi_{11100}\zeta_1.$$

This determines γ and ψ_{20100}. In particular, we obtain

$$\begin{aligned} 3\gamma &= \langle 4\mathbf{R}_{2,0}(\overline{\zeta_0}, \Psi_{02000}) + 4\mathbf{R}_{2,0}(\zeta_1, \Psi_{01100}) + 6\mathbf{R}_{3,0}(\zeta_1, \zeta_1, \overline{\zeta_0}), \zeta_1^* \rangle \\ &\quad - \langle 2\mathbf{R}_{2,0}(\zeta_0, \Psi_{01010}) + 2\mathbf{R}_{2,0}(\overline{\zeta_1}, \Psi_{11000}) + 2\mathbf{R}_{2,0}(\zeta_1, \Psi_{10010}), \zeta_1^* \rangle \\ &\quad - \langle 6\mathbf{R}_{3,0}(\zeta_0, \zeta_1, \overline{\zeta_1}) - 2\widetilde{\Psi}_{20010} + \widetilde{\Psi}_{11100}, \zeta_1^* \rangle. \end{aligned}$$

Finally the equation (D.44) gives

$$\begin{aligned}2\psi_{11100} = {} & \langle 2\mathbf{R}_{2,0}(\zeta_1, \Psi_{01010}) + 2\mathbf{R}_{2,0}(\overline{\zeta_1}, \Psi_{02000}) + 3\mathbf{R}_{3,0}(\zeta_1, \zeta_1, \overline{\zeta_1}), \zeta_1^* \rangle \\ & - \langle \widetilde{\Psi}_{11010} + \widetilde{\Psi}_{02100}, \zeta_1^* \rangle,\end{aligned}$$

and completes the computation of the normal form.

References

1. R.A. Adams. *Sobolev spaces*. Pure and Applied Mathematics, Vol. 65. Academic Press, New York-London, 1975.
2. S. Agmon. *Lectures on elliptic boundary value problems*. Van Nostrand Mathematical Studies, No. 2, D. Van Nostrand Co., Inc., Princeton, NJ, Toronto–London, 1965.
3. A. Andronov, E. Leontovich. Some cases of dependence of limit cycles on a parameter. (Russian) *J. State Univ. Gorki* 6 (1937), 3–24.
4. V.I. Arnold. *Geometrical methods in the theory of ordinary differential equations*. Grundlehren der Mathematischen Wissenschaften [Fundamental Principles of Mathematical Science], 250. Springer-Verlag, New York–Berlin, 1983.
5. M. Barrandon. Résonance 1 : 2 pour les champs de vecteurs réversibles. *C. R. Math. Acad. Sci. Paris* 334 (2002), 1, 7–10.
6. J.T. Beale. Exact solitary water waves with capillary ripples at infinity. *Comm. Pure Appl. Math.* 44 (1991), 2, 211–257.
7. G.R. Belickii. Normal forms in relation to the filtering action of a group. (Russian) *Trudy Moskov. Mat. Obshch.* 40 (1979), 3–46.
8. D. Bensimon, P. Kolodner, C.M. Surko, H. Williams, V. Croquette. Competing and coexisting dynamical states of travelling wave convection in an annulus. *J. Fluid Mech.* 217 (1990), 441–467.
9. D. Brézis. Perturbations singulières et problèmes d'évolution avec défaut d'ajustement. *C. R. Acad. Sci. Paris Sér. A-B* 276 (1973), A1597–A1600.
10. B. Buffoni, J. Toland. *Analytic theory of global bifurcation. An introduction*. Princeton Series in Applied Mathematics. Princeton University Press, Princeton, NJ, 2003.
11. J.M. Burgers. A mathematical model illustrating the theory of turbulence. *Adv. Appl. Mech.* 1 (1948), 171–199.
12. J. Carr. *Applications of centre manifold theory*. Applied Mathematical Sciences, 35. Springer-Verlag, New York-Berlin, 1981.
13. A.R. Champneys. Homoclinic orbits in reversible systems and their applications in mechanics, fluids and optics. Time-reversal symmetry in dynamical systems (Coventry, 1996). *Phys. D* 112 (1998), 1-2, 158–186.
14. S. Chandrasekhar. *Hydrodynamic and hydromagnetic stability*. The International Series of Monographs on Physics. Clarendon Press, Oxford, 1961.
15. P. Chossat, G. Iooss. *The Couette–Taylor problem*. Applied Mathematical Sciences, 102. Springer-Verlag, New York, 1994.
16. P. Chossat, R. Lauterbach. *Methods in equivariant bifurcations and dynamical systems*. Advanced Series in Nonlinear Dynamics, 15. World Scientific Publishing Co., Inc., River Edge, NJ, 2000.
17. S.N. Chow, J. K. Hale. *Methods of bifurcation theory*. Grundlehren der Mathematischen Wissenschaften [Fundamental Principles of Mathematical Science], 251. Springer-Verlag, New York-Berlin, 1982.

M. Haragus, G. Iooss, *Local Bifurcations, Center Manifolds, and Normal Forms in Infinite-Dimensional Dynamical Systems*, Universitext,
DOI 10.1007/978-0-85729-112-7,

18. P. Collet, J.P. Eckmann. *Instabilities and fronts in extended systems*. Princeton Series in Physics. Princeton University Press, Princeton, NJ, 1990.
19. P. Coullet, G.Iooss. Instabilities of one-dimensional cellular patterns. *Phys. Rev. Lett.* 64 (1990), 8, 866–869.
20. R. Cushman, J.A. Sanders. Splitting algorithm for nilpotent normal forms. *Dynam. Stability Systems* 2 (1987), 3-4, 235–246.
21. G. Da Prato, P. Grisvard. Sommes d'opérateurs linéaires et équations différentielles opérationnelles. *J. Math. Pures Appl.* (9) 54 (1975), 3, 305–387.
22. F. Dias, G. Iooss. Capillary-gravity solitary waves with damped oscillations. *Phys. D* 65 (1993), 4, 399–423.
23. F. Dias, G. Iooss. Water-waves as a spatial dynamical system. *Handbook of mathematical fluid dynamics*, Vol. II 443–499, North-Holland, Amsterdam, 2003.
24. C. Elphick, G. Iooss, E. Tirapegui. Normal form reduction for time-periodically driven differential equations. *Phys. Lett. A* 120 (1987), 9, 459–463.
25. C. Elphick, E. Tirapegui, M.E. Brachet, P. Coullet, G. Iooss, A simple global characterization for normal forms of singular vector fields. *Phys. D* 29 (1987), 1-2, 95–127.
26. J.-M. Gambaudo. Perturbation of a Hopf bifurcation by an external time-periodic forcing. *J. Differential Equations* 57 (1985), 2, 172–199.
27. S.A. van Gils, M. Krupa, W.F. Langford. Hopf bifurcation with nonsemisimple 1 : 1 resonance. *Nonlinearity* 3 (1990), 3, 825–850.
28. L.Yu. Glebsky, L.M. Lerman. On small stationary localized solutions for the generalized 1-D Swift-Hohenberg equation. *Chaos* 5 (1995), 2, 424–431.
29. M. Golubitsky, D.G. Schaeffer. *Singularities and groups in bifurcation theory*. Vol. I. Applied Mathematical Sciences, 51. Springer-Verlag, New York, 1985.
30. M. Golubitsky, I. Stewart, D.G. Schaeffer. *Singularities and groups in bifurcation theory*. Vol. II. Applied Mathematical Sciences, 69. Springer-Verlag, New York, 1988.
31. R. Grimshaw, G. Iooss. Solitary waves of a coupled Korteweg-de Vries system. Nonlinear waves: computation and theory, II (Athens, GA, 2001). *Math. Comput. Simulation* 62 (2003), 1-2, 31–40.
32. D.M. Grobman. Homeomorphism of systems of differential equations. (Russian) *Dokl. Akad. Nauk SSSR* 128 (1959), 880–881.
33. M.D. Groves. Three-dimensional travelling gravity-capillary water waves. *GAMM-Mitt.* 30 (2007), 1, 8–43.
34. M.D. Groves, M. Haragus. A bifurcation theory for three-dimensional oblique travelling gravity-capillary water waves. *J. Nonlinear Sci.* 13 (2003), 4, 397–447.
35. M.D. Groves, A. Mielke. A spatial dynamics approach to three-dimensional gravity-capillary steady water waves. *Proc. Roy. Soc. Edinburgh Sect. A* 131 (2001), 1, 83–136.
36. M.D. Groves, B. Sandstede. A plethora of three-dimensional periodic travelling gravity-capillary water waves with multipulse transverse profiles. *J. Nonlinear Sci.* 14 (2004), 3, 297–340.
37. M.D. Groves, S.M. Sun. Fully localised solitary-wave solutions of the three-dimensional gravity-capillary water-wave problem. *Arch. Ration. Mech. Anal.* 188 (2008), 1, 1–91.
38. J. Guckenheimer, P. Holmes. *Nonlinear oscillations, dynamical systems, and bifurcations of vector fields*. Applied Mathematical Sciences, 42. Springer-Verlag, New York, 1983.
39. J.K. Hale, H. Koçak. *Dynamics and bifurcations*. Texts in Applied Mathematics, 3. Springer-Verlag, New York, 1991.
40. M. Haragus, T. Kapitula. Spots and stripes in NLS-type equations with nearly one-dimensional potentials. *Math. Methods Appl. Sci.* (to appear).
41. M. Haragus, K. Kirchgässner. Three-dimensional steady capillary-gravity waves. *Ergodic theory, analysis, and efficient simulation of dynamical systems*, 363–397, Springer, Berlin, 2001.
42. M. Haragus, A. Scheel. Corner defects in almost planar interface propagation. *Ann. Inst. H. Poincaré, Anal. Non Linéaire* 23 (2006), 3, 283–329.
43. M. Haragus, A. Scheel. Almost planar waves in anisotropic media. *Comm. Partial Differential Equations* 31 (2006), 4-6, 791–815.

44. M. Haragus, A. Scheel. A bifurcation approach to non-planar traveling waves in reaction-diffusion systems. *GAMM-Mitt.* 30 (2007), 1, 75–95.
45. M. Haragus, A. Scheel. Interfaces between rolls in the Swift-Hohenberg equation. *Int. J. Dyn. Syst. Differ. Equ.* 1 (2007), 2, 89–97.
46. P. Hartman. A lemma in the theory of structural stability of differential equations. *Proc. Amer. Math. Soc.* 11 (1960), 610–620.
47. D. Henry. *Geometric theory of semilinear parabolic equations*. Lecture Notes in Mathematics, 840. Springer-Verlag, Berlin-New York, 1981.
48. M.W. Hirsch, C.C. Pugh. Stable manifolds and hyperbolic sets. Global Analysis. *Proc. Symp. Pure Math., Vol. XIV, Berkeley, CA* (1968) pp. 133–163; Amer. Math. Soc., Providence, R.I., 1970.
49. A.J. Homburg, B. Sandstede. Homoclinic and heteroclinic bifurcations in vector fields. *Handbook of Dynamical Systems*, Vol. III (to appear).
50. E. Hopf. Abzweigung einer periodischen Lösung von einer stationären Lösung eines Differentialsystems. *Ber. Verh. Sächs. Akad. Wiss. Leipzig. Math.-Nat. Kl.* 95 (1943), 1, 3–22.
51. K. Ikeda, K. Murota. *Imperfect Bifurcation in Structures and Materials*. Applied Math. Sci., 149, Springer, New York, 2002.
52. G. Iooss. Théorie non linéaire de la stabilité des écoulements laminaires dans le cas de "l'échange des stabilités". *Arch. Rational Mech. Anal.* 40 (1970/1971), 166–208.
53. G. Iooss. Global characterization of the normal form for a vector field near a closed orbit. *J. Diff. Equ.* 76 (1988), 1, 47–76.
54. G. Iooss. A codimension 2 bifurcation for reversible vector fields. Normal forms and homoclinic chaos (Waterloo, ON, 1992), *Fields Inst. Com.* 4 (1995), 201–217.
55. G. Iooss. Travelling waves in the Fermi–Pasta–Ulam lattice. *Nonlinearity* 13 (2000), 3, 849–866.
56. G. Iooss, M. Adelmeyer. *Topics in bifurcation theory and applications*. Second ed. Advanced Series in Nonlinear Dynamics, 3. World Scientific Publishing, River Edge, NJ, 1998.
57. G. Iooss, G. James. Localized waves in nonlinear oscillator chains. *Chaos* 15 (2005), 1, 015113, 15 pp.
58. G. Iooss, D.D. Joseph. *Elementary stability and bifurcation theory*. Second ed. Undergraduate Texts in Mathematics. Springer-Verlag, New York, 1990.
59. G. Iooss, K. Kirchgässner. Water waves for small surface tension: an approach via normal form. *Proc. Roy. Soc. Edinburgh A* 122 (1992), 3-4, 267–299.
60. G. Iooss, K. Kirchgässner. Travelling waves in a chain of coupled nonlinear oscillators. *Comm. Math. Phys.* 211 (2000), 2, 439–464.
61. G. Iooss, E. Lombardi. Normal forms with exponentially small remainder: application to homoclinic connections for the reversible $0^{2+}i\omega$ resonance. *C. R. Math. Acad. Sci. Paris* 339 (2004), 12, 831–838.
62. G. Iooss, E. Lombardi. Polynomial normal forms with exponentially small remainder for analytic vector fields. *J. Diff. Equ.* 212 (2005), 1, 1–61.
63. G. Iooss, E. Lombardi. Approximated invariant manifold up to exponentially small terms. *J. Diff. Equ.* 248 (2010), 6, 1410–1431.
64. G. Iooss, J. Los. Bifurcation of spatially quasi-periodic solutions in hydrodynamic stability problems. *Nonlinearity* 3 (1990), 851–871.
65. G. Iooss, A. Mielke. Bifurcating time-periodic solutions of Navier-Stokes equations in infinite cylinders. *J. Nonlinear Sci.* 1 (1991), 107–146.
66. G. Iooss, A. Mielke. Time-periodic Ginzburg-Landau equations for one-dimensional patterns with large wave length. *Z. Angew. Math. Phys.* 43 (1992), 1, 125–138.
67. G. Iooss, A. Mielke, Y. Demay. Theory of steady Ginzburg-Landau equation, in hydrodynamic stability problems. *European J. Mech. B Fluids* 8 (1989), 3, 229–268.
68. G. Iooss, M.-C. Pérouème. Perturbed homoclinic solutions in reversible 1 : 1 resonance vector fields. *J. Diff. Equ.* 102 (1993), 1, 62–88.
69. G. Iooss, P. Plotnikov. Small divisor problem in the theory of three-dimensional water gravity waves. *Mem. Amer. Math. Soc.* 200 (2009), 940, 1–128.

70. G. Iooss, M. Rossi. Hopf bifurcation in the presence of spherical symmetry: analytical results. *SIAM J. Math. Anal.* 20 (1989), 3, 511–532.
71. C.G.J. Jacobi. Über die Figur des Gleichgewichts. *Ann. Phys. Chem.* 33 (1834) 229–233.
72. G. James. Centre manifold reduction for quasilinear discrete systems. *J. Nonlinear Sci.* 13 (2003), 1, 27–63.
73. G. James. Internal travelling waves in the limit of a discontinuously stratified fluid. *Arch. Ration. Mech. Anal.* 160 (2001), 1, 41–90.
74. A. Joets. Apollonios, premier géomètre des singularités. *Quadrature* 66 (2007), 37–41
75. D.D. Joseph. *Stability of fluid motions*. I and II. Springer Tracts in Natural Philosophy, Vol. 27 and 28. Springer-Verlag, Berlin-New York, 1976.
76. T. Kato. *Perturbation theory for linear operators*. Reprint of the 1980 ed. Classics in Mathematics. Springer-Verlag, Berlin, 1995.
77. A. Kelley. The stable, center-stable, center, center-unstable, unstable manifolds. *J. Differential Equations* 3 (1967), 546–570.
78. H. Kielhöfer, K. Kirchgässner. Stability and bifurcation in fluid dynamics. Rocky Mountain Consortium Symposium on Nonlinear Eigenvalue Problems (Santa Fe, N.M., 1971). *Rocky Mountain J. Math.* 3 (1973), 275–318.
79. H. Kielhöfer. *Bifurcation theory. An introduction with applications to PDEs*. Applied Mathematical Sciences, 156. Springer-Verlag, New York, 2004.
80. K. Kirchgässner. Wave-solutions of reversible systems and applications. *J. Diff. Equ.* 45 (1982), 1, 113–127.
81. K. Kirchgässner. Nonlinearly resonant surface waves and homoclinic bifurcation. *Adv. Appl. Mech.* 26 (1988), 135–181.
82. P. Kirrmann. Reduktion nichtlinearer elliptischer Systeme in Zylindergebieten unter Verwendung von optimaler Regularität in Hölder Räumen. PhD thesis, Universität Stuttgart, 1991.
83. E.L. Koschmieder. *Bénard cells and Taylor vortices*. Cambridge Monographs on Mechanics and Applied Mathematics. Cambridge University Press, New York, 1993.
84. Y.A. Kuznetsov. *Elements of applied bifurcation theory*. Third ed. Applied Mathematical Sciences, 112. Springer-Verlag, New York, 2004.
85. O.A. Ladyzhenskaya. *The mathematical theory of viscous incompressible flow*. Revised English ed. Translated from the Russian, Gordon and Breach Science Pub., New York-London 1963.
86. L. Landau. On the problem of turbulence. *C. R. (Doklady) Acad. Sci. URSS (N.S.)* 44 (1944), 311–314.
87. O.E. Lanford. Bifurcation of periodic solutions into invariant tori: the work of Ruelle and Takens. *Nonlinear problems in the physical sciences and biology*. Proceedings of a Battelle Summer Institute, Seattle, Wash., July 3–28, 1972. I. Stakgold, D. D. Joseph and D. H. Sattinger eds. Lecture Notes in Mathematics, Vol. 322. Springer-Verlag, Berlin-New York, 1973.
88. W.F. Langford, G. Iooss. Interactions of Hopf and pitchfork bifurcations. Bifurcation problems and their numerical solution (Proc. Workshop, Univ. Dortmund, Dortmund, 1980), *Internat. Ser. Numer. Math.* 54 (1980), 103–134.
89. V.G. LeBlanc, W.F. Langford. Classification and unfoldings of 1 : 2 resonant Hopf bifurcation. *Arch. Rational Mech. Anal.* 136 (1996), 4, 305–357.
90. T. Levi-Civita. Détermination rigoureuse des ondes permanentes d'ampleur finie. *Math. Ann.* 93 (1925), 1, 264–314.
91. E. Lombardi. Orbits homoclinic to exponentially small periodic orbits for a class of reversible systems. Application to water waves. *Arch. Rational Mech. Anal.* 137 (1997), 3, 227–304.
92. E. Lombardi. Non-persistence of homoclinic connections for perturbed integrable reversible systems. *J. Dyn. Diff. Equ.* 11 (1999), 1, 129–208.
93. E. Lombardi. *Oscillatory integrals and phenomena beyond all algebraic orders. With applications to homoclinic orbits in reversible systems*. Lecture Notes in Mathematics, 1741. Springer-Verlag, Berlin, 2000.
94. J.E. Marsden, M. McCracken. *The Hopf bifurcation and its applications*. Applied Mathematical Sciences, Vol. 19. Springer-Verlag, New York, 1976.

95. A. Mielke. A reduction principle for nonautonomous systems in infinite-dimensional spaces. *J. Diff. Equ.* 65 (1986), 1, 68–88.
96. A. Mielke. Über maximale L^p-Regularität für Differentialgleichungen in Banach- und Hilbert-Räumen. *Math. Ann.* 277 (1987), 1, 121–133.
97. A. Mielke. Reduction of quasilinear elliptic equations in cylindrical domains with applications. *Math. Methods Appl. Sci.* 10 (1988), 1, 51–66.
98. A. Mielke. On nonlinear problems of mixed type: a qualitative theory using infinite-dimensional center manifolds. *J. Dynam. Diff. Equ.* 4 (1992), 3, 419–443.
99. L.A. Ostrowsky. Nonlinear internal waves in a rotating ocean. *Okeanologia* 18 (1978), 181–191.
100. R. L. Pego, J. R. Quintero. A host of traveling waves in a model of three-dimensional water-wave dynamics. *J. Nonlinear Sci.* 12 (2002), 1, 59–83.
101. V. Pliss. A reduction principle in the theory of stability of motion. (Russian) *Izv. Akad. Nauk SSSR Ser. Mat.* 28 (1964) 1297–1324.
102. H. Poincaré. *Les méthodes nouvelles de la mécanique céleste*. Tome I. Reprint of the 1892 original. Les Grands Classiques Gauthier-Villars. Librairie Scientifique et Technique Albert Blanchard, Paris, 1987.
103. H. Poincaré. Sur l'équilibre d'une masse fluide animée d'un mouvement de rotation. *Acta Math.* 7 (1885), 1, 259–380.
104. B. van der Pol. Forced oscillations in a circuit with nonlinear resistance. *Phil. Mag.* 3 (1927), 65–80.
105. Y. Pomeau, A. Ramani, B. Grammaticos. Structural stability of the Korteweg-de Vries solitons under a singular perturbation. *Phys. D* 31 (1988), 1, 127–134
106. I. Prigogine, R. Lefever. Symmetry breaking instabilities in dissipative systems. 2. *J. Chem. Phys.* 48 (1968), 4, 1695–1700.
107. C. Rorres. Completing Book II of Archimedes's on floating bodies. *Math. Intelligencer* 26 (2004), 3, 32–42.
108. D. Ruelle, F. Takens. On the nature of turbulence. *Comm. Math. Phys.* 20 (1971), 167–192, and 23 (1971), 343–344.
109. D.H. Sattinger. *Group-theoretic methods in bifurcation theory*. Lecture Notes in Mathematics, 762. Springer, Berlin, 1979.
110. S.H. Strogatz. *Nonlinear dynamics and chaos: With applications to physics, biology, chemistry, and engineering*. Studies in Nonlinearity, Perseus Books, Cambridge, 1994.
111. M.B. Sevryuk. *Reversible systems*. Lecture Notes in Mathematics, 1211. Springer-Verlag, Berlin, 1986.
112. J. Sijbrand. Studies in nonlinear stability and bifurcation theory. PhD Thesis, University of Utrecht, 1981.
113. V.S. Sorokin. On steady motions in a fluid heated from below. (Russian) *Akad. Nauk SSSR. Prikl. Mat. Meh.* 18 (1954), 197–204.
114. S.J.van Strien. Center manifolds are not C^∞. *Math. Z.* 166 (1979), 2, 143–145.
115. S.M. Sun. Existence of a generalized solitary wave solution for water with positive Bond number less than $1/3$. *J. Math. Anal. Appl.* 156 (1991), 2, 471–504.
116. S.M. Sun, M.C. Shen. Exponentially small estimate for the amplitude of capillary ripples of a generalized solitary wave. *J. Math. Anal. Appl.* 172 (1993), 2, 533–566.
117. R. Tagg. The Couette–Taylor problem. *Nonlinear Science Today* 4 (1994), 2–25; see also: Couette–Taylor Reference Base, 1999, http://carbon.cudenver.edu/˜rtagg/Special/SpecialWebMaterials.html.
118. R. Temam. *Navier–Stokes equations. Theory and numerical analysis*. Studies in Mathematics and its Applications, Vol. 2. North-Holland Publishing Co., Amsterdam-New York-Oxford, 1977.
119. M.R. Ukhovskii, V.I. Yudovich. On the equations of steady-state convection. *Prikl. Mat. Meh.* 27, 295–300 (Russian); translated as *J. Appl. Math. Mech.* 27 (1963), 432–440.
120. A. Vanderbauwhede. Centre manifolds, normal forms and elementary bifurcations. *Dynamics reported*, Vol. 2, 89–169, Dynam. Report. Ser. Dynam. Systems Appl., 2, Wiley, Chichester, 1989.

121. A. Vanderbauwhede, S.A. van Gils. Center manifolds and contractions on a scale of Banach spaces. *J. Funct. Anal.* 72 (1987), 2, 209–224.
122. A. Vanderbauwhede, G. Iooss. Center manifold theory in infinite dimensions. *Dynamics reported: expositions in dynamical systems*, 125–163, Dynam. Report. Expositions Dynam. Systems (N.S.), 1, Springer, Berlin, 1992.
123. W. Velte. Stabilitätsverhalten und Verzweigung stationärer Lösungen der Navier-Stokesschen Gleichungen. *Arch. Rational Mech. Anal.* 16 (1964), 97–125.
124. L.-J. Wang. Homoclinic and heteroclinic orbits for the 0^2 or $0^2 i\omega$ singularity in the presence of two reversibility symmetries. *Quart. Appl. Math.* 67 (2009), 1–38.
125. L.-J. Wang. Studies on 0^2 or $0^2 i\omega$ singularity in the presence of two reversibility symmetries and on dynamics of the Swift–Hohenberg equation with Dirichlet–Neumann boundary conditions. PhD thesis, Université de Toulouse, 2008.
126. V.I. Yudovich. Secondary flows and fluid instability between rotating cylinders. *Prikl. Mat. Meh.* 30, 688–698 (Russian); translated as *J. Appl. Math. Mech.* 30 (1966), 822–833.
127. V.I. Yudovich. On the origin of convection. *Prikl. Mat. Meh.* 30, 1000–1005 (Russian); translated as *J. Appl. Math. Mech.* 30 (1966), 1193–1199.
128. V.I. Yudovich. Free convection and bifurcation. *Prikl. Mat. Meh.* 31, 101–111 (Russian); translated as *J. Appl. Math. Mech.* 31 (1967), 103–114.
129. V.I. Yudovich. *The linearization method in hydrodynamical stability theory*. Translations of Mathematical Monographs, 74. American Mathematical Society, Providence, RI, 1989.

Index

M. Haragus, G. Iooss, *Local Bifurcations, Center Manifolds, and Normal Forms in Infinite-Dimensional Dynamical Systems*, Universitext,
DOI 10.1007/978-0-85729-112-7, © EDP Sciences 2011